$19XOON

Student's Solutions Manual

ALGEBRA FOR COLLEGE STUDENTS

Robert F. Blitzer
Miami–Dade Community College

Jack C. Gill
Miami–Dade Community College

D0080590

Prepared by
Robert F. Blitzer and Kevin E. Calderon

Problems solved by
Robert F. Blitzer, Alina Botero, Karl Brisson, Kevin E. Calderon, Carlo Castano, Opal Chin, Karl Dhana, Darryl Ho, Humberto Jaramillo, Hugo Leonard, Cristobal Martinez, Richard Orellana, Jeronimo Reverte, Ana Serna, Yuk Yin Sham, Julio Vidal

Additional Keyboarding and Formatting By Richard Unzueta and Javier Unzueta, Javi Graphics, Miami, Florida

MACMILLAN PUBLISHING COMPANY
New York

MAXWELL MACMILLAN CANADA, INC.
Toronto

MAXWELL MACMILLAN INTERNATIONAL
New York   Oxford   Singapore   Sydney

Copyright © 1992, Macmillan Publishing Company,
a division of Macmillan, Inc.

Macmillan Publishing Company is part of the
Maxwell Communications Group of Companies.

Printed in the United States of America

All rights reserved. No part of this book may be reproduced or
transmitted in any form or by any means, electronic or mechanical,
including photocopying, recording, or any information storage and
retrieval system, without permission in writing from the Publisher.

Macmillan Publishing Company
866 Third Avenue, New York, New York 10022

Maxwell Macmillan Canada, Inc.
1200 Eglinton Avenue East
Suite 200
Don Mills, Ontario M3C 3N1

Printing: 2 3 4 5 6 7   Year: 3 4 5 6 7 8

ISBN   0-02-343031-1

Preface

This manual has been prepared to accompany <u>Algebra for College Students</u> by Robert F. Blitzer and Jack C. Gill. It contains detailed solutions to many odd-numbered problems in the problem sets (particularly the harder problems that may cause difficulty), the three review problems at the end of each problem set, all review problems at the conclusion of each chapter, and all review problems appearing in the final appendix.

One intended outcome of your work in this course is to develop problem solving skills and to enhance your ability to think critically. Critical thinking can be increased with practice, and included in the problem sets are many questions that will give you something to think about. Consequently, the solutions manual is not intended to be used as a substitute for working the problems. If you have difficulty with a particular problem, you should refer to the solution given to see the way it can be solved, but do this only after you have made a diligent attempt to work the problem yourself. Having the manual show you how to work a problem is not the same as working the problem yourself.

Some of the problems in the text can be solved in more than one way, although in most instances we have not presented alternate solutions. The methods shown here follow the same methods as those in the text. It is anticipated that sometimes you will devise a strategy for solving a problem that requires fewer steps than the one shown. If this occurs, you may feel pleased with your ingenuity.

The key to success in algebra is working problems. The more problems you work, the better you will become at working them. The more problems you work, the more you will strengthen your ability to think critically, a skill that will benefit you academically and professionally.

Robert F. Blitzer
Kevin E. Calderon
Miami-Dade Community College

CONTENTS

6 Quadratic Equations

7 Graphing and Systems of Equations

8 Algebraic Solutions to Systems of Equations

9 Functions: Exponential and Logarithmic Functions

10 Sequences and Series

Problem set 1.1.1

17. $\sqrt{1} = 1$ and $\sqrt{4} = 2$, so the set is $\{1, 2\}$.

31. true; Every set is a subset of itself. In general, $A \subseteq A$.

33. false; True statements include, $1 \in \{1, 2, 5\}$ and $\{1\} \subseteq \{1, 2, 5\}$.

37. false; The number 26 is an element of the set of natural numbers.

39. false; The true statement is : Every integer is a rational number. $\left(4 = \frac{4}{1}, -7 = \frac{-7}{1}, \text{etc.}\right)$ Observe that $\frac{2}{3}$ is a rational number, but is not an integer.

43. true; All the irrational numbers are real numbers, but they are not rational numbers.

47. true; π is an irrational number. Only rational numbers can be expressed as the quotient of two integers. π is only approximately equal to $\frac{22}{7}$.

49. false; $\frac{3}{0}$ is undefined. In general, division by zero is undefined.

53. false; Prime numbers must be greater than one.

55. $\{0, 1, 2, 3\} \not\subseteq N$ because 0 is not a natural number.

59. $\{-2, 2\} \not\subseteq W$ because -2 is not a whole number.

61. $\left\{\frac{1}{2}, 2\right\} \not\subseteq I$ because $\frac{1}{2}$ is not an integer.

63. $W \not\subseteq N$ because 0 belongs to the set of whole numbers, but not to the set of natural numbers.

Problem set 1.1.2

25. The property is the multiplication principle because both sides are multiplied by $\frac{1}{4}$.

37. false; True statements include
$-(7 + 4) = (-7) + (-4)$;
$-(7 \cdot 4) = 7(-4)$
$-(7 \cdot 4) = (-7)(4)$.

39. false; True statements include
$2(3 \cdot 8) = (2 \cdot 3)8$;
$2(3 + 8) = (2 \cdot 3) + (2 \cdot 8)$.

47. $7 + (11 + 3x) = (7 + 11) + 3x = 18 + 3x$

51. $3(9y) = (3 \cdot 9)y = 27y$

53. $\frac{1}{7}(7y) = \left(\frac{1}{7} \cdot 7\right)y = 1y = y$

57. $\frac{7}{10}\left(\frac{10}{7}y\right) = \left(\frac{7}{10} \cdot \frac{10}{7}\right)y = 1y = y$

67. $\frac{3}{4}(12x + 16) = \frac{3}{4} \cdot 12x + \frac{3}{4} \cdot 16 = 9x + 12$

73. $4(3y+5) + 80 = 4 \cdot 3y + 4 \cdot 5 + 80 = 12y + 20 + 80 = 12y + 100$

77. $6 + (2 + 4x)3 = 6 + 2 \cdot 3 + 4x \cdot 3 = 6 + 6 + 12x = 12 + 12x$

79. $7(3x + 2y + 5) + 60 =$
$7 \cdot 3x + 7 \cdot 2y + 7 \cdot 5 + 60 =$
$21x + 14y + 35 + 60 = 21x + 14y + 95$

87. The numbers are 1 and -1. $(1)(1) = 1$ (the identity of multiplication) and $(-1)(-1)$ $= 1$. Observe that multiplication of a number by its inverse brings us back to one, the multiplicative identity.

97. If $x + 5 = 17$, then $x + 5 + (-5) = 17 + (-5)$. By the addition principle, -5 is added to each side.

103.

Statements	Reasons
1. $a = b$	1. Given
2. ac is a real number	2. Closure of multiplication.
3. $ac = ac$	3. Reflexive property
4. $ac = bc$	4. Substitution

(a, on the right in step 3, is replaced by b)

105.

Statements	Reasons
1. $a + c = b + c$	1. Given
2. $(a + c) + (-c) =$ $(b + c) + (-c)$	2. Addition principle (The number $(-c)$ is added to each side.)
3. $(a + c) + (-c) = a$ $(b + c) + (-c) = b$	3. Problem 104
4. $a = b$	4. Substitution

(a is substituted for $(a + c) + (-c)$ and b is substituted for $(b + c) + (-c)$ in step 2.)

107. a. Each operation $(S \circ S, S \circ L, S \circ A, \text{etc.})$ results in a unique element that is a member of $\{S,L,A,R\}$.

b. $(L \circ A) \circ R = R \circ R = A$
$L \circ (A \circ R) = L \circ L = A$
This illustrates the associative property.

c. $L \circ R = S$ and $R \circ L = S$, illustrating the commutative property.

d. S, the identity

e. $S \circ S = S; L \circ R = S$
(R is the inverse of L);
$A \circ A = S; R \circ L = S$
This illustrates the inverse property.

109. $a \circ b = aa + bb$
$b \circ a = bb + aa = aa + bb$
Since $a \circ b = b \circ a$, the operation is commutative.

110. $\{3, 5, 7, 11, 13\}$

111. 13 is a natural number. 0 and 13 are whole numbers. -10, 0 and 13 are integers. -10, $-\frac{3}{2}$, 0, and 13 are rational numbers. $\sqrt{5}$ is the only irrational number.

112. False; True statements include, $\{3\} \subseteq \{1, 2, 3, 4, 5\}$ and $3 \in \{1, 2, 3, 4, 5\}$.

Problem set 1.2.1

15. Since $\frac{4}{5} = \frac{8}{10}$, $\frac{4}{5}$ $\left(\text{or } \frac{8}{10} \right) \le \frac{7}{10}$ is false.

17. true; 3 is greater than 1 and 3 is also less than 12.

21. false; $2(1+3) = 2(4) = 8$ and $7 + 2 = 9$
8 is not greater than or equal to 9.

23. false; True statements include $3(5+7) =$
$3 \cdot 5 + 3 \cdot 7$ and $3(5+7) \geq 3 \cdot 5 + 3 \cdot 7$.

53. x is a least 2 means that x is 2 or greater.
x is at most 5 means that x is 5 or less
than 5. Thus, $\{x \mid 2 \leq x \leq 5\}$.

57. x is negative means that x is less than
zero ($x < 0$). x is at least -2 means that
x is -2 or greater ($x \geq$ -2). Thus,
$\{ x \mid $ -2 $\leq x < 0\}$.

63. Commutative of addition; The order of
addends is reversed. $4 + 3 = 3 + 4$

64. The addition principle; The real number
(-7) is added to each side.

65. true; 1,362 is a natural number.

Problem set 1.2.2

9. $-|6| = -6$ Notice that the negative sign is
outside the absolute value bars. Do not
confuse this with $|-6| = 6$.

13. $-|-6| = -6$ (Since $|-6| = 6$, and we
must copy the negative sign outside the
absolute value bars. Do not confuse this
with $-(-6) = 6$.)

19. $\left|\frac{3}{5}\right| + \left|-\frac{2}{5}\right| - |-1| = \frac{3}{5} + \frac{2}{5} - 1 =$
$\frac{3}{5} + \frac{2}{5} - \frac{5}{5} = 0$

21. $\left|1\frac{2}{3}\right| + \left|-3\frac{1}{6}\right| - \left|2\frac{1}{2}\right| =$
$\frac{5}{3} + \frac{19}{6} - \frac{5}{2} = \frac{10}{6} + \frac{19}{6} - \frac{15}{6} =$
$\frac{14}{6} = \frac{7}{3}$

23. $x = 5$ or $x = -5$ since $|5| = 5$ and
$|-5| = 5$.

43. false; $|-7| = 7$ and $|-15| = 15$;
7 is not greater than 15.

45. false; A true statement is $-12 < -3 < 1$.
Equivalently, $-12 < -3 < |-1|$.

47. $|-12| = 12; 4(2+1) = 4(3) = 12$
The statement $12 > 12$ is false.

51. false; suppose a is negative. Let $a = -3$.
$a \geq |a|$ becomes $-3 \geq |-3|$.
$-3 \geq 3$ is false.

53. false; Let $a = 4$. $|-a| = -|a|$ becomes
$|-4| = -|4|$; $4 = -4$ is false.

55. No values satisfy $|x| < -1$. If $x = 0$,
$|0| = 0$, which is not less than -1.
The absolute value of any other number
results in a positve number.
No positive number is less than -1.

59. $\{-7 < x \leq 13\}$

60. Inverse property for addition

61. True; $-13 = -\frac{13}{1}$, which is an element
of the set of rational numbers.

Problem set 1.3.1

13. The difference between 7.415 and 3.2
is 4.215. Thus $(-7.415) + (3.2) = -4.215$.

27. $-\dfrac{5}{8} - \left(-\dfrac{1}{6}\right) = -\dfrac{5}{8} + \left(\dfrac{1}{6}\right) =$
$-\dfrac{15}{24} + \left(\dfrac{4}{24}\right) = -\dfrac{11}{24}.$

33. $11 - (-5) - (-3) = 11 + (5) - (-3) =$
$16 - (-3) = 16 + (3) = 19$

39. $17 + (-5) - (-3 + 11 - 4) =$
$17 + (-5) - (4) = 12 + (-4) = 8$

43. $-\dfrac{5}{7} + \dfrac{5}{14} - \left(-\dfrac{3}{4}\right) = -\dfrac{20}{28} + \dfrac{10}{28} + \left(\dfrac{21}{28}\right) =$
$-\dfrac{10}{28} + \left(\dfrac{21}{28}\right) = \dfrac{11}{28}$

49. $17 - \left[13 - (9 - 19)\right] =$
$17 - \left[13 - (-10)\right] = 17 - \left[13 + (10)\right] =$
$17 - 23 = -6$

51. $(-13) + (-15) + \left[-20 - (13 - 18)\right] =$
$(-28) + \left[-20 - (-5)\right] =$
$(-28) + \left[-20 + (5)\right] =$
$(-28) + (-15) = -43$

59. $|-12| + |3| - (|9| + |-11|) =$
$12 + 3 - (9 + 11) = 15 - 20 = -5$

61. $\left[7.5 + (-4)\right] - \left[8.5 + (-6)\right] =$
$3.5 - 2.5 = 1$

63. $2\dfrac{1}{9} + \left[\dfrac{4}{9} - \left|-\dfrac{5}{9}\right|\right] = \dfrac{19}{9} + \left[\dfrac{4}{9} - \dfrac{5}{9}\right] =$
$\dfrac{19}{9} + \left(-\dfrac{1}{9}\right) = \dfrac{18}{9} = 2$

65. $(-7) - (-5) = (-7) + (5) = -2$
$(-7) - (-9) = (-7) + (9) = 2$
As these examples show, the difference
between two negative numbers
is sometimes a negative number.

71. $9400 - 600 + 1030 - 550 = 9280$
Its new altitude is 9,280 meters.

73. Population changed by $1500 - 850 = 650$
people. Since there was an increase
in 650 people, at the beginning there
were 650 fewer people.
Thus, $19,150 - 650 = 18,500.$
There were 18,500 people at the beginning.

79. false; Let $a = 4$ and $b = 9$
$$|a - b| \overset{?}{=} |a| - |b|$$
$$|4 - 9| \overset{?}{=} |4| - |9|$$
$$|-5| \overset{?}{=} |4| - |9|$$
$$5 \overset{?}{=} 4 - 9$$
$$5 \neq -5$$

81.

82. $\{x|\ x > 4 \text{ or } x < -4\}$

83. $\{x\,|-4 < x < 4\}$

Problem set 1.3.2

17. $(-3)(-1)(-2)\left(-\frac{1}{2}\right)(-4)=-12;$

   The product is negative with an odd
   number of negative factors (five).

25. $(-5)^2=(-5)(-5)=25$

27. $-5^2=-(5\cdot 5)=-25$

33. $\left(\frac{2}{5}\right)^4=\left(\frac{2}{5}\right)\left(\frac{2}{5}\right)\left(\frac{2}{5}\right)\left(\frac{2}{5}\right)=\frac{16}{625}$

39. $(-2)^3(-4)^2=(-8)(16)=-128$

41. $\left(-\frac{2}{3}\right)^3\cdot 9^2=\left(-\frac{8}{27}\right)\cdot\left(\frac{81}{1}\right)=-24$

47. $-(-5)^3=-(-125)=125$

49. $\left[-(-5)\right]^3=5^3=125$

55. $15\div\left(-\frac{1}{5}\right)=15\cdot(-5)=-75$

57. $-\frac{3}{8}\div\left(-\frac{2}{5}\right)=-\frac{3}{8}\cdot\left(-\frac{5}{2}\right)=\frac{15}{16}$

63. $1\frac{2}{3}\div\left(-6\frac{2}{3}\right)=\frac{5}{3}\div\left(-\frac{20}{3}\right)=$
   $\frac{5}{3}\cdot\left(-\frac{3}{20}\right)=-\frac{1}{4}$

67. $\dfrac{a}{b+c}=\dfrac{a}{b}+\dfrac{a}{c}$ is false.

   $\dfrac{a}{b+c}\overset{?}{=}\dfrac{a}{b}+\dfrac{a}{c}$ Let $a=1,\ b=1,\ c=1$

   $\dfrac{1}{1+1}\overset{?}{=}\dfrac{1}{1}+\dfrac{1}{1}$

   $\dfrac{1}{2}\overset{?}{=}1+1$

   $\dfrac{1}{2}\neq 2$

81. $(-b)^n$ and $-b^n$ are <u>sometimes</u> equal (true).
   If $b=2$ and $n=3$:
   $(-b)^n=(-2)^3=(-2)(-2)(-2)=-8$
   $-b^n=-2^3=-(2)(2)(2)=-8$
   Notice that $(-b)^n$ and $-b^n$ are not always
   equal.
   If $b=2$ and $n=2$
   $(-b)^n=(-2)^2=(-2)(-2)=4$
   $-b^n=-2^2=-(2)(2)=-4$

85. $\{x\mid 0<x\le 7\}$

86.

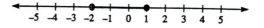

87. Substitution

Problem set 1.3.3

7. $3-5(-4-2)=3-5(-6)=$
   $3+30=33$

15. $6(3-5)^3 - 2(1-3)^3 =$

$6(-2)^3 - 2(-2)^3 = 6(-8) - 2(-8) =$
$-48 + 16 = -32$

17. $8^2 - 16 \div 2^2 \cdot 4 - 3 =$

$64 - 16 \div 4 \cdot 4 - 3 = 64 - 4 \cdot 4 - 3 =$
$64 - 16 - 3 = 48 - 3 = 45$

23. $\left(\frac{1}{2}\right)^2 + \left(\frac{6-4}{5}\right)^2 + \left(\frac{5+2}{10}\right)^2 =$

$\left(\frac{1}{2}\right)^2 + \left(\frac{2}{5}\right)^2 + \left(\frac{7}{10}\right)^2 =$

$\frac{1}{4} + \frac{4}{25} + \frac{49}{100} =$

$\frac{25}{100} + \frac{16}{100} + \frac{49}{100} = \frac{90}{100} = \frac{9}{10}$

27. $\left(\frac{1}{2} - \frac{7}{4}\right) \div \left(1 - \frac{3}{8}\right) =$

$\left(\frac{2}{4} - \frac{7}{4}\right) \div \left(\frac{8}{8} - \frac{3}{8}\right) =$

$-\frac{5}{4} \div \frac{5}{8} = -\frac{5}{4} \cdot \frac{8}{5} = -2$

29. $\frac{1}{4} - 6(2 + 8) \div \left(-\frac{1}{3}\right)\left(-\frac{1}{9}\right) =$

$\frac{1}{4} - 6(10) \div \left(-\frac{1}{3}\right)\left(-\frac{1}{9}\right) =$

$\frac{1}{4} - 60 \div \left(-\frac{1}{3}\right)\left(-\frac{1}{9}\right) =$

$\frac{1}{4} - 60 \cdot (-3)\left(-\frac{1}{9}\right) =$

$\frac{1}{4} - (-180)\left(-\frac{1}{9}\right) =$

$\frac{1}{4} - 20 = \frac{1}{4} - \frac{80}{4} = -\frac{79}{4}$

35. $8 - 3\left[-2(2 - 5) - 4(8 - 6)\right] =$

$8 - 3\left[-2(-3) - 4(2)\right] =$

$8 - 3[6 - 8] = 8 - 3(-2) =$
$8 + 6 = 14$

39. $10 - (-8)\left[\frac{2(-3) - 5(6)}{7 - (-1)}\right] =$

$10 - (-8)\left[\frac{-6 - 30}{8}\right] =$

$10 - (-8)\left(-\frac{36}{8}\right) = 10 - 36 = -26$

41. $6 - (-12)\left[\frac{2 - 4(3 - 7)}{-4 - 5(1 - 3)}\right] =$

$6 - (-12)\left[\frac{2 - 4(-4)}{-4 - 5(-2)}\right] =$

$6 - (-12)\left[\frac{-2 + 16}{-4 + 10}\right] =$

$6 - (-12)\left(\frac{18}{6}\right) = 6 - (-12)(3) =$

$6 - (-36) = 6 + 36 = 42$

45. $-\frac{4}{5}\left[8(-3) + (-50)\left(-\frac{1}{2}\right)\right] =$

$-\frac{4}{5}[-24 + 25] = -\frac{4}{5}(1) = -\frac{4}{5}$

47. $\dfrac{-7\left(\frac{3 \cdot 10}{5 - (-2)}\right) - 8\left(\frac{-8 - 3}{7 - 6}\right)}{2\left(\frac{-2 - 9}{5 \cdot 3 - 2^2}\right) + 9\left(\frac{2 - 6 - 5}{2 - (-1)}\right)} =$

$\dfrac{-7\left(-\frac{7}{7}\right) - 8\left(\frac{-11}{1}\right)}{2\left(\frac{-11}{11}\right) + 9\left(\frac{-9}{3}\right)} =$

$\frac{-7(-1) - 8(-11)}{2(-1) + 9(-3)} = \frac{7 + 88}{-2 - 27} = \frac{-95}{29}$

51. $t = 1970 - 1955 = 15$

$I = 0.03(15)^2 - 0.13(15) + 5.05 =$
$0.03(225) - 0.13(15) + 5.05 =$
$6.75 - 1.95 + 5.05 = 9.85$
Median income for 1970 is
$(9.85)(1000) = \$9,850$

55. $N = \left[14,400 + 120(2) + 100(2)^2\right] \div$

$\left(144 + 2^2\right) = (14,400 + 240 + 400) \div$

$(148) = 15,040 \div 148 = 101.6$
There are approximately 102
bacteria present.

57. $T(age\ 30) = 3(30-20)^2 \div 50 + 10 =$
$3(100) \div 50 + 10 = 300 \div 50 + 10 =$
$6 + 10 = 16$ seconds

$T(age\ 40) = 3(40-20)^2 \div 50 + 10 =$
$3(400) \div 50 + 10 = 24 + 10 =$
34 seconds

fraction for percent increase =

$\dfrac{\text{amount of increase}}{\text{original amount}} = \dfrac{18}{16} = \dfrac{9}{8} = 1.125$

The percent increase is 112.5%.

59. Cost $= 10(20) + 12(37) + 15(54) =$
$200 + 444 + 810 = 1454$ The cost
    is \$14.54.

63. b is not true. The set of whole numbers
$\{0, 1, 2, 3, \dots\}$ is a subset of the set of
integers $\{\dots, -3, -2, -1, 0, 1, 2, 3, \dots\}$.
Thus, a true statement is
$\{\text{whole numbers}\} \subseteq \{\text{integers}\}$.

64. $\{4, 6\}$

65. $\{x \mid x \geq 6 \text{ or } x \leq -6\}$

Problem set 1.4.1

21. $-3(y^2 - 1) - (y^2 - 7) =$
$-3y^2 + 3 - y^2 + 7 = -4y^2 + 10$

27. $-2(x^2 - 1) - 4(3x^2 - 5) =$
$-2x^2 + 2 - 12x^2 + 20 = -14x^2 + 22$

33. $-3(5x - z) - (z - 4y) + 5(x + 3y - z) =$
$-15x + 3z - z + 4y + 5x + 15y - 5z =$
$-10x + 19y - 3z$

37. $4[x - 2(x - 3y)] = 4[x - 2x + 6y] =$
$4[-x + 6y] = -4x + 24y$

41. $4 - 3(4x - 3y) - 3(-2x + 3y) =$
$4 - 12x + 9y + 6x - 9y = -6x + 4$

43. $\frac{1}{3}[8a - 2(a - 12) + 3] =$
$\frac{1}{3}[8a - 2a + 24 + 3] = \frac{1}{3}[6a + 27] =$
$2a + 9$

47. $3[x^2 + 5(x^2 + 3)] - 6 =$
$3[x^2 + 5x^2 + 15] - 6 =$
$3[6x^2 + 15] - 6 =$
$18x^2 + 45 - 6 = 18x^2 + 39$

49. $4x^2 - [2(x^2 - y^2) - 2(x^2 + y^2)] =$
$4x^2 - [2x^2 - 2y^2 - 2x^2 - 2y^2] =$
$4x^2 - (-4y^2) = 4x^2 + 4y^2$

51. $10x - 6x \cdot 3 + 15y^2 \div 5 \cdot 3 =$
$10x - 18x + 15y^2 \div 5 \cdot 3 =$
$10x - 18x + 3y^2 \cdot 3 =$
$10x - 18x + 9y^2 = -8x + 9y^2$

53. $9y + 6y \div 2y - 8 \cdot 3 = 9y + 3 - 24 =$
$9y - 21$

55. $\left(-\frac{4}{5}\right) \div \left(-\frac{2}{5}x + 6x\right) =$
$\left(-\frac{4}{5}\right) \div \left(-\frac{2}{5}x + \frac{30}{5}x\right) =$
$\left(-\frac{4}{5}\right) \div \left(\frac{28}{5}x\right) =$
$\left(-\frac{4}{5}\right) \cdot \left(\frac{5}{28x}\right) = -\frac{1}{7x}$

57. $4x + 8x \div 2x - 6x \cdot (-2) + 5 =$
$4x + 4 + 12x + 5 = 16x + 9$

61. $-4\left(2a^2 - b\right) - 5\left(b - a^2\right) =$
$-8a^2 + 4b - 5b + 5a^2 = -3a^2 - b =$
$-3(-2)^2 - (-1) = -3(4) - (-1) =$
$-12 + 1 = -11$

65. $3\left[x - 2(x + 2y)\right] = 3\left[x - 2x - 4y\right] =$
$3\left[-x - 4y\right] = -3x - 12y =$
$-3\left(-\frac{1}{3}\right) - 12\left(\frac{1}{6}\right) = 1 - 2 = -1$

67. $N = 3(x + 19x - y) + 7\left[y - (7x - 3y)\right] =$
$3(x + 19x - y) + 7\left[y - 7x + 3y\right] =$
$3(20x - y) + 7(4y - 7x) =$
$60x - 3y + 28y - 49x = 11x + 25y$

75. $6x + 3x \div \frac{1}{3} = 6x + 3x \cdot 3 =$
$6x + 9x = 15x$

77. $x;\ 2x;\ 2x + 9;\ (2x + 9) + x;$
$\left[(2x + 9) + x\right] \div 3;\ \left[(2x + 9) + x\right] \div 3 + 4;$
$\left[(2x + 9) + x\right] \div 3 + 4 - x$
$\left[(2x + 9) + x\right] \div 3 + 4 - x =$
$(3x + 9) \cdot \frac{1}{3} + 4 - x = x + 3 + 4 - x = 7$

78. $1978;\ t = 1978 - 1970 = 8$
$1983;\ t = 1983 - 1970 = 13$
At $t = 8,\ I = 200(1 + 8) + 1300 =$
$200(9) + 1300 = 3,100$
At $t = 13,\ I = 100(1 + 13)^2 = 100(14)^2 =$
$100(196) = 19,600$

fraction for percent increase =
$\frac{\text{amount of increase}}{\text{original amount}} = \frac{19,600 - 3,100}{3,100} =$
$\frac{16,500}{3,100} \approx 5.32 = 532\%$
The percent increase is approximately 532%.

79. Symmetric property of equality
(If a = b then b = a)

80. $6 - 4\left[-2(3 - 6) - 5(19 - 16)\right] =$
$6 - 4\left[-2(-3) - 5(3)\right] =$
$6 - 4[6 - 15] = 6 - 4(-9) =$
$6 + 36 = 42$

Problem set 1.4.2

7. $\left(-5x^3y^2\right)\left(2xy^{17}\right) = -10x^{3+1}y^{2+17} =$
$-10x^4y^{19}$

15. $\left(2xy^2\right)^3 = 2^3x^3\left(y^2\right)^3 = 8x^3y^6$

23. $(2x)^3(-3xy) + 25x^4y =$
$\left(8x^3\right)(-3xy) + 25x^4y =$
$-24x^4y + 25x^4y = x^4y$

31. $\left(\frac{-5x^3}{2y}\right)^3 = \frac{\left(-5x^3\right)^3}{(2y)^3} = \frac{(-5)^3\left(x^3\right)^3}{2^3y^3} =$
$\frac{-125x^9}{8y^3}$

41. $(-4)^{-3} = \frac{1}{(-4)^3} = \frac{1}{-64} = -\frac{1}{64}$

45. $-4^{-2} = -\frac{1}{4^2} = -\frac{1}{16}$

47. $\left(\frac{3}{4}\right)^{-2} = \frac{1}{\left(\frac{3}{4}\right)^2} = \frac{1}{\frac{9}{16}} = \frac{16}{9}$

53. $\frac{1}{-3^{-4}} = \frac{1}{-1} \cdot 3^4 = -3^4 = -81$

59. $\dfrac{-100x^{18}}{25x^{17}} = \dfrac{-100}{25} \cdot x^{18-17} = -4x$

61. $\dfrac{20x^4y^3}{5xy^3} = 4x^{4-1}y^{3-3} = 4x^3y^0 =$

$4x^3(1) = 4x^3$

65. $\dfrac{20x^3}{-5x^4} = -4x^{-1} = \dfrac{-4}{x}$

69. $\dfrac{20a^3b^8}{2ab^{13}} = 10a^{3-1}b^{8-13} = 10a^2b^{-5} =$

$\dfrac{10a^2}{b^5}$

75. $\left(2a^5\right)\left(-3a^{-7}\right) = -6a^{5-7} = -6a^{-2} =$

$-\dfrac{6}{a^2}$

83. $\dfrac{x^3}{x^{-7}} = x^{3-(-7)} = x^{10}$

Alternate solution: $\dfrac{x^3}{x^{-7}} =$

$x^3 \cdot x^7 = x^{10}$

87. $\dfrac{4x^6}{-20x^{-11}} = \left(\dfrac{4}{-20}\right)x^{6-(-11)} =$

$-\dfrac{1}{5}x^{17} \text{ or } -\dfrac{x^{17}}{5}$

91. $\dfrac{x^{-7}}{x^{-3}} = x^{-7-(-3)} = x^{-7+3} = x^{-4} =$

$\dfrac{1}{x^4}$

95. $\dfrac{25a^{-8}b^2}{-75a^{-3}b^4} = \dfrac{25}{-75}a^{-8-(-3)}b^{2-4} =$

$-\dfrac{1}{3}a^{-5}b^{-2} = -\dfrac{1}{3a^5b^2}$

99. $\left(\dfrac{15a^4b^2}{-5a^{10}b^{-3}}\right)^3 =$

$\left(-3a^{4-10}b^{2-(-3)}\right)^3 = \left(-3a^{-6}b^5\right)^3 =$

$(-3)^3\left(a^{-6}\right)^3\left(b^5\right)^3 = -27a^{-18}b^{15} =$

$\dfrac{-27b^{15}}{a^{18}}$

101. $\left(\dfrac{10y^2}{y}\right) + \left(\dfrac{4xy^4}{xy^3}\right) = 10y + 4y = 14y$

103. $P = \left(2x^2y\right)^3 - 46,500 =$

$8x^6y^3 - 46,500$

$P = 8(3)^6(2)^3 - 46,500 =$

$8(729)(8) - 46,500 =$

$46,656 - 46,500 = 156$

The profit is \$156

105. $2^2 + 2^3 \neq 2^{2+3}$

$4 + 8 \neq 2^5$

$12 \neq 32$

107. $(3+4)^2 \neq 3^2 + 4^2$

$7^2 \neq 9 + 16$

$49 \neq 25$

109. $(-2)^{-3} \neq \dfrac{1}{2^3}$

$\dfrac{1}{(-2)^3} \neq \dfrac{1}{2^3}$

$-\dfrac{1}{8} \neq \dfrac{1}{8}$

111. $A = LW$

$20x^2y^3 = L\left(5xy^2\right)$

$L = \dfrac{20x^2y^3}{5xy^2} = 4xy$

113. $A = \pi r^2$ Since $d = 10x^6$, $r = \frac{1}{2}\left(10x^6\right) =$

$5x^6$; $A = \pi\left(5x^6\right)^2 = \pi \cdot 5^2\left(x^6\right)^2 =$

$25\pi x^{12}$

115. $\left[6(xy-1)-2xy\right] -$

$4\left[(2xy+3)-2(xy+2)\right] =$

$\left[6xy-6-2xy\right] - 4\left[2xy+3-2xy-4\right] =$

$4xy-6-4(-1) = 4xy-6+4 = 4xy-2$

116. $-\frac{4}{3}\left[\left(-\frac{3}{4}\right)\left(-\frac{4}{3}\right) = 1, \begin{array}{l}\text{the multiplicative}\\ \text{identity}\end{array}\right]$

117. $\dfrac{3}{7^2+(-7)(-5)} - \left[\left(\frac{2}{3}+\frac{1}{4}\right)\cdot \frac{6}{7}\right] =$

$\dfrac{3}{49+35} - \left[\left(\frac{8}{12}+\frac{3}{12}\right)\cdot \frac{6}{7}\right] =$

$\dfrac{3}{84} - \left[\frac{11}{12}\cdot \frac{6}{7}\right] = \dfrac{1}{28} - \dfrac{11}{14} =$

$\dfrac{1}{28} - \dfrac{22}{28} = \dfrac{-21}{28} = -\dfrac{3}{4}$

Review Problems: Chapter 1

1. $\{4, 8, 12, 16, \ldots\}$

2. $\{0\}$

3. $90 = 10 \cdot 9 = 5 \cdot 2 \cdot 3 \cdot 3$

    $\{2, 3, 5\}$

4. false; 3 is not an element of the given set.

5. true; $\{13\}$ is a subset of the given set .

6. true

7. false; Every integer can be expressed as the quotient of integers $\left(7 = \frac{7}{1}, -9 = \frac{-9}{1}, \text{etc.}\right)$. Thus, all integers are rational numbers.

8. Commutative of addition $(a + b = b + a)$

9. Associative of multiplication

    $\left[(a \cdot b) \cdot c = a \cdot (b \cdot c)\right]$

10. Symmetric of equality

    (If $a = b$, then $b = a$)

11. Identity of addition $(a + 0 = a)$

12. Inverses of multiplication

    $\left(\frac{1}{a} \cdot a = 1, \ a \neq 0\right)$

13. Distributive $\left[a\,(b+c) = ab + ac\right]$

14. Addition principle

    (If $a = b$, then $a + c = b + c$)

15. Double negative property $\left[-(-a) = a\right]$

16. -8 because $8 + (-8) = 0$,

    the additive identity

17. 13 because $\frac{1}{13} \cdot 13 = 1$,

    the multiplicative identity

18.

19.

20.

21. $\{x \mid -3 < x < 3\}$

     ($|x| < c$ is equivalent to $-c < x < c$)

22. $\{x \mid x \geq 1 \text{ or } x \leq -1\}$

     ($|x| \geq c$ is equivalent to $x \geq c \text{ or } x \leq -c$)

23. $\{x \mid -3 \leq x < 6\}$

24. $\{x \mid x \leq 12\}$

25. $|-23| + |17| - |-6| = 23 + 17 - 6$
     $40 - 6 = 34$

26. If $|x| = 6$, then $x = 6$ or $x = -6$.

27. $3 + (-17) + (-25) = (-14) + (-25) =$
     -39

28. $16 - (-14) = 16 + (14) = 30$

29. $-11 - \left[-17 + (-3)\right] = -11 - (-20) =$
     $-11 + (20) = 9$

30. $|-17| + |3| - (|10| + |-13|) =$
     $17 + 3 - (10 + 13) = 20 - 23 = -3$

31. $(-0.2)(-0.5) = 0.1$

32. $-\frac{1}{2}(-16)(-3) = 8(-3) = -24$

33. $-12 \div \frac{1}{4} = -12 \cdot 4 = -48$

34. $\left(-\frac{1}{2}\right)^3 \cdot 2^4 = -\frac{1}{8} \cdot 16 = -2$

35. $-\frac{2}{7} \div \left(-\frac{3}{7}\right) = -\frac{2}{7} \cdot \left(-\frac{7}{3}\right) = \frac{2}{3}$

36. $-3[4 - (6 - 8)] = -3[4 - (-2)] =$
     $-3[4 + 2] = -3(6) = -18$

37. $8^2 - 36 \div 3^2 \cdot 4 - (-7) =$
     $64 - 36 \div 9 \cdot 4 - (-7) =$
     $64 - 4 \cdot 4 - (-7) = 64 - 16 - (-7) =$
     $48 - (-7) = 48 + 7 = 55$

38. $\dfrac{(-2)^4 + (-3)^2}{2^2 - (-21)} = \dfrac{16 + 9}{4 - (-21)} =$
     $\dfrac{16 + 9}{4 + 21} = \dfrac{25}{25} = 1$

39. $25 \div \left(\dfrac{8+9}{2^3 - 3}\right) - 6 = 25 \div \left(\dfrac{17}{5}\right) - 6 =$
     $25 \cdot \dfrac{5}{17} - 6 = \dfrac{125}{17} - 6 = \dfrac{125}{17} - \dfrac{102}{17} =$
     $\dfrac{23}{17}$

40. $6 - (-20)\left[\dfrac{6 - 1(6 - 10)}{14 - 3(6 - 8)}\right] =$
     $6 - (-20)\left[\dfrac{6 - 1(-4)}{14 - 3(-2)}\right] =$
     $6 - (-20)\left[\dfrac{6 + 4}{14 + 6}\right] = 6 - (-20)\left(\dfrac{10}{20}\right) =$
     $6 - (-20)\left(\dfrac{1}{2}\right) = 6 - (-10) =$
     $6 + 10 = 16$

41. L for $16 = 3(16 \div 2 - 2)^2 = 3(8 - 2)^2 =$
     $3(6)^2 = 3(36) = 108$
     L for $12 = 3(12 \div 2 - 2)^2 = 3(6 - 2)^2 =$
     $3(4)^2 = 3(16) = 48$
     The difference is $108 - 48 = 60$.

42. $V = 1200(1 - 0.07 \cdot 4) =$
     $1200(1 - 0.28) = 1200(0.72) = 864$
     The value is \$864.

43. $D = 2(70) +$

$3[70 + 0.02(18,000 - 15,0000)] +$

$10[160 + 0.03(40,000 - 30,000)] =$

$140 + 3[70 + 0.02(3,000)] +$

$10[160 + 0.03(10,000)] =$

$140 + 3[70 + 60] + 10[160 + 300] =$

$140 + 3(130) + 10(460) =$

$140 + 390 + 4600 = 5130$

Total dues are $5,130

44. $6(2x - 3) - 5(3x - 2)$

$12x - 18 - 15x + 10 = -3x - 8$

45. $6[b - 3(a - 6b)] = 6[b - 3a + 18b] =$

$6[19b - 3a] = 114b - 18a$

46. $3x - [y - (2x - 3y)] = 3x - [y - 2x + 3y] =$

$3x - [-2x + 4y] = 3x + 2x - 4y =$

$5x - 4y$

47. $-x^2 - \left[3y^2 - \left(2x^2 - y^2\right)\right] =$

$-x^2 - \left[3y^2 - 2x^2 + y^2\right] =$

$-x^2 - \left[-2x^2 + 4y^2\right] =$

$-x^2 + 2x^2 - 4y^2 = x^2 - 4y^2$

48. $8x - 2x \cdot 5 + 6x^2 \div 3 \cdot 2 =$

$8x - 10x + 2x^2 \cdot 2 = 8x - 10x + 4x^2 =$

$4x^2 - 2x$

49. $\left[6(xy - 1) - 2xy\right] -$

$\left[4(2xy + 3) - 2(xy + 2)\right] =$

$\left[6xy - 6 - 2xy\right] - \left[8xy + 12 - 2xy - 4\right] =$

$\left[4xy - 6\right] - \left[6xy + 8\right] =$

$4xy - 6 - 6xy - 8 = -2xy - 14$

50. $\left(-3y^7\right)\left(-8y^6\right) = 24y^{7+6} = 24y^{13}$

51. $\left(7x^3y\right)^2 = 7^2\left(x^3\right)^2 y^2 = 49x^6y^2$

52. $(-3xy)\left(2x^2\right)^3 = (-3xy)\left(2^3\left(x^2\right)^3\right) =$

$(-3xy)\left(8x^6\right) = -24x^7y$

53. $\left(\frac{2}{3}\right)^{-2} = \frac{1}{\left(\frac{2}{3}\right)^2} = \frac{1}{\frac{4}{9}} = \frac{9}{4}$

Alternate solution: $\left(\frac{2}{3}\right)^{-2} = \frac{2^{-2}}{3^{-2}} =$

$\frac{3^2}{2^2} = \frac{9}{4}$

54. $(-6xy)\left(-3x^2y\right) - 25x^3y^2 =$

$18x^3y^2 - 25x^3y^2 = -7x^3y^2$

55. $\frac{16y^3}{-2y^{10}} = -8y^{3-10} = -8y^{-7} = -\frac{8}{y^7}$

56. $\left(-3x^4\right)\left(-4x^{-11}\right) = 12x^{4-11} =$

$12x^{-7} = \frac{12}{x^7}$

57. $\frac{12x^7}{-4x^{-3}} = -3x^{7-(-3)} = -3x^{10}$

58. $\frac{-10a^5b^6}{20a^{-3}b^{11}} = -\frac{1}{2}a^{5-(-3)}b^{6-11} =$

$-\frac{1}{2}a^8b^{-5} = -\frac{a^8}{2b^5}$

59. $(-2)^{-3} + 2^{-2} + \frac{1}{2}x^0 =$

$$\frac{1}{(-2)^3} + \frac{1}{2^2} + \frac{1}{2} \cdot 1 =$$

$$\frac{1}{-8} + \frac{1}{4} + \frac{1}{2} = -\frac{1}{8} + \frac{2}{8} + \frac{4}{8} = \frac{5}{8}$$

60. $\left(\dfrac{-2}{ab}\right)^5 = \dfrac{(-2)^5}{(ab)^5} = \dfrac{-32}{a^5 b^5}$

Problem set 2.1.1

11.
$$8x - 3.1 = 11x + 9.02$$
$$8x - 3.1 + (-11x) = 11x + 9.02 + (-11x)$$
$$-3x - 3.1 = 9.02$$
$$-3x - 3.1 + 3.1 = 9.02 + 3.1$$
$$-3x = 12.12$$
$$\left(-\frac{1}{3}\right)(-3x) = (12.12)\left(-\frac{1}{3}\right)$$
$$x = -4.04 \; ; \quad \{-4.04\}$$

13.
$$\frac{1}{8}y + \frac{1}{4} = \frac{1}{2}$$
$$8\left(\frac{1}{8}y + \frac{1}{4}\right) = 8\left(\frac{1}{2}\right)$$
$$y + 2 = 4$$
$$y = 2 \; ; \quad \{2\}$$

17.
$$2 - (7y + 5) = 13 - 3y$$
$$2 - 7y - 5 = 13 - 3y$$
$$-7y - 3 + 3y = 13 - 3y + 3y$$
$$-4y - 3 = 13$$
$$-4y - 3 + 3 = 13 + 3$$
$$-4y = 16$$
$$\left(-\frac{1}{4}\right)(-4y) = \left(-\frac{1}{4}\right)(16)$$
$$y = -4 \; ; \quad \{-4\}$$

21.
$$4(2y + 1) - 29 = 3(2y - 5)$$
$$8y + 4 - 29 = 6y - 15$$
$$8y - 25 = 6y - 15$$
$$2y - 25 = -15$$
$$2y = 10$$
$$y = 5 \; ; \quad \{5\}$$

23.
$$2(3x - 7) - (6 - x) = 4 - (-x + 6)$$
$$6x - 14 - 6 + x = 4 + x - 6$$
$$7x - 20 = x - 2$$
$$6x - 20 = -2$$
$$6x = 18$$
$$x = 3 \; ; \quad \{3\}$$

25.
$$3(2x - 1) = 5\left[3x - (2 - x)\right]$$
$$6x - 3 = 5\left[3x - 2 + x\right]$$
$$6x - 3 = 5(4x - 2)$$
$$6x - 3 = 20x - 10$$
$$-14x = -7$$
$$x = \frac{-7}{-14} = \frac{1}{2} \; ; \quad \left\{\frac{1}{2}\right\}$$

29.
$$-2\{7 - \left[4 - 2(1 - x) + 3\right]\} =$$
$$10 - \left[4x - 2(x - 3)\right]$$
$$-2\{7 - \left[4 - 2 + 2x + 3\right]\} =$$
$$10 - \left[4x - 2x + 6\right]$$
$$-2\{7 - \left[2x + 5\right]\} = 10 - \left[2x + 6\right]$$
$$-2\{7 - 2x - 5\} = 10 - 2x - 6$$
$$-2\{-2x + 2\} = -2x + 4$$
$$4x - 4 = -2x + 4$$
$$6x - 4 = 4$$
$$6x = 8$$
$$x = \frac{8}{6} = \frac{4}{3} \; ; \quad \left\{\frac{4}{3}\right\}$$

33. a. $G = \frac{3}{20}(140 - 100) = \frac{3}{20} \cdot \frac{40}{1} = 6$

    b. $\frac{17}{4} = \frac{3}{20}(I - 100)$
$$20\left(\frac{17}{4}\right) = 20 \cdot \frac{3}{20}(I - 100)$$
$$85 = 3(I - 100)$$
$$85 = 3I - 300$$
$$385 = 3I$$
$$I = \frac{385}{3} = 128\frac{1}{3} \quad \text{The predicted IQ is } 128\frac{1}{3}.$$

37.
$$116.124 = 0.672I + 113.1$$
$$116.124 + (-113.1) =$$
$$0.672I + 113.1 + (-113.1)$$
$$3.024 = 0.672I$$
$$I = \frac{3.024}{0.672} = 4.5$$

The income is $ 4.5 billion.

39.

$$42.16 = f\frac{12(12+1)}{36(36+1)}$$

$$42.16 = f \cdot \frac{1}{3} \cdot \frac{13}{37}$$

$$42.16 = f \cdot \frac{13}{111}$$

$$(111)(42.16) = f \cdot \frac{13}{111}(111)$$

$$4679.76 = 13f$$

$$f = \frac{4679.76}{13} \approx 359.98$$

Finance charge was approximately $359.98.

41. $400 = C + 273$

$127 = C$

$$F = \frac{9}{5}(127) + 32 = \frac{1143}{5} + 32 =$$

$228.6 + 32 = 260.6°F$

The reaction will take place at 261°F.

45. $R = -\frac{35}{2}L + 195$

$$2R = 2\left(-\frac{35}{2}L + 195\right)$$

$$2R = -35L + 390$$

$$35L = 390 - 2R$$

$$L = \frac{1}{35}(390 - 2R)$$

$$L = \frac{1}{35}(390 - 2 \cdot 125) = \frac{1}{35}(390 - 250) =$$

$$\frac{1}{35}(140) = 4$$

The length is 4 cm.

This approach is more efficient than the one in problem 34 since we can now immediately determine L for a given value of R.

49.

$$ax - bx = 13$$

$$x(a - b) = 13$$

$$x(a - b) \cdot \frac{1}{(a-b)} = 13 \cdot \frac{1}{(a-b)}$$

$$x = \frac{13}{(a-b)}$$

51.

$$ax + 13 = bx - 12$$

$$ax - bx = -25$$

$$x(a - b) = -25$$

$$x(a - b) \cdot \frac{1}{(a-b)} = -25 \cdot \frac{1}{(a-b)}$$

$$x = \frac{-25}{(a-b)}$$

Alternate solution:

$$ax + 13 = bx - 12$$

$$25 = bx - ax$$

$$25 = x(b - a)$$

$$25 \cdot \frac{1}{(b-a)} = x(b - a) \cdot \frac{1}{(b-a)}$$

$$\frac{25}{(b-a)} = x$$

Note: The two answers are the same. Multiplying numerator and denominator of either answer by (-1) produces the other form of the answer.

$$\frac{-25}{(a-b)} \cdot \frac{(-1)}{(-1)} = \frac{25}{-a+b} = \frac{25}{(b-a)}$$

53.

$$D = 0.8(L - S) + S$$

$$D = 0.8L - 0.8S + S$$

$$D = 0.8L + 0.2S$$

$$D - 0.8L = 0.2S$$

$$\frac{1}{0.2}(D - 0.8L) = \frac{1}{0.2}(0.2S)$$

$$\frac{D}{0.2} - 4L = S$$

$$S = \frac{D}{0.2} - 4L$$

55.

$$P = C + MC$$

$$P - C = MC$$

$$\frac{1}{C}(P - C) = MC\left(\frac{1}{C}\right)$$

$$\frac{P - C}{C} = M$$

$$\left(\text{Equivalently, } M = \frac{1}{C} \cdot P - \frac{1}{C} \cdot C = \frac{P}{C} - 1\right)$$

59. $\{x|\ |x| > 2\}$

$\{x|\ x > 2 \text{ or } x < -2\}$

60. $4x^2 - \left[6y^2 - \left(3y^2 - 8x^2\right)\right]$

$= 4x^2 - \left[3y^2 + 8x^2\right] = -4x^2 - 3y^2$

61. $\dfrac{-30x^2y^8}{10x^{-4}y^{11}} = -3x^{2-(-4)}y^{8-11} =$

$-3x^6y^{-3} = \dfrac{-3x^6}{y^3}$

Problem set 2.1.2

5. $7x - 2\left[3(1-x)\right] = -(4+2-9x-4x)$

$7x - 2\left[3 - 3x\right] = -(6 - 13x)$

$7x - 6 + 6x = -6 + 13x$

$13x - 6 = -6 + 13x$

$-6 = -6$; true for all x $\{x|\ x \in \Re\}$

7. $8x - \left[5x + 2(x-3) - (3x+4)\right] = 30$

$8x - \left[5x + 2x - 6 - 3x - 4\right] = 30$

$8x - \left[4x - 10\right] = 30$

$8x - 4x + 10 = 30$

$4x = 20$

$x = 5$; $\{5\}$

9. $12\left(\tfrac{1}{3}y - \tfrac{1}{2}\right) = 8\left(\tfrac{1}{2}y - 1\right)$

$4y - 6 = 4y - 8$

$-6 = -8$; no solution $\varnothing$

11. $2(3y+4) - 4 = 9y + 4 - 3y$

$6y + 8 - 4 = 6y + 4$

$6y + 4 = 6y + 4$

$4 = 4$;true for all values of y $\{y|\ y \in \Re\}$

13. $-11p + 4(p-3) + 6p = 4p - 12$

$-11p + 4p - 12 + 6p = 4p - 12$

$-p - 12 = 4p - 12$

$-5p - 12 = -12$

$-5p = 0$

$\left(-\tfrac{1}{5}\right)(-5p) = \left(-\tfrac{1}{5}\right)(0)$

$p = 0$; $\{0\}$

15. $3(7 + 4m) = -4\left[6 - (-2 + 3m)\right]$

$21 + 12m = -4\left[6 + 2 - 3m\right]$

$21 + 12m = -4\left[8 - 3m\right]$

$21 + 12m = -32 + 12m$

$21 = -32$; no solution $\varnothing$

16. $4.5 = \tfrac{1}{30}A + \tfrac{1}{2}$

$30(4.5) = 30\left(\tfrac{1}{30}A + \tfrac{1}{2}\right)$

$135 = A + 15$

$120 = A$

Supply must be 120 millon pounds.

17. $\left(-8x^3\right)\left(-9x^{-13}\right) = 72x^{3-13} =$

$72x^{-10} = \dfrac{72}{x^{10}}$

18. $|-18| + |2| - (|-11| + |-17|) =$

$18 + 2 - (11 + 17) = 20 - 28 = -8$

Problem set 2.1.3

3. $\frac{2}{5}x + \frac{1}{2}(4x+3) = \frac{x}{10} - 2$

$\frac{2}{5}x + 2x + \frac{3}{2} = \frac{x}{10} - 2$

$10\left(\frac{2}{5}x + 2x + \frac{3}{2}\right) = 10\left(\frac{x}{10} - 2\right)$

$4x + 20x + 15 = x - 20$

$24x + 15 = x - 20$

$23x = -35$

$x = -\frac{35}{23}\ ;\ \left\{-\frac{35}{23}\right\}$

11. $6\left(\frac{4y-3}{3} - 6\right) = 6\left(\frac{3y}{2} - 8\right)$

$2(4y-3) - 36 = 3(3y) - 48$

$8y - 6 - 36 = 9y - 48$

$8y - 42 = 9y - 48$

$6 = y\ ;\ \{6\}$

13. $18\left(\frac{2y-3}{9} + \frac{y-3}{2}\right) = 18\left(\frac{y+5}{6} - 1\right)$

$2(2y-3) + 9(y-3) = 3(y+5) - 18$

$4y - 6 + 9y - 27 = 3y + 15 - 18$

$13y - 33 = 3y - 3$

$10y = 30$

$y = 3\ ;\ \{3\}$

17. $36\left(\frac{y-3}{12} + \frac{y-1}{6}\right) = 36\left(\frac{y+2}{9} - 1\right)$

$3(y-3) + 6(y-1) = 4(y+2) - 36$

$3y - 9 + 6y - 6 = 4y + 8 - 36$

$9y - 15 = 4y - 28$

$5y = -13$

$y = -\frac{13}{5}\ ;\ \left\{-\frac{13}{5}\right\}$

19. $5\left(\frac{x}{5} - 6\right) = 5\left(6 + \frac{1}{5}x\right)$

$x - 30 = 30 + x$

$-30 = 30\ \ \ \ \text{No solution}\ \varnothing$

23. $30\left(\frac{y-2}{6}\right) = 30\left(\frac{1}{15} - \frac{2y+1}{10}\right)$

$5(y-2) = 2 - 3(2y+1)$

$5y - 10 = 2 - 6y - 3$

$5y - 10 = -6y - 1$

$11y = 9$

$y = \frac{9}{11}\ ;\ \left\{\frac{9}{11}\right\}$

25. $24(1-3p) = 24\left[\frac{3}{8}(2-p) - \frac{5}{6}\left(3p - \frac{1}{4}\right)\right]$

$24 - 72p = 9(2-p) - 20\left(3p - \frac{1}{4}\right)$

$24 - 72p = 18 - 9p - 60p + 5$

$24 - 72p = -69p + 23$

$1 = 3p$

$\frac{1}{3} = p\ ;\ \left\{\frac{1}{3}\right\}$

27. $B = \frac{1}{7}ac$

$7B = 7\left(\frac{1}{7}ac\right)$

$7B = ac$

$(7B)\frac{1}{c} = (ac)\frac{1}{c}$

$\frac{7B}{c} = a$

29.
$$a = \frac{d}{7}(B+x)$$
$$7a = 7\left[\frac{d}{7}(B+x)\right]$$
$$7a = d(B+x)$$
$$7a = dB + dx$$
$$7a - dx = dB$$
$$\frac{1}{d}(7a - dx) = \frac{1}{d}(dB)$$
$$\frac{7a - dx}{d} = B$$

Note: The answer can also be written as:

$$B = \frac{1}{d}(7a - dx) = \frac{1}{d}(7a) - \frac{1}{d}(dx) = \frac{7a}{d} - x$$

31.
$$\frac{2}{7}(c+d) = 4(c-a)$$
$$7\left[\frac{2}{7}(c+d)\right] = 7 \cdot 4(c-a)$$
$$2(c+d) = 28(c-a)$$
$$2c + 2d = 28c - 28a$$
$$28a + 2d = 26c$$
$$\frac{28a + 2d}{26} = c$$

The answer can be simplified using the distributive property.

$$c = \frac{28a + 2d}{26} = \frac{2(14a + d)}{2 \cdot 13} = \frac{14a + d}{13}$$

Alternate solution: $\frac{2}{7}(c+d) = 4(c-a)$

$$\frac{2}{7}c + \frac{2}{7}d = 4c - 4a$$
$$7\left(\frac{2}{7}c + \frac{2}{7}d\right) = 7(4c - 4a)$$
$$2c + 2d = 28c - 28a$$

At this point, continue as shown above.

35.
$$z\left(\frac{4xy}{z}\right) = z(-5)$$
$$4xy = -5z$$
$$\frac{1}{4y}(4xy) = \frac{1}{4y}(-5z)$$
$$x = \frac{-5z}{4y}$$

37.
$$6\left[\frac{1}{3}(c - dx) + 2d\right] = 6\left[\frac{1}{6}(6d - x)\right]$$
$$2(c - dx) + 12d = 6d - x$$
$$2c - 2dx + 12d = 6d - x$$
$$2c + 6d = 2dx - x$$
$$2c + 6d = x(2d - 1)$$
$$(2c + 6d) \cdot \frac{1}{2d - 1} = x(2d - 1) \cdot \frac{1}{2d - 1}$$
$$\frac{2c + 6d}{2d - 1} = x$$

Note: An alternate solution involves distributing first and then multiplying by the LCM.

39.
$$15\left(\frac{acx}{3} + \frac{4c}{5}\right) = 15\left(\frac{2acx}{3}\right)$$
$$5acx + 12c = 10acx$$
$$12c = 5acx$$
$$\frac{1}{5ac} \cdot 12c = \frac{1}{5ac} \cdot 5acx$$
$$\frac{12}{5a} = x$$

41.
$$2h = 2\left(V_0 t + \frac{1}{2}at^2\right)$$
$$2h = 2V_0 t + at^2$$
$$2h - 2V_0 t = at^2$$
$$\frac{2h - 2V_0 t}{t^2} = a$$

42. $\{x| \ x \ge 7\}$

43. c is not true. True statements include $\{3\} \subseteq \{1, 2, 3, \ldots\}$ and $3 \in \{1, 2, 3, \ldots\}$.

44. $\{3, 6, 9, 12, 15\}$

Problem set 2.2.1

7. Let x = the number

$$3(x - 2) + x = 2x - 4$$
$$3x - 6 + x = 2x - 4$$
$$4x - 6 = 2x - 4$$
$$2x = 2$$
$$x = 1 \quad \text{The number is 1.}$$

9. Let x = the number

$$5x - 7 = 5(x + 3)$$
$$5x - 7 = 5x + 15$$
$$-7 = 15 \quad \text{No solution}$$

No numbers satisfy the given conditions.

13. Integers are represented by x, x + 2, x + 4.

$$x + x + 2 + x + 4 = 198$$
$$3x = 192$$
$$x = 64 \quad \text{Integers are } 64, 66, 68.$$

17. Integers are represented by x, x + 1, x + 2.

$$x + x + 1 + x + 2 = 25$$
$$3x + 3 = 25$$
$$3x = 22$$
$$x = \frac{22}{3} \quad \text{Since } \frac{22}{3} \text{ is not an}$$

integer (it is a rational number), no integers satisfy the given conditions.

19. Integers are represented by x, x + 2, x + 4.

$$x + 4 = 2x - 6$$
$$10 = x \quad \text{The integers are } 10, 12, 14.$$

21. Integers are represented by x, x + 2, x + 4.

$$x + (x + 4) = \tfrac{1}{2}(x + 2) + 15$$
$$2(2x + 4) = 2\left[\tfrac{1}{2}(x + 2) + 15\right]$$
$$4x + 8 = x + 2 + 30$$
$$4x + 8 = x + 32$$
$$3x = 24$$
$$x = 8 \quad \text{Integers are } 8, 10, 12.$$

23. Integers are represented by x, x + 2, x + 4.

$$2(x + 2) = x + (x + 4)$$
$$2x + 4 = 2x + 4$$
$$4 = 4 \quad \text{The conditions are true for}$$

any three consecutive even integers.

25. Integers are represented by x, x + 2, x + 4. The mean is the sum of the three integers dividedby 3.

$$\frac{x + x + 2 + x + 4}{3} = 20$$
$$3\left(\frac{3x + 6}{3}\right) = 3(20)$$
$$3x + 6 = 60$$
$$3x = 54$$
$$x = 18$$

The integers are $18, 20, 22$.

27.
$$ay = 3r - by$$
$$ay + by = 3r$$
$$y(a + b) = 3r$$
$$y = \frac{3r}{a + b}$$

28. Commutative of both addition and multiplication.

29. $12\left(\dfrac{3x-2}{4}-\dfrac{x-2}{3}\right)=12\left(\dfrac{13}{4}-\dfrac{10x-8}{12}\right)$

$$3(3x-2)-4(x-2)=3(13)-(10x-8)$$
$$9x-6-4x+8=39-10x+8$$
$$5x+2=-10x+47$$
$$15x=45$$
$$x=3\,;\qquad\{3\}$$

Problem set 2.2.2

3. $x=$ Jim's age; $3x=$ Bob's age;
$x+5=$ Susan's age
$x+3x+x+5=60;\ x=11$
Susan's age is $11+5=16$

5. $x=$ Dale's share;
$2x+5{,}000=$ Jackie's share;
$6x-15{,}000=$ Maggie's share.
$x+(2x+5{,}000)+(6x-15{,}000)=80{,}000;$
$x=10{,}000$
Dale: $x=\$10{,}000;$ Jackie: $2x+5{,}000=$
$2(10{,}000)+5{,}000=\$25{,}000;$
Maggie: $6x-15{,}000=6(10{,}000)-15{,}000=$
$\$45{,}000.$

9. Barry's present age: x
Simon's present age: $4x$
Barry's age in 4 years: $x+4$
Simon's age in 4 years: $4x+4$
$4x+4=2(x+4);\ x=2$
Barry is 2 and Simon is $4\cdot 2=8$

11. Let $x=$ the number of years ago
Rita's age x years ago: $23-x$
Joel's age x years ago: $15-x$
$23-x=3(15-x);\ x=11$
11years ago

13. x: amount at 14% $2x$: amount at 12%
Interest from 14% investment + Interest
from 12% investment $=256.5$
$.14x+.12(2x)=256.5$
$.14x+.24x=256.5$
$.38x=256.5$
$x=\dfrac{256.5}{.38}=675$
Amount at 14% : $\$675$
Amount at 12% $=2(675)=\$1350$

15. x: amount at 10% $2x+300$: amount at 13%
$.1x+.13(2x+300)=338.52$
$.1x+.26x+39=338.52$
$.36x=299.52$
$x=\dfrac{299.52}{.36}=832$
Amount at 10%: $\$832$
Amount at 13% $=2(832)+300=\$1{,}964$

17. x: amount at 9% $1200-x$: amount at 12%
$.09x+.12(1200-x)=129;\ x=500$
$\$500$ at 9% and $1200-500=\$700$ at 12%

19. total sum: x　　amount at 9%: $\frac{1}{2}$x

　　　at 12%: $\frac{1}{3}$x　　at 15%: $x-\left(\frac{1}{2}x+\frac{1}{3}x\right)=\frac{1}{6}x$

$$.09\left(\frac{1}{2}x\right)+.12\left(\frac{1}{3}x\right)+.15\left(\frac{1}{6}x\right)=957$$

$$6\left[.09\left(\frac{1}{2}x\right)+.12\left(\frac{1}{3}x\right)+.15\left(\frac{1}{6}x\right)\right]=6(957)$$

$$.27x+.24x+.15x=5742$$

$$.66x=5742$$

$$x=\frac{5742}{.66}=8700$$

The total sum invested is $8,700.00.

21. number of dimes : x

number of quarters: $20-x$

(no. of dimes)(value of 1 dime) +

(no. of quarters)(value of 1 quarter) = 320

$x(10)+(20-x)(25)=320;\ x=12$

Jane has 12 dimes and $20-12=8$ quarters.

25. number of half dollars: x

number of quarters: $x-3$

number of dimes: $3x+2$

50(no. of half dollars) +

　　　+ 25(no. of quarters)

　　　　　+ 10(no. of dimes) = 680

$50x+25(x-3)+10(3x+2)=680;\ x=7$

7 half dollars, $7-3=4$ quarters, $3\cdot 7+2=$

23 dimes

29. number of lbs of 60¢ tea: x

number of lbs of 75¢ tea: $100-x$

(no. of lbs of 60¢ tea)(cost per lb) +

$\left(\text{no. of lbs of 75¢ tea}\right)(\text{cost per lb})=$

(no. of lbs of mixture)(cost per lb)

$$60x+75(100-x)=72(100)$$

$$60x+7500-75x=7200$$

$$-15x=-300$$

$$x=20$$

20 lbs of 60¢ tea and

$100-20=80$ lbs of 75¢ tea.

31. Gallons of solution A (30% acid) : x

Gallons of solution B (60% acid) : $60-x$

Acid in solution A + Acid in solution B =

Acid in mixture

$.3x+.6(60-x)=.5(60);\ x=20$

20 gallons of solution A and

$60-20=40$ gallons of solution B.

33. (This one is tricky!)

Amount of "filler" with neither bluegrass

nor rye: x

$(80)(.3)=24$ lbs bluegrass in original

mixture

Since bluegrass seed is not being added:

Amount of bluegrass in new mixture is

still 24 lbs.

$.2(80+x)=24;\ x=40$

Add 40 lbs of seed containing neither

bluegrass nor rye.

35. x: number of hours
Distanced covered by faster plane +
Distance covered by slower plane = 2500
$300x + 200x = 2500$; $x = 5$; in 5 hours

37. x: speed of slower truck
x + 5: speed of faster truck
Sum of the distances covered by the trucks
in 5 hours = 600 miles.
$5x + 5(x + 5) = 600$; $x = 57.5$
Speed of slower truck: 57.5 m/h
faster truck: $57.5 + 5 = 62.5$ m/h

45. x: time on outgoing trip
$10 - x$: time on return trip
The distance on outgoing trip = The distance
on return trip.
(Distance = Rate · Time)
$20x = 30(10 - x)$; $x = 6$
She took 6 hours on the outgoing trip and
$10 - 6 = 4$ hours on the return trip.
Distance = $20x = 20 \cdot 6 =$
120 miles each way

49. x + 3: time of first traveler
x: time of second traveler
At the moment the second traveler overtakes
the first:
Distance covered by first traveler = Distance
covered by second traveler
$50(x + 3) = 60x$; $x = 15$
It will take the second traveler 15 hours.

53. x: Rate of speed of rowing in still water
x + 3: Rate of speed of rowing with the
current (downstream)
x − 3: Rate of speed of rowing against the
current (upstream)
Distance traveled downstream = Distance
traveled upstream
(D = RT)
$3(x + 3) = 5(x - 3)$; $x = 12$
Distance is either $3(x + 3)$ or $5(x - 3)$ with
$x = 12$.
Distance is $3(x + 3) = 3(12 + 3) = 45$ miles.

55. x: speed of the wind
$300 - x$: speed of plane against the wind
$300 + x$: speed of plane with the wind
Outgoing distance against the wind = return
distance with the wind
$4(300 - x) = 3(300 + x)$; $x = \dfrac{300}{7} = 42\dfrac{6}{7}$
The speed of the wind is $42\dfrac{6}{7}$ m/h.

61. Canada: x   USA: 9x − 1   Australia: x − 2
England: $10(9x - 1) - 9 = 90x - 10 - 9 =$
$90x - 19$
England exceeds sum of the others by 537.
$90x - 19 = x + (9x - 1) + (x - 2) + 537$;
$x = 7$
Population density per square mile:
Canada: 7   USA: $9(7) - 1 = 62$
Australia: $7 - 2 = 5$
England: $90(7) - 19 = 611$

65. A: x B: $x+2$ C: $3x-4$ D: $x-4$
4(no. of A's)$+3$(no. of B's)$+$
2(no. of C's)$+1$(no. of D's)$=50$
$4x+3(x+2)+2(3x-4)+(x-4)=50$;
$x=4$
A: 4 B: $4+2=6$ C: $3(4)-4=8$
D: $4-4=0$

67. First no.: x Second no.: $2x+1$
Third no.: $3x-4$
$\dfrac{x+2x+1+3x-4}{3}=31$; $x=16$
First no.: 16 Second no.: $2(16)+1=33$
Third no.: $3(16)-4=44$

69. x: Year Sunset Cove opened
$x+15$: Year Bay Cove opened
$1988-x$: Age of Sunset Cove in 1988
$1988-(x+15)$: Age of Bay Cove in 1988
In 1988, twice the age of Bay Cove =
the age of Sunset Cove.
$2[1988-(x+15)]=(1988-x)$
$\quad 2(1973-x)=1988-x$
$\quad 3946-2=1988-x$
$\quad\quad\quad 1958=x$
The Bay Cove opened in $x+15=$
$1958+15=1973$.
It will celebrate its centenial anniversary in
$1973+100=2073$.

73. x: last four digits of her phone number
(Sophie is 30.)
$30+70=3x-5939$; $x=2013$
Her phone number is $279-2013$.

77. (This is a hard one!)
x: original number of goods
first stop: $\frac{1}{3}x$ relinquished
Thus, $x-\frac{1}{3}x=\frac{3x}{3}-\frac{1x}{3}=\frac{2x}{3}$ remains.
second stop: $\frac{1}{4}\left(\frac{2x}{3}\right)=\frac{1}{6}x$ relinquished.
Thus, $\frac{2x}{3}-\frac{x}{6}=\frac{4x}{6}-\frac{x}{6}=\frac{3x}{6}=\frac{1}{2}x$ remains.
third stop: $\frac{1}{5}\left(\frac{1}{2}x\right)=\frac{1}{10}x$ relinquished.
Sum of the amount relinquished on all 3
stops $=24$
$\quad \frac{1}{3}x+\frac{1}{6}x+\frac{1}{10}x=24$
$\quad 30\left(\frac{1}{3}x+\frac{1}{6}x+\frac{1}{10}x\right)=30(24)$
$\quad\quad 10x+5x+3x=720$; $x=40$
The original number of goods was 40.

79. x: no. of hours worked
$30-x$: no. of hours not worked
$4.40x=6.60(30-x)$; $x=18$
The tutor worked 18 hours.

81. x: Carla's original amount of money
$4[3(2x-30)-54]-72=48$; $x=29$
Carla's originally had $29.

83. x: weight of Kay's orange
$\dfrac{9}{10}x+\dfrac{9}{10}=x$; $x=9$
The orange weighs (an incredible)
9 pounds.

87. (This one is similar to problem 77, which we know you enjoyed!)

x: number originally stolen

To first guard, he gives $\frac{1}{2}x+2$.

He still has $x-\left(\frac{1}{2}x+2\right)=\frac{1}{2}x-2$

To the second guard, he gives

$\frac{1}{2}\left(\frac{1}{2}x-2\right)+2=\frac{1}{4}x+1$.

He still has $\frac{1}{2}x-2-\left(\frac{1}{4}x+1\right)=\frac{1}{4}x-3$.

To the third guard, he gives

$\frac{1}{2}\left(\frac{1}{4}x-3\right)+2=\frac{1}{8}x+\frac{1}{2}$.

He still has $\frac{1}{4}x-3-\left(\frac{1}{8}x+\frac{1}{2}\right)=\frac{1}{8}x-\frac{7}{2}$.

Since he leaves with one plant,

$$\frac{1}{8}x-\frac{7}{2}=1$$
$$8\left(\frac{1}{8}x-\frac{7}{2}\right)=8(1)$$
$$x-28=8$$
$$x=36$$

The thief originally stole 36 plants.

88. $4\left[\dfrac{4(y-1)}{4}\right]=(y+4)4$

$4y-4=4y+16$

$-4=16$; No solution $\varnothing$

89.

90. $P=80(4)+$

$3\left[900-(6-4)^2+(12-2\cdot 5)^3\right]$

$=320+3\left[900-2^2+2^3\right]$

$=320+3(900-4+8)$

$=320+3(904)=320+2712=3032$

The profit is \$3,032.00

3. One number: 3x   Other number: 7x

$7x=3x+12;\ x=3$

Numbers are $3(3)=9$ and $7(3)=21$.

9. $2\frac{1}{2}x=\left(5\frac{1}{4}\right)\left(3\frac{1}{8}\right)$

$\frac{5}{2}x=\frac{21}{4}\cdot\frac{25}{8}$

$\frac{5}{2}x=\frac{525}{32}$

$32\left(\frac{5}{2}x\right)=32\left(\frac{525}{32}\right)$

$80x=525$

$x=\frac{525}{80}=\frac{105}{16}$ $\left\{\frac{105}{16}\right\}$

15. $\dfrac{5-4y}{3y+10}=\dfrac{-3}{5}$

$5(5-4y)=-3(3y+10)$

$25-20y=-9y-30$

$-11y=-55$

$y=5$ $\{5\}$

17. $-\dfrac{2w+6}{72}=\dfrac{w-7}{-12}$

$-12(-2w-6)=72(w-7)$

$24w+72=72w-504$

$576=48w$

$12=w$ $\{12\}$

Alternate solution: First multiply both sides of the proportion by -1.

$\dfrac{2w+6}{72}=\dfrac{w-7}{12}$

Now solve as above.

21. $\dfrac{\text{salary}}{\text{hours}}$: $\dfrac{47}{18}=\dfrac{93}{x}$

$47x=(18)(93)$

$x\approx 35.6\approx 36$

Approximately 36 hours

27. $\dfrac{\text{tagged}}{\text{total}}$: $\dfrac{50}{x} = \dfrac{27}{108}$

$$27x = (50)(108)$$
$$27x = 5400$$
$$x = \frac{5400}{27} = 200 \; ; \; 200 \text{ bass}$$

29. $\dfrac{\text{people}}{\text{time}}$: $\dfrac{147}{24} = \dfrac{x}{78}$

$$24x = (147)(78)$$
$$24x = 11,466$$
$$x = \frac{11,466}{24} \approx 478; \; 478 \text{ people}$$

31.
$$\frac{r_E{}^2}{r_M{}^2} = \frac{d_E{}^3}{d_M{}^3}$$
$$\frac{(365)^2}{(687)^2} = \frac{(93)^3}{d_M{}^3}$$
$$133,225d_M{}^3 = (687)^2(93)^3$$
$$d_M{}^3 = 2,849,552$$
$$d_M \approx 141.8 \; ; \; 141.8 \text{ millon miles}$$

33.
$$\frac{P}{S} = \frac{1}{4}$$
$$\frac{P}{P+C} = \frac{1}{4}$$
$$4P = P + C$$
$$3P = C$$

Find $\dfrac{P}{C}$.

$$3\frac{P}{C} = 1$$
$$\frac{1}{3}\left(3\frac{P}{C}\right) = \frac{1}{3}(1)$$
$$\frac{P}{C} = \frac{1}{3}$$

34. $4\big[y - (3+y)\big] = 4\left[5 + \dfrac{2(y-2)}{4}\right]$

$$4\big[y - 3 - y\big] = 20 + 2(y-2)$$
$$-12 = 20 + 2y - 4$$
$$-12 = 16 + 2y$$
$$-28 = 2y$$
$$-14 = y \qquad \{-14\}$$

35.
$$2w = 2\left(\frac{11}{2}h - 220\right)$$
$$2w = 11h - 440$$
$$2w + 440 = 11h$$
$$h = \frac{2w + 440}{11}$$
$$h = \frac{2(154) + 440}{11} = 68;$$

Height should be 68 in.

36. $-4x^2 - \left[6y^2 - \left(5x^2 - 2y^2\right)\right]$
$$= -4x^2 - \left[6y^2 - 5x^2 + 2y^2\right]$$
$$= -4x^2 - \left[8y^2 - 5x^2\right]$$
$$= -4x^2 - 8y^2 + 5x^2 = x^2 - 8y^2$$

Problem set 2.2.4

3. width: x length: $2x + 6$
$$2W + 2L = \text{perimeter}$$
$$2x + 2(2x + 6) = 288; \; x = 36$$
width: 36 ft.   length: $2(36) + 6 = 78$ ft.

5. shorter side: x longer side: $x + 2$
$$2(x+2) + \frac{1}{2}x + 5 = 44; \; x = 14$$
Sides measure 14 and 16 cm.
$$\text{Area} = LW = (14)(16) = 224 \text{ cm}^2$$

9. Height: x   Each nonparallel side: $2x - 3$
Shorter Base: $x + 3$ Longer Base: $3x + 1$
Perimeter $= 30$
$2(2x - 3) + (x + 3) + (3x + 1) = 30$
$\quad 4x - 6 + x + 3 + 3x + 1 = 30$
$\qquad\qquad\qquad\quad 8x = 32; \ x = 4$
$A = \frac{1}{2}h(B + b) = \frac{1}{2} \cdot \ 4(7 + 13) = 40 \ \text{in}^2$

15. side a: x   side b: $\frac{2}{3}x + 8$
side c: $\frac{1}{4}\left(x + \frac{2}{3}x + 8\right) - 2$
Perimeter $= 11$ yd $= (3)(11) = 33$ ft
$x + \frac{2}{3}x + 8 + \frac{1}{4}\left(x + \frac{2}{3}x + 8\right) - 2 = 33$
$\quad x + \frac{2}{3}x + 8 + \frac{1}{4}x + \frac{1}{6}x + 2 - 2 = 33$
$\quad 12\left(x + \frac{2}{3}x + 8 + \frac{1}{4}x + \frac{1}{6}x\right) = 12(33)$
$\quad 12x + 8x + 96 + 3x + 2x = 396 \ ; \ x = 12$
side a: 12 ft   side b: $\frac{2}{3}(12) + 8 = 16$ ft
side c: $\frac{1}{4}\left(12 + \frac{2}{3} \cdot \ 12 + 8\right) - 2 = \frac{1}{4}(28) - 2 =$
5 ft

21. Represent angles by x, 5x, 3x.
$x + 5x + 3x = 180°; \ x = 20°$
Largest angle: $5x = 5(20) = 100°$

25. One angle : 5x   Second angle: 2x
$5x - 2x = 15; \ x = 5$
One angle : $5(5) = 25°$
Second angle: $2(5) = 10°$
Third angle: $180° - (25° + 10°) = 145°$
The largest angle measures 145°.

27. $\dfrac{\text{original area}}{\text{new area}} = \dfrac{\text{original cost}}{\text{new cost}} \quad \left(A = \pi r^2\right)$
$\dfrac{\pi \cdot 4^2}{\pi \cdot 8^2} = \dfrac{15}{x}$
$\dfrac{16}{64} = \dfrac{15}{x}$
$16x = (15)(64)$
$x = 60$
Cost is \$60.

29. $\dfrac{\text{original volume}}{\text{new volume}} = \dfrac{\text{original no. of marbles}}{\text{new no. of marbles}}$
$\left(V = \pi r^2 h\right)$
$\dfrac{\pi \cdot 2^2 \cdot 1}{\pi \cdot 4^2 \cdot 3} = \dfrac{400}{x}$
$\dfrac{4}{48} = \dfrac{400}{x}$
$\dfrac{1}{12} = \dfrac{400}{x}$
$x = (12)(400) = 4800$
4,800 marbles

31. Original width: x   Original length: $x + 6$
New width: $x + 3$ New length: $2(x + 6)$
Perimeter of new rectangle = Perimeter of
original rectangle $+ 26$
$2(x + 3) + 2\left[2(x + 6)\right] = 2x + 2(x + 6) + 26$
$x = 4$
Original width: 4cm
Original length: $4 + 6 = 10$cm

35. x: size of one side of the square that is cut
from corners
Length of box: $16 - 2x$ (16 minus x cut
from each end)
Width of box: $12 - 2x$
$16 - 2x = 3(12 - 2x); \ x = 5$
Squares must be 5m by 5m.

37. Area of wall $= (30)(10) = 300 \text{ ft}^2$

Area painted: x   Area papered: $300 - x$

$1.4x + .6(300 - x) = 324$

$1.4x + 180 - .6x = 324$

$.8x = 144$

$x = \dfrac{144}{.8} = 180$

Area painted: 180 ft^2

39. One original side: $4x$

Second original side: $4(x + 1)$

$2(4x + 5) + 2[4(x + 1) + 5] = 140;\ x = 7$

Original dimensions: $4x = 4(7) = 28m$

$4(x + 1) = 4(7 + 1) = 32m$

40. $P = \dfrac{100(300 - 50)}{(300)(-50)} = \dfrac{(100)(250)}{(300)(-50)} =$

$\dfrac{(1)(-5)}{(3)(1)} = \dfrac{-5}{3}$

Power is $-\dfrac{5}{3}$

41. $\quad D = A(n - 1)$

$D = An - A$

$D + A = An$

$\dfrac{D + A}{A} = n$

42. $10^3 + 10^0 - 10^{-1}$

$= 1000 + 1 - \dfrac{1}{10}$

$= 1001 - \dfrac{1}{10} = 1000.9$

Problem set 2.3.1

7. $8x + 3 > 3(2x + 1) + x + 5$

$8x + 3 > 6x + 3 + x + 5$

$8x + 3 > 7x + 8$

$x + 3 > 8$

$x > 5$

$\{x \mid x > 5\}$

13. $4\left[\dfrac{1}{4}(x + 3)\right] < 4[4x - 2(x - 3)]$

$x + 3 < 16x - 8(x - 3)$

$x + 3 < 16x - 8x + 24$

$x + 3 < 8x + 24$

$-7x < 21$

$x > -3$

$\{x \mid x > -3\}$

15. $7(y + 4) - 13 < 12 + 13(3 + y)$

$7y + 28 - 13 < 12 + 39 + 13y$

$7y + 15 < 51 + 13y$

$-6y < 36$

$y > -6$

$\{y \mid y > -6\}$

17. $3[3(y + 5) + 8y + 7] +$

$\qquad 5[3(y - 6) - 2(3y - 5)] < 2(4y + 3)$

$3[3y + 15 + 8y + 7] +$

$\qquad 5[3y - 18 - 6y + 10] < 8y + 6$

$3(11y + 22) + 5(-3y - 8) < 8y + 6$

$33y + 66 - 15y - 40 < 8y + 6$

$18y + 26 < 8y + 6$

$10y < -20$

$y < -2$

$\{y \mid y < -2\}$

19. $4(3y+2)-3y < 3(1+3y)-7$
$12y+8-3y < 3+9y-7$
$9y+8 < 9y-4$
$8 < -4$
contradiction; No Solution $\varnothing$

21. $\frac{5}{6}\left[3(y+2)-2(y+1)\right] <$
$$\frac{2}{3}\left[\frac{1}{2}(y-4)-2(y-9)\right]$$
$\frac{5}{2}(y+2)-\frac{5}{3}(y+1) < \frac{1}{3}(y-4)-\frac{4}{3}(y-9)$
$6\left[\frac{5}{2}(y+2)-\frac{5}{3}(y+1)\right] <$
$$6\left[\frac{1}{3}(y-4)-\frac{4}{3}(y-9)\right]$$
$15y+30-10y-10 < 2y-8-8y+72$
$5y+20 < -6y+64$
$11y < 44$
$y < 4$
$\{y \mid y < 4\}$

25. $8x-1 \le 29+3x$
$5x \le 30$
$x \le 6;\quad \{x \mid x \le 6\}$

29. Mia: x　Eleanor: $2x+3$
$x+(2x+3) \ge 24$
$3x \ge 21$
$x \ge 7$
Mia's age: ≥ 7
Eleanor's age: $\ge 2(7)+3$
Mia is at least 7. Eleanor is at least 17.
　A is false. Mia can be 8
$\longrightarrow$ B is true. Eleanor is at least 17,
　　so she can be 19.
　C is false. Since Eleanor is at least 17,
　　she can be 17.
　D is false. Mia is at least 7.
　　She cannot be 6.

31. x: Monthly sales
$.3(x-1000) > 700$
$.3x-300 > 700$
$.3x > 1000$
$x > \frac{1000}{.3}$
$x > 3333.\overline{3}$
Monthly sales greater than $3,333.34

35. x: number of tickets
$10+3x \le 25;\ x \le 5$
The maximum number of tickets is 5.

39. x: time spent working out
First Club Costs: $500+x$
Second Club Costs: $440+1.75x$
$500+x < 440+1.75x$
$-.75x < -60$
$x > \frac{-60}{-.75}$
$x > 80$
More than 80 hours working out

43. $3\left[y-(4y-2)\right]+12 = 0$
$3\left[y-4y+2\right]+12 = 0$
$3(-3y+2)+12 = 0$
$-9y+6+12 = 0$
$-9y = -18$
$y = 2$　　$\{2\}$

44. Original angle: x　Second angle: $3x+3$
Third angle: $(x+3x+3)+78$
$x+3x+3+(x+3x+3)+78 = 180;\ x = 12$
Original angle: $12°$
Second angle: $3(12)+3 = 39°$
Third angle: $12+3(12)+3+78 = 129°$

45. $\dfrac{\text{Pounds}}{\text{Dollars}} : \dfrac{1.57}{1} = \dfrac{87.5}{x}$

$1.57x = 87.5$

$x = \dfrac{87.5}{1.57} \approx 57.73$

Billing charge is \$57.73

Problem set 2.3.2

1. $5x < 20 \qquad\qquad 3x > -18$

$\qquad x < -4 \qquad\qquad\quad x > -6$

$\qquad \{x \mid -6 < x < -4\}$

5. $2x > 5x - 15 \qquad 7x > 2x + 10$

$\qquad -3x > -15 \qquad\quad 5x > 10$

$\qquad\quad x < 5 \qquad\qquad\quad x > 2$

$\qquad \{x \mid 2 < x < 5\}$

7. $4(1-y) < -6 \qquad 5\left(\dfrac{y-7}{5}\right) \le 5(-2)$

$\qquad 4 - 4y < -6 \qquad\qquad y - 7 \le -10$

$\qquad\quad -4y < -10 \qquad\qquad\quad y \le -3$

$\qquad\qquad y > \dfrac{5}{2}$

$\qquad \emptyset$

11. $\dfrac{9+4x}{3} > -5 \qquad\quad \dfrac{x}{3} + 4 < 3$

$\qquad 9 + 4x > -15 \qquad\quad x + 12 < 9$

$\qquad\qquad 4x > -24 \qquad\qquad x < -3$

$\qquad\qquad\quad x > -6$

$\qquad \{x \mid -6 < x < -3\}$

15. $-5 \le 2x - 7 < 9$

$\qquad 2 \le 2x < 16$

$\qquad 1 \le x < 8$

$\qquad \{x \mid 1 \le x < 8\}$

19. $-2 < -3y \le 3$

$\qquad \dfrac{-2}{-3} > y \ge \dfrac{-3}{-3}$

$\qquad \dfrac{2}{3} > y \ge -1$

Equivalently $-1 \le y < \dfrac{2}{3}$

$\qquad \left\{y \mid -1 \le y < \dfrac{2}{3}\right\}$

23. $-2 \le -\dfrac{1}{2}y + 3 \le 6$

$\qquad 2(-2) \le 2\left(-\dfrac{1}{2}y + 3\right) \le 2(6)$

$\qquad -4 \le -y + 6 \le 12$

$\qquad -10 \le -y \le 6$

$\qquad \dfrac{-10}{-1} \ge \dfrac{-y}{-1} \ge \dfrac{6}{-1}$

$\qquad 10 \ge y \ge -6$

$\qquad \{y \mid -6 \le y \le 10\}$

25. $-7 \le 8 - 3y \le 20 \qquad -7 < 6y - 1 < 41$

$\qquad -15 \le -3y \le 12 \qquad\quad -6 < 6y < 42$

$\qquad\quad 5 \ge y \ge -4 \qquad\qquad\quad -1 < y < 7$

$\qquad -4 \le y \le 5$

$\qquad \{y \mid -1 < y \le 5\}$

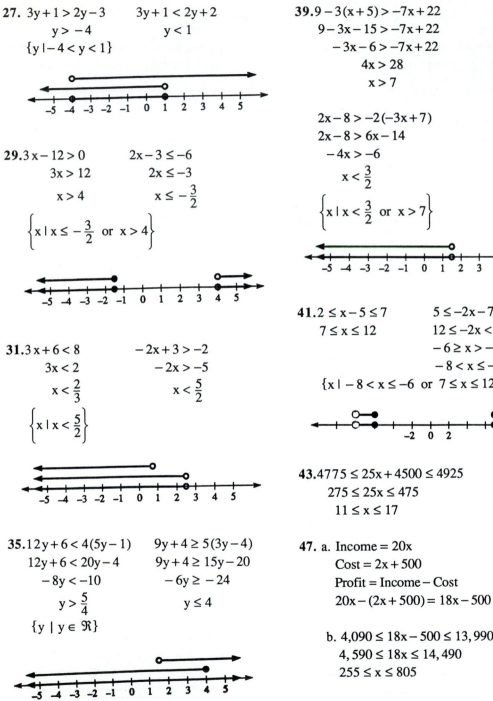

27. $3y+1>2y-3$ $3y+1<2y+2$
$y>-4$ $y<1$
$\{y\,|-4<y<1\}$

29. $3x-12>0$ $2x-3\le-6$
$3x>12$ $2x\le-3$
$x>4$ $x\le-\frac{3}{2}$
$\left\{x\mid x\le-\frac{3}{2}\ \text{or}\ x>4\right\}$

31. $3x+6<8$ $-2x+3>-2$
$3x<2$ $-2x>-5$
$x<\frac{2}{3}$ $x<\frac{5}{2}$
$\left\{x\mid x<\frac{5}{2}\right\}$

35. $12y+6<4(5y-1)$ $9y+4\ge5(3y-4)$
$12y+6<20y-4$ $9y+4\ge15y-20$
$-8y<-10$ $-6y\ge-24$
$y>\frac{5}{4}$ $y\le4$
$\{y\mid y\in\Re\}$

39. $9-3(x+5)>-7x+22$
$9-3x-15>-7x+22$
$-3x-6>-7x+22$
$4x>28$
$x>7$

$2x-8>-2(-3x+7)$
$2x-8>6x-14$
$-4x>-6$
$x<\frac{3}{2}$
$\left\{x\mid x<\frac{3}{2}\ \text{or}\ x>7\right\}$

41. $2\le x-5\le7$ $5\le-2x-7<9$
$7\le x\le12$ $12\le-2x<16$
$-6\ge x>-8$
$-8<x\le-6$
$\{x\mid-8<x\le-6\ \text{or}\ 7\le x\le12\}$

43. $4775\le25x+4500\le4925$
$275\le25x\le475$
$11\le x\le17$

47. a. Income $=20x$
Cost $=2x+500$
Profit $=$ Income $-$ Cost
$20x-(2x+500)=18x-500$

b. $4,090\le18x-500\le13,990$
$4,590\le18x\le14,490$
$255\le x\le805$

49. Width: x   Length : $3x - 11$

Perimeter: $2x + 2(3x - 11) = 8x - 22$

$10 \le 8x - 22 \le 114$

$32 \le 8x \le 136$

$4 \le x \le 17$

Maximum width: 17in

Minimum width: 4 in

51. x: grade on sixth exam

$80 \le \dfrac{90 + 70 + 82 + 80 + 90 + x}{6} \le 85$

$80 \le \dfrac{412 + x}{6} \le 85$

$6(80) \le 6\left(\dfrac{412 + x}{6}\right) \le 6(85)$

$480 \le 412 + x \le 510$

$68 \le x \le 98$

53. x: time in which the cars will meet

Distance traveled by first car: 50x

Distance traveled by second car: 40x

(D = RT)

$90 \le 50x + 40x \le 120$

$90 \le 90x \le 120$

$1 \le x \le 1\frac{1}{3}$

Cars will meet between 1 hour and $1\frac{1}{3}$ hours.

57. $18 \le 2(y + 2) + y \le 22$

$18 \le 2y + 4 + y \le 22$

$18 \le 3y + 4 \le 22$

$14 \le 3y \le 18$

$4\frac{2}{3} \le y \le 6$

59. $s = -16t^2 + vt$

$435 = -16(5)^2 + v(5)$

$435 = -400 + 5v$

$835 = 5v$

$167 = v$

Velocity should be 167 ft/sec.

60. $60 = \frac{5}{11}d + 15$

$11(60) = 11\left(\frac{5}{11}d + 15\right)$

$660 = 5d + 165$

$495 = 5d$

$99 = d$ Depth is 99 ft.

61. Gloria : x   Mike: $x + 7$

$x + 7 - 10 = 2(x - 10) - 5$

$x - 3 = 2x - 25$

$-x = -22$

$x = 22$

Gloria is 22. Mike is 22+7=29.

Problem set 2.4.1

3. $\left|\dfrac{4y - 2}{3}\right| = 2$

$\dfrac{4y - 2}{3} = 2$ or $-\left(\dfrac{4y - 2}{3}\right) = 2$

$4y - 2 = 6$ $-(4y - 2) = 6$

$4y = 8$ $-4y + 2 = 6$

$y = 2$ $-4y = 4$

$y = -1$

check: check:

$\left|\dfrac{4 \cdot 2 - 2}{3}\right| = 2$ $\left|\dfrac{4 \cdot (-1) - 2}{3}\right| = 2$

$\left|\dfrac{6}{3}\right| = 2$ $\left|\dfrac{-6}{3}\right| = 2$

$|2| = 2$ $|-2| = 2$

$2 = 2$ $2 = 2$

$\{-1, 2\}$

5. x: number

$|2x - 3| = 13$

$2x - 3 = 13$ or $-(2x - 3) = 13$

$2x = 16$ $-2x = 10$

$x = 8$ $x = -5$

The numbers are 8 and -5.

11. x: number

$$\left|\frac{1}{6}(5x-1)\right| = 4$$

$$\frac{1}{6}(5x-1) = 4 \qquad \text{or} \quad -\frac{1}{6}(5x-1) = 4$$

$$6\left[\frac{1}{6}(5x-1)\right] = 6(4) \quad 6\left[-\frac{1}{6}(5x-1)\right] = 6(4)$$

$$5x - 1 = 24 \qquad\qquad -(5x-1) = 24$$

$$5x = 25 \qquad\qquad -5x + 1 = 24$$

$$x = 5 \qquad\qquad -5x = 23$$

$$x = -\frac{23}{5}$$

Numbers are 5 or $-\frac{23}{5}$ $\left(-4\frac{3}{5}\right)$.

13. Numbers: x and $5x - 3$

$$|x - (5x-3)| = 11$$

$$|-4x + 3| = 11$$

$$-4x + 3 = 11 \quad \text{or} \quad -(-4x+3) = 11$$

$$-4x = 8 \qquad\qquad 4x - 3 = 11$$

$$x = -2 \qquad\qquad 4x = 14$$

$$x = \frac{7}{2}$$

Numbers are -2 and $5(-2) - 3 = -13$, or

numbers are $\frac{7}{2}$ and $5\left(\frac{7}{2}\right) - 3 = \frac{35}{2} - \frac{6}{2} = \frac{29}{2}$.

15. $|4x + 3| = -5$

No value of x satisfies this equation, since the absolute value of any number can never be negative.

$|x| = c$ has no solution if $c < 0$. $\varnothing$

17. $|2y + 2| = |y + 2|$

$$2y + 2 = y + 2 \quad \text{or} \quad 2y + 2 = -(y+2)$$

$$y = 0 \qquad\qquad 2y + 2 = -y - 2$$

$$3y = -4$$

$$\left\{-\frac{4}{3}, 0\right\} \qquad\qquad y = -\frac{4}{3}$$

21. $|2x - 4| = |2x + 3|$

$$2x - 4 = 2x + 3 \quad \text{or} \quad 2x - 4 = -(2x+3)$$

$$-4 = 3 \qquad\qquad 2x - 4 = -2x - 3$$

$$\varnothing \qquad\qquad 4x = 1$$

$$\left\{\frac{1}{4}\right\} \qquad\qquad x = \frac{1}{4}$$

23. $\left|\frac{2}{3}x - 2\right| = \left|\frac{1}{3}x + 3\right|$

$$\frac{2}{3}x - 2 = \frac{1}{3}x + 3$$

$$3\left(\frac{2}{3}x - 2\right) = 3\left(\frac{1}{3}x + 3\right)$$

$$2x - 6 = x + 9$$

$$x = 15$$

or

$$\frac{2}{3}x - 2 = -\left(\frac{1}{3}x + 3\right)$$

$$3\left(\frac{2}{3}x - 2\right) = -3\left(\frac{1}{3}x + 3\right)$$

$$2x - 6 = -x - 9$$

$$3x = -3$$

$$x = -1 \qquad\qquad \{-1, 15\}$$

27. Consecutive odd integers: x, $x + 2$, $x + 4$

$$2(x + x + 2) = 4(x + 4) - 12$$

$$4x + 4 = 4x + 16 - 12$$

$$4x + 4 = 4x + 4 \qquad \{x \mid x \in \Re\}$$

Any three consecutive odd integers will satisfy the given conditions.

28. x: amount invested at 6%

$x + 10,000$: amount invested at 8%

$$.06x + .08(x + 10,000) = 12,000$$

$$.14x + 800 = 12,000$$

$$.14x = 11,200$$

$$x = 80,000$$

$80,000 at 6%; $90,000 at 8%

29. Base: x   Each equal side : $2x+1$

$$x+2(2x+1)=37$$
$$5x+2=37$$
$$5x=35$$
$$x=7$$

Base: 7 cm

Each equal side: $2(7)+1 = 15cm$

Problem set 2.4.2

3. $|2(x+1)+3| \le 5$

$$|2x+5| \le 5$$
$$-5 \le 2x+5 \le 5$$
$$-10 \le 2x \le 0$$
$$-5 \le x \le 0 \qquad \{x|-5 \le x \le 0\}$$

5. $\left|\dfrac{2(3x-1)}{3}\right| < \dfrac{1}{6}$

$$-\dfrac{1}{6} < \dfrac{2(3x-1)}{3} < \dfrac{1}{6}$$
$$6\left(-\dfrac{1}{6}\right) < 6\left[\dfrac{2(3x-1)}{3}\right] < 6\left(\dfrac{1}{6}\right)$$
$$-1 < 4(3x-1) < 1$$
$$-1 < 12x-4 < 1$$
$$3 < 12x < 5$$
$$\dfrac{1}{4} < x < \dfrac{5}{12}$$
$$\left\{x|\dfrac{1}{4} < x < \dfrac{5}{12}\right\}$$

7. $|3x-1| > 13$

$$3x-1 > 13 \quad \text{or} \quad 3x-1 < -13$$
$$3x > 14 \qquad\qquad 3x < -12$$
$$x > \dfrac{14}{3} \qquad\qquad x < -4$$
$$\left\{x \,|\, x > \dfrac{14}{3} \text{ or } x < -4\right\}$$

9. $|2(x-3)+3(x+2)-17| \ge 13$

$$|2x-6+3x+6-17| \ge 13$$
$$|5x-17| \ge 13$$
$$5x-17 \ge 13 \quad \text{or} \quad 5x-17 \le -13$$
$$5x \ge 30 \qquad\qquad 5x \le 4$$
$$x \ge 6 \qquad\qquad x \le \dfrac{4}{5}$$
$$\left\{x \,|\, x \ge 6 \text{ or } x \le \dfrac{4}{5}\right\}$$

13. $\left|\dfrac{7x-2}{4}\right| \ge \dfrac{5}{4}$

$$\dfrac{7x-2}{4} \ge \dfrac{5}{4} \quad \text{or} \quad \dfrac{7x-2}{4} \le -\dfrac{5}{4}$$
$$7x-2 \ge 5 \qquad\qquad 7x-2 \le -5$$
$$7x \ge 7 \qquad\qquad 7x \le -3$$
$$x \ge 1 \qquad\qquad x \le -\dfrac{3}{7}$$
$$\left\{x \,|\, x \ge 1 \text{ or } x \le -\dfrac{3}{7}\right\}$$

17. x: number

$$|2x-1|-2 < 3$$
$$|2x-1| < 5$$
$$-5 < 2x-1 < 5$$
$$-4 < 2x < 6$$
$$-2 < x < 3 \qquad \{x|-2 < x < 3\}$$

21. x: number

$|2x - 1| > 1$

$2x - 1 > 1$ or $2x - 1 < -1$

$2x > 2$ $2x < 0$

$x > 1$ $x < 0$

$\{x|\ x > 1 \text{ or } x < 0\}$

25. $|x - 2,560,000| \le 135,000$

$-135,000 \le x - 2,560,000 \le 135,000$

$2,425,000 \le x \le 2,695,000$

high: 2,695,000 barrels

low: 2,425,000 barrels

27. $|p - 0.35\%| \le 0.16\%$

$-0.16\% \le p - 0.35\% \le 0.16\%$

$0.19\% \le p \le .51\%$

100,000 products; N: number of defective products

$(.19\%)(100,000) \le N \le (.51\%)(100,000)$

$(.0019)(100,000) \le N \le (.0051)(100,000)$

$190 \le N \le 510$

\$6 refund for each product;

R: cost of refunds

$6(190) \le R \le 6(510)$

$1140 \le R \le 3060$

$\{R \mid \$1140 \le R \le \$3060\}$

29. $|2x - 1| + 8 \ge 10$

$|2x - 1| \ge 2$

$2x - 1 \ge 2$ or $2x - 1 \le -2$

$x \ge \dfrac{3}{2}$ $x \le -\dfrac{1}{2}$

$\left\{x \mid x \ge \dfrac{3}{2} \text{ or } x \le -\dfrac{1}{2}\right\}$

31. $|2x + 4| < -1$

The absolute value of a number cannot result in a negative number, and a number less than -1 is negative. $\varnothing$

33. $|2x - b| \le c$

$-c \le 2x - b \le c$

$b - c \le 2x \le b + c$

$\dfrac{b-c}{2} \le x \le \dfrac{b+c}{2}$

$\left\{x \;\middle|\; \dfrac{b-c}{2} \le x \le \dfrac{b+c}{2}\right\}$

35. $|10 - x| > 5$

$10 - x > 5$ or $10 - x < -5$

$-x > -5$ $-x < -15$

$x < 5$ $x > 15$

$\{x \mid x < 5 \text{ or } x > 15 \}$

37. no. of quarters: x dimes: $2x$

nickles: $\dfrac{5}{2}(2x) = 5x$

$25x + 10(2x) + 5(5x) = 1120$

$70x = 1120$

$x = 16$

16 quarters, $2(16) = 32$ dimes,

$5(16) = 80$ nickles

38. x: time on outgoing trip

$6 - x$: time on return trip

Distance on outgoing trip = distance on return trip $(RT = D)$

$45x = 55(6 - x)$

$45x = 330 - 55x$

$100x = 330$

$x = 3.3$

Distance in one direction =

$45(3.3) = 148.5$ miles

39. x: number of ounces of the 4% alcohol solution

alcohol in 4% solution + alcohol in 40% solution = alcohol in mixture

$.04x + (.4)(70) = .18(x + 70)$

$.04x + 28 = .18x + 12.6$

$15.4 = .14x$

$x = \dfrac{15.4}{.14} = 110$

110 ounces must be added

Review Problems: Chapter 2

1. $2(y-2) = 2[y - 5(1-y)]$

$2y - 4 = 2[y - 5 + 5y]$

$2y - 4 = 2(6y - 5)$

$2y - 4 = 12y - 10$

$-10y = -6$

$$y = \frac{3}{5} \qquad \left\{ \frac{3}{5} \right\}$$

2. $8\left(\frac{x}{4} - \frac{1}{2}x + \frac{1}{8} \right) = 8\left[\frac{1}{8}(5x - 3) \right]$

$2x - 4x + 1 = 5x - 3$

$-2x + 1 = 5x - 3$

$4 = 7x$

$$\frac{4}{7} = x \qquad \left\{ \frac{4}{7} \right\}$$

3. $-3\{5 - [1 - (1-x) + 2]\} =$

$\qquad\qquad\qquad 7 - [9x - 3(x+1)]$

$-3\{5 - x - 2\} = 7 - (6x - 3)$

$-9 + 3x = 10 - 6x$

$9x = 19$

$$x = \frac{19}{9} \qquad \left\{ \frac{19}{9} \right\}$$

4. $2(x+6) + 3(x+1) = 4x + 10 + x + 5$

$2x + 12 + 3x + 3 = 5x + 15$

$5x + 15 = 5x + 15$

$15 = 15 \qquad \{ x \mid x \in \mathfrak{R} \}$

5. $3(x-1) + 7(x-3) = 5(2x-5)$

$3x - 3 + 7x - 21 = 10x - 25$

$10x - 24 = 10x - 25$

$-24 = -25 \qquad\qquad \varnothing$

6. $6\left[\frac{1}{2}(3c - x) \right] = 6\left[\frac{1}{3}(5c + x) + 4c \right]$

$3(3c - x) = 2(5c + x) + 24c$

$9c - 3x = 10c + 2x + 24c$

$9c - 3x = 34c + 2x$

$-25c = 5x$

$-5c = x \qquad \{-5c\}$

7. $6\left(\frac{ay}{3} - \frac{by}{2} \right) = \left(\frac{17}{6} \right)6$

$2ay - 3by = 17$

$y(2a - 3b) = 17$

$$y = \frac{17}{2a - 3b} \qquad \left\{ \frac{17}{2a - 3b} \right\}$$

8. $3(z+5) \le 6(z+1)$

$3z + 15 \le 6z + 6$

$-3z \le -9$

$z \ge 3 \qquad \{ z \mid z \ge 3 \}$

9. $4\left[\frac{1}{2}(x-1) + \frac{1}{4}(x-3) \right] > 4(2)$

$2(x-1) + x - 3 > 8$

$2x - 2 + x - 3 > 8$

$3x - 5 > 8$

$3x > 13$

$$x > \frac{13}{3}$$

$$\left\{ x \mid x > \frac{13}{3} \right\}$$

10. $2(3x+2) > 3(2x+1)$

$6x + 4 > 6x + 3$

$4 > 3 \qquad \{ x \mid x \in \mathfrak{R} \}$

11. $7x + 3(x + 2) + 2 < 5(2x - 3) + 22$

$\qquad 7x + 3x + 6 + 2 < 10x - 15 + 22$

$\qquad\qquad 10x + 8 < 10x + 7$

$\qquad\qquad\qquad 8 < 7 \qquad\qquad \varnothing$

12. $\left|\dfrac{3x+4}{2}\right| = 5$

$\qquad \dfrac{3x+4}{2} = 5 \quad$ or $\quad -\left(\dfrac{3x+4}{2}\right) = 5$

$\qquad 3x + 4 = 10 \qquad\qquad 3x + 4 = -10$

$\qquad\quad 3x = 6 \qquad\qquad\qquad 3x = -14$

$\qquad\qquad x = 2 \qquad\qquad\qquad\quad x = -\dfrac{14}{3}$

$$\left\{-\dfrac{14}{3},\ 2\right\}$$

13. $|6x + 5| > 29$

$\qquad 6x + 5 > 29 \quad$ or $\quad 6x + 5 < -29$

$\qquad\quad 6x > 24 \qquad\qquad\quad 6x < -34$

$\qquad\qquad x > 4 \qquad\qquad\qquad x < -\dfrac{17}{3}$

$$\left\{x \mid x > 4 \text{ or } x < -\dfrac{17}{3}\right\}$$

14. $|7x + 3| \leq 4$

$\qquad -4 \leq 7x + 3 \leq 4$

$\qquad -7 \leq 7x \leq 1$

$\qquad -1 \leq x \leq \dfrac{1}{7}$

$$\left\{x \mid -1 \leq x \leq \dfrac{1}{7}\right\}$$

15. x: number

$\qquad 7x - 1 = 5x + 9$

$\qquad\quad 2x = 10$

$\qquad\quad\ x = 5 \qquad$ The number is 5.

16. x: number

$\qquad\qquad |4x - 1| = 19$

$\qquad 4x - 1 = 19 \quad$ or $\quad -(4x - 1) = 19$

$\qquad\quad 4x = 20 \qquad\qquad -4x + 1 = 19$

$\qquad\qquad x = 5 \qquad\qquad\qquad -4x = 18$

$\qquad\qquad\qquad\qquad\qquad\qquad x = \dfrac{18}{-4} = -\dfrac{9}{2}$

The numbers are 5 and $-\dfrac{9}{2}$.

17. x: number

$\qquad\qquad |3x + 2| > 5$

$\qquad 3x + 2 > 5 \quad$ or $\quad 3x + 2 < -5$

$\qquad\quad 3x > 3 \qquad\qquad\quad 3x < -7$

$\qquad\qquad x > 1 \qquad\qquad\qquad x < -\dfrac{7}{3}$

All numbers greater than 1 or all numbers less than $-\dfrac{7}{3}$.

18. x: number

$\qquad .23x = 3.4891$

$\qquad\quad x = \dfrac{3.4891}{.23} = 15.17$

The number is 15.17.

19. x: smaller number

$\qquad 3x + 26$: larger number

$\qquad x + 3x + 26 = 174$

$\qquad\qquad 4x = 148$

$\qquad\qquad\ x = 37$

The larger number is $3(37) + 26 = 137$.

20. Brother's age: x

Vera's age: $x - 4$

$$x - 4 + 10 = \frac{4}{5}(x + 10)$$

$$x + 6 = \frac{4}{5}(x + 10)$$

$$5(x + 6) = 5\left[\frac{4}{5}(x + 10)\right]$$

$$5x + 30 = 4x + 40$$

$$x = 10$$

Vera is $10 - 4 = 6$.

21. $\dfrac{\text{Inches}}{\text{miles}} : \dfrac{\frac{2}{3}}{400} = \dfrac{1\frac{3}{5}}{x}$

$$\frac{2}{3}x = (400)\left(\frac{8}{5}\right)$$

$$\frac{2}{3}x = 640$$

$$2x = 1920$$

$$x = 960 \qquad 960 \text{ miles.}$$

22. $\dfrac{\text{Tagged}}{\text{Total}} : \dfrac{133}{x} = \dfrac{70}{150}$

$$7x = (133)(15)$$

$$x = \frac{1995}{7} = 285 \qquad 285 \text{ deer.}$$

23. Consecutive even integers: $x, x + 2, x + 4$

$$x + (x + 4) = 3(x + 2)$$

$$2x + 4 = 3x + 6$$

$$-2 = x$$

Integers are $-2, 0, 2$.

24. x: no. of \$5 bills

$2x + 1$: no. of \$10 bills

$3x - 5$: no. of \$20 bills

$$5x + 10(2x + 1) + 20(3x - 5) = 165$$

$$5x + 20x + 10 + 60x - 100 = 165$$

$$85x = 255$$

$$x = 3$$

no. of \$5 bills : 3

no. of \$10 bills : $2(3) + 1 = 7$

no. of \$20 bills : $3(3) - 5 = 4$

25. x: pounds of almonds

$600 - x$: pounds of cashews

Value of almonds + value of cashews =

value of mixture

$$.4x + .7(600 - x) = .45(600)$$

$$.4x + 420 - .7x = 270$$

$$-.3x = -150$$

$$x = \frac{-150}{-.3} = 500$$

500 pounds of almonds and 100 pounds

of cashews

26. x: time $\qquad$ (RT = D)

Distance of car 1 + distance of car 2 = 660

$$50x + 60x = 660$$

$$110x = 660$$

$$x = 6$$

6 hours

27. speed of slower train: x

faster train: $x + 20$

Distance of slower + distance of faster =

500 $\quad$ (RT = D)

$$5x + 5(x + 20) = 500 \;\; ; \; x = 40$$

slower train: 40 m/h

faster train: 60 m/h

28. x: time in which planes meet

When the planes meet, the sum of their

distances = 3000.

$$650x + 550x = 3000 \; ; \;\; x = 2.5$$

They will meet in $2\frac{1}{2}$ hours.

29. x: amount at 13%

$13,800 - x$: amount at 17%

$$.13x + .17(13,800 - x) = 2138$$

$$.13x + 2346 - .17x = 2138$$

$$-.04x = -208$$

$$x = \frac{-208}{-.04} = 5200$$

\$5,200 at 13%;

$13,800 - 5,200 = \$8,600$ at 17%

30. $3x - 6 = 90$; $x = 32$

$2x = 2(32) = 64°$; $3x - 6 = 90°$;

$x - 5 = 32 - 5 = 27°$

$64° + 90° + 27° = 181°$

Thus, $\angle$ DOA is not a straight angle.

31. $(3x + 6) + (4x - 12) + (2x + 24) = 180$

$x = 18$

Angle 1: $3(18) + 6 = 60°$

Angle 2: $4(18) - 12 = 60°$

Angle 3: $2(18) + 24 = 60°$

32. $2[x - (1 - 3x)] - 1 = 3(x + 4)$

$2[4x - 1] - 1 = 3x + 12$

$8x - 3 = 3x + 12$

$5x = 15$

$x = 3$

One side: $3(x + 4) = 3(3 + 4) = 21$ yd

Perimeter: $4(21) = 84$ yd

Area: $(21)(21) = 441$ yd^2

33. x: side of square   2x: length of rectangle

$x - 1$: width of rectangle

perimeter of rectangle = perimeter of

square $+ 8$

$2(2x) + 2(x - 1) = 4x + 8$

$6x - 2 = 4x + 8$; $x = 5$

side of square: 5yd

34. Height : x   Non-parallel side: $x - 2$

Shorter base: $x + 2$

Longer base: $3x - 8$

Perimeter = 26yd

$2(x - 2) + (x + 2) + (3x - 8) = 26$

$2x - 4 + x + 2 + 3x - 8 = 26$

$6x = 36$

$x = 6$

Height: 6 yd   Shorter base: $6 + 2 = 8$ yd

Longer base: $3(6) - 8 = 10$ yd

$A = \frac{1}{2}h(B + b) = \frac{1}{2} \cdot \frac{6}{1}(8 + 10) =$

$3(18) = 54$ yd^2

35. Width : x   Length: $x - 3$

New width: $x - 5$

New length: $x - 3 + 4 = x + 1$

Area of new rectangle = area of original

rectangle $- 15$

$(x - 5)(x + 1) = x(x - 3) - 15$

$x^2 + x - 5x - 5 = x^2 - 3x - 15$

$x^2 - 4x - 5 = x^2 - 3x - 15$

$-4x - 5 = -3x - 15$

$10 = x$

Width: 10 ft   Length : $10 - 3 = 7$ ft

36. $24\left(\frac{y+2}{6} - \frac{y-3}{4}\right) = 24\left(\frac{y}{8}\right)$

$4(y + 2) - 6(y - 3) = 3y$

$4y + 8 - 6y + 18 = 3y$

$-2y + 26 = 3y$

$26 = 5y$

$\frac{26}{5} = y$ $\left\{\frac{26}{5}\right\}$

37.
$$R = 1.63t + 9.85$$
$$32.67 = 1.63t + 9.85$$
$$22.82 = 1.63t$$
$$t = \frac{22.82}{1.63} = 14$$
14 years after 1965 ; 1979

38. $400(3x - 1) + 700 = 1250x$
$$1200x + 300 = 1250x$$
$$300 = 50x$$
$$6 = x$$
a. 6 units; $1250x > 400(3x - 1) + 700$; $x > 6$
b. more than 6 units

39. $F = \frac{1}{4}C + 40$
$$4F = 4\left(\frac{1}{4}C + 40\right)$$
$$4F = C + 160$$
a. $4F - 160 = C$
b. $4F - 160 = 200$
$$4F = 360$$
$$F = 90; \text{ control temperature at } 90°$$

40. $600 - 5x < 505$
$$-5x < -95$$
$$x > 19 ; \text{ after 19 days}$$

41. x: no. of checks
method 1: $.06x + 11$ method 2: $.2x + 4$
Method 1 < Method 2
$$.06x + 11 < .2x + 4$$
$$-.14x < -7$$
$$x > 50 ; \text{ more than 50 checks}$$

42. $15 \leq \frac{5}{9}(F - 32) \leq 35$
$$9(15) \leq 9\left[\frac{5}{9}(F - 32)\right] \leq 9(35)$$
$$135 \leq 5F - 160 \leq 315$$
$$295 \leq 5F \leq 475$$
$$59 \leq F \leq 95$$
$59°$ to $95°$, inclusively

43.
$$|4x - 3| = |7x + 9|$$

$4x - 3 = 7x + 9$ $4x - 3 = -(7x + 9)$
$-3x = 12$ $4x - 3 = -7x - 9$
$x = -4$ $11x = -6$
$$x = -\frac{6}{11}$$

$$\left\{-4, \ -\frac{6}{11}\right\}$$

44. $|x - 73| \leq 67$
$$-67 \leq x - 73 \leq 67$$
$$6 \leq x \leq 140$$
Daily high: 140   Daily low: 6
Five-day high: $(140)(5) = 700$
Five-day low: $(6)(5) = 30$

45. $|p - .3\%| \leq .2\%$
$$-.002 \leq (p - .003) \leq .002$$
$$.001 \leq p \leq .005$$
Number of defective products
$$(100,000)(.001) \leq N \leq (100,000)(.005)$$
$$100 \leq N \leq 500$$
R: refund
$$(100)(5) \leq R \leq (500)(5)$$
$$\{R \mid \$500 \leq R \leq \$2,500\}$$

46. $3(5 - y) > 3\left(\frac{y - 1}{3}\right)$ $3\left(\frac{2y}{3} + 5\right) > 3(7)$
$15 - 3y > y - 1$ $2y + 15 > 21$
$-4y > -16$ $2y > 6$
$y < 4$ $y > 3$
$$\{y \mid 3 < y < 4\}$$

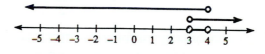

47. $-4 \le \dfrac{4-x}{3} < 0$

$$3(-4) \le 3\left(\dfrac{4-x}{3}\right) < 3(0)$$

$$-12 \le 4-x < 0$$

$$-16 \le -x < -4$$

$$16 \ge x > 4$$

$$\{x \mid 4 < x \le 16\}$$

48.

$5x+3 \le 3x-1$	$4-x > 5x-6$
$2x \le -4$	$-6x > -10$
$x \le -2$	$x < \dfrac{5}{3}$

$$\{x \mid x \le -2\}$$

49.

$3(3-y) > 3\left(\dfrac{2y-6}{3}\right)$	$3\left(\dfrac{5y}{3}+6\right) < 3(1)$
$9-3y > 2y-6$	$5y+18 < 3$
$-5y > -15$	$5y < -15$
$y < 3$	$y < -3$

$$\{y \mid y < 3\}$$

50.

$7x+5 < 2(3x-1)$	$7x+2 \ge 3(2x-1)$
$7x+5 < 6x-2$	$7x+2 \ge 6x-3$
$x < -7$	$x \ge -5$

$$\{x \mid x < -7 \ \text{ or } \ x \ge -5\}$$

Problem set 3.1

9. $-7r^3 + 8r^2 + 3r - 2$

$\underline{\qquad\quad -3r^2 + 7r + 4\qquad}$

$-7r^3 + 5r^2 + 10r + 2$

15. $\frac{1}{3}x^9y^2 - \frac{1}{5}x^5y \qquad\qquad + \frac{1}{2}x^2y^3 + 7$

$\underline{-\frac{1}{5}x^9y^2 + \frac{3}{5}x^5y + \frac{1}{4}x^4y - \frac{3}{4}x^2y^3 - \frac{1}{2}}$

$\left(\frac{1}{3} - \frac{1}{5}\right)x^9y^2 + \left(-\frac{1}{5} + \frac{3}{5}\right)x^5y + \frac{1}{4}x^4y +$

$\qquad\left(\frac{1}{2} - \frac{3}{4}\right)x^2y^3 + \left(7 - \frac{1}{2}\right)$

$= \left(\frac{5}{15} - \frac{3}{15}\right)x^9y^2 + \frac{2}{5}x^5y + \frac{1}{4}x^4y +$

$\qquad\left(\frac{2}{4} - \frac{3}{4}\right)x^2y^3 + 6\frac{1}{2}$

$= \frac{2}{15}x^9y^2 + \frac{2}{5}x^5y + \frac{1}{4}x^4y -$

$\qquad\frac{1}{4}x^2y^3 + \frac{13}{2}$

19. $(.07 + 1.8 - .11)x^4 + (.11 - .01 + .37)x^3 +$

$\qquad(-.2 + 1 + .001 + 10)x^2 + (5)x +$

$\qquad\qquad(.17 + .85 - .03)$

$= 1.76x^4 + 0.47x^3 + 10.801x^2 + 5x + 0.99$

23. $(1 - 5)y^{2n} + (7 + 3)y^n + (-3 + 8)$

$= -4y^{2n} + 10y^n + 5$

27. $13r^5 + 9r^4 - 5r^2 + 3r + 6 -$

$\qquad\left(-9r^5 - 7r^3 + 8r^2 + 11\right)$

$= (13 + 9)r^5 + 9r^4 + 7r^3 + (-5 - 8)r^2 +$

$\qquad\qquad 3r + (6 - 11)$

$= 22r^5 + 9r^4 + 7r^3 - 13r^2 + 3r - 5$

29. $-6x^3y^2 - 8x^2y + 11xy - 3 -$

$\qquad\left(7x^3y^2 - 5x^2y + 9xy - 3\right)$

$= (-6 - 7)x^3y^2 + (-8 + 5)x^2y +$

$\qquad(11 - 9)xy + (-3 + 3)$

$= -13x^3y^2 - 3x^2y + 2xy$

33. $-x^4y^3 - x^3y^2 + xy - 1 -$

$\qquad\left(x^4y^3 - x^3y^2 - xy + 1\right)$

$= (-1 - 1)x^4y^3 + (-1 + 1)x^3y^2 +$

$\qquad(1 + 1)xy + (-1 - 1)$

$= -2x^4y^3 + 2xy - 2$

35. $\left(3x^{2n} + 7x^n - 4\right) - \left(-2x^{2n} + 5x^n - 4\right)$

$= (3 + 2)x^{2n} + (7 - 5)x^n + (-4 + 4)$

$= 5x^{2n} + 2x^n$

39. $\left(5x^2 - 7x + 8\right) + \left(2x^2 - 3x + 7\right) -$

$\qquad\qquad\left(x^2 - 4x - 3\right)$

$= (5 + 2 - 1)x^2 + (-7 - 3 + 4)x + (8 + 7 + 3)$

$= 6x^2 - 6x + 18$

45. $\left(y^{3n} - 7y^{2n} + 3\right) - \left(-3y^{3n} - 2y^{2n} - 1\right) +$

$\qquad\qquad\left(6y^{3n} - y^{2n} + 1\right)$

$= (1 + 3 + 6)y^{3n} + (-7 + 2 - 1)y^{2n} +$

$\qquad\qquad(3 + 1 + 1)$

$= 10y^{3n} - 6y^{2n} + 5$

47. $\left(12a^2b^2 + 15a^2 + 3\right) +$

$\qquad \left(19a^2b^2 - 7a^2 + 1\right) -$

$\qquad\qquad \left(-11a^2b^2 - 5a^2 - 2\right)$

$\quad = (12 + 19 + 11)a^2b^2 + (15 - 7 + 5)a^2 +$

$\qquad (3 + 1 + 2)$

$\quad = 42a^2b^2 + 13a^2 + 6$

49. $\left(9 - 2a^3b^2 - a^2b^3 - ab^4\right) +$

$\qquad \left(4 - 3a^3b^2 - 7a^2b^3 + 5ab^4\right) -$

$\qquad\qquad \left(13 - 8a^2b^3 - 5a^3b^2 - 4ab^4\right)$

$\quad = (9 + 4 - 13) + (-2 - 3 + 5)a^3b^2 +$

$\qquad (-1 - 7 + 8)a^2b^3 + (-1 + 5 + 4)ab^4$

$\quad = 8ab^4$

55. $h_{moon} - h_{earth} = -2.7t^2 + 48t + 6 -$

$\qquad \left(-16t^2 + 48t + 6\right) = (-2.7 + 16)t^2 +$

$\qquad (48 - 48)t + (6 - 6) = 13.3t^2$

59. $P(x) = -2x^2 + 4x - 3;$

$\qquad P(-1) = -2(-1)^2 + 4(-1) - 3$

$\qquad = -2 - 4 - 3 = -9$

$\qquad P(2) = -2(2)^2 + 4(2) - 3 = -8 + 8 - 3 = -3$

61. $P(-1) = 2(-1)^4 - 2(-1)^2 - 5(-1)$

$\qquad = 2(1) - 2(1) - 5(-1) = 2 - 2 + 5 = 5$

$\qquad P(2) = 2 \cdot 2^4 - 2 \cdot 2^2 - 5 \cdot 2 = 32 - 8 - 10$

$\qquad = 14$

65. $\left[5x^2 - \left(8xy + 11y^2\right)\right] -$

$\qquad \left[6x^2 - \left(-8xy - 9y^2\right)\right]$

$\quad = \left[5x^2 - 8xy - 11y^2\right] -$

$\qquad\qquad \left[6x^2 + 8xy + 9y^2\right]$

$\quad = (5 - 6)x^2 + (-8 - 8)xy + (-11 - 9)y^2$

$\quad = -x^2 - 16xy - 20y^2$

69. Desired polynomial: $P(x)$

$\qquad P(x) - \left(4x^2 + 2x - 3\right) = 5x^2 - 5x + 8$

$\qquad P(x) - 4x^2 - 2x + 3 = 5x^2 - 5x + 8$

$\qquad P(x) = 5x^2 - 5x + 8 + 4x^2 + 2x - 3$

$\qquad P(x) = 9x^2 - 3x + 5$

71. $\left(y^3 + 8y - 3\right) - \left(y^3 - 6y + 5\right) = -1$

$\qquad y^3 + 8y - 3 - y^3 + 6y - 5 = -1$

$\qquad\qquad 14y - 8 = -1$

$\qquad\qquad\qquad 14y = 7$

$\qquad\qquad\qquad y = \frac{7}{14} = \frac{1}{2} \quad \left\{\frac{1}{2}\right\}$

73. True; $17 = 17x^0$

77. $12\left(\frac{1}{2} - \frac{7y}{4}\right) \geq 12\left(\frac{3y}{4} - \frac{5}{6}\right)$

$\qquad 6 - 21y \geq 9y - 10$

$\qquad -30y \geq -16$

$\qquad y \leq \frac{-16}{-30}$

$\qquad y \leq \frac{8}{15} \quad \left\{y \,\middle|\, y \leq \frac{8}{15}\right\}$

78. x: amount invested at 7%

$\quad$ 11,200 − x: amount invested at 9%

$\quad .07x + .09(11,200 − x) = 924$

$\quad .07x + 1008 − .09x = 924$

$\quad\quad\quad\quad −.02x = −84$

$\quad\quad\quad\quad\quad\quad x = 4200$

$\quad$ \$4,200 at 7%;

$\quad$ 11,200 − 4,200 = \$7,000 at 9%

79. x : number of liters of 45% salt solution

$\quad .45x + .6(40 − x) = (40)(.48)$

$\quad .45x + 24 − .6x = 19.2$

$\quad\quad\quad −.15x = −4.8$

$\quad\quad\quad\quad\quad x = 32 \; ; \; 32$ liters

Problem set 3.2.1

7. $\left(-3xy^2z^5\right)\left(2xy^7z^4\right)$

$\quad = -6x^{1+1}y^{2+7}z^{5+4} = -6x^2y^9z^9$

13. $\frac{1}{3}x^3y^7\left(\frac{1}{2}xy^6 + \frac{2}{5}x^4y^2 + 6\right)$

$\quad = \frac{1}{3}x^3y^7\left(\frac{1}{2}xy^6\right) + \frac{1}{3}x^3y^7\left(\frac{2}{5}x^4y^2\right) +$

$\quad \frac{1}{3}x^3y^7(6)$

$\quad = \frac{1}{6}x^4y^{13} + \frac{2}{15}x^7y^9 + 2x^3y^7$

17. $\left(6uv^3w - 8uv + w^4\right)\left(-5u^5v^3w\right)$

$\quad = \left(6uv^3w\right)\left(-5u^5v^3w\right) +$

$\quad (-8uv)\left(-5u^5v^3w\right) + w^4\left(-5u^5v^3w\right)$

$\quad = -30u^6v^6w^2 + 40u^6v^4w - 5u^5v^3w^5$

21. $y^{n-3}\left(y^{2n+7} - 3y^4 - 1\right)$

$\quad = \left(y^{n-3}\right)\left(y^{2n+7}\right) + \left(y^{n-3}\right)\left(-3y^4\right)$

$\quad + \left(y^{n-3}\right)(-1)$

$\quad = y^{n-3+2n+7} - 3y^{n-3+4} - y^{n-3}$

$\quad = y^{3n-4} - 3y^{n+1} - y^{n-3}$

23. $\left(2x^{n-4}y^{5n+3}\right) \cdot$

$\quad\quad \left(-6x^{3n+4}y^{-5n-3} - x^4y^5 + 2\right)$

$\quad = \left(2x^{n-4}y^{5n+3}\right)\left(-6x^{3n+4}y^{-5n-3}\right) +$

$\quad \left(2x^{n-4}y^{5n+3}\right)\left(-x^4y^5\right) +$

$\quad \left(2x^{n-4}y^{5n+3}\right)(2)$

$\quad = -12x^{n-4+3n+4}y^{5n+3-5n-3} -$

$\quad\quad 2x^{n-4+4}y^{5n+3+5} + 4x^{n-4}y^{5n+3}$

$\quad = -12x^{4n}y^0 - 2x^ny^{5n+8} +$

$\quad\quad 4x^{n-4}y^{5n+3}$

$\quad = -12x^{4n} - 2x^ny^{5n+8} + 4x^{n-4}y^{5n+3}$

25. $(x+y)\left(x^2 - xy + y^2\right) = x\left(x^2 - xy + y^2\right)$

$\quad + y\left(x^2 - xy + y^2\right) = x^3 - x^2y + xy^2 +$

$\quad x^2y - xy^2 + y^3 = x^3 + y^3$

29. $\left(a^2b^4 + 3\right)\left(a^4b^8 - 3a^2b^4 + 9\right)$

$\quad = a^2b^4\left(a^4b^8 - 3a^2b^4 + 9\right) +$

$\quad 3\left(a^4b^8 - 3a^2b^4 + 9\right)$

$\quad = a^6b^{12} - 3a^4b^8 + 9a^2b^4 + 3a^4b^8 -$

$\quad 9a^2b^4 + 27 = a^6b^{12} + 27$

31. $\left(9r^4 - r^3 - r^2 + r - 1\right)\left(3r^2 + 5r - 1\right)$

$= 3r^2\left(9r^4 - r^3 - r^2 + r - 1\right) +$

$\quad 5r\left(9r^4 - r^3 - r^2 + r - 1\right) -$

$\quad 1\left(9r^4 - r^3 - r^2 + r - 1\right)$

$= 27r^6 - 3r^5 - 3r^4 + 3r^3 - 3r^2 + 45r^5 -$

$\quad 5r^4 - 5r^3 + 5r^2 - 5r - 9r^4 + r^3 + r^2 - r + 1$

$= 27r^6 + 42r^5 - 17r^4 - r^3 + 3r^2 - 6r + 1$

33. $2y\left(5x^2y - 3y + 4x + x^3y - xy^2\right) +$

$\quad 3x^2y\left(5x^2y - 3y + 4x + x^3y - xy^2\right) -$

$\quad 4x\left(5x^2y - 3y + 4x + x^3y - xy^2\right)$

$= 10x^2y^2 - 6y^2 + 8xy + 2x^3y^2 - 2xy^3 +$

$\quad 15x^4y^2 - 9x^2y^2 + 12x^3y + 3x^5y^2 -$

$\quad 3x^3y^3 - 20x^3y + 12xy - 16x^2 - 4x^4y$

$\quad + 4x^2y^2$

$= 3x^5y^2 + 15x^4y^2 - 4x^4y - 3x^3y^3 +$

$\quad 2x^3y^2 - 8x^3y + 5x^2y^2 - 16x^2 +$

$\quad 20xy - 6y^2$

35. $\left(9x^2 - 4\right)\left(3x^2 + 5\right) = \left(9x^2\right)\left(3x^2\right) +$

$\quad \left(9x^2\right)(5) + \left(3x^2\right)(-4) + (-4)(5)$

$= 27x^4 + 45x^2 - 12x^2 - 20$

$= 27x^4 + 33x^2 - 20$

39. $\left(3x^3 + 2y^2\right)(5x + 4y) = \left(3x^3\right)(5x) +$

$\quad \left(3x^3\right)(4y) + \left(2y^2\right)(5x) + \left(2y^2\right)(4y)$

$= 15x^4 + 12x^3y + 10xy^2 + 8y^3$

47. $(3a - 2b)^2 = (3a - 2b)(3a - 2b)$

$= 9a^2 - 6ab - 6ab + 4b^2$

$= 9a^2 - 12ab + 4b^2$

55. $\left(9y^n - 2\right)\left(y^n + 4\right) = \left(9y^n\right)\left(y^n\right) +$

$\quad \left(9y^n\right)(4) + (-2)\left(y^n\right) + (-2)(4)$

$= 9y^{2n} + 36y^n - 2y^n - 8$

$= 9y^{2n} + 34y^n - 8$

57. $\left(4x^{2a} - y^{5a}\right)\left(3x^{5a} - y^{2a}\right)$

$= 12x^{7a} - 4x^{2a}y^{2a} - 3x^{5a}y^{5a} + y^{7a}$

59. $\left(4x^2y + 5x\right)\left(4x^2y - 5x\right)$

$= \left(4x^2y\right)^2 - (5x)^2 = 16x^4y^2 - 25x^2$

61. $\left(-3a^4b^2 + 5c\right)\left(3a^4b^2 + 5c\right)$

$= \left(5c - 3a^4b^2\right)\left(5c + 3a^4b^2\right)$

$= (5c)^2 - \left(3a^4b^2\right)^2 = 25c^2 - 9a^8b^4$

67. $\left(7x^{n-1} + y^{3n}\right)\left(7x^{n-1} - y^{3n}\right)$

$= \left(7x^{n-1}\right)^2 - \left(y^{3n}\right)^2$

$= 49x^{2n-2} - y^{6n}$

71. $\left(x^n + 4\right)^2 = \left(x^n\right)^2 + 2 \cdot x^n \cdot 4 + 4^2$

$= x^{2n} + 8x^n + 16$

73. $\left(x^{2n} - 6y^n\right)^2 = \left(x^{2n}\right)^2 - 2 \cdot x^{2n} \cdot 6y^n$

$+ \left(6y^n\right)^2 = x^{4n} - 12x^{2n}y^n + 36y^{2n}$

77. $\left(5m^2n^3 - 3mn^2\right)^2 = \left(5m^2n^3\right)^2 -$

$2\left(5m^2n^3\right)\left(3mn^2\right) + \left(3mn^2\right)^2$

$= 25m^4n^6 - 30m^3n^5 + 9m^2n^4$

79. $(3x + 7 + 5y)(3x + 7 - 5y)$

$= (3x + 7)^2 - (5y)^2 = (3x)^2 + 2(3x)(7)$

$+ 7^2 - (5y)^2 = 9x^2 + 42x + 49 - 25y^2$

81. $\left[5y - (2x + 3)\right]\left[5y + (2x + 3)\right]$

$= (5y)^2 - (2x + 3)^2 = 25y^2 -$

$\left[(2x)^2 + 2 \cdot (2x) \cdot 3 + 3^2\right]$

$= 25y^2 - 4x^2 - 12x - 9$

83. $(2x + y + 1)^2 = \left[(2x + y) + 1\right]^2 = (2x + y)^2$

$+ 2 \cdot (2x + y) \cdot 1 + 1^2$

$= 4x^2 + 4xy + y^2 + 4x + 2y + 1$

87. $\left[(3x - 1) + y\right]\left[(3x - 1) - y\right]$

$= (3x - 1)^2 - y^2 = 9x^2 - 6x + 1 - y^2$

89. $(x + y - 3)(x - y + 3)$

$= \left[x + (y - 3)\right]\left[x - (y - 3)\right] = x^2 - (y - 3)^2$

$= x^2 - \left(y^2 - 6y + 9\right) = x^2 - y^2 + 6y - 9$

91. $(2a + b)^3 = (2a + b)(2a + b)^2$

$= (2a + b)\left(4a^2 + 4ab + b^2\right) = 8a^3 + 8a^2b$

$+ 2ab^2 + 4a^2b + 4ab^2 + b^3$

$= 8a^3 + 12a^2b + 6ab^2 + b^3$

95. $\left(x^4 - 2x^5\right)^3 = \left(x^4 - 2x^5\right)\left(x^4 - 2x^5\right)^2$

$= \left(x^4 - 2x^5\right)\left(x^8 - 4x^9 + 4x^{10}\right)$

$= x^{12} - 4x^{13} + 4x^{14} - 2x^{13}$

$+ 8x^{14} - 8x^{15}$

$= -8x^{15} + 12x^{14} - 6x^{13} + x^{12}$

99. Profit = Revenue − Cost

$= 25x + 20y -$

$\left[300 + 10x + 13y + .01(x + y)^2\right]$

$= 25x + 20y -$

$\left[300 + 10x + 13y + .01x^2 + .02xy + .01y^2\right]$

$= 25x + 20y - 300 - 10x - 13y - .01x^2$

$- .02xy - .01y^2$

$= -.01x^2 + 15x + 7y - .02xy$

$- .01y^2 - 300$

103. $A_1 = 1x = x \qquad A_3 = 1x = x$

$A_2 = (x + 3)(1 + x + 1) = (x + 3)(x + 2)$

$= x^2 + 5x + 6$

$A = A_1 + A_2 + A_3$

$= x + x + x^2 + 5x + 6 = x^2 + 7x + 6$

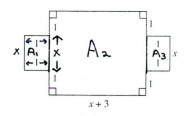

105. $A_1 = \frac{1}{2}(2x)x = x^2$

$A_2 = (2x)(x+10) = 2x^2 + 20x$

$A_3 = (x+2)(4x-2x) = (x+2)(2x)$

$\quad = 2x^2 + 4x$

$A_4 = \frac{1}{2}(6x-4x)(x+2) = \frac{1}{2}(2x)(x+2)$

$\quad = x^2 + 2x$

$A = A_1 + A_2 + A_3 + A_4 = 6x^2 + 26x$

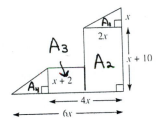

107. Volume of entire solid with " slit " on top
filled in = LWH

$= (2x+1)(x+5)(x+2)$

$= (2x+1)\left(x^2+7x+10\right)$

$= 2x^3 + 15x^2 + 27x + 10$

Volume of solid " slit " on top = LWH

$= x(x+5)3 = 3x^2 + 15x$

Desired volume = volume of entire solid
− volume of slit

$= 2x^3 + 15x^2 + 27x + 10 - \left(3x^2 + 15x\right)$

$= 2x^3 + 12x^2 + 12x + 10$

111. $P(x)$: desired polynomial

$\dfrac{P(x)}{x+1} = x^2 - x + 1$

$P(x) = (x+1)\left(x^2 - x + 1\right)$

$= x^3 - x^2 + x + x^2 - x + 1 = x^3 + 1$

113. $(11-5)^2 = (11-5)(11-5)$

$= 11^2 - 2 \cdot 11 \cdot 5 - 5^2$

$= 121 - 110 + 25$

114. $|4-3x| \geq 10$

$\quad 4-3x \geq 10 \quad$ or $\quad 4-3x \leq -10$

$\quad\quad -3x \geq 6 \quad\quad\quad\quad -3x \leq -14$

$\quad\quad\quad x \leq -2 \quad\quad\quad\quad\quad x \geq \dfrac{14}{3}$

$$\left\{ x \mid x \leq -2 \ \text{ or } \ x \geq \dfrac{14}{3}\right\}$$

115. $\left[\left(2y^2 - 3y - 1\right) + \left(3y^2 + y - 1\right)\right] -$

$$\left(y^2 + 3y - 1\right)$$

$= \left(5y^2 - 2y - 2\right) - y^2 - 3y + 1$

$= 4y^2 - 5y - 1$

116. measure of $\angle DBC$: x

measure of $\angle ABD$: $12x - 2$

$x + 12x - 2 = 180$

$\quad\quad\quad 13x = 182$

$\quad\quad\quad\quad x = 14$

measure of $\angle ABD = 12(14) - 2 = 166°$

measure of $\angle DBC$: $14°$

Problem set 3.2.2

3. x: first integer

$x+2$: next consecutive odd integer

$x(x+2) = (x+2)^2 - 30$

$x^2 + 2x = x^2 + 4x + 4 - 30$

$\quad\quad 2x = 4x - 26$

$\quad -2x = -26$

$\quad\quad\quad x = 13 \quad$ The integers are 13 and 15.

5. consecutive integers: $x, x+1, x+2$

$$x(x+1)(x+2) = (x+1)^3 - 18$$

$$x\left(x^2+3x+2\right) = (x+1)(x+1)^2 - 18$$

$$x^3 + 3x^2 + 2x = (x+1)\left(x^2+2x+1\right) - 18$$

$$x^3 + 3x^2 + 2x = x^3 + 3x^2 + 3x - 17$$

$$2x = 3x - 17$$

$$17 = x$$

The integers are 17,18,19.

9.

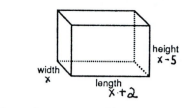

Area of top + area of bottom = sum of areas of two sides + 112

$$x(x+2) + x(x+2) = x(x-5) + $$
$$x(x-5) + 112$$

$$2x^2 + 4x = 2x^2 - 10x + 112$$

$$14x = 112$$

$$x = 8$$

width: $x = 8$ cm

length: $x + 2 = 8 + 2 = 10$ cm

height: $x - 5 = 8 - 5 = 3$ cm

11. x: width of picture

$x+2$: length of picture

Area of picture + frame =

area of picture + 108

$$(x+6)(x+8) = x(x+2) + 108$$

$$x^2 + 14x + 48 = x^2 + 2x + 108$$

$$12x = 60$$

$$x = 5$$

width of picture : $x = 5$cm

length of picture: $x + 2 = 5 + 2 = 7$cm

13. x: edge of middle cube

$x+2$: edge of largest cube

$x-3$: edge of smallest cube

Surface area of largest cube = Surface area of smallest cube + 210

$$6(x+2)^2 = 6(x-3)^2 + 210$$

$$6\left(x^2+4x+4\right) = 6\left(x^2-6x+9\right) + 210$$

$$6x^2 + 24x + 24 = 6x^2 - 36x + 54 + 210$$

$$24x + 24 = -36x + 264$$

$$60x = 240$$

$$x = 4$$

edge of smallest cube: $x - 3 = 4 - 3 = 1$ cm

edge of middle cube: $x = 4$ cm

edge of largest cube: $x + 2 = 4 + 2 = 6$ cm

15.

$y - 1 \le 7y - 1$	$4y - 7 < 3 - y$
$-6y \le 0$	$5y < 10$
$y \ge 0$	$y < 2$

$$\{y \mid 0 \le y < 2\}$$

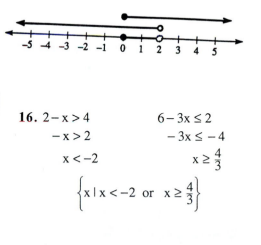

16.

$2 - x > 4$	$6 - 3x \le 2$
$-x > 2$	$-3x \le -4$
$x < -2$	$x \ge \frac{4}{3}$

$$\left\{x \mid x < -2 \text{ or } x \ge \frac{4}{3}\right\}$$

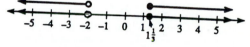

17. Pennies: x Dimes: $2x+1$

Quarters: $x-3$

Total value = 303 ¢

$1x + 10(2x+1) + 25(x-3) = 303$

$46x - 65 = 303$

$46x = 368$

$x = 8$

Pennies: 8   Dimes: $2(8)+1 = 17$

Quarters: $8 - 3 = 5$

Problem set 3.3.1

7. $\dfrac{-54x^7y^4z^5}{3x^3yz^2} = -18x^{7-3}y^{4-1}z^{5-2} =$

$-18x^4y^3z^3$

9. $\dfrac{24x^7 - 15x^4 + 18x^3}{3x} = \dfrac{24x^7}{3x} - \dfrac{15x^4}{3x} +$

$\dfrac{18x^3}{3x} = 8x^6 - 5x^3 + 6x^2$

13. $\dfrac{64x^7 - 20x^3 + 24x - 80}{-4x^3} = \dfrac{64x^7}{-4x^3} - \dfrac{20x^3}{-4x^3} +$

$\dfrac{24x}{-4x^3} - \dfrac{80}{-4x^3} = -16x^4 + 5x^0 - 6x^{-2} +$

$20x^{-3} = -16x^4 + 5 - \dfrac{6}{x^2} + \dfrac{20}{x^3}$

17. $\dfrac{16x^3y^2 - 28x^2y^3 - 20xy^5}{4x^2y} =$

$\dfrac{16x^3y^2}{4x^2y} - \dfrac{28x^2y^3}{4x^2y} - \dfrac{20xy^5}{4x^2y}$

$= 4xy - 7y^2 - 5x^{-1}y^4$

$= 4xy - 7y^2 - \dfrac{5y^4}{x}$

19. $\dfrac{36x^4y^3 - 18x^5z - 12xz^2}{6x^2yz} =$

$\dfrac{36x^4y^3z^0}{6x^2yz} - \dfrac{18x^5y^0z}{6x^2yz} - \dfrac{12xy^0z^2}{6x^2yz}$

$= 6x^2y^2z^{-1} - 3x^3y^{-1}z^0 - 2x^{-1}y^{-1}z^2$

$= \dfrac{6x^2y^2}{z} - \dfrac{3x^3}{y} - \dfrac{2z^2}{xy}$

25.

$$\begin{array}{r} b+2 \\ b-6\overline{)b^2-4b-12} \\ \underline{b^2-6b} \\ 2b-12 \\ \underline{2b-12} \\ 0 \end{array}$$

27.

$$\begin{array}{r} c+12 \\ -c+2\overline{)-c^2-10c+24} \\ \underline{-c^2+2c} \\ -12c+24 \\ \underline{-12c+24} \\ 0 \end{array}$$

31.

$$\begin{array}{r} 2b^2+3b+5 \\ 3b+4\overline{)6b^3+17b^2+27b+20} \\ \underline{6b^3+8b^2} \\ 9b^2+27b \\ \underline{9b^2+12b} \\ 15b+20 \\ \underline{15b+20} \\ 0 \end{array}$$

35.

$$4x + 3 + \frac{2}{3x-2}$$

$$3x-2 \,\big)\, 12x^2 + x - 4$$

$$\underline{12x^2 - 8x}$$
$$9x - 4$$
$$\underline{9x - 6}$$
$$2$$

39.

$$4x^3 + 16x^2 + 60x + 246 + \frac{984}{x-4}$$

$$x-4 \,\big)\, 4x^4 + 0x^3 - 4x^2 + 6x + 0$$

$$\underline{4x^4 - 16x^3}$$
$$16x^3 - 4x^2$$
$$\underline{16x^3 - 64x^2}$$
$$60x^2 + 6x$$
$$\underline{60x^2 - 240x}$$
$$246x + 0$$
$$\underline{246x - 984}$$
$$984$$

41.

$$x^2 + x + 1$$

$$x-1 \,\big)\, x^3 + 0x^2 + 0x - 1$$

$$\underline{x^3 - x^2}$$
$$x^2 + 0x$$
$$\underline{x^2 - x}$$
$$x - 1$$
$$\underline{x - 1}$$
$$0$$

43.

$$2a + 5$$

$$3a^2 - a - 3 \,\big)\, 6a^3 + 13a^2 - 11a - 15$$

$$\underline{6a^3 - 2a^2 - 6a}$$
$$15a^2 - 5a - 15$$
$$\underline{15a^2 - 5a - 15}$$
$$0$$

47.

$$6y^2 + 3y - 1 + \frac{-3y+1}{3y^2+1}$$

$$3y^2 + 1 \,\big)\, 18y^4 + 9y^3 + 3y^2 + 0y + 0$$

$$\underline{18y^4 \qquad\quad + 6y^2}$$
$$9y^3 - 3y^2 + 0y$$
$$\underline{9y^3 \qquad\quad + 3y}$$
$$-3y^2 - 3y + 0$$
$$\underline{-3y^2 \qquad\quad - 1}$$
$$-3y + 1$$

49.

$$x + y$$

$$x-y \,\big)\, x^2 + 0xy - y^2$$

$$\underline{x^2 - xy}$$
$$xy - y^2$$
$$\underline{xy - y^2}$$
$$0$$

51.

$$a^3 + a^2 b + ab^2 + b^3$$

$$a-b \,\big)\, a^4 + 0a^3b + 0a^2b^2 + 0ab^3 - b^4$$

$$\underline{a^4 - a^3b}$$
$$a^3b + 0a^2b^2$$
$$\underline{a^3b - a^2b^2}$$
$$a^2b^2 + 0ab^3$$
$$\underline{a^2b^2 - ab^3}$$
$$ab^3 - b^4$$
$$\underline{ab^3 - b^4}$$
$$0$$

53.

$$x^2 - x + 2 \overline{\smash{\big)}\ x^5 + 0x^4 + 0x^3 + 0x^2 + 0x - 1} \quad \left[\ x^3 + x^2 - x - 3 + \dfrac{-x+5}{x^2-x+2}\right]$$

$$\underline{x^5 - x^4 + 2x^3}$$
$$x^4 - 2x^3 + 0x^2$$
$$\underline{x^4 - x^3 + 2x^2}$$
$$-x^3 - 2x^2 + 0x$$
$$\underline{-x^3 + x^2 - 2x}$$
$$-3x^2 + 2x - 1$$
$$\underline{-3x^2 + 3x - 6}$$
$$-x + 5$$

55. $V = LWH$

$$11y^2 + 6y^3 + 6 - 19y = (2y-1)(y+3)H$$

$$6y^3 + 11y^2 - 19y + 6 = \left(2y^2 + 5y - 3\right)H$$

$$\dfrac{6y^3 + 11y^2 - 19y + 6}{2y^2 + 5y - 3} = H$$

$$2y^2 + 5y - 3 \overline{\smash{\big)}\ 6y^3 + 11y^2 - 19y + 6} \quad [\ 3y - 2\ ]$$
$$\underline{6y^3 + 15y^2 - 9y}$$
$$-4y^2 - 10y + 6$$
$$\underline{-4y^2 - 10y + 6}$$
$$0$$

$$H = 3y - 2$$

59.

$$4y + 3 \overline{\smash{\big)}\ 20y^3 + 23y^2 - 10y + K} \quad [\ 5y^2 + 2y - 4\ ]$$
$$\underline{20y^3 + 15y^2}$$
$$8y^2 - 10y$$
$$\underline{8y^2 + 6y}$$
$$-16y + K$$
$$\underline{-16y - 12}$$
$$K + 12$$

Remainder $K + 12 = 0$ $K = -12$

61.

$$3y^n - 1 \overline{\smash{\big)}\ 27y^{3n} + 0y^{2n} + 0y^n - 1} \quad [\ 9y^{2n} + 3y^n + 1\ ]$$
$$\underline{27y^{3n} - 9y^{2n}}$$
$$9y^{2n} + 0y^n$$
$$\underline{9y^{2n} - 3y^n}$$
$$3y^n - 1$$
$$\underline{3y^n - 1}$$
$$0$$

63. Note: $2 + \dfrac{2}{3}$

$$3\ \overline{|\ 8\ } \qquad \text{and, thus, } 3\left(2 + \dfrac{2}{3}\right) = 8$$

Similarly:

$$P(x) \overline{|\ 2x^2 - 7x + 9\ } \quad \left[\ 2x - 3 + \dfrac{3}{P(x)}\ \right] \quad \text{means that}$$

$$P(x)\left[2x - 3 + \dfrac{3}{P(x)}\right] = 2x^2 - 7x + 9$$

$$2x\,P(x) - 3\,P(x) + 3 = 2x^2 - 7x + 9$$

$$P(x)(2x - 3) = 2x^2 - 7x + 6$$

$$P(x) = \dfrac{2x^2 - 7x + 6}{2x - 3}$$

$$2x - 3 \overline{\smash{\big)}\ 2x^2 - 7x + 6} \quad [\ x - 2\ ]$$
$$\underline{2x^2 - 3x}$$
$$-4x + 6$$
$$\underline{-4x + 6}$$
$$0$$

$$P(x) = x - 2$$

64. x: number

$$\left|\dfrac{x}{2} - 4\right| \leq 2$$

$$-2 \leq \dfrac{x}{2} - 4 \leq 2$$

$$-4 \leq x - 8 \leq 4$$

$$4 \leq x \leq 12 \qquad \{x \mid 4 \leq x \leq 12\}$$

65. $\dfrac{\text{Original cost}}{\text{New cost}} = \dfrac{\text{Original Area}}{\text{New Area}} : \left(A = \pi\, r^2\right)$

$\dfrac{5}{x} = \dfrac{\pi \cdot 6^2}{\pi \cdot 12^2}$ $(d = 2 \text{ ft}; r = 1 \text{ ft} = 12 \text{ in})$

$\dfrac{5}{x} = \dfrac{36}{144}$

$\dfrac{5}{x} = \dfrac{1}{4}$

$x = 5(4)$

$x = 20$ Cost is \$ 20

66. $3b^2 - b\{5b - 2[6 - (b-4)]\}$

$= 3b^2 - b\{5b - 2[6 - b + 4]\}$

$= 3b^2 - b\{5b - 2[10 - b]\}$

$= 3b^2 - b\{5b - 20 + 2b\}$

$= 3b^2 - b\{7b - 20\}$

$= 3b^2 - 7b^2 + 20b$

$= -4b^2 + 20b$

Problem set 3.3.2

3. $\dfrac{3x^2 + 7x - 20}{x + 5}$

$\underline{-5|}\ \ 3\quad\ \ 7\quad -20$
$\qquad\qquad -15\quad\ \ 40$
$\qquad\ \ 3\quad -8\quad\ \ 20$

$= 3x - 8 + \dfrac{20}{x+5}$

5. $\dfrac{4x^3 - 3x^2 + 3x - 1}{x - 1}$

$\underline{1|}\ \ 4\quad -3\quad\ \ 3\quad -1$
$\qquad\qquad\ \ 4\quad\ \ 1\quad\ \ 4$
$\qquad\ \ 4\quad\ \ 1\quad\ \ 4\quad\ \ 3$

$= 4x^2 + x + 4 + \dfrac{3}{x-1}$

7. $\dfrac{6y^5 + 0y^4 - 2y^3 + 4y^2 - 3y + 1}{y - 2}$

$\underline{2|}\ \ 6\quad\ \ 0\quad -2\quad\ \ 4\quad -3\quad\ \ 1$
$\qquad\qquad\ \ 12\quad\ \ 24\quad\ 44\quad\ 96\quad\ 186$
$\qquad\ \ 6\quad\ 12\quad\ \ 22\quad\ 48\quad\ 93\quad\ 187$

$= 6y^4 + 12y^3 + 22y^2 + 48y + 93 + \dfrac{187}{y-2}$

11. $\dfrac{z^5 + 0z^4 + z^3 + 0z^2 + 0z - 2}{z - 1}$

$\underline{1|}\ \ 1\quad\ \ 0\quad\ \ 1\quad\ \ 0\quad\ \ 0\quad -2$
$\qquad\qquad\ \ 1\quad\ \ 1\quad\ \ 2\quad\ \ 2\quad\ \ 2$
$\qquad\ \ 1\quad\ \ 1\quad\ \ 2\quad\ \ 2\quad\ \ 2\quad\ \ 0$

$= z^4 + z^3 + 2z^2 + 2z + 2$

13. $\dfrac{y^4 + 0y^3 + 0y^2 + 0y - 256}{y - 4}$

$\underline{4|}\ \ 1\quad\ \ 0\quad\ \ 0\quad\ \ 0\quad -256$
$\qquad\qquad\ \ 4\quad\ \ 16\quad\ 64\quad\ \ 256$
$\qquad\ \ 1\quad\ \ 4\quad\ \ 16\quad\ 64\quad\ \ 0$

$= y^3 + 4y^2 + 16y + 64$

17. $\underline{5|}\ \ 2\quad -8\quad -7\quad -15$
$\qquad\qquad\ \ 10\quad\ \ 10\quad\ \ 15$
$\qquad\ \ 2\quad\ \ 2\quad\ \ 3\quad\ \ 0$

Since the remainder is 0, 5 is a solution

to $2x^3 - 8x^2 - 7x - 15 = 0$.

21. $\dfrac{1}{3}\Big|\ 3\quad -1\quad\ \ 9\quad -1\quad\ \ 0$
$\qquad\qquad\ \ 1\quad\ \ 0\quad\ \ 3\quad\ \ \dfrac{2}{3}$
$\qquad\ \ 3\quad\ \ 0\quad\ \ 9\quad\ \ 2\quad\ \ \dfrac{2}{3}$

Since the remainder is not 0, $\dfrac{1}{3}$ is not a

solution to $3x^4 - x^3 + 9x^2 - x = 0$.

23.

$$3 \,|\, 5 \quad 2 \quad 0 \quad -4$$
$$ 15 \quad 51 \quad 153$$
$$ 5 \quad 17 \quad 51 \quad 149$$

$P(3) = 149$

27. Find $C(200)$. This will be the remainder.

$$200 \,|\, -.01 \quad 6 \quad 400$$
$$ -2 \quad 800$$
$$ -.01 \quad 4 \quad 1200$$

$C(200) = 1200$. It will cost $1,200.00.

29. $\dfrac{-2x^3 + 3x^2 + x + K}{x+1}$:

$$-1 \,|\, -2 \quad 3 \quad 1 \quad K$$
$$ 2 \quad -5 \quad 4$$
$$ -2 \quad 5 \quad -4 \quad K+4$$

$K + 4 = 3$

$K = -1$

31. $\dfrac{15y^3 - 5y^2}{5y^2} - \dfrac{12y^4 - 6y^2}{6y^2}$

$= 3y - 1 - \left(2y^2 - 1 \right)$

$= 3y - 1 - 2y^2 + 1 = -2y^2 + 3y$

32. x: no. of ml of water to be added

No. of ml of water in original solution +
No. of ml of water to be added = No. of
ml of water in the final solution

$(.6)(500) + x = .75(500 + x)$

$300 + x = 375 + .75x$

$.25x = 75$

$x = 300$

300 ml of water must be added.

33. Bases: x and 3x

$\dfrac{1}{2} \cdot \dfrac{8}{1}(x + 3x) = 96$

$4(4x) = 96$

$16x = 96$

$x = 6$

Bases have lengths of 6 cm and 18 cm.

Problem set 3.4.1

21. GCF $= 12xy^2$

$\dfrac{24xy^3 - 36x^3y^2 + 12x^2y^4}{12xy^2} =$

$ 2y - 3x^2 + xy^2$

$24xy^3 - 36x^3y^2 + 12x^2y^4 =$

$ 12xy^2\left(2y - 3x^2 + xy^2 \right)$

23. $26 = 2 \cdot 13 \qquad 52 = 2 \cdot 26 = 2 \cdot 2 \cdot 13$

$39 = 3 \cdot 13$

The GCF has a coefficient of 13.

GCF $= 13x^5y^2$ (Have you noticed that
the variable factors in the GCF contain
the variable raised to the <u>smallest power</u>
that appears in the terms ?)

$\dfrac{26x^5y^3 + 52x^7y^2 - 39x^8y^5}{13x^5y^2} =$

$ 2y + 4x^2 - 3x^3y^3$

$26x^5y^3 + 52x^7y^2 - 39x^8y^5 =$

$ 13x^5y^2\left(2y + 4x^2 - 3x^3y^3 \right)$

27. $55 = 11 \cdot 5 \qquad 77 = 11 \cdot 7$

Coefficient of GCF $= 11$

Variable factors are x^2, y^2, and z since
2 is the smallest power of x, 2 is the
smallest power of y, and 1 is the smallest
power of z in the terms.

GCF $= 11x^2y^2z$

$55x^2y^2z^4 - 77x^3y^2z =$

$ 11x^2y^2z\left(5z^3 - 7x \right)$

31. $70 = 7 \cdot 5 \cdot 2 \qquad 42 = 7 \cdot 3 \cdot 2$

$ 28 = 7 \cdot 2 \cdot 2 \qquad 84 = 7 \cdot 3 \cdot 2 \cdot 2$

Coefficient of GCF $= 7 \cdot 2 = 14$

Variable factors are x and y.

GCF $= 14xy$

$70x^3y + 42x^2y - 28x^2y^2 - 84xy^3 =$

$$14xy\left(5x^2 + 3x - 2xy - 6y^2\right)$$

35. $-2a(x+7y) + 4c(x+7y)$

$= (x+7y)(-2a+4c)$

$= (x+7y) \cdot 2(-a+2c)$

$= 2(x+7y)(2c-a)$

37. $4x(a+b-c) - 2y(a+b-c)$

$= (a+b-c)(4x-2y)$

$= (a+b-c) \cdot 2(2x-y)$

$= 2(a+b-c)(2x-y)$

41. $(x-4y)z^2 + (x-4y)z + (4y-x)$

$= (x-4y)z^2 + (x-4y)z + (-4y+x)(-1)$

$= (x-4y)z^2 + (x-4y)z + (x-4y)(-1)$

$= (x-4y)\left(z^2 + z - 1\right)$

43. $5x^3(2a-7b) + 15x^2(2a-7b)$

GCF $= 5x^2(2a-7b)$

$= 5x^2(2a-7b)(x+3)$

A second method involves first factoring out $(2a-7b)$.

$5x^3(2a-7b) + 15x^2(2a-7b)$

$= (2a-7b)\left(5x^3 + 15x^2\right)$

$= (2a-7b) \cdot 5x^2(x+3)$

$= 5x^2(2a-7b)(x+3)$

49. $(7x-4y)a - b(4y-7x)$

$= (7x-4y)a - b(-1)(-4y+7x)$

$= (7x-4y)a + b(7x-4y)$

$= (7x-4y)(a+b)$

51. $11a(7x-4y) + (4y-7x)$

$= 11a(7x-4y) + (-1)(-4y+7x)$

$= 11a(7x-4y) - (7x-4y)$

$= (7x-4y)(11a-1)$

53. $a^2c + 5ac + 2a + 10$

$= ac(a+5) + 2(a+5)$

$= (a+5)(ac+2)$

55. $3Y^2 + 4YZ + 24Y + 32Z$

$= Y(3Y+4Z) + 8(3Y+4Z)$

$= (3Y+4Z)(Y+8)$

59. $c + d + 2c^2 + 2cd$

$= (c+d) + 2c(c+d)$

$= (c+d)(1+2c)$

61. $4x^2y + 16xy - x - 4$

$= 4xy(x+4) - (x+4)$

$= (x+4)(4xy-1)$

65. $4x^2 + 4xy - 3xy^2 - 3y^3$

$= 4x(x+y) - 3y^2(x+y)$

$= (x+y)\left(4x - 3y^2\right)$

69. $ab - c - ac + b$

$= ab + b - ac - c$

$= b(a+1) - c(a+1)$

$= (a+1)(b-c)$

71. $9a^3 + 10b^2 - 6a^2b - 15ab$

$= 9a^3 - 6a^2b + 10b^2 - 15ab$

$= 3a^2(3a - 2b) + 5b(2b - 3a)$

$= 3a^2(3a - 2b) + 5b(-1)(-2b + 3a)$

$= 3a^2(3a - 2b) - 5b(3a - 2b)$

$= (3a - 2b)\left(3a^2 - 5b\right)$

73. $ax + ay + az - bx - by - bz + cx + cy + cz$

$= a(x + y + z) - b(x + y + z) + c(x + y + z)$

$= (x + y + z)(a - b + c)$

75. $GCF = x^{n+2}$

$$\dfrac{4x^{n+3} - 5x^{n+2}}{x^{n+2}}$$

$= 4x^{n+3-(n+2)} - 5x^{n+2-(n+2)}$

$= 4x - 5x^0 = 4x - 5$

$4x^{n+3} - 5x^{n+2} = x^{n+2}(4x - 5)$

77. $GCF = 3x^{3n}$

(x^{3n} involves the lowest power of x in all three terms.)

$$\dfrac{3x^{5n} - 9x^{4n} + 6x^{3n}}{3x^{3n}}$$

$= x^{5n-3n} - 3x^{4n-3n} + 2x^{3n-3n}$

$= x^{2n} - 3x^n + 2$

$3x^{5n} - 9x^{4n} + 6x^{3n} =$
$$3x^{3n}\left(x^{2n} - 3x^n + 2\right)$$

79. $GCF = x^n y^m$

$$\dfrac{7x^{2n}y^m + 3x^n y^{m+1}}{x^n y^m}$$

$= 7x^{2n-n}y^{m-m} + 3x^{n-n}y^{m+1-m}$

$= 7x^n + 3y$

$7x^{2n}y^m + 3x^n y^{m+1} =$
$$x^n y^m\left(7x^n + 3y\right)$$

85. $A = P + Pr + (P + Pr)r = (P + Pr)(1 + r)$

$= P(1 + r)(1 + r) = P(1 + r)^2$

89. x: number of hours

Distance covered by faster car + Distance covered by slower car = 400

$$(D = RT)$$

$$60x + 40x = 400$$

$$100x = 400$$

$$x = 4 \text{ ; in 4 hours}$$

90. $\left|\dfrac{3x-8}{2}\right| \geq 4$

$\dfrac{3x-8}{2} \geq 4$ or $\dfrac{3x-8}{2} \leq -4$

$3x - 8 \geq 8$ $3x - 8 \leq -8$

$3x \geq 16$ $3x \leq 0$

$x \geq \dfrac{16}{3}$ $x \leq 0$

$\left\{x \mid x \geq \dfrac{16}{3} \text{ or } x \leq 0\right\}$

91. x: pounds of tea costing 70¢ per pound
$.70x + .90(30) = .85(x+30)$; $x = 10$
10 pounds

Problem set 3.4.2

9. $t^2 + 6t + 9 = (t+3)(t+3) = (t+3)^2$

25. $-12x + 35 + x^2 = x^2 - 12x + 35$
$= (x-7)(x-5)$

35. $2x^3 + 6x^2 + 4x = 2x\left(x^2 + 3x + 2\right)$
$= 2x(x+2)(x+1)$

47. $9y^2 - 30y + 25 = (3y-5)(3y-5)$
$= (3y-5)^2$

61. $15w^3 - 25w^2 + 10w$
$= 5w\left(3w^2 - 5w + 2\right) = 5w(3w+1)(w-2)$

63. $-x^2 + xy + 6y^2 = -\left(x^2 - xy - 6y^2\right)$
$= -(x-3y)(x+2y)$

65. $4a^4b - 24a^3b - 64a^2b$
$= 4a^2b\left(a^2 - 6a - 16\right)$
$= 4a^2b(a+2)(a-8)$

67. $36x^3y - 6x^2y - 20xy$
$= 2xy\left(18x^2 - 3x - 10\right)$
$= 2xy(6x-5)(3x+2)$

77. $6a^3b^3 + 12a^2b^2 - 90ab$
$= 6ab\left(a^2b^2 + 2ab - 15\right)$
$= 6ab(ab+5)(ab-3)$

79. $13x^3y^3 + 39x^3y^2 - 52x^3y$
$= 13x^3y\left(y^2 + 3y - 4\right)$
$= 13x^3y(y+4)(y-1)$

81. $y^4 + 5y^2 + 6$
$= x^2 + 5x + 6\left(\text{with } x = y^2\right)$
$= (x+3)(x+2) = \left(y^2+3\right)\left(y^2+2\right)$

83. $5m^4 + m^2 - 6$
$= 5x^2 + x - 6\left(\text{with } x = m^2\right)$
$= (5x+6)(x-1) = \left(5m^2+6\right)\left(m^2-1\right)$
$= \left(5m^2+6\right)(m+1)(m-1)$

85. $2n^4 + mn^2 - 6m^2$
$= 2x^2 + mx - 6m^2\left(\text{with } x = n^2\right)$
$= (2x-3m)(x+2m)$
$= \left(2n^2-3m\right)\left(n^2+2m\right)$

87. $y^6 - 9y^3 - 36$
$= x^2 - 9x - 36\left(\text{with } x = y^3\right)$
$= (x-12)(x+3)$
$= \left(y^3-12\right)\left(y^3+3\right)$

89. $y^8 + 10y^4 - 39$
$= x^2 + 10x - 39\left(\text{with } x = y^4\right)$
$= (x+13)(x-3)$
$= \left(y^4+13\right)\left(y^4-3\right)$

91. $(a-3b)^2 - 5(a-3b) - 36$

$\quad = x^2 - 5x - 36 \text{ (with } x = a-3b)$

$\quad = (x+4)(x-9)$

$\quad = (a-3b+4)(a-3b-9)$

93. $5(a+b)^2 + 12(a+b) + 7$

$\quad = 5x^2 + 12x + 7 \text{ (with } x = a+b)$

$\quad = (5x+7)(x+1)$

$\quad = [5(a+b)+7](a+b+1)$

$\quad = (5a+5b+7)(a+b+1)$

95. $18(x+y)^2 - 3(x+y)b - 28b^2$

$\quad = 18w^2 - 3wb - 28b^2 \text{ (with } w = x+y)$

$\quad = (6w+7b)(3w-4b)$

$\quad = [6(x+y)+7b][3(x+y)-4b]$

$\quad = (6x+6y+7b)(3x+3y-4b)$

97. $6a^2(a-b)^2 - 13ab(a-b) + 6b^2$

$\quad = 6a^2x^2 - 13abx + 6b^2 \text{ (} x = a-b)$

$\quad = (3ax-2b)(2ax-3b)$

$\quad = [3a(a-b)-2b][2a(a-b)-3b]$

$\quad = \left(3a^2 - 3ab - 2b\right)\left(2a^2 - 2ab - 3b\right)$

105. $y^{3n} + 10y^{2n} + 16y^n$

$\quad = y^n\left(y^{2n} + 10y^n + 16\right)$

$\quad = y^n\left(y^n + 8\right)\left(y^n + 2\right)$

107. $-16t^2 + 16t + 32$

$\quad = -16\left(t^2 - t - 2\right)$

$\quad = -16(t-2)(t+1)$

109. $x^2 + Kx - 6$ K is the coefficient of x.

$\quad (x-3)(x+2) \ K = -1; (x-6)(x+1) \ K = -5$

$\quad (x+3)(x-2) \ K = 1; (x+6)(x-1) \ K = 5$

111. $4x^2 + Kx - 1$

$\quad (4x-1)(x+1) \ K = 3$

$\quad (4x+1)(x-1) \ K = -3$

$\quad (2x-1)(2x+1) \ K = 0$

113. $4x^2 + 4x + 1 = (2x+1)(2x+1)$

Since $ax + b$ is a factor, $a = 2$ and $b = 1$.

Thus, $ax + b = 2x + 1$. Since $ax + b = 2x + 1$

is a factor of $2x^2 - 5x + c$,

$2x^2 - 5x + c = (2x+1)\underline{\quad ?\quad}$

$\qquad\qquad\qquad = (2x+1)(x-3)$

Only a second factor of $(x-3)$ will result

in $2x^2$ when the first terms are multiplied

and $-5x$ when the sum of the outside

and inside terms is taken.

Thus, $2x^2 - 5x + c = (2x+1)(x-3) =$

$2x^2 - 5x - 3$, which means that $c = -3$.

114. $\dfrac{16x^5y^4 - 8x^3y^3 - 6x^2y^2}{-2x^2y^2}$

$\quad = \dfrac{16x^5y^4}{-2x^2y^2} - \dfrac{8x^3y^3}{-2x^2y^2} - \dfrac{6x^2y^2}{-2x^2y^2}$

$\quad = -8x^3y^2 + 4xy + 3$

115.

$$6x^2 - 7x + 10 - \frac{18}{x+2}$$

$$x+2 \overline{\big)\, 6x^3 + 5x^2 - 4x + 2}$$

$$\underline{6x^3 + 12x^2}$$

$$-7x^2 - 4x$$

$$\underline{-7x^2 - 14x}$$

$$10x + 2$$

$$\underline{10x + 20}$$

$$-18$$

$$\begin{array}{r|rrrr} -2 & 6 & 5 & -4 & 2 \\ & & -12 & 14 & -20 \\ \hline & 6 & -7 & 10 & -18 \end{array}$$

116. $P = \frac{2}{5}w\left(1 + \frac{n}{50}\right)$

$720 = \frac{2}{5}(1000)\left(1 + \frac{n}{50}\right)$

$720 = 400\left(1 + \frac{n}{50}\right)$

$720 = 400 + 8n$

$320 = 8n$

$40 = n \; ; \; 40 \text{ years}$

Problem set 3.4.3

3. $25 - a^2 = 5^2 - a^2 = (5+a)(5-a)$

5. $36x^2 - 49 = (6x)^2 - 7^2 = (6x+7)(6x-7)$

7. $36x^2 - 49y^2 = (6x)^2 - (7y)^2$
$= (6x+7y)(6x-7y)$

9. $x^2y^2 - a^2b^2 = (xy)^2 - (ab)^2$
$= (xy+ab)(xy-ab)$

11. $x^2y^6 - a^4b^2 = \left(xy^3\right)^2 - \left(a^2b\right)^2$
$= \left(xy^3 + a^2b\right)\left(xy^3 - a^2b\right)$

13. $4x^2y^6 - 25a^4b^2 = \left(2xy^3\right)^2 - \left(5a^2b\right)^2$
$= \left(2xy^3 + 5a^2b\right)\left(2xy^3 - 5a^2b\right)$

15. $81a^2b^4c^6 - 49x^8y^2$
$= \left(9ab^2c^3\right)^2 - \left(7x^4y\right)^2$
$= \left(9ab^2c^3 + 7x^4y\right)\left(9ab^2c^3 - 7x^4y\right)$

19. $(x+y)^2 - 36 = (x+y)^2 - 6^2$
$= (x+y+6)(x+y-6)$

23. $16y^2 - (3x-1)^2 = (4y)^2 - (3x-1)^2$
$= [(4y)+(3x-1)][(4y)-(3x-1)]$
$= (4y+3x-1)(4y-3x+1)$

27. $(2x-1)^2 - (3x+2)^2$
$= [(2x-1)+(3x+2)][(2x-1)-(3x+2)]$
$= (5x+1)(-x-3)$

29. $a^{14} - 9 = \left(a^7\right)^2 - 3^2$
$= \left(a^7+3\right)\left(a^7-3\right)$

33. $-16 + x^2 = x^2 - 16 = (x+4)(x-4)$

35. $-25A^2 + x^2y^2 = (xy)^2 - (5A)^2$
$= (xy+5A)(xy-5A)$

37. $y^4 - 1 = \left(y^2\right)^2 - 1^2 = \left(y^2 + 1\right)\left(y^2 - 1\right)$

$\quad = \left(y^2 + 1\right)(y + 1)(y - 1)$

39. $1 - 81b^2 = 1^2 - (9b)^2 = (1 + 9b)(1 - 9b)$

41. $x^2 y^3 - 16y = y\left(x^2 y^2 - 16\right)$

$\quad = y\left[(xy)^2 - 4^2\right] = y(xy + 4)(xy - 4)$

43. $3x^3 - 3x = 3x\left(x^2 - 1\right) = 3x(x + 1)(x - 1)$

45. $3x^3 y - 12xy = 3xy\left(x^2 - 4\right)$

$\quad = 3xy(x + 2)(x - 2)$

47. $x^{2n} - 144 = \left(x^n\right)^2 - 12^2$

$\quad = \left(x^n + 12\right)\left(x^n - 12\right)$

49. $x^{4n} - y^{4n} = \left(x^{2n}\right)^2 - \left(y^{2n}\right)^2$

$\quad = \left(x^{2n} + y^{2n}\right)\left(x^{2n} - y^{2n}\right)$

$\quad = \left(x^{2n} + y^{2n}\right)\left[\left(x^n\right)^2 - \left(y^n\right)^2\right]$

$\quad = \left(x^{2n} + y^{2n}\right)\left(x^n + y^n\right)\left(x^n - y^n\right)$

53. $27x^3 - 64y^3 = (3x)^3 - (4y)^3$

$\quad = (3x - 4y)\left[(3x)^2 + (3x)(4y) + (4y)^2\right]$

$\quad = (3x - 4y)\left(9x^2 + 12xy + 16y^2\right)$

55. $27R^3 - 1 = (3R)^3 - 1^3$

$\quad = (3R - 1)\left[(3R)^2 + (3R)(1) + 1^2\right]$

$\quad = (3R - 1)\left(9R^2 + 3R + 1\right)$

57. $125b^3 + 64 = (5b)^3 + 4^3$

$\quad = (5b + 4)\left[(5b)^2 - (5b)\cdot 4 + 4^2\right]$

$\quad = (5b + 4)\left(25b^2 - 20b + 16\right)$

61. $8a^3 b^3 - 27 = (2ab)^3 - 3^3$

$\quad = (2ab - 3)\left[(2ab)^2 + (2ab)(3) + 3^2\right]$

$\quad = (2ab - 3)\left(4a^2 b^2 + 6ab + 9\right)$

65. $8x^3 + 27y^{12} = (2x)^3 + \left(3y^4\right)^3$

$\quad = \left(2x + 3y^4\right)\cdot$

$\qquad \left[(2x)^2 - (2x)\left(3y^4\right) + \left(3y^4\right)^2\right]$

$\quad = \left(2x + 3y^4\right)\left(4x^2 - 6xy^4 + 9y^8\right)$

67. $a^3 b^6 - c^6 d^{12} = \left(ab^2\right)^3 - \left(c^2 d^4\right)^3$

$\quad = \left(ab^2 - c^2 d^4\right)\cdot$

$\qquad \left[\left(ab^2\right)^2 + \left(ab^2\right)\left(c^2 d^4\right) + \left(c^2 d^4\right)^2\right]$

$\quad = \left(ab^2 - c^2 d^4\right)\cdot$

$\qquad \left(a^2 b^4 + ab^2 c^2 d^4 + c^4 d^8\right)$

71. $x^6 + 1 = \left(x^2\right)^3 + 1^3$

$\quad = \left(x^2 + 1\right)\left[\left(x^2\right)^2 - x^2 \cdot 1 + 1^2\right]$

$\quad = \left(x^2 + 1\right)\left(x^4 - x^2 + 1\right)$

73. $(x+y)^3 + (x-y)^3$

$\quad = \left[(x+y) + (x-y)\right] \cdot$

$\qquad \left[(x+y)^2 - (x+y)(x-y) + (x-y)^2\right]$

$\quad = 2x \cdot$

$\quad \left(x^2 + 2xy + y^2 - x^2 + y^2 + x^2 - 2xy + y^2\right)$

$\quad = 2x\left(x^2 + 3y^2\right)$

75. $(2x-y)^3 - (2x+y)^3$

$\quad = \left[(2x-y) - (2x+y)\right] \cdot$

$\quad \left[(2x-y)^2 + (2x-y)(2x+y) + (2x+y)^2\right]$

$\quad = (-2y)(4x^2 - 4xy + y^2 + 4x^2 - y^2$

$\qquad\qquad\qquad + 4x^2 + 4xy + y^2)$

$\quad = (-2y)\left(12x^2 + y^2\right)$

77. $x^{3n} - 8 = \left(x^n\right)^3 - 2^3$

$\quad = \left(x^n - 2\right)\left[\left(x^n\right)^2 + x^n \cdot 2 + 2^2\right]$

$\quad = \left(x^n - 2\right)\left(x^{2n} + 2x^n + 4\right)$

79. $125x^{3n} + 27y^{12m} = \left(5x^n\right)^3 + \left(3y^{4m}\right)^3$

$\quad = \left(5x^n + 3y^{4m}\right) \cdot$

$\quad \left[\left(5x^n\right)^2 - \left(5x^n\right)\left(3y^{4m}\right) + \left(3y^{4m}\right)^2\right]$

$\quad = \left(5x^n + 3y^{4m}\right) \cdot$

$\qquad \left(25x^{2n} - 15x^n y^{4m} + 9y^{8m}\right)$

81. $1000x^3 y^{15} - 64w^6 z^9$

$\quad = 8\left(125x^3 y^{15} - 8w^6 z^9\right)$

$\quad = 8\left[\left(5xy^5\right)^3 - \left(2w^2 z^3\right)^3\right]$

$\quad = 8\left(5xy^5 - 2w^2 z^3\right) \cdot$

$\quad \left[\left(5xy^5\right)^2 + \left(5xy^5\right)\left(2w^2 z^3\right) + \left(2w^2 z^3\right)^2\right]$

$\quad = 8\left(5xy^5 - 2w^2 z^3\right) \cdot$

$\quad \left(25x^2 y^{10} + 10xy^5 w^2 z^3 + 4w^4 z^6\right)$

83. $y^2 + 4y + 4 = y^2 + 2 \cdot 2y + 2^2$

$\quad = (y+2)^2$

87. $9x^2 - 12xy + 4y^2$

$\quad = (3x)^2 - 2(3x)(2y) + (2y)^2$

$\qquad \left[a^2 - 2ab + b^2\right]$

$\quad = (3x - 2y)^2$

$\qquad \left[(a-b)^2, \text{ with } a = 3x \text{ and } b = 2y\right]$

89. $25a^2 b^2 - 20ab + 4$

$\quad = (5ab)^2 - 2(5ab)(2) + 2^2$

$\qquad \left[A^2 - 2AB + B^2\right]$

$\quad = (5ab - 2)^2$

$\qquad \left[(A-B)^2, \text{ with } A = 5ab \text{ and } B = 2\right]$

93. $4x^4 - 20x^2y^2 + 25y^4$

$= \left(2x^2\right)^2 - 2\left(2x^2\right)\left(5y^2\right) + \left(5y^2\right)^2$

$= \left(2x^2 - 5y^2\right)^2$

$\left[(A-B)^2, \text{ with } A = 2x^2 \text{ and } B = 5y^2\right]$

95. $(x+y)^2 + 2(x+y) + 1$

$= \left[(x+y)+1\right]^2 = (x+y+1)^2$

97. $(v-w)^2 + 4(v-w) + 4$

$= \left[(v-w)+2\right]^2 = (v-w+2)^2$

99. $x^2 - 6x + 9 - y^2$

$= (x-3)^2 - y^2$

$= (x-3+y)(x-3-y)$

101. $9x^2 - 30x + 25 - 36y^2$

$= (3x-5)^2 - (6y)^2$

$= (3x-5+6y)(3x-5-6y)$

103. $r^2 - \left(16s^2 - 24s + 9\right)$

$= r^2 - (4s-3)^2$

$= \left[r+(4s-3)\right]\left[r-(4s-3)\right]$

$= (r+4s-3)(r-4s+3)$

105. $y^2 - x^2 - 4x - 4$

$= y^2 - \left(x^2 + 4x + 4\right) = y^2 - (x+2)^2$

$= \left[y+(x+2)\right]\left[y-(x+2)\right]$

$= (y+x+2)(y-x-2)$

107. $z^2 - x^2 + 4xy - 4y^2$

$= z^2 - \left(x^2 - 4xy + 4y^2\right)$

$= z^2 - (x-2y)^2$

$= \left[z+(x-2y)\right]\left[z-(x-2y)\right]$

$= (z+x-2y)(z-x+2y)$

109. $x^3 - y^3 - x + y$

$= \left(x^3 - y^3\right) - (x-y)$

$= (x-y)\left(x^2 + xy + y^2\right) - (x-y)$

$= (x-y)\left(x^2 + xy + y^2 - 1\right)$

111. $x^3 + y^3 - 3x - 3y$

$= \left(x^3 + y^3\right) - 3(x+y)$

$= (x+y)\left(x^2 - xy + y^2\right) - 3(x+y)$

$= (x+y)\left(x^2 - xy + y^2 - 3\right)$

113. $4x^{2n} + 20x^ny^m + 25y^{2m}$

$= \left(2x^n\right)^2 + 2\left(2x^n\right)\left(5y^m\right) + \left(5y^m\right)^2$

$= \left(2x^n + 5y^m\right)^2$

115. $x^2 - 12x + K$: In order to obtain x^2 and $-12x$, we must have $(x-6)(x-6)$ or $(x-6)^2$.

Thus $x^2 - 12x + K = (x-6)^2$

$= x^2 - 12x + 36.$ $K = 36$

117. $Kx^2 + 8xy + y^2$: In order to obtain $8xy$ and y^2, we must have $(4x+y)(4x+y)$ or $(4x+y)^2$.

Thus $Kx^2 + 8xy + y^2 = (4x+y)^2$

$= 16x^2 + 8xy + y^2$. $K = 16$

119. Area of shaded region $=$ area of large square $-$ area of small square.

$= x^2 - y^2 = (x+y)(x-y)$

121. Area of shaded region $=$ area of large square $- 4$(area of each small square)

$= x^2 - 4y^2 = (x+2y)(x-2y)$

123. $10^2 - 9^2 = (10+9)(10-9)$

$= (19)(1) = 19$

125. $x^6 - y^6 = \left(x^3\right)^2 - \left(y^3\right)^2$

$= \left(x^3 + y^3\right)\left(x^3 - y^3\right)$

$= (x+y)\left(x^2 - xy + y^2\right)(x-y) \cdot$

$\left(x^2 + xy + y^2\right)$

$x^6 - y^6 = \left(x^2\right)^3 - \left(y^2\right)^3$

$= \left(x^2 - y^2\right)\left[\left(x^2\right)^2 + x^2 y^2 + \left(y^2\right)^2\right]$

$= (x+y)(x-y)\left(x^4 + x^2 y^2 + y^4\right)$

Since both factorizations involve $(x+y) \cdot$ $(x-y)$, we can conclude that

$\left(x^2 - xy + y^2\right)\left(x^2 + xy + y^2\right) =$ $x^4 + x^2 y^2 + y^4$.

127. $(c+d)^3 = c^3 + 3c^2 d + 3cd^2 + d^3$

$27a^3 + 27a^2 b + 9ab^2 + b^3$

implies that $c = 3a$ and $d = b$

$= (3a)^3 + 3(3a)^2 b + 3(3a)b^2 + b^3$

$= (c+d)^3 = (3a+b)^3$

128. $\left(x^3 - 2x^2 + 5x\right)\left(3x^4 - 4x - 1\right)$

$= x^3\left(3x^4 - 4x - 1\right) - 2x^2\left(3x^4 - 4x - 1\right)$

$\quad + 5x\left(3x^4 - 4x - 1\right)$

$= 3x^7 - 4x^4 - x^3 - 6x^6 + 8x^3 + 2x^2$

$\quad + 15x^5 - 20x^2 - 5x$

$= 3x^7 - 6x^6 + 15x^5 - 4x^4 + 7x^3$

$\quad - 18x^2 - 5x$

129. $24\left[\frac{1}{6}(2x+4)\right] = 24\left[\frac{1}{4}(2x+4) - \frac{5}{8}x\right]$

$4(2x+4) = 6(2x+4) - 3(5x)$

$8x + 16 = 12x + 24 - 15x$

$8x + 16 = -3x + 24$

$11x = 8$

$x = \frac{8}{11}$ $\left\{\frac{8}{11}\right\}$

130. $3(x-2) - 2x > -6$ $9 - x > 5$

$3x - 6 - 2x > -6$ $-x > -4$

$x > 0$ $x < 4$

$\{x \mid 0 < x < 4\}$

Problem set 3.4.4

1. $c^3 - 16c = c\left(c^2 - 16\right) = c(c+4)(c-4)$

3. $3x^2 + 18x + 27 = 3\left(x^2 + 6x + 9\right)$

$= 3(x+3)^2$

5. $81x^3 - 3 = 3\left(27x^3 - 1\right) = 3\left[(3x)^3 - 1^3\right]$

$= 3(3x-1)\left(9x^2 + 3x + 1\right)$

7. $B^2C - 16C + 32 - 2B^2$

$= C\left(B^2 - 16\right) - 2\left(B^2 - 16\right)$

$= \left(B^2 - 16\right)(C-2)$

$= (B+4)(B-4)(C-2)$

9. $-x^2 + 12x - 27 = -\left(x^2 - 12x + 27\right)$

$= -(x-9)(x-3)$

11. $4a^2b - 2ab - 30b = 2b\left(2a^2 - a - 15\right)$

$= 2b(2a+5)(a-3)$

13. $a\left(y^2 - 4\right) - 4\left(y^2 - 4\right) = \left(y^2 - 4\right)(a-4)$

$= (y+2)(y-2)(a-4)$

15. $11x^5 - 11xy^2 = 11x\left(x^4 - y^2\right)$

$= 11x\left(x^2 + y\right)\left(x^2 - y\right)$

17. $3x^2 + 3x + 3y - 3y^2 = 3\left(x^2 - y^2 + x + y\right)$

$= 3\left[(x+y)(x-y) + (x+y)\right]$

$= 3\left[(x+y)(x-y+1)\right] = 3(x+y)(x-y+1)$

23. $s^2 - 12s + 36 - 49t^2 = (s-6)^2 - (7t)^2$

$= (s-6+7t)(s-6-7t)$

27. $9s^2t^2 - 36t^2 = 9t^2\left(s^2 - 4\right)$

$= 9t^2(s+2)(s-2)$

33. $20a^3b - 245ab^3 = 5ab\left(4a^2 - 49b^2\right)$

$= 5ab(2a-7b)(2a+7b)$

35. $63y^2 + 30y - 72 = 3\left(21y^2 + 10y - 24\right)$

$= 3(7y-6)(3y+4)$

41. $100x^4 + 120x^3y + 36x^2y^2$

$= 4x^2\left(25x^2 + 30xy + 9y^2\right)$

$= 4x^2(5x+3y)^2$

45. $71bx^4 - 71b = 71b\left(x^4 - 1\right)$

$= 71b\left(x^2 + 1\right)\left(x^2 - 1\right)$

$= 71b\left(x^2 + 1\right)(x+1)(x-1)$

49. $\left(r^3 - s^3\right) + (r-s)$

$= (r-s)\left(r^2 + rs + s^2\right) + (r-s)$

$= (r-s)\left(r^2 + rs + s^2 + 1\right)$

51. $x^{2n} - y^{2m} = \left(x^n\right)^2 - \left(y^m\right)^2$

$= \left(x^n + y^m\right)\left(x^n - y^m\right)$

53. $a^2 + 4a + 4 - 16b^2 = (a+2)^2 - (4b)^2$

$= (a+2+4b)(a+2-4b)$

55. $27r^3s + 72r^2s^2 + 48rs^3$

$= 3rs\left(9r^2 + 24rs + 16s^2\right)$

$= 3rs(3r + 4s)^2$

61. $(3x - y)^2 - 100a^2 = (3x - y)^2 - (10a)^2$

$= (3x - y + 10a)(3x - y - 10a)$

67. $48y^4 - 243 = 3\left(16y^4 - 81\right)$

$= 3\left(4y^2 + 9\right)\left(4y^2 - 9\right)$

$= 3\left(4y^2 + 9\right)(2y + 3)(2y - 3)$

69. $20bx^4 + 220bx^2y + 605by^2$

$= 5b\left(4x^4 + 44x^2y + 121y^2\right)$

$= 5b\left(2x^2 + 11y\right)^2$

73. $4x^7 + 32xy^3 = 4x\left(x^6 + 8y^3\right)$

$= 4x\left[\left(x^2\right)^3 + (2y)^3\right]$

$= 4x\left(x^2 + 2y\right)\left[\left(x^2\right)^2 - x^2 \cdot 2y + (2y)^2\right]$

$= 4x\left(x^2 + 2y\right)\left(x^4 - 2x^2y + 4y^2\right)$

75. $36(c - d)y^2 - 6(c - d)yx - 20(c - d)x^2$

Let $z = (c - d)$

$36zy^2 - 6zyx - 20zx^2$

$= 2z\left(18y^2 - 3xy - 10x^2\right)$

$= 2z(6y - 5x)(3y + 2x)$

$= 2(c - d)(6y - 5x)(3y + 2x)$

77. $x^8 - y^{12} = \left[\left(x^4\right)^2 - \left(y^6\right)^2\right]$

$= \left(x^4 + y^6\right)\left(x^4 - y^6\right)$

$= \left(x^4 + y^6\right)\left[\left(x^2\right)^2 - \left(y^3\right)^2\right]$

$= \left(x^4 + y^6\right)\left(x^2 + y^3\right)\left(x^2 - y^3\right)$

79. $4x^2y + 5 - 20x^2 - y$

$= 4x^2y - y - 20x^2 + 5$

$= y\left(4x^2 - 1\right) - 5\left(4x^2 - 1\right)$

$= (y - 5)\left(4x^2 - 1\right)$

$= (2x + 1)(2x - 1)(y - 5)$

83. $4r^2s^2 - 4r^2 - 9s^2 + 9$

$= 4r^2\left(s^2 - 1\right) - 9\left(s^2 - 1\right)$

$= \left(4r^2 - 9\right)\left(s^2 - 1\right)$

$= (2r + 3)(2r - 3)(s + 1)(s - 1)$

85. $y^{2n+1} + 2y^{n+1} + y$

$= y\left(y^{2n} + 2y^n + 1\right) = y\left(y^n + 1\right)^2$

87. $rs^2 - 2a - r + 2as^2 = rs^2 - r + 2as^2 - 2a$

$= r\left(s^2 - 1\right) + 2a\left(s^2 - 1\right)$

$= (r + 2a)\left(s^2 - 1\right)$

$= (r + 2a)(s + 1)(s - 1)$

93. $\left(4x^2 - 12xy + 9y^2\right) + (72by - 48bx)$

$$- 25b^2$$

$= (2x - 3y)^2 - 24b(2x - 3y) - 25b^2$

Let $z = (2x - 3y)$

$= z^2 - 24bz - 25b^2 = (z - 25b)(z + b)$

$= (2x - 3y - 25b)(2x - 3y + b)$

95. $16a^3b + 4a^2b^2 - 42ab^3$

$= 2ab\left(8a^2 + 2ab - 21b^2\right)$

$= 2ab(4a + 7b)(2a - 3b)$

97. $2x^{n+2} - 7x^{n+1} + 3x^n =$

$x^n\left(2x^2 - 7x + 3\right) = x^n(2x - 1)(x - 3)$

99. $a^6b^6 - a^3b^3 = a^3b^3\left(a^3b^3 - 1\right)$

$= a^3b^3\left[(ab)^3 - 1^3\right]$

$= a^3b^3(ab - 1)\left(a^2b^2 + ab + 1\right)$

101. $10x^3 - 6x^2 - 21x = x\left(10x^2 - 6x - 21\right)$

107. $\dfrac{y^3 - 5y^2 + 10y - 8}{y - 2} = y^2 - 3y + 4$

$$\underline{2\,|}\ \ 1\ \ -5\ \ \ \ 10\ \ \ -8$$
$$\underline{\qquad\ \ \ \ 2\ \ \ -6\ \ \ \ \ 8}$$
$$1\ \ -3\ \ \ \ 4\ \ \ \ \ 0$$

$y^3 - 5y^2 + 10y - 8 = (y - 2)\left(y^2 - 3y + 4\right)$

109.

$$
\begin{array}{r}
x^2 + x - 1 \\
x^2 + 3x + 1\,\overline{\smash{\big)}\,x^4 + 4x^3 + 3x^2 - 2x - 1} \\
\underline{x^4 + 3x^3 + x^2} \\
x^3 + 2x^2 - 2x \\
\underline{x^3 + 3x^2 + x} \\
-x^2 - 3x - 1 \\
\underline{-x^2 - 3x - 1} \\
0
\end{array}
$$

$x^4 + 4x^3 + 3x^2 - 2x - 1$

$\quad = \left(x^2 + 3x + 1\right)\left(x^2 + x - 1\right)$

111. $x^4 + 2x^2y^2 + 9y^4$

$= x^4 + 6x^2y^2 + 9y^4 - 4x^2y^2$

$= \left(x^2 + 3y^2\right)^2 - (2xy)^2$

$= \left(x^2 + 3y^2 + 2xy\right)\left(x^2 + 3y^2 - 2xy\right)$

119. $-(7 - z) = 2\left[3z + 4(z - 1)\right]$

$-7 + z = 2\left[3z + 4z - 4\right]$

$-7 + z = 14z - 8$

$-13z = -1$

$z = \dfrac{1}{13} \qquad \left\{\dfrac{1}{13}\right\}$

120. x: number

$2.5x - 3.8 = -7.9$

$2.5x = -7.9 + 3.8$

$2.5x = -4.1$

$x = \dfrac{-4.1}{2.5} = -1.64$

The number is -1.64.

121. odd integers: x, $x+2$, $x+4$

$$x+(x+2) = 3(x+4)-27$$
$$2x+2 = 3x-15$$
$$17 = x$$

Integers are 17, 19, 21.

Review Problems: Chapter Three

1. a. trinomial, degree 3

    b. monomial, degree $4+3+1 = 8$

    c. binomial, degree $1+6 = 7$

    d. (five terms), degree 5

2. a. $\left(4x^2-5xy+3y^2\right)$

$$+\left(7xy-10y^2+13x^2\right)$$
$$= (4+13)x^2+(-5+7)xy+(3-10)y^2$$
$$= 17x^2+2xy-7y^2$$

   b. $\left(5y^2-8xy+7x^2\right)$

$$-\left(13xy-12y^2+11x^2\right)$$
$$= 5y^2-8xy+7x^2-13xy+12y^2-11x^2$$
$$= (7-11)x^2+(-8-13)xy+(5+12)y^2$$
$$= -4x^2-21xy+17y^2$$

   c. $\left(4x^{2n}-5x^n+3\right)+\left(7x^{2n}-6x^n-8\right)$

$$-\left(2x^{2n}-4x^n-9\right)$$
$$= 4x^{2n}-5x^n+3+7x^{2n}-6x^n-8$$
$$\quad -2x^{2n}+4x^n+9$$
$$= (4+7-2)x^{2n}+(-5-6+4)x^n$$
$$\quad +(3-8+9)$$
$$= 9x^{2n}-7x^n+4$$

3. $\left(6+3a^4b^3+4ab^3\right)$

$$+\left(-5-2a^4b^3-3ab^3\right)$$
$$-\left(10-6a^4b^3-2ab^3\right)$$
$$= 6+3a^4b^3+4ab^3-5-2a^4b^3-3ab^3$$
$$\quad -10+6a^4b^3+2ab^3$$
$$= (3-2+6)a^4b^3+(4-3+2)ab^3$$
$$\quad +(6-5-10)$$
$$= 7a^4b^3+3ab^3-9$$

4. $P(-2) = 4(-2)^2-8(-2)-3$
$$= 16+16-3 = 29$$

$P(5) = 4(5)^2-8(5)-3$
$$= 100-40-3 = 57$$

5. $\left(4x^2yz^5\right)\left(-3x^4yz^2\right)$

$$= -12x^{2+4}y^{1+1}z^{5+2}$$
$$= -12x^6y^2z^7$$

6. $-4a^3b^2c\left(3a^2b^3c^2-\frac{1}{2}abc^5-2a^4b^2c\right)$

$$= \left(-4a^3b^2c\right)\left(3a^2b^3c^2\right)$$
$$\quad +\left(-4a^3b^2c\right)\left(-\frac{1}{2}abc^5\right)$$
$$\quad +\left(-4a^3b^2c\right)\left(-2a^4b^2c\right)$$
$$= -12a^5b^5c^3+2a^4b^3c^6+8a^7b^4c^2$$

7. $(4x-2)(3x-5) = 12x^2 - 20x - 6x + 10$

$\qquad = 12x^2 - 26x + 10$

8. $\left(2a^2b + c\right)^2$

$\qquad = \left(2a^2b\right)^2 + 2\left(2a^2b\right)(c) + c^2$

$\qquad = 4a^4b^2 + 4a^2bc + c^2$

9. $\left(3x^2y + 2y\right)\left(2x^2y - 3y\right)$

$\qquad = 6x^4y^2 - 9x^2y^2 + 4x^2y^2 - 6y^2$

$\qquad = 6x^4y^2 - 5x^2y^2 - 6y^2$

10. $\left(4x^3y^2 + 1\right)\left(4x^3y^2 - 1\right)$

$\qquad = \left(4x^3y^2\right)^2 - 1^2$

$\qquad = 16x^6y^4 - 1$

11. $(2xy + 2)\left(x^2y - 3y + 4\right)$

$\qquad = 2xy\left(x^2y - 3y + 4\right) + 2\left(x^2y - 3y + 4\right)$

$\qquad = 2x^3y^2 - 6xy^2 + 8xy + 2x^2y - 6y + 8$

$\qquad = 2x^3y^2 + 2x^2y - 6xy^2 + 8xy - 6y + 8$

12. $\left(3x^{2n} - y^{5n}\right)\left(4x^{2n} + 2y^{5n}\right)$

$\qquad = 12x^{4n} + 6x^{2n}y^{5n} - 4x^{2n}y^{5n} - 2y^{10n}$

$\qquad = 12x^{4n} + 2x^{2n}y^{5n} - 2y^{10n}$

13. $\left[5y - (2x + 7)\right]\left[5y + (2x + 7)\right]$

$\qquad = (5y)^2 - (2x + 7)^2$

$\qquad = 25y^2 - \left(4x^2 + 28x + 49\right)$

$\qquad = 25y^2 - 4x^2 - 28x - 49$

14. $(3x + y + 1)^2 = (3x + y + 1)(3x + y + 1)$

$\qquad = 3x(3x + y + 1) + y(3x + y + 1)$

$\qquad\qquad\qquad\qquad + 1(3x + y + 1)$

$\qquad = 9x^2 + 3xy + 3x + 3xy + y^2 + y$

$\qquad\qquad\qquad\qquad + 3x + y + 1$

$\qquad = 9x^2 + 6xy + 6x + 2y + y^2 + 1$

15. Integers : x, x+1, x+2

$\qquad x(x+1)(x+2) = (x+1)^3 - 12$

$\qquad x\left(x^2 + 3x + 2\right) = (x+1)(x+1)^2 - 12$

$\qquad x^3 + 3x^2 + 2x = (x+1)\left(x^2 + 2x + 1\right) - 12$

$\qquad x^3 + 3x^2 + 2x = x^3 + 2x^2 + x$

$\qquad\qquad\qquad\qquad + x^2 + 2x + 1 - 12$

$\qquad x^3 + 3x^2 + 2x = x^3 + 3x^2 + 3x - 11$

$\qquad\qquad\qquad\qquad 2x = 3x - 11$

$\qquad\qquad\qquad\qquad 11 = x$

$\qquad$ Integers : 11, 12, 13

16. x: length of cube's edge

$\qquad$ x+4 : length of box

$\qquad$ x−3 : width of box

$\qquad$ x: height of box

$\qquad$ Surface area of cube = Surface area of box

$\qquad 6x^2 = 2(x+4)(x-3) + 2(x-3)x + 2(x+4)x$

$\qquad 6x^2 = 2x^2 + 2x - 24 + 2x^2 - 6x + 2x^2 + 8x$

$\qquad 6x^2 = 6x^2 + 4x - 24$

$\qquad\qquad 24 = 4x$

$\qquad\qquad 6 = x$

$\qquad$ length: 6+4 = 10ft　　width: 6−3 = 3ft

17. $\dfrac{16x^4y^3z^7}{-8x^2yz^5} = -2x^{4-2}y^{3-1}z^{7-5}$

$= -2x^2y^2z^2$

18. $\dfrac{16a^4b^2 - 8a^3b^2 + 20ab}{4ab}$

$= \dfrac{16a^4b^2}{4ab} - \dfrac{8a^3b^2}{4ab} + \dfrac{20ab}{4ab}$

$= 4a^3b - 2a^2b + 5$

19.
$$3y^2 + 7y + 2 - \dfrac{26}{5y-2}$$

$$
\require{enclose}
\begin{array}{r}
3y^2 + 7y + 2 - \frac{26}{5y-2} \\
5y-2 \enclose{longdiv}{15y^3 + 29y^2 - 4y - 30} \\
\underline{15y^3 - 6y^2} \\
35y^2 - 4y \\
\underline{35y^2 - 14y} \\
10y - 30 \\
\underline{10y - 4} \\
-26
\end{array}
$$

20.
$$
\begin{array}{r}
x^3 + x^2y + xy^2 + y^3 \\
x-y \enclose{longdiv}{x^4 + 0x^3y + 0x^2y^2 + 0xy^3 - y^4} \\
\underline{x^4 - x^3y} \\
x^3y + 0x^2y^2 \\
\underline{x^3y - x^2y^2} \\
x^2y^2 + 0xy^3 \\
\underline{x^2y^2 - xy^3} \\
xy^3 - y^4 \\
\underline{xy^3 - y^4} \\
0
\end{array}
$$

21. $\dfrac{25x^2y^3z^4 - 45xy^4z^3}{5xyz^2}$

$= \dfrac{25x^2y^3z^4}{5xyz^2} - \dfrac{45xy^4z^3}{5xyz^2}$

$= 5xy^2z^2 - 9y^3z$

22.
$$
\begin{array}{r}
2x^2 + 2x + 4 \\
2x^2+3 \enclose{longdiv}{4x^4 + 4x^3 + 14x^2 + 6x + 12} \\
\underline{4x^4 \qquad\quad + 6x^2} \\
4x^3 + 8x^2 + 6x \\
\underline{4x^3 \qquad + 6x} \\
8x^2 \qquad + 12 \\
\underline{8x^2 \qquad + 12} \\
0
\end{array}
$$

23. $\dfrac{4x^3 - 3x^2 - 2x + 1}{x+1}$

$$
\begin{array}{r|rrrr}
-1 & 4 & -3 & -2 & 1 \\
   &   & -4 & 7 & -5 \\
\hline
   & 4 & -7 & 5 & -4
\end{array}
$$

$= 4x^2 - 7x + 5 - \dfrac{4}{x+1}$

24. $\dfrac{3y^4+0y^3-2y^2-10y+0}{y-2}$

$$\begin{array}{r|rrrrr} 2 & 3 & 0 & -2 & -10 & 0 \\ & & 6 & 12 & 20 & 20 \\ \hline & 3 & 6 & 10 & 10 & 20 \end{array}$$

$= 3y^3+6y^2+10y+10+\dfrac{20}{y-2}$

25.
$$\begin{array}{r|rrrrr} -5 & -3 & 2 & 5 & -9 & 10 \\ & & 15 & -85 & 400 & -1955 \\ \hline & -3 & 17 & -80 & 391 & -1945 \end{array}$$
$$\uparrow$$
(The remainder is not 0.)

-5 is <u>not</u> a solution of

$-3x^4+2x^3+5x^2-9x+10=0$

26.
$$\begin{array}{r|rrrr} -4 & 3 & 4 & -3 & 2 \\ & & -12 & 32 & -116 \\ \hline & 3 & -8 & 29 & -114 \end{array}$$
$P(-4)=-114$

27. $5a^4b^2c^3-20a^3b^3c^4+15a^2b^2c^3$

$= 5a^2b^2c^3\left(a^2-4abc+3\right)$

28. $21x^{3n+1}-35x^{3n}$

$= 7x^{3n}(3x-5)$

29. $(x+5y)z^2+(x+5y)z-42(x+5y)$

$= (x+5y)\left(z^2+z-42\right)$

$= (x+5y)(z+7)(z-6)$

30. $bc-d-bd+c$

$= bc+c-bd-d$

$= c(b+1)-d(b+1)$

$= (b+1)(c-d)$

31. $(a-4b)x^2+(a-4b)x+(4b-a)$

$= (a-4b)x^2+(a-4b)x+(a-4b)(-1)$

$= (a-4b)\left(x^2+x-1\right)$

32. $3x^2+15x-2xy-10y$

$= 3x(x+5)-2y(x+5)$

$= (x+5)(3x-2y)$

33. $a^2+37a+36$

$= (a+36)(a+1)$

34. $-2x^3+36x^2-64x$

$= -2x\left(x^2-18x+32\right)$

$= -2x(x-2)(x-16)$

35. $8y^4-14y^2-15$

$= \left(4y^2+3\right)\left(2y^2-5\right)$

36. $x^2\left(b^2-9\right)-25\left(b^2-9\right)$

$= \left(b^2-9\right)\left(x^2-25\right)$

$= (b+3)(b-3)(x+5)(x-5)$

37. $4(a+b)^2-27(a+b)+18$ (Let $c=a+b$)

$= 4c^2-27c+18$

$= (4c-3)(c-6)$

$= [4(a+b)-3][(a+b)-6]$

$= (4a+4b-3)(a+b-6)$

38. $x^{2n}-5x^n-36$

$= \left(x^n+4\right)\left(x^n-9\right)$

39. $10x^2 - 160$

$= 10\left(x^2 - 16\right)$

$= 10(x+4)(x-4)$

40. $(x+4)^2 - (3x-1)^2$

$= \left[(x+4)+(3x-1)\right]\left[(x+4)-(3x-1)\right]$

$= (4x+3)(-2x+5)$

41. $9x^2y^6 - 25a^4b^2$

$= \left(3xy^3\right)^2 - \left(5a^2b\right)^2$

$= \left(3xy^3 + 5a^2b\right)\left(3xy^3 - 5a^2b\right)$

42. $81x^4 - 100y^4$

$= \left(9x^2\right)^2 - \left(10y^2\right)^2$

$= \left(9x^2 + 10y^2\right)\left(9x^2 - 10y^2\right)$

43. $1 - 64y^3 = 1^3 - (4y)^3$

$= (1-4y)\left[1^2 + 1(4y) + (4y)^2\right]$

$= (1-4y)\left(1 + 4y + 16y^2\right)$

44. $9x^2 - 21xy + 10y^2$

$= (3x-2y)(3x-5y)$

45. $4a^3 + 32$

$= 4\left(a^3 + 8\right) = 4\left(a^3 + 2^3\right)$

$= 4(a+2)\left(a^2 - a \cdot 2 + 2^2\right)$

$= 4(a+2)\left(a^2 - 2a + 4\right)$

46. $x^2 + 6x + 9 - 4a^2$

$= (x+3)^2 - (2a)^2$

$= (x+3+2a)(x+3-2a)$

47. $x^{3n} + y^{3m}$

$= \left(x^n\right)^3 + \left(y^m\right)^3$

$= \left(x^n + y^m\right) \cdot$

$\quad\left[\left(x^n\right)^2 - \left(x^n\right)\left(y^m\right) + \left(y^m\right)^2\right]$

$= \left(x^n + y^m\right)\left(x^{2n} - x^n y^m + y^{2m}\right)$

48. $(x+y)^3 - (2x+y)^3$

$= \left[(x+y) - (2x+y)\right] \cdot$

$\quad\left[(x+y)^2 + (x+y)(2x+y) + (2x+y)^2\right]$

$= (-x)(x^2 + 2xy + y^2 + 2x^2 + 3xy + y^2$

$\qquad\qquad + 4x^2 + 4xy + y^2)$

$= -x\left(7x^2 + 9xy + 3y^2\right)$

49. $9x^2 + 30xy + 25y^2$

$= (3x+5y)^2$

50. $x^3 + y + y^3 + x$

$= x^3 + y^3 + x + y$

$= (x+y)\left(x^2 - xy + y^2\right) + (x+y)$

$= (x+y)\left(x^2 - xy + y^2 + 1\right)$

51. prime

52. $2xy - 2x^{10}y$

$= 2xy\left(1 - x^9\right) = 2xy\left[1^3 - \left(x^3\right)^3\right]$

$= 2xy\left(1 - x^3\right)\left[1^2 + 1\left(x^3\right) + \left(x^3\right)^2\right]$

$= 2xy\left(1 - x^3\right)\left(1 + x^3 + x^6\right)$

$= 2xy(1 - x)\left(1 + x + x^2\right)\left(1 + x^3 + x^6\right)$

53. $x^{2n} - 1$

$= \left(x^n\right)^2 - 1^2$

$= \left(x^n + 1\right)\left(x^n - 1\right)$

54. $x^4 - 6x^2 + 9$

$= \left(x^2 - 3\right)^2$

55. $-x^2 + 4x + 21$

$= -\left(x^2 - 4x - 21\right)$

$= -(x - 7)(x + 3)$

56. $6x^3y^2 - 6xy^4 - 6x^2y^2 + 6xy^3$

$= 6xy^2\left(x^2 - y^2 - x + y\right)$

$= 6xy^2\left[(x + y)(x - y) - (x - y)\right]$

$= 6xy^2\left[(x - y)(x + y - 1)\right]$

$= 6xy^2(x - y)(x + y - 1)$

57. $x^4 - x^3 - x + 1$

$= x^3(x - 1) - (x - 1)$

$= (x - 1)\left(x^3 - 1\right)$

$= (x - 1)(x - 1)\left(x^2 + x + 1\right)$

$= (x - 1)^2\left(x^2 + x + 1\right)$

58. $27x^{3n} - 8 = \left(3x^n\right)^3 - 2^3$

$= \left(3x^n - 2\right)\left[\left(3x^n\right)^2 + \left(3x^n\right)(2) + 2^2\right]$

$= \left(3x^n - 2\right)\left(9x^{2n} + 6x^n + 4\right)$

59. $9x^{2n} + 24x^n y^m + 16y^{2m}$

$= \left(3x^n\right)^2 + 2\left(3x^n\right)\left(4y^m\right) + \left(4y^m\right)^2$

$$\left[A^2 + 2AB + B^2\right]$$

$= \left(3x^n + 4y^m\right)^2 \qquad \left[(A + B)^2\right]$

60. $a^{4m} - b^{4m}$

$= \left(a^{2m}\right)^2 - \left(b^{2m}\right)^2$

$= \left(a^{2m} + b^{2m}\right)\left(a^{2m} - b^{2m}\right)$

$= \left(a^{2m} + b^{2m}\right)\left[\left(a^m\right)^2 - \left(b^m\right)^2\right]$

$= \left(a^{2m} + b^{2m}\right)\left(a^m + b^m\right)\left(a^m - b^m\right)$

61. $r^2 - 4rs + 4s^2 - 13r + 26s + 12$

$= \left(r^2 - 4rs + 4s^2\right) + (-13r + 26s) + 12$

$= (r - 2s)^2 - 13(r - 2s) + 12$

(Let $x = r - 2s$)

$= x^2 - 13x + 12$

$= (x - 12)(x - 1)$

$= (r - 2s - 12)(r - 2s - 1)$

62. $27b^3 - 125c^3$

$= (3b)^3 - (5c)^3$

$= (3b - 5c)\left[(3b)^2 + (3b)(5c) + (5c)^2\right]$

$= (3b - 5c)\left(9b^2 + 15bc + 25c^2\right)$

63. $x + y + 3x^2 + 3xy$

$= 1(x + y) + 3x(x + y)$

$= (x + y)(1 + 3x)$

64. $a^4b + 9 - 9a^4 - b$

$= a^4b - 9a^4 + 9 - b$

$= a^4(b - 9) - 1(b - 9)$

$= (b - 9)\left(a^4 - 1\right)$

$= (b - 9)\left(a^2 + 1\right)\left(a^2 - 1\right)$

$= (b - 9)\left(a^2 + 1\right)(a + 1)(a - 1)$

65. $\dfrac{4x^3 - 16x^2 - 9x + 36}{x - 4}$

$$
\begin{array}{r|rrrr}
4 & 4 & -16 & -9 & 36 \\
  &   & 16 & 0 & -36 \\
\hline
  & 4 & 0 & -9 & 0
\end{array}
$$

$= 4x^2 + 0x - 9$

$4x^3 - 16x^2 - 9x + 36 = (x - 4)\left(4x^2 - 9\right)$

$= (x - 4)(2x + 3)(2x - 3)$

66.

$$
\begin{array}{r}
2x^2 + 9x - 18 \\
2x + 5 \overline{\smash{\big)}\, 4x^3 + 28x^2 + 9x - 90} \\
\underline{4x^3 + 10x^2} \\
18x^2 + 9x \\
\underline{18x^2 + 45x} \\
-36x - 90 \\
\underline{-36x - 90} \\
0
\end{array}
$$

$4x^3 + 28x^2 + 9x - 90 =$

$(2x + 5)\left(2x^2 + 9x - 18\right)$

$= (2x + 5)(2x - 3)(x + 6)$

Problem set 4.1

11. Neither factor in the denominator can equal 0.

$x - 1 \neq 0 \qquad 2x + 6 \neq 0$

$x \neq 1 \qquad 2x \neq -6$

$\qquad\qquad x \neq -3$

$\{x \mid x \neq 1 \text{ and } x \neq -3\}$

13. $\dfrac{n+4}{n^2-25} = \dfrac{n+4}{(n+5)(n-5)}$

$\{n \mid n \neq -5 \text{ and } n \neq 5\}$

15. $\dfrac{3y+2}{4y-5} \qquad 4y - 5 \neq 0$

$\qquad\qquad\qquad 4y \neq 5$

$\qquad\qquad\qquad y \neq \frac{5}{4} \quad \left\{ y \mid y \neq \frac{5}{4} \right\}$

17. $\dfrac{-3p+17}{p^2+p-12} = \dfrac{-3p+17}{(p-3)(p+4)}$

$\{p \mid p \neq 3 \text{ and } p \neq -4\}$

19. a. $C = \dfrac{4(80)}{100-80} = \dfrac{320}{20} = 16$

Cost is $ 16,000.

b. $C = \dfrac{4(95)}{100-95} = \dfrac{380}{5} = 76$

Cost is $ 76,000.

c. $x = 100$ is not permissible since this value causes 0 in the denominator.

d. The denominator becomes very small and the value of $\dfrac{4x}{100-x}$ gets very large.

The cost of removing almost all pollutants from the stream is extremely high.

23. $\dfrac{3a+3}{a+3} = \dfrac{3(a+3)}{a+3} = 3$

25. $\dfrac{a-c}{a^2-ac} = \dfrac{a-c}{a(a-c)} = \dfrac{1}{a}$

29. $\dfrac{3b}{3b+3c} = \dfrac{3b}{3(b+c)} = \dfrac{b}{b+c}$

31. $\dfrac{12ab^2}{6ab^3-6ab^4} = \dfrac{12ab^2}{6ab^3(1-b)} = \dfrac{2}{b(1-b)}$

35. $\dfrac{5a^2b+5a^2c}{15ab+15ac} = \dfrac{5a^2(b+c)}{15a(b+c)} = \dfrac{a}{3}$

37. $\dfrac{x^2-4}{2x-4} = \dfrac{(x+2)(x-2)}{2(x-2)} = \dfrac{x+2}{2}$

43. $\dfrac{x^2-2xy+y^2}{x^2-y^2} = \dfrac{(x-y)^2}{(x+y)(x-y)} = \dfrac{x-y}{x+y}$

47. $\dfrac{y^2-4y-5}{y^2+5y+4} = \dfrac{(y-5)(y+1)}{(y+4)(y+1)} = \dfrac{y-5}{y+4}$

49. $\dfrac{6b^2-b-2}{3b^2+4b-4} = \dfrac{(3b-2)(2b+1)}{(3b-2)(b+2)} = \dfrac{2b+1}{b+2}$

53. $\dfrac{a^3+4^3}{a^2-4^2} = \dfrac{(a+4)\left(a^2-4a+16\right)}{(a+4)(a-4)}$

$= \dfrac{a^2-4a+16}{a-4}$

57. $\dfrac{x^3+x^2-20x}{x^3+2x^2-15x} = \dfrac{x\left(x^2+x-20\right)}{x\left(x^2+2x-15\right)}$

$= \dfrac{x(x+5)(x-4)}{x(x+5)(x-3)} = \dfrac{x-4}{x-3}$

63. $\dfrac{x^2-5x+6}{x^2-7x-18}=\dfrac{(x-2)(x-3)}{(x-9)(x+2)}$

Although numerator and denominator can be factored, with no identical factors in numerator and denominator, the algebraic fraction cannot be simplified.

65. $\dfrac{a^2-16}{a^2-4a+3ab-12b}=\dfrac{(a+4)(a-4)}{a(a-4)+3b(a-4)}$

$=\dfrac{(a+4)(a-4)}{(a+3b)(a-4)}=\dfrac{a+4}{a+3b}$

67. $\dfrac{3m^2+9m-mx-3x}{m^3+27}$

$=\dfrac{3m(m+3)-x(m+3)}{m^3+3^3}$

$=\dfrac{(3m-x)(m+3)}{(m+3)\left(m^2-3m+9\right)}=\dfrac{3m-x}{m^2-3m+9}$

69. $\dfrac{3-y}{y-3}=\dfrac{-1(-3+y)}{y-3}=\dfrac{-1(y-3)}{y-3}=-1$

You can immediately obtain -1 using the property that the quotient of two polynomials that have exactly opposite signs and are additive inverses is -1.

71. $\dfrac{a^2-4}{2-a}=\dfrac{(a+2)(a-2)}{2-a}=-1(a+2)$

73. $\dfrac{3-x}{x^2-7x+12}=\dfrac{3-x}{(x-4)(x-3)}=\dfrac{-1}{x-4}$

The answer can be equivalently expressed by attaching the negative sign to the denominator rather than the numerator.

$\dfrac{-1}{x-4}=\dfrac{1}{-(x-4)}=\dfrac{1}{-x+4}=\dfrac{1}{4-x}$

75. $\dfrac{x^{2n}-2x^n-3}{x^{2n}+x^n-12}=\dfrac{\left(x^n-3\right)\left(x^n+1\right)}{\left(x^n-3\right)\left(x^n+4\right)}$

$=\dfrac{x^n+1}{x^n+4}$

77. $\dfrac{\left(x^n\right)^2-\left(y^n\right)^2}{x^{2n}+2x^ny^n+y^{2n}}$

$=\dfrac{\left(x^n+y^n\right)\left(x^n-y^n\right)}{\left(x^n+y^n\right)^2}=\dfrac{x^n-y^n}{x^n+y^n}$

79. $\dfrac{a^{2n}-b^{2n}}{a^{2n}+a^nb^n}=\dfrac{\left(a^n+b^n\right)\left(a^n-b^n\right)}{a^n\left(a^n+b^n\right)}$

$=\dfrac{a^n-b^n}{a^n}$

81. $\dfrac{\left(x^3-y^3\right)\left(x^3-xy^2\right)}{x^3+x^2y+xy^2}$

$=\dfrac{(x-y)\left(x^2+xy+y^2\right)x\left(x^2-y^2\right)}{x\left(x^2+xy+y^2\right)}$

$=(x-y)\left(x^2-y^2\right)$ or $(x-y)^2(x+y)$

83. $\dfrac{\left(x^2+9\right)\left(54-2x^3\right)}{2x^4-162}=\dfrac{\left(x^2+9\right)2\left(3^3-x^3\right)}{2\left(x^4-81\right)}$

$=\dfrac{\left(x^2+9\right)2(3-x)\left(9+3x+x^2\right)}{2\left(x^2+9\right)\left(x^2-9\right)}$

$=\dfrac{(3-x)\left(9+3x+x^2\right)}{(x+3)(x-3)}=\dfrac{-\left(9+3x+x^2\right)}{x+3}$

85. $\dfrac{c^4+16c^2d^2+64d^4-16c^2d^2}{c^2+4cd+8d^2}$

$= \dfrac{\left(c^2+8d^2\right)^2-(4cd)^2}{c^2+4cd+8d^2}$

$= \dfrac{\left(c^2+8d^2+4cd\right)\left(c^2+8d^2-4cd\right)}{c^2+4cd+8d^2}$

$= c^2+8d^2-4cd$

89. $\dfrac{150x}{250x-50} = \dfrac{150x}{50(5x-1)} = \dfrac{3x}{5x-1}$

$x = 70$

$\dfrac{3x}{5x-1} = \dfrac{3(70)}{5(70)-1} = \dfrac{210}{349} \approx .6$

Cost is $(.6)(1,000,000)$, approximately $600,000.

91. $\dfrac{x^2+7}{7} \neq x^2+1$ We cannot cancel

identical terms in numerator and denominator. We can cancel identical

factors. $\left(\dfrac{x^27}{7} = x^2\right)$

If $x=1$: $\dfrac{1^2+7}{7} \neq 1^2+1$

$\dfrac{8}{7} \neq 2$

95. $dx-2x = d^2-4d+4$

$x(d-2) = d^2-4d+4$

$x = \dfrac{d^2-4d+4}{d-2}$ $d \neq 2$

$x = \dfrac{(d-2)^2}{d-2}$

$x = d-2$ $(d \neq 2)$

97. $d^2(x-1) = 5d+10+4(x-1)$

$d^2x-d^2 = 5d+10+4x-4$

$d^2x-4x = d^2+5d+6$

$x\left(d^2-4\right) = d^2+5d+6$

$x = \dfrac{d^2+5d+6}{d^2-4}$

Since $d^2-4 \neq 0$, $(d+2)(d-2) \neq 0$, so $d \neq -2$ and $d \neq 2$.

$x = \dfrac{(d+2)(d+3)}{(d+2)(d-2)} = \dfrac{d+3}{d-2}$ $(d \neq -2$ and $d \neq 2)$

99. $x^7-x = x\left(x^6-1\right) = x\left[\left(x^3\right)^2-1^2\right]$

$= x\left(x^3+1\right)\left(x^3-1\right)$

$= x(x+1)\left(x^2-x+1\right)(x-1)\left(x^2+x+1\right)$

method 2 : $x^7-x = x\left(x^6-1\right)$

$= \left[\left(x^2\right)^3-1^3\right]$

$= x\left(x^2-1\right)\left[\left(x^2\right)^2+x^2 \cdot 1+1^2\right]$

$= x(x+1)(x-1)\left(x^4+x^2+1\right)$

The two answers are identical. See chapter 3, problem set 3.4.4, problems 111 and 112, for a factoring technique that shows

$x^4+x^2+1 = \left(x^2-x+1\right)\left(x^2+x+1\right).$

100. x: side of smaller triangle

$x+10$: side of larger triangle

$3x+3(x+10) = 186$

$6x+30 = 186$

$6x = 156; \ x = 26$

Side of larger triangle: $26+10 = 36$ m.

101. $\left(3x^{\,n}-5y^{\,n}\right)\left(6x^{\,n}+y^{\,n}\right)$

$= 18x^{\,2n}+3x^{\,n}y^{\,n}-30x^{\,n}y^{\,n}-5y^{\,2n}$

$= 18x^{\,2n}-27x^{\,n}y^{\,n}-5y^{\,2n}$

35. $\dfrac{(x-3)\left(x^{\,2}+3x+9\right)\left(3-x^{\,2}\right)}{x\left(x^{\,2}-3\right)(x-3)}$

$= \dfrac{-\left(x^{\,2}+3x+9\right)}{x}$

Problem set 4.2.1

In most cases, numerators and denominators are immediately expressed in factored form.

41. $\dfrac{p(r-s)+q(r-s)}{p(r+s)+q(r+s)} \cdot \dfrac{m(r+s)-n(r+s)}{m(r-s)+n(r-s)}$

$= \dfrac{(r-s)(p+q)(r+s)(m-n)}{(r+s)(p+q)(r-s)(m+n)} = \dfrac{m-n}{m+n}$

11. $\dfrac{(y+2)(y-2)}{y-2} \cdot \dfrac{y-2}{(y+3)(y-2)} = \dfrac{y+2}{y+3}$

15. $\dfrac{b(b+1)(b+3)(b+2)}{(b+2)(b-2)(b+1)(b-1)} = \dfrac{b(b+3)}{(b-2)(b-1)}$

43. $\dfrac{x^{\,2}(x-4)+(x-4)}{2x^{\,2}(x-4)+(x-4)} \cdot \dfrac{2x^{\,2}(x+1)+(x+1)}{x^{\,3}(x-1)+x(x-1)}$

$= \dfrac{(x-4)\left(x^{\,2}+1\right)(x+1)\left(2x^{\,2}+1\right)}{(x-4)\left(2x^{\,2}+1\right)(x-1)\left(x^{\,3}+x\right)}$

17. $\dfrac{(m+2)(m-2)2(m+2)}{(m+2)^{\,2}(m+3)(m-2)} = \dfrac{2}{m+3}$

$= \dfrac{(x-4)\left(x^{\,2}+1\right)(x+1)\left(2x^{\,2}+1\right)}{(x-4)\left(2x^{\,2}+1\right)(x-1)x\left(x^{\,2}+1\right)}$

$= \dfrac{x+1}{x(x-1)}$

21. $\dfrac{(x-2)\left(x^{\,2}+2x+4\right)(x+2)}{(x+2)(x-2)3x} = \dfrac{x^{\,2}+2x+4}{3x}$

23. $\dfrac{(2a+1)(a-7)(a-2)(a+1)}{(a-7)(a+1)(2a+1)(a-3)} = \dfrac{a-2}{a-3}$

45. $\dfrac{y^{\,n}(y-3)y\left(y^{\,n}+2\right)}{y^{\,n}\left(y^{\,n}+2\right)y(y-3)} = 1$

25. $\dfrac{(2y-5)(y+7)(3y+1)(y+3)}{(2y-5)(3y+1)(y+7)(y+3)} = 1$

47. $\dfrac{\left(y^{\,n}+1\right)\left(y^{\,n}-1\right)\left(y^{\,n}-3\right)\left(y^{\,n}+2\right)}{\left(y^{\,n}+2\right)\left(y^{\,n}+1\right)\left(y^{\,n}+4\right)\left(y^{\,n}-3\right)}$

$= \dfrac{y^{\,n}-1}{y^{\,n}+4}$

27. $\dfrac{(y+5)^{\,2}(y-4)(y+3)}{(y-4)(y+5)(y+3)} = y+5$

29. $\dfrac{(m+4)(m-3)(m+3)(m+2)(m+6)(m+1)}{(m+6)(m-5)(m-3)(m+1)(m+3)}$

$= \dfrac{(m+4)(m+2)}{m-5}$

33. $\dfrac{(x+4)(x+1)(3-x)}{(x+4)(x-3)(x+1)} = \dfrac{3-x}{x-3} = -1$

49. $\dfrac{a(x-y)+3(x-y)}{x^3+y^3} \cdot \dfrac{-\left(x^2-xy+y^2\right)}{b(a+3)+c(a+3)}$

$= \dfrac{-(x-y)(a+3)\left(x^2-xy+y^2\right)}{(x+y)\left(x^2-xy+y^2\right)(a+3)(b+c)}$

$= \dfrac{-(x-y)}{(x+y)(b+c)}$

55. $V = 4 \cdot \dfrac{a^2-b^2}{10x^3} \cdot \dfrac{5x^2}{2a+2b}$

$= \dfrac{4(a+b)(a-b)5x^2}{10x^3 \cdot 2(a+b)} = \dfrac{a-b}{x}$

56. rate of first car : x

     rate of second car: $x+10$

     Distance covered by first car + distance

     covered by second car = 300

$$(RT = D)$$

$$3x + 3(x+10) = 300$$

$$6x + 30 = 300$$

$$6x = 270$$

$$x = 45$$

     first car: 45 m/h

     second car: $45 + 10 = 55$ m/h

57. $(3a-5b)\left(6a^2+4ab-9b^2\right)$

$= 3a\left(6a^2+4ab-9b^2\right)$

$\qquad\qquad - 5b\left(6a^2+4ab-9b^2\right)$

$= 18a^3 + 12a^2b - 27ab^2 - 30a^2b$

$\qquad\qquad - 20ab^2 + 45b^3$

$= 8a^3 - 18a^2b - 47ab^2 + 45b^3$

58.

$$3x+y \enclose{longdiv}{3x^3+x^2y-6x-2y} \quad x^2-2$$

$$\underline{3x^3+x^2y}$$

$$-6x-2y$$

$$\underline{-6x-2y}$$

$$0$$

$$3x^3+x^2y-6x-2y = (3x+y)\left(x^2-2\right)$$

Problem set 4.2.2

In all solutions, the division is immediately
expressed as multiplication by the reciprocal
of the divisor.

3. $\dfrac{x+y}{x^2-xy} \cdot \dfrac{x-y}{3x+3y}$

$= \dfrac{(x+y)(x-y)}{x(x-y)3(x+y)} = \dfrac{1}{3x}$

7. $\dfrac{x^3-27}{a^3+8} \cdot \dfrac{a+2}{x-3}$

$= \dfrac{(x-3)\left(x^2+3x+9\right)(a+2)}{(a-2)\left(a^2-2a+4\right)(x-3)}$

$= \dfrac{x^2+3x+9}{a^2-2a+4}$

13. $\dfrac{9x^2-12x+4}{2x^2+3x-5} \cdot \dfrac{2x^2+7x+5}{3x^2-8x+4}$

$= \dfrac{(3x-2)^2(2x+5)(x+1)}{(2x+5)(x-1)(3x-2)(x-2)}$

$= \dfrac{(3x-2)(x+1)}{(x-1)(x-2)}$

15. $b(x+y)+3(x+y) \cdot \dfrac{b+3}{x^2-y^2}$

$= \dfrac{(x+y)(b+3)(b+3)}{(x+y)(x-y)} = \dfrac{(b+3)^2}{x-y}$

17. $\dfrac{x-2}{(2x)^3-3^3} \cdot \dfrac{3-2x}{a(x-2)+b(x-2)}$

$= \dfrac{(x-2)(3-2x)}{(2x-3)\left(4x^2+6x+9\right)(x-2)(a+b)}$

$= \dfrac{-1}{\left(4x^2+6x+9\right)(a+b)}$

19. $\dfrac{2y^2+13y+20}{6y^2-13y-5} \cdot \dfrac{9y^2-3y-2}{(-1)\left(3y^2+10y-8\right)}$

$= \dfrac{(2y+5)(y+4)(3y+1)(3y-2)}{-(3y+1)(2y-5)(3y-2)(y+4)}$

$= \dfrac{-(2y+5)}{2y-5}$

21. $\dfrac{4y^3-12y^2}{4y^n+8} \cdot \dfrac{y^{2n}-4}{y^{n+1}-2y}$

$= \dfrac{4y^2(y-3)\left(y^n+2\right)\left(y^n-2\right)}{4\left(y^n+2\right)y\left(y^n-2\right)}$

$= y(y-3)$

23. $\dfrac{y^{2n}-1}{2y^{2n}+y^n-3} \cdot \dfrac{2y^{2n}-y^n-6}{y^{2n}-y^n-2}$

$= \dfrac{\left(y^n+1\right)\left(y^n-1\right)\left(2y^n+3\right)\left(y^n-2\right)}{\left(2y^n+3\right)\left(y^n-1\right)\left(y^n-2\right)\left(y^n+1\right)} = 1$

25. $\dfrac{y-1}{y^2-3y} \cdot \dfrac{y^3-9y}{1}$

$= \dfrac{(y-1)y(y+3)(y-3)}{y(y-3)} = (y-1)(y+3)$

27. $\dfrac{x^3(x-y)^3}{x^3-y^3} \cdot \dfrac{x^2+xy+y^2}{x^3-2x^2y+xy^2}$

$= \dfrac{x^3(x-y)^3\left(x^2+xy+y^2\right)}{(x-y)\left(x^2+xy+y^2\right)x(x-y)^2}$

$= x^2$

29. $\left(\dfrac{a-b}{4c} \cdot \dfrac{c}{b-a}\right) \div \dfrac{a-b}{c^2}$

$= \dfrac{-1}{4} \cdot \dfrac{c^2}{a-b} = \dfrac{-c^2}{4(a-b)}$

31. $\dfrac{a^2-8a+15}{2a^3-10a^2} \cdot \dfrac{2a^2+3a}{3a^3-27a} \cdot \dfrac{a^2-6a-27}{14a+21}$

$= \dfrac{(a-5)(a-3)a(2a+3)(a-9)(a+3)}{2a^2(a-5)3a(a+3)(a-3)7(2a+3)}$

$= \dfrac{a-9}{42a^2}$

33. Area of rectangle =

$\dfrac{1}{x^2-9} \cdot x^2+6x+9 = \dfrac{(x+3)^2}{(x+3)(x-3)} = \dfrac{x+3}{x-3}$

Area of triangle =

$\dfrac{1}{2} \cdot \dfrac{1}{x^2+6x+9} \cdot 2(x+3) = \dfrac{2(x+3)}{2(x+3)^2}$

$= \dfrac{1}{x+3}$

Area of rectangle ÷ Area of triangle =

$\dfrac{x+3}{x-3} \div \dfrac{1}{x+3} = \dfrac{x+3}{x-3} \cdot \dfrac{x+3}{1} = \dfrac{(x+3)^2}{x-3}$

35. $y^n - \left[3y^n - \left(1 - 2y^n\right)\right]$

$= y^n - \left[3y^n - 1 + 2y^n\right]$

$= y^n - \left(5y^n - 1\right) = y^n - 5y^n + 1$

$= -4y^n + 1$

36. First piece: x   Second: x + 3   Third: 2x

$x + (x + 3) + 2x = 11$

$4x = 8$

$x = 2$

First piece: 2 ft   Second: 2 + 3 = 5 ft

Third: 2(2) = 4 ft

37. $\left|\dfrac{3y-5}{2}\right| < 10$

$-10 < \dfrac{3y-5}{2} < 10$

$-20 < 3y - 5 < 20$

$-15 < 3y < 25$

$-5 < y < \dfrac{25}{3}$

$\left\{ y \,\middle|\, -5 < y < \dfrac{25}{3} \right\}$

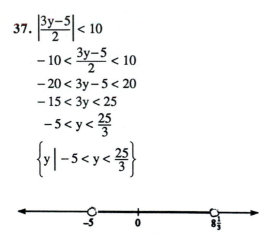

Problem set 4.3.1

9. $\dfrac{y + 5y + 6y}{6} = \dfrac{12y}{6} = 2y$

17. $\dfrac{4x + 8y}{x + 2y} = \dfrac{4(x + 2y)}{x + 2y} = 4$

21. $\dfrac{x + x - 5}{2x - 5} = \dfrac{2x - 5}{2x - 5} = 1$

23. $\dfrac{3x - y + x + 3y}{2} = \dfrac{4x + 2y}{2} = \dfrac{2(2x + y)}{2}$

$= 2x + y$

25. $\dfrac{4b - 3c + 2b + 9c}{b + c} = \dfrac{6b + 6c}{b + c}$

$= \dfrac{6(b + c)}{b + c} = 6$

29. $\dfrac{a^2 + 7a + 3 + 2a + 6}{a^2 + 9a + 9} = \dfrac{a^2 + 9a + 9}{a^2 + 9a + 9} = 1$

35. $\dfrac{4a - a}{6y} = \dfrac{3a}{6y} = \dfrac{a}{2y}$

37. $\dfrac{x^2 - y^2}{x + y} = \dfrac{(x + y)(x - y)}{x + y} = x - y$

41. $\dfrac{x^2 - 25}{x^2 + 2x - 15} = \dfrac{(x + 5)(x - 5)}{(x + 5)(x - 3)} = \dfrac{x - 5}{x - 3}$

43. $\dfrac{5x + 5y - (3x + y)}{2x} = \dfrac{5x + 5y - 3x - y}{2x}$

$= \dfrac{2x + 4y}{2x} = \dfrac{2(x + 2y)}{2x} = \dfrac{x + 2y}{x}$

45. $\dfrac{y - (x + y)}{x + y} = \dfrac{y - x - y}{x + y} = \dfrac{-x}{x + y}$

47. $\dfrac{a^2 - 4a - (a - 6)}{a^2 - a - 6} = \dfrac{a^2 - 4a - a + 6}{a^2 - a - 6}$

$= \dfrac{a^2 - 5a + 6}{a^2 - a - 6} = \dfrac{(a - 3)(a - 2)}{(a - 3)(a + 2)} = \dfrac{a - 2}{a + 2}$

51. $\dfrac{3a+b+2a+3b-(4a+3b)}{a+b} = \dfrac{a+b}{a+b} = 1$

53. $\dfrac{3a^3+4b^3-\left(5b^3+2a^3\right)}{a^2-b^2} = \dfrac{a^3-b^3}{a^2-b^2}$

$= \dfrac{(a-b)\left(a^2+ab+b^2\right)}{(a+b)(a-b)} = \dfrac{a^2+ab+b^2}{a+b}$

55. $\dfrac{b-a}{ac+ad-bc-bd} = \dfrac{b-a}{a(c+d)-b(c+d)}$

$= \dfrac{b-a}{(a-b)(c+d)} = \dfrac{-1}{c+d}$

57. $\dfrac{2y}{y-5} - \left(\dfrac{2+y-2}{y-5}\right) = \dfrac{2y}{y-5} - \dfrac{y}{y-5}$

$= \dfrac{2y-y}{y-5} = \dfrac{y}{y-5}$

59. R: desired rational expression

$\dfrac{6y-2}{5-3y} + R = \dfrac{2y-3}{5-3y}$

$R = \dfrac{2y-3}{5-3y} - \dfrac{6y-2}{5-3y} = \dfrac{2y-3-(6y-2)}{5-3y}$

$R = \dfrac{-4y-1}{5-3y}$

61. $\dfrac{1}{a-b} + \dfrac{1}{b-a} = \dfrac{1}{a-b} + \dfrac{1}{b-a} \cdot \dfrac{(-1)}{(-1)}$

$= \dfrac{1}{a-b} + \dfrac{-1}{a-b} = \dfrac{1-1}{a-b} = \dfrac{0}{a-b} = 0$

62. x: number of orchestra tickets

$8-x$: number of balcony tickets

$57x + 43(8-x) = 386$

$57x + 344 - 43x = 386$

$14x = 42$

$x = 3$

Orchestra tickets: 3

Balcony tickets: $8-3 = 5$

63. $-(5+3x)-1 \le x-18$

$-5-3x-1 \le x-18$

$-6-3x \le x-18$

$-4x \le -12$

$x \ge 3 \quad \{x \mid x \ge 3\}$

64. First Rectangle: length: $3x$ width: x

Second Rectangle: length : $3x+4$

width: $x-1$

Area of first rectangle = area of second

$3x \cdot x = (3x+4)(x-1)$

$3x^2 = 3x^2 + x - 4$

$4 = x$

First Rectangle: length : $3(4) = 12$ m

                  width : 4 m

Problem set 4.3.2

1. $\dfrac{3}{6x^3} - \dfrac{2}{9x^2} \quad \left(\text{LCD is } 18x^3\right)$

$= \dfrac{3}{6x^3} \cdot \dfrac{3}{3} - \dfrac{2}{9x^2} \cdot \dfrac{2x}{2x}$

$= \dfrac{9}{18x^3} - \dfrac{4x}{18x^3} = \dfrac{9-4x}{18x^3}$

3. $\dfrac{2a}{3c^2} - \dfrac{3b}{4cd}$ $\left(\text{LCD is } 12c^2d\right)$

$= \dfrac{2a}{3c^2} \cdot \dfrac{4d}{4d} - \dfrac{3b}{4cd} \cdot \dfrac{3c}{3c}$

$= \dfrac{8ad}{12c^2d} - \dfrac{9bc}{12c^2d} = \dfrac{8ad-9bc}{12c^2d}$

5. $\dfrac{3a}{2c^2} - \dfrac{2a}{3cd} + \dfrac{a}{6d^2}$ $\left(\text{LCD is } 6c^2d^2\right)$

$= \dfrac{3a}{2c^2} \cdot \dfrac{3d^2}{3d^2} - \dfrac{2a}{3cd} \cdot \dfrac{2cd}{2cd} + \dfrac{a}{6d^2} \cdot \dfrac{c^2}{c^2}$

$= \dfrac{9ad^2}{6c^2d^2} - \dfrac{4acd}{6c^2d^2} + \dfrac{ac^2}{6c^2d^2}$

$= \dfrac{9ad^2-4acd+ac^2}{6c^2d^2}$

7. $\dfrac{2b-2c}{b^2c} + \dfrac{b-c}{bc^2}$ $\left(\text{LCD is } b^2c^2\right)$

$= \dfrac{2b-2c}{b^2c} \cdot \dfrac{c}{c} + \dfrac{b-c}{bc^2} \cdot \dfrac{b}{b}$

$= \dfrac{(2b-2c)c}{b^2c^2} + \dfrac{(b-c)b}{b^2c^2}$

$= \dfrac{2bc-2c^2+b^2-bc}{b^2c^2} = \dfrac{b^2+bc-2c^2}{b^2c^2}$

9. $\dfrac{b-2y}{4b^2y} + \dfrac{2b+y}{6by^2}$ $\left(\text{LCD is } 12b^2y^2\right)$

$= \dfrac{b-2y}{4b^2y} \cdot \dfrac{3y}{3y} + \dfrac{2b+y}{6by^2} \cdot \dfrac{2b}{2b}$

$= \dfrac{3y(b-2y)}{12b^2y^2} + \dfrac{2b(2b+y)}{12b^2y^2}$

$= \dfrac{3by-6y^2+4b^2+2by}{12b^2y^2}$

$= \dfrac{5by-6y^2+4b^2}{12b^2y^2}$

11. $\dfrac{4x-3y}{6xy} - \dfrac{x-4z}{8xz} - \dfrac{3y-z}{4yz}$ $(\text{LCD is } 24xyz)$

$= \dfrac{4x-3y}{6xy} \cdot \dfrac{4z}{4z} - \dfrac{x-4z}{8xz} \cdot \dfrac{3y}{3y} - \dfrac{3y-z}{4yz} \cdot \dfrac{6x}{6x}$

$= \dfrac{4z(4x-3y)-3y(x-4z)-6x(3y-z)}{24xyz}$

$= \dfrac{16xz-12yz-3xy+12yz-18xy+6xz}{24xyz}$

$= \dfrac{22xz-21xy}{24xyz} = \dfrac{x(22z-21y)}{24xyz}$

$= \dfrac{22z-21y}{24yz}$

13. $\dfrac{10}{x+4} - \dfrac{2}{x-6}$ $\left[\text{LCD is } (x+4)(x-6)\right]$

$= \dfrac{10}{x+4} \cdot \dfrac{x-6}{x-6} - \dfrac{2}{x-6} \cdot \dfrac{x+4}{x+4}$

$= \dfrac{10(x-6)-2(x+4)}{(x+4)(x-6)}$

$= \dfrac{10x-60-2x-8}{(x+4)(x-6)} = \dfrac{8x-68}{(x+4)(x-6)}$

15. $\dfrac{b}{b-c} - \dfrac{c}{b+c}$ $\left[\text{LCD is } (b-c)(b+c)\right]$

$= \dfrac{b}{b-c} \cdot \dfrac{b+c}{b+c} - \dfrac{c}{b+c} \cdot \dfrac{b-c}{b-c}$

$= \dfrac{b(b+c)-c(b-c)}{(b-c)(b+c)} = \dfrac{b^2+bc-bc+c^2}{(b-c)(b+c)}$

$= \dfrac{b^2+c^2}{(b-c)(b+c)}$

17. $\dfrac{3}{a+1} - \dfrac{3}{a}$ $\left[\text{LCD is } a(a+1)\right]$

$= \dfrac{3}{a+1} \cdot \dfrac{a}{a} - \dfrac{3}{a} \cdot \dfrac{a+1}{a+1}$

$= \dfrac{3a-3(a+1)}{a(a+1)} = \dfrac{3a-3a-3}{a(a+1)} = \dfrac{-3}{a(a+1)}$

19. $\dfrac{5x}{x-2} - \dfrac{x-1}{x+2}$ $\left[\text{LCD is } (x-2)(x+2)\right]$

$= \dfrac{5x}{x-2} \cdot \dfrac{x+2}{x+2} - \dfrac{x-1}{x+2} \cdot \dfrac{x-2}{x-2}$

$= \dfrac{5x(x+2)-(x-1)(x-2)}{(x-2)(x+2)}$

$= \dfrac{5x^2+10x-x^2+3x-2}{(x-2)(x+2)}$

$= \dfrac{4x^2+13x-2}{(x-2)(x+2)}$

23. $\dfrac{a-b}{a+b} - \dfrac{a+b}{a-b}$ $\left[\text{LCD is } (a+b)(a-b)\right]$

$= \dfrac{a-b}{a+b} \cdot \dfrac{a-b}{a-b} - \dfrac{a+b}{a-b} \cdot \dfrac{a+b}{a+b}$

$= \dfrac{(a-b)(a-b)-(a+b)(a+b)}{(a+b)(a-b)}$

$= \dfrac{a^2-2ab+b^2-a^2-2ab-b^2}{(a+b)(a-b)}$

$= \dfrac{-4ab}{(a+b)(a-b)}$

25. $\dfrac{4}{x+2} - \dfrac{3}{x+1} + \dfrac{2}{x}$ $\left[\text{LCD is } x(x+1)(x+2)\right]$

$= \dfrac{4}{x+2} \cdot \dfrac{x}{x} \cdot \dfrac{x+1}{x+1} - \dfrac{3}{x+1} \cdot \dfrac{x}{x} \cdot \dfrac{x+2}{x+2}$

$\qquad\qquad + \dfrac{2}{x} \cdot \dfrac{x+1}{x+1} \cdot \dfrac{x+2}{x+2}$

$= \dfrac{4x(x+1)-3x(x+2)+2(x+1)(x+2)}{x(x+1)(x+2)}$

$= \dfrac{4x^2+4x-3x^2-6x+2x^2+6x+4}{x(x+1)(x+2)}$

$= \dfrac{3x^2+4x+4}{x(x+1)(x+2)}$

27. $\dfrac{5}{2b-8} + \dfrac{3}{4b-2} = \dfrac{5}{2(b-4)} + \dfrac{3}{2(2b-1)}$

$\qquad \left[\text{LCD is } 2(b-4)(2b-1)\right]$

$= \dfrac{5}{2(b-4)} \cdot \dfrac{2b-1}{2b-1} + \dfrac{3}{2(2b-1)} \cdot \dfrac{b-4}{b-4}$

$= \dfrac{5(2b-1)+3(b-4)}{2(b-4)(2b-1)} = \dfrac{10b-5+3b-12}{2(b-4)(2b-1)}$

$= \dfrac{13b-17}{2(b-4)(2b-1)}$

29. $\dfrac{4}{x^2+6x+9} + \dfrac{4}{x+3} = \dfrac{4}{(x+3)^2} + \dfrac{4}{x+3}$

$\qquad \left[\text{LCD is } (x+3)^2\right]$

$= \dfrac{4}{(x+3)^2} + \dfrac{4}{x+3} \cdot \dfrac{x+3}{x+3}$

$= \dfrac{4+4(x+3)}{(x+3)^2} = \dfrac{4x+16}{(x+3)^2}$

31. $\dfrac{c}{c^2-10c+25} - \dfrac{c-4}{2c-10} = \dfrac{c}{(c-5)^2} - \dfrac{c-4}{2(c-5)}$

$\qquad \left[\text{LCD is } 2(c-5)^2\right]$

$= \dfrac{c}{(c-5)^2} \cdot \dfrac{2}{2} - \dfrac{c-4}{2(c-5)} \cdot \dfrac{c-5}{c-5}$

$= \dfrac{2c-(c-4)(c-5)}{2(c-5)^2} = \dfrac{2c-c^2+9c-20}{2(c-5)^2}$

$= \dfrac{-c^2+11c-20}{2(c-5)^2}$

33. $\dfrac{a-b}{3a+3b} + \dfrac{a+b}{2a-2b} = \dfrac{a-b}{3(a+b)} + \dfrac{a+b}{2(a-b)}$

$\qquad \left[\text{LCD is } 6(a+b)(a-b)\right]$

$\qquad = \dfrac{a-b}{3(a+b)} \cdot \dfrac{2(a-b)}{2(a-b)} + \dfrac{a+b}{2(a-b)} \cdot \dfrac{3(a+b)}{3(a+b)}$

$\qquad = \dfrac{2(a-b)(a-b) + 3(a+b)(a+b)}{6(a+b)(a-b)}$

$\qquad = \dfrac{2a^2 - 4ab + 2b^2 + 3a^2 + 6ab + 3b^2}{6(a+b)(a-b)}$

$\qquad = \dfrac{5a^2 + 2ab + 5b^2}{6(a+b)(a-b)}$

35. $\dfrac{b+2}{b^2+b-2} + \dfrac{2}{b^2-1}$

$\qquad = \dfrac{b+2}{(b+2)(b-1)} + \dfrac{2}{(b+1)(b-1)}$

$\qquad = \dfrac{1}{b-1} + \dfrac{2}{(b+1)(b-1)}$

$\qquad \left[\text{LCD is } (b+1)(b-1)\right]$

$\qquad = \dfrac{1}{b-1} \cdot \dfrac{b+1}{b+1} + \dfrac{2}{(b+1)(b-1)}$

$\qquad = \dfrac{b+3}{(b+1)(b-1)}$

37. $\dfrac{y+3}{y^2-y-2} - \dfrac{y-1}{y^2+2y+1}$

$\qquad = \dfrac{y+3}{(y+1)(y-2)} - \dfrac{y-1}{(y+1)^2}$

$\qquad \left[\text{LCD is } (y+1)^2(y-2)\right]$

$\qquad = \dfrac{y+3}{(y+1)(y-2)} \cdot \dfrac{y+1}{y+1} - \dfrac{y-1}{(y+1)^2} \cdot \dfrac{y-2}{y-2}$

$\qquad = \dfrac{(y+3)(y+1)}{(y+1)^2(y-2)} - \dfrac{(y-1)(y-2)}{(y+1)^2(y-2)}$

$\qquad = \dfrac{y^2+4y+3 - \left(y^2-3y+2\right)}{(y+1)^2(y-2)}$

$\qquad = \dfrac{7y+1}{(y+1)^2(y-2)}$

39. $\dfrac{x^2+x+2}{x^3-1} - \dfrac{1}{x-1}$

$\qquad = \dfrac{x^2+x+2}{(x-1)\left(x^2+x+1\right)} - \dfrac{1}{x-1}$

$\qquad \left[\text{LCD is } (x-1)\left(x^2+x+1\right)\right]$

$\qquad = \dfrac{x^2+x+2}{(x-1)\left(x^2+x+1\right)} - \dfrac{1}{x-1} \cdot \dfrac{x^2+x+1}{x^2+x+1}$

$\qquad = \dfrac{x^2+x+2 - \left(x^2+x+1\right)}{(x-1)\left(x^2+x+1\right)}$

$\qquad = \dfrac{1}{(x-1)\left(x^2+x+1\right)}$

41. $\dfrac{1}{y-x} + \dfrac{1}{x-y} = \dfrac{1}{y-x} + \dfrac{1}{x-y} \cdot \dfrac{-1}{-1}$

$\qquad = \dfrac{1}{y-x} + \dfrac{-1}{-x+y} = \dfrac{1}{y-x} - \dfrac{1}{y-x}$

$\qquad = \dfrac{1-1}{y-x} = \dfrac{0}{y-x} = 0$

43. $\dfrac{y^2}{y-7} + \dfrac{6y+7}{7-y} = \dfrac{y^2}{y-7} + \dfrac{6y+7}{7-y} \cdot \dfrac{-1}{-1}$

$\qquad = \dfrac{y^2}{y-7} + \dfrac{(-1)(6y+7)}{y-7}$

$\qquad = \dfrac{y^2-6y-7}{y-7} = \dfrac{(y-7)(y+1)}{y-7} = y+1$

45. $\dfrac{x}{1} - \dfrac{3}{x-2} = \dfrac{x}{1} \cdot \dfrac{x-2}{x-2} - \dfrac{3}{x-2}$

$\qquad = \dfrac{x(x-2)-3}{x-2} = \dfrac{x^2-2x-3}{x-2}$

47. $\dfrac{x+3y}{(x-4y)(x-3y)} - \dfrac{x-3y}{(x-4y)(x+3y)}$

$\left[\text{LCD is } (x-4y)(x-3y)(x+3y)\right]$

$= \dfrac{x+3y}{(x-4y)(x-3y)} \cdot \dfrac{(x+3y)}{(x+3y)}$

$\qquad - \dfrac{x-3y}{(x-4y)(x+3y)} \cdot \dfrac{(x-3y)}{(x-3y)}$

$= \dfrac{(x+3y)(x+3y)-(x-3y)(x-3y)}{(x-4y)(x-3y)(x+3y)}$

$= \dfrac{x^2+6xy+9y^2-\left(x^2-6xy+9y^2\right)}{(x-4y)(x-3y)(x+3y)}$

$= \dfrac{12xy}{(x-4y)(x-3y)(x+3y)}$

49. $\dfrac{y^n}{y^{2n}-1} + \dfrac{2}{y^n-1}$

$= \dfrac{y^n}{\left(y^n+1\right)\left(y^n-1\right)} + \dfrac{2}{y^n-1}$

$\left[\text{LCD is } \left(y^n+1\right)\left(y^n-1\right)\right]$

$= \dfrac{y^n}{\left(y^n+1\right)\left(y^n-1\right)} + \dfrac{2}{y^n-1} \cdot \dfrac{y^n+1}{y^n+1}$

$= \dfrac{y^n+2\left(y^n+1\right)}{\left(y^n+1\right)\left(y^n-1\right)} = \dfrac{3y^n+2}{\left(y^n+1\right)\left(y^n-1\right)}$

53. $\dfrac{y}{y-4} + \dfrac{y}{y+4} - \dfrac{16}{(y+4)(y-4)}$

$\left[\text{LCD is } (y+4)(y-4)\right]$

$= \dfrac{y}{y-4} \cdot \dfrac{y+4}{y+4} + \dfrac{y}{y+4} \cdot \dfrac{y-4}{y-4} - \dfrac{16}{(y+4)(y-4)}$

$= \dfrac{y(y+4)+y(y-4)-16}{(y+4)(y-4)} = \dfrac{2y^2-16}{(y+4)(y-4)}$

55. $\dfrac{a+2}{a} - \dfrac{a+1}{a+3} + 1 \qquad \left[\text{LCD is } a(a+3)\right]$

$= \dfrac{a+2}{a} \cdot \dfrac{a+3}{a+3} - \dfrac{a+1}{a+3} \cdot \dfrac{a}{a} + 1 \cdot \dfrac{a(a+3)}{a(a+3)}$

$= \dfrac{(a+2)(a+3)-a(a+1)+a(a+3)}{a(a+3)}$

$= \dfrac{a^2+5a+6-a^2-a+a^2+3a}{a(a+3)}$

$= \dfrac{a^2+7a+6}{a(a+3)} \quad \left[\text{or } \dfrac{(a+6)(a+1)}{a(a+3)}\right]$

57. $\dfrac{y-3}{y-2} + \dfrac{7-4y}{(2y-5)(y-2)} - \dfrac{y+1}{2y-5}$

$\left[\text{LCD is } (2y-5)(y-2)\right]$

$= \dfrac{y-3}{y-2} \cdot \dfrac{2y-5}{2y-5} + \dfrac{7-4y}{(2y-5)(y-2)}$

$\qquad - \dfrac{y+1}{2y-5} \cdot \dfrac{y-2}{y-2}$

$= \dfrac{(y-3)(2y-5)+7-4y-(y+1)(y-2)}{(2y-5)(y-2)}$

$= \dfrac{2y^2-11y+15+7-4y-y^2+y+2}{(2y-5)(y-2)}$

$= \dfrac{y^2-14y+24}{(2y-5)(y-2)} = \dfrac{(y-12)(y-2)}{(2y-5)(y-2)}$

$= \dfrac{y-12}{2y-5}$

59. $\dfrac{1}{c(c-a)-b(c-a)} - \dfrac{1}{b(b-a)-c(b-a)}$

$\qquad\qquad + \dfrac{1}{a(a-b)-c(a-b)}$

$= \dfrac{1}{(c-a)(c-b)} - \dfrac{1}{(b-a)(b-c)} + \dfrac{1}{(a-b)(a-c)}$

$\qquad \left[\text{LCD is } (a-b)(a-c)(b-c)\right]$

$= \dfrac{1}{(c-a)(c-b)} \cdot \dfrac{-1}{-1} \cdot \dfrac{-1}{-1} \cdot \dfrac{a-b}{a-b}$

$\qquad - \dfrac{1}{(b-a)(b-c)} \cdot \dfrac{-1}{-1} \cdot \dfrac{a-c}{a-c}$

$\qquad + \dfrac{1}{(a-b)(a-c)} \cdot \dfrac{b-c}{b-c}$

$= \dfrac{a-b-(-1)(a-c)+b-c}{(a-b)(a-c)(b-c)}$

$= \dfrac{a-b+a-c+b-c}{(a-b)(a-c)(b-c)} = \dfrac{2a-2c}{(a-b)(a-c)(b-c)}$

$= \dfrac{2(a-c)}{(a-b)(a-c)(b-c)} = \dfrac{2}{(a-b)(b-c)}$

61. $\dfrac{3x}{2y} + \dfrac{4y}{y^2} \cdot \dfrac{3y}{4} = \dfrac{3x}{2y} + \dfrac{12y^2}{4y^2} = \dfrac{3x}{2y} + \dfrac{3}{1}$

$\qquad \left[\text{LCD is } 2y\right]$

$= \dfrac{3x}{2y} + \dfrac{3}{1} \cdot \dfrac{2y}{2y} = \dfrac{3x+6y}{2y} \quad \left[\text{or } \dfrac{3(x+2y)}{2y}\right]$

63. $\dfrac{6}{5a} \cdot \dfrac{2}{10a^2} + \dfrac{4}{a} \div \dfrac{8}{3a^2}$

$= \dfrac{6 \cdot 2}{5a \cdot 10a^2} + \dfrac{4}{a} \cdot \dfrac{3a^2}{8} = \dfrac{6}{25a^3} + \dfrac{3a}{2}$

$\qquad \left[\text{LCD is } 50a^3\right]$

$= \dfrac{6}{25a^3} \cdot \dfrac{2}{2} + \dfrac{3a}{2} \cdot \dfrac{25a^3}{25a^3} = \dfrac{12+75a^4}{50a^3}$

65. $\dfrac{6}{(x+4)(x-4)} \cdot \dfrac{x+4}{12} - \dfrac{1}{2} = \dfrac{1}{2(x-4)} - \dfrac{1}{2}$

$\qquad \left[\text{LCD is } 2(x-4)\right]$

$= \dfrac{1}{2(x-4)} - \dfrac{1}{2} \cdot \dfrac{x-4}{x-4} = \dfrac{1-(x-4)}{2(x-4)}$

$= \dfrac{-x+5}{2(x-4)}$

67. $\left(2 - \dfrac{6}{y+1}\right)\left(1 + \dfrac{3}{y-2}\right)$

$= \left[2\dfrac{(y+1)}{(y+1)} - \dfrac{6}{y+1}\right]\left[1\dfrac{(y-2)}{(y-2)} + \dfrac{3}{y-2}\right]$

$= \left(\dfrac{2y-4}{y+1}\right)\left(\dfrac{y+1}{y-2}\right) = \dfrac{2(y-2)(y+1)}{(y+1)(y-2)} = 2$

69. $\dfrac{3y^2}{(y+3)(y+2)} \cdot \dfrac{y+3}{y}$

$\qquad + \dfrac{(y+3)(y-3)}{6y} \div \dfrac{(y-3)^2}{3y^2}$

$= \dfrac{3y^2(y+3)}{y(y+3)(y+2)} + \dfrac{(y+3)(y-3)3y^2}{6y(y-3)^2}$

$= \dfrac{3y}{y+2} + \dfrac{y(y+3)}{2(y-3)} \quad \left[\text{LCD is } 2(y-3)(y+2)\right]$

$= \dfrac{3y}{y+2} \cdot \dfrac{2(y-3)}{2(y-3)} + \dfrac{y(y+3)}{2(y-3)} \cdot \dfrac{y+2}{y+2}$

$= \dfrac{6y(y-3)+y(y+3)(y+2)}{2(y-3)(y+2)}$

$= \dfrac{6y^2-18y+y^3+5y^2+6y}{2(y-3)(y+2)}$

$= \dfrac{y^3+11y^2-12y}{2(y-3)(y+2)}$

71. $\left(\dfrac{1}{b+2} + \dfrac{1}{b+4}\right) \cdot \left(b^2+6b+8\right)$

$= \left(\dfrac{1}{b+2} \cdot \dfrac{b+4}{b+4} + \dfrac{1}{b+4} \cdot \dfrac{b+2}{b+2}\right) \cdot$

$\qquad\qquad\qquad\qquad \left(b^2+6b+8\right)$

$= \dfrac{2b+6}{(b+2)(b+4)} \cdot \dfrac{(b+2)(b+4)}{1}$

$= 2b+6$

73.
$$\left[\frac{1}{(a-b)\left(a^2+ab+b^2\right)} \cdot \frac{(c+d)(a-b)}{1}\right]$$
$$-\frac{c-d}{a^2+ab+b^2}$$
$$=\frac{c+d}{a^2+ab+b^2} - \frac{c-d}{a^2+ab+b^2}$$
$$=\frac{(c+d)-(c-d)}{a^2+ab+b^2} = \frac{2d}{a^2+ab+b^2}$$

75. $\left(\dfrac{1}{x^2} - \dfrac{1}{y^2}\right)\left(x-y+\dfrac{2y^2}{x+y}\right)$
$$\div\left(x^3 - \frac{x^3y+y^4}{x+y}\right)$$
$$=\left(\frac{1}{x^2}\cdot\frac{y^2}{y^2} - \frac{1}{y^2}\cdot\frac{x^2}{x^2}\right)$$
$$\cdot\left(x\cdot\frac{x+y}{x+y} - y\cdot\frac{x+y}{x+y} + \frac{2y^2}{x+y}\right)$$
$$\div\left(x^3\cdot\frac{x+y}{x+y} - \frac{x^3y+y^4}{x+y}\right)$$
$$=\left(\frac{y^2-x^2}{x^2y^2}\right)\left(\frac{x^2+y^2}{x+y}\right)\div\left(\frac{x^4-y^4}{x+y}\right)$$
$$=\frac{(y-x)(y+x)\left(x^2+y^2\right)(x+y)}{\left(x^2y^2\right)(x+y)\left(x^4-y^4\right)}$$
$$=\frac{(y-x)\left(x^2+y^2\right)(x+y)}{x^2y^2\left(x^2+y^2\right)(x+y)(x-y)}$$
$$=-\frac{1}{x^2y^2}$$

77. $\left(\dfrac{1}{x+h} - \dfrac{1}{x}\right)\div h = \left(\dfrac{1}{x+h}\cdot\dfrac{x}{x} - \dfrac{1}{x}\cdot\dfrac{x+h}{x+h}\right)\div h$
$$=\left[\frac{x-x-h}{x(x+h)}\right]\div\frac{h}{1} = \frac{-h}{x(x+h)}\cdot\frac{1}{h}$$
$$=\frac{-1}{x(x+h)}$$

85. $\dfrac{1}{C_1} + \dfrac{1}{C_2} + \dfrac{1}{C_3}$
$$=\frac{1}{C_1}\cdot\frac{C_2C_3}{C_2C_3} + \frac{1}{C_2}\cdot\frac{C_1C_3}{C_1C_3}$$
$$+\frac{1}{C_3}\cdot\frac{C_2C_1}{C_2C_1}$$
$$=\frac{C_2C_3+C_1C_3+C_2C_1}{C_1C_2C_3}$$

87. $\dfrac{300x}{x+2} + \dfrac{140y}{y+4}$
$$=\frac{300x}{x+2}\cdot\frac{y+4}{y+4} + \frac{140y}{y+4}\cdot\frac{x+2}{x+2}$$
$$=\frac{300x(y+4)+140y(x+2)}{(x+2)(y+4)}$$
$$=\frac{440xy+1200x+280y}{(x+2)(y+4)}$$

89. $\dfrac{(n+2)(n-2)}{10n^3}\left[6\cdot\dfrac{n+2}{n+2} + \dfrac{20}{n+2}\right]$
$$=\frac{(n+2)(n-2)}{10n^3}\cdot\frac{6n+32}{n+2}$$
$$=\frac{(n+2)(n-2)2(3n+16)}{10n^3(n+2)}$$
$$=\frac{(n-2)(3n+16)}{5n^3}$$

91. $R = x(x+1)\left[3\cdot\dfrac{20(x+1)}{20(x+1)} - \dfrac{x}{20(x+1)}\right]$
$$=x(x+1)\left[\frac{60x+60-x}{20(x+1)}\right]$$
$$=\frac{x(x+1)(59x+60)}{20(x+1)} = \frac{x(59x+60)}{20}$$

93. Area of trapezoid:

$$\frac{1}{2} \cdot \frac{1}{y}\left[\left(\frac{1}{y}+2-\frac{1}{y}\right)+\left(2-\frac{1}{y}\right)\right]$$

$$= \frac{1}{2y}\left(4-\frac{1}{y}\right) = \frac{1}{2y}\left(\frac{4y-1}{y}\right) = \frac{4y-1}{2y^2}$$

Area of right triangle:

$$\frac{1}{2}\left[5y-(y+1)\right] \cdot \frac{1}{y} = \frac{1}{2} \cdot \frac{4y-1}{y}$$

$$= \frac{4y-1}{2y}$$

Sum of areas:

$$\frac{4y-1}{2y^2}+\frac{4y-1}{2y} = \frac{4y-1}{2y^2}+\frac{4y-1}{2y} \cdot \frac{y}{y}$$

$$= \frac{4y-1+4y^2-y}{2y^2} = \frac{4y^2+3y-1}{2y^2}$$

95. $\dfrac{1}{a}+\dfrac{1}{b} = \dfrac{1}{a} \cdot \dfrac{b}{b}+\dfrac{1}{b} \cdot \dfrac{a}{a} = \dfrac{b+a}{ab}$

97. $\dfrac{1}{x}+7 = \dfrac{1}{x}+7 \cdot \dfrac{x}{x} = \dfrac{1+7x}{x}$

99. $\dfrac{a+bx}{a} = \dfrac{a}{a}+\dfrac{bx}{a} = 1+\dfrac{b}{a}x$

101. $\dfrac{a}{x}+\dfrac{a}{b} = \dfrac{a}{x} \cdot \dfrac{b}{b}+\dfrac{a}{b} \cdot \dfrac{x}{x} = \dfrac{ab+ax}{bx}$

102. $ay-bx = 2cy+dx$

$$ay-2cy = bx+dx$$

$$y(a-2c) = bx+dx$$

$$y = \frac{bx+dx}{a-2c}$$

103.

$$
\require{enclose}
\begin{array}{r}
y^2+2y+3 \\
y^3-2\,\enclose{longdiv}{y^5+2y^4+3y^3-2y^2-4y-6} \\
\underline{y^5-2y^2} \\
2y^4+3y^3-4y \\
\underline{2y^4-4y} \\
3y^3-6 \\
\underline{3y^3-6} \\
0
\end{array}
$$

104. $y-2(-y-3) = 3(y-5)+2$

$$y+2y+6 = 3y-15+2$$

$$3y+6 = 3y-13$$

$$6 = -13 \quad \text{No Solution} \quad \varnothing$$

Problem set 4.4

1. $\dfrac{\frac{3}{y}}{y-\frac{1}{y}} \cdot \dfrac{y}{y} = \dfrac{\frac{3}{y} \cdot y}{y \cdot y-\frac{1}{y} \cdot y} = \dfrac{3}{y^2-1}$

5. $\dfrac{\frac{1}{y}+\frac{1}{y^2}}{1+\frac{1}{y}} \cdot \dfrac{y^2}{y^2} = \dfrac{\frac{1}{y} \cdot y^2+\frac{1}{y^2} \cdot y^2}{1 \cdot y^2+\frac{1}{y} \cdot y^2}$

$$= \dfrac{y+1}{y^2+y} = \dfrac{y+1}{y(y+1)} = \dfrac{1}{y}$$

7. $\dfrac{\frac{x}{y}+\frac{y}{x}}{\frac{1}{y}+\frac{1}{x}} \cdot \dfrac{xy}{xy} = \dfrac{\frac{x}{y} \cdot xy+\frac{y}{x} \cdot xy}{\frac{1}{y} \cdot xy+\frac{1}{x} \cdot xy} = \dfrac{x^2+y^2}{x+y}$

11. $\dfrac{\dfrac{x}{y}-\dfrac{y}{x}}{\dfrac{x^2}{y}-y} \cdot \dfrac{xy}{xy} = \dfrac{x^2-y^2}{x^3-xy^2}$

$= \dfrac{x^2-y^2}{x\left(x^2-y^2\right)} = \dfrac{1}{x}$

13. $\dfrac{y+5+\dfrac{6}{y}}{y-\dfrac{9}{y}} \cdot \dfrac{y}{y} = \dfrac{y^2+5y+6}{y^2-9}$

$= \dfrac{(y+3)(y+2)}{(y+3)(y-3)} = \dfrac{y+2}{y-3}$

19. $\dfrac{b^2-c^2}{\dfrac{1}{b}+\dfrac{1}{c}} \cdot \dfrac{bc}{bc} = \dfrac{bc\left(b^2-c^2\right)}{c+b}$

$= \dfrac{bc(b+c)(b-c)}{c+b} = bc(b-c)$

23. $\dfrac{b-3}{b-\dfrac{3}{b-2}} \cdot \dfrac{b-2}{b-2} = \dfrac{(b-3)(b-2)}{b(b-2)-3}$

$= \dfrac{(b-3)(b-2)}{b^2-2b-3} = \dfrac{(b-3)(b-2)}{(b+1)(b-3)}$

$= \dfrac{b-2}{b+1}$

27. $\dfrac{\dfrac{3}{y-2}-\dfrac{4}{y+2}}{\dfrac{7}{(y+2)(y-2)}} \cdot \dfrac{(y+2)(y-2)}{(y+2)(y-2)}$

$= \dfrac{3(y+2)-4(y-2)}{7} = \dfrac{-y+14}{7}$

29. $\dfrac{\dfrac{4}{b-2}+1}{\dfrac{3}{(b+2)(b-2)}+1} \cdot \dfrac{(b+2)(b-2)}{(b+2)(b-2)}$

$= \dfrac{4(b+2)+(b+2)(b-2)}{3+(b+2)(b-2)}$

$= \dfrac{(b+2)\left[4+(b-2)\right]}{3+b^2-4} = \dfrac{(b+2)(b+2)}{b^2-1}$

$= \dfrac{(b+2)^2}{b^2-1}$

31. $\dfrac{\dfrac{6}{(x+5)(x-3)}-\dfrac{1}{x-3}}{\dfrac{1}{x+5}+1} \cdot \dfrac{(x+5)(x-3)}{(x+5)(x-3)}$

$= \dfrac{6-(x+5)}{x-3+(x+5)(x-3)} = \dfrac{6-x-5}{(x-3)\left[1+(x+5)\right]}$

$= \dfrac{-x+1}{(x-3)(x+6)}$

33. $\dfrac{\dfrac{3}{x+2y}-\dfrac{2y}{x(x+2y)}}{\dfrac{3y}{x(x+2y)}+\dfrac{5}{x}} \cdot \dfrac{x(x+2y)}{x(x+2y)}$

$= \dfrac{3x-2y}{3y+5(x+2y)} = \dfrac{3x-2y}{3y+5x+10y}$

$= \dfrac{3x-2y}{5x+13y}$

35. $\dfrac{\dfrac{1}{(x-5)\left(x^2+5x+25\right)}}{\dfrac{1}{(x+5)(x-5)}-\dfrac{1}{x^2+5x+25}}$

$\dfrac{\left[\dfrac{(x+5)(x-5)\left(x^2+5x+25\right)}{(x+5)(x-5)\left(x^2+5x+25\right)}\right]}{}$

$= \dfrac{x+5}{x^2+5x+25-(x+5)(x-5)}$

$= \dfrac{x+5}{x^2+5x+25-x^2+25} = \dfrac{x+5}{5x+50}$

37. $2+\dfrac{1}{1+\dfrac{2}{x+\dfrac{1}{x}}\cdot\dfrac{x}{x}}=2+\dfrac{1}{1+\dfrac{2x}{x^2+1}}\cdot\dfrac{x^2+1}{x^2+1}$

$=2+\dfrac{x^2+1}{x^2+1+2x}=2+\dfrac{x^2+1}{x^2+2x+1}$

$=2\cdot\dfrac{x^2+2x+1}{x^2+2x+1}+\dfrac{x^2+1}{x^2+2x+1}$

$=\dfrac{2x^2+4x+2+x^2+1}{x^2+2x+1}$

$=\dfrac{3x^2+4x+3}{x^2+2x+1}$

39. $\dfrac{\dfrac{1}{a^2+b^2}}{1-\dfrac{a}{a+\dfrac{b^2}{a-b}}\cdot\dfrac{a-b}{a-b}}=\dfrac{\dfrac{1}{a^2+b^2}}{1-\dfrac{a(a-b)}{a(a-b)+b^2}}$

$=\dfrac{\dfrac{1}{a^2+b^2}}{1-\dfrac{a^2-ab}{a^2-ab+b^2}}$

$=\dfrac{\dfrac{1}{a^2+b^2}}{1\cdot\dfrac{a^2-ab+b^2}{a^2-ab+b^2}-\dfrac{a^2-ab}{a^2-ab+b^2}}$

$=\dfrac{\dfrac{1}{a^2+b^2}}{\dfrac{b^2}{a^2-ab+b^2}}=\dfrac{1}{a^2+b^2}\cdot\dfrac{a^2-ab+b^2}{b^2}$

$=\dfrac{a^2-ab+b^2}{b^2\left(a^2+b^2\right)}$

41. $\left[x(x-1)^{-1}+1\right]\left[3\left(x^2-1\right)^{-1}+4\right]^{-1}$

$=\left[\dfrac{x}{(x-1)^1}+1\right]\left[\dfrac{3}{\left(x^2-1\right)^1}+4\right]^{-1}$

$=\left[\dfrac{x}{x-1}+1\right]\cdot\dfrac{1}{\left[\dfrac{3}{x^2-1}+4\right]^1}$

$=\left(\dfrac{x}{x-1}+1\right)\cdot\dfrac{1}{\dfrac{3}{x^2-1}+4}\cdot\dfrac{x^2-1}{x^2-1}$

$=\left(\dfrac{x}{x-1}+1\right)\cdot\left[\dfrac{x^2-1}{3+4\left(x^2-1\right)}\right]$

$=\left(\dfrac{x}{x-1}+\dfrac{x-1}{x-1}\right)\left(\dfrac{x^2-1}{3+4x^2-4}\right)$

$=\left(\dfrac{2x-1}{x-1}\right)\left(\dfrac{x^2-1}{4x^2-1}\right)$

$=\dfrac{(2x-1)(x+1)(x-1)}{(x-1)(2x+1)(2x-1)}=\dfrac{x+1}{2x+1}$

43. $y+\dfrac{y}{y+\dfrac{1}{y}}\cdot\dfrac{y}{y}=y+\dfrac{y^2}{y^2+1}$

$=y\cdot\dfrac{y^2+1}{y^2+1}+\dfrac{y^2}{y^2+1}$

$=\dfrac{y\left(y^2+1\right)+y^2}{y^2+1}=\dfrac{y^3+y^2+y}{y^2+1}$

45. $1-\dfrac{1}{1-\dfrac{1}{x-2}}\cdot\dfrac{x-2}{x-2}=1-\dfrac{x-2}{x-2-1}$

$=1-\dfrac{x-2}{x-3}=\dfrac{x-3}{x-3}-\dfrac{x-2}{x-3}=\dfrac{x-3-(x-2)}{x-3}$

$=-\dfrac{1}{x-3}$

47. $\dfrac{\dfrac{2y}{y-\dfrac{1}{y}}-\dfrac{1}{y}}{2y+\dfrac{2y}{1-\dfrac{1}{y}}} = \dfrac{\dfrac{2y}{y-\dfrac{1}{y}}\cdot\dfrac{y}{y}-\dfrac{1}{y}}{2y+\dfrac{2y}{1-\dfrac{1}{y}}\cdot\dfrac{y}{y}}$

$= \dfrac{\dfrac{2y^2}{y^2-1}-\dfrac{1}{y}}{2y+\dfrac{2y^2}{y-1}}$

$= \dfrac{\dfrac{2y^2}{(y+1)(y-1)}-\dfrac{1}{y}}{2y+\dfrac{2y^2}{y-1}}\cdot\dfrac{y(y+1)(y-1)}{y(y+1)(y-1)}$

$= \dfrac{2y^3-(y+1)(y-1)}{2y^2(y+1)(y-1)+2y^3(y+1)}$

$= \dfrac{2y^3-\left(y^2-1\right)}{2y^2(y+1)\left[(y-1)+y\right]}$

$= \dfrac{2y^3-y^2+1}{2y^2(y+1)(2y-1)}$

49. $\dfrac{2d}{\dfrac{d}{r_1}+\dfrac{d}{r_2}}\cdot\dfrac{r_1r_2}{r_1r_2} = \dfrac{2dr_1r_2}{dr_2+dr_1}$

$= \dfrac{2dr_1r_2}{d(r_2+r_1)} = \dfrac{2r_1r_2}{r_2+r_1}$

$\dfrac{2r_1r_2}{r_2+r_1} = \dfrac{2(30)(20)}{30+20} = \dfrac{1200}{50} = 24$ m/h

55. $\dfrac{4x^2-y^2}{2x^2-5xy-3y^2} = \dfrac{(2x+y)(2x-y)}{(2x+y)(x-3y)}$

$= \dfrac{2x-y}{x-3y}$

56. $\dfrac{y}{y+4}-\dfrac{y}{y-4}-\dfrac{32}{(y+4)(y-4)}$

$\left[\text{LCD is }(y+4)(y-4)\right]$

$= \dfrac{y}{y+4}\cdot\dfrac{y-4}{y-4}-\dfrac{y}{y-4}\cdot\dfrac{y+4}{y+4}-\dfrac{32}{(y+4)(y-4)}$

$= \dfrac{y(y-4)-y(y+4)-32}{(y+4)(y-4)}$

$= \dfrac{y^2-4y-y^2-4y-32}{(y+4)(y-4)}$

$= \dfrac{-8y-32}{(y+4)(y-4)} = \dfrac{-8(y+4)}{(y+4)(y-4)} = \dfrac{-8}{y-4}$

57. $6\left[\dfrac{2(y-3)}{3}-\dfrac{3y-4}{6}\right] = 6\left[\dfrac{1}{3}-\dfrac{y+2}{2}\right]$

$4(y-3)-(3y-4) = 2-3(y+2)$

$4y-12-3y+4 = 2-3y-6$

$y-8 = -3y-4$

$4y = 4$

$y = 1 \qquad \{1\}$

Problem set 4.5

1. $12x\left(\dfrac{2}{3x}+\dfrac{1}{4}\right) = 12x\left(\dfrac{11}{6x}-\dfrac{1}{3}\right)$

$12x\cdot\dfrac{2}{3x}+12x\cdot\dfrac{1}{4} = 12x\cdot\dfrac{11}{6x}-12x\cdot\dfrac{1}{3}$

$8+3x = 22-4x$

$7x = 14$

$x = 2 \qquad \{2\}$

5. $(y-3)(y+3)\left[\dfrac{2}{y-3}+\dfrac{3y+1}{y+3}\right]$

$= (y-3)(y+3)\cdot 3$

$2(y+3)+(3y+1)(y-3) = 3(y-3)(y+3)$

$2y+6+3y^2-8y-3 = 3y^2-27$

$-6y+3 = -27$

$-6y = -30$

$y = 5 \qquad \{5\}$

9. $(c+1)(c-1)\dfrac{3c}{c+1} = (c+1)(c-1)\left[\dfrac{5c}{c-1}-2\right]$

$3c(c-1) = 5c(c+1)-2(c+1)(c-1)$

$3c^2-3c = 5c^2+5c-2c^2+2$

$3c^2-3c = 3c^2+5c+2$

$-3c = 5c+2$

$-8c = 2$

$c = \dfrac{2}{-8} = -\dfrac{1}{4} \qquad \left\{-\dfrac{1}{4}\right\}$

11. $\dfrac{x+5}{(x+2)(x-2)} - \dfrac{3}{2(x-2)} = \dfrac{1}{2(x+2)}$

$2(x+2)(x-2)\left[\dfrac{x+5}{(x+2)(x-2)} - \dfrac{3}{2(x-2)}\right]$

$\qquad\qquad = 2(x+2)(x-2)\cdot \dfrac{1}{2(x+2)}$

$2(x+5)-3(x+2) = x-2$

$2x+10-3x-6 = x-2$

$-x+4 = x-2$

$6 = 2x$

$3 = x \qquad\qquad \{3\}$

15. $\dfrac{y+5}{y+1} - \dfrac{y}{y+2} = \dfrac{4y+1}{(y+1)(y+2)}$

$(y+1)(y+2)\left[\dfrac{y+5}{y+1} - \dfrac{y}{y+2}\right]$

$\qquad\qquad = (y+1)(y+2)\dfrac{4y+1}{(y+1)(y+2)}$

$(y+2)(y+5)-y(y+1) = 4y+1$

$y^2+7y+10-y^2-y = 4y+1$

$6y+10 = 4y+1$

$2y = -9$

$y = \dfrac{-9}{2} \qquad \left\{-\dfrac{9}{2}\right\}$

17. $5(c+4)\dfrac{c}{c+4} = 5(c+4)\left[\dfrac{2}{5}-\dfrac{4}{c+4}\right]$

$5c = 2(c+4)-20$

$5c = 2c+8-20$

$5c = 2c-12$

$3c = -12$

$c = -4$

The value -4 makes two denominators in the original equation equal to zero. Thus, -4 is not a solution. The equation has no solution. $\varnothing$

19. $(y-3)\left[\dfrac{3y}{y-3} - \dfrac{5y}{y-3}\right] = (y-3)(-2)$

$3y-5y = -2y+6$

$-2y = -2y+6$

$0 = 6$

This contradiction indicates that the equation has no solution. $\varnothing$

21. $\dfrac{1}{x-2} - \dfrac{1}{x} = \dfrac{2}{x(x-2)}$

$x(x-2)\left[\dfrac{1}{x-2} - \dfrac{1}{x}\right] = x(x-2)\cdot \dfrac{2}{x(x-2)}$

$x-(x-2) = 2$

$x-x+2 = 2$

$2 = 2$

All real numbers other than those that cause zero in the equation's denominators (namely 0 and 2) satisfy the equation.

$\{x \mid x \neq 0, \; x \neq 2\}$

25. $.453 = .45+\dfrac{.75}{x}$

$.453x = .45x+.75$

$.003x = .75$

$x = \dfrac{.75}{.003} = 250 \; ; \; 250 \text{ gallons}$

27. P = 19, since P is given in thousands.

$$19 = 20 - \frac{4}{t+1}$$

$$(t+1)\,19 = (t+1)\left[20 - \frac{4}{t+1}\right]$$

$$19t + 19 = 20(t+1) - 4$$

$$19t + 19 = 20t + 16$$

$$3 = t$$

The population will be 19,000 in 3 years after 1990, or in 1993.

31. $d \cdot \dfrac{D}{d} = d\left(q + \dfrac{R}{d}\right)$

$$D = dq + R$$

$$D - R = dq$$

$$\frac{D-R}{q} = d$$

33.

$$W = \frac{10x}{150-x}$$

$$(150-x)W = (150-x)\frac{10x}{150-x}$$

$$150W - Wx = 10x$$

$$150W = Wx + 10x$$

$$150W = x(W+10)$$

$$\frac{150W}{W+10} = x$$

If W = 5,

$$x = \frac{150(5)}{5+10} = \frac{750}{15} = 50$$

50% can be raised in 5 weeks.

35. $(RR_1 R_2)\left(\dfrac{1}{R}\right) = (RR_1 R_2)\left(\dfrac{1}{R_1} + \dfrac{1}{R_2}\right)$

$$R_1 R_2 = RR_2 + RR_1$$

$$R_1 R_2 - RR_2 = RR_1$$

$$R_2(R_1 - R) = RR_1$$

$$R_2 = \frac{RR_1}{R_1 - R}$$

37. $(1-r)S = (1-r)\dfrac{a}{1-r}$

$$S - Sr = a$$

$$S - a = Sr$$

$$\frac{S-a}{S} = r$$

41. $S = \dfrac{rL - a}{r-1}$

$$(r-1)S = (r-1)\left(\frac{rL-a}{r-1}\right)$$

$$Sr - S = rL - a$$

$$Sr - rL = S - a$$

$$r(S - L) = S - a$$

$$r = \frac{S-a}{S-L}$$

47. $\dfrac{1}{f} = (n-1)\left(\dfrac{1}{R_1} - \dfrac{1}{R_2}\right)$

$$\frac{1}{f} = \frac{1}{R_1} \cdot (n-1) - \frac{1}{R_2} \cdot (n-1)$$

$$(fR_1 R_2)\frac{1}{f}$$

$$= (fR_1 R_2)\left[\frac{1}{R_1} \cdot (n-1) - \frac{1}{R_2} \cdot (n-1)\right]$$

$$R_1 R_2 = fR_2(n-1) - fR_1(n-1)$$

$$R_1 R_2 = fnR_2 - fR_2 - fnR_1 + fR_1$$

$$R_1 R_2 + fnR_1 - fR_1 = fnR_2 - fR_2$$

$$R_1(R_2 + fn - f) = fnR_2 - fR_2$$

$$R_1 = \frac{fnR_2 - fR_2}{R_2 + fn - f}$$

49. a. $(y+5)(y+1)\dfrac{3y+1}{y+5}$

$$= (y+5)(y+1)\left[\dfrac{y-1}{y+1}+2\right]$$

$(y+1)(3y+1) = (y+5)(y-1)$
$$\qquad\qquad\qquad + 2(y+5)(y+1)$$

$3y^2+4y+1 = y^2+4y-5+2y^2+12y+10$

$3y^2+4y+1 = 3y^2+16y+5$

$\qquad\quad 4y+1 = 16y+5$

$\qquad\quad -4 = 12y$

$\qquad\quad -\dfrac{4}{12} = y$

$\qquad\qquad y = -\dfrac{1}{3} \qquad \left\{-\dfrac{1}{3}\right\}$

51. We are told that $x = -6$.

$$\dfrac{7x+4}{b}+13 = x$$

$$\dfrac{7(-6)+4}{b}+13 = -6$$

$$b\left(\dfrac{-38}{b}+13\right) = b(-6)$$

$$-38+13b = -6b$$

$$19b = 38$$

$$b = 2$$

53. $2\,|\,1-2y\,|-3 < 1$

$\qquad 2\,|\,1-2y\,| < 4$

$\qquad |\,1-2y\,| < 2$

$\qquad -2 < 1-2y < 2$

$\qquad -3 < -2y < 1$

$\qquad \dfrac{3}{2} > y > -\dfrac{1}{2}$

Equivalently: $-\dfrac{1}{2} < y < \dfrac{3}{2}$

$$\left\{y \,\middle|\, -\dfrac{1}{2} < y < \dfrac{3}{2}\right\}$$

54. $\left[1-(-1)\right]^{-2}\left[2(1)+(-1)\right]$

$\quad = 2^{-2} \cdot 1 = \dfrac{1}{2^2} = \dfrac{1}{4}$

55. $\left(x^2-x-1\right)\left(x^3-x^2+1\right)$

$\quad = x^2\left(x^3-x^2+1\right)-x\left(x^3-x^2+1\right)$

$\qquad\qquad\qquad\qquad -1\left(x^3-x^2+1\right)$

$\quad = x^5-x^4+x^2-x^4+x^3-x$

$\qquad\qquad\qquad -x^3+x^2-1$

$\quad = x^5-2x^4+2x^2-x-1$

Problem set 4.6

1. x: number

$$\dfrac{3}{4}x = \dfrac{1}{2}x+11$$

$$4\left(\dfrac{3}{4}x\right) = 4\left(\dfrac{1}{2}x+11\right)$$

$$3x = 2x+44$$

$$x = 44 \qquad\text{The number is 44.}$$

3. x: number

$$\dfrac{3+x}{8+x} = \dfrac{2}{3}$$

$$3(3+x) = 2(8+x)$$

$$9+3x = 16+2x$$

$$x = 7 \qquad\text{The number is 7.}$$

5. Numbers : x and 3x

$$\dfrac{1}{x}+\dfrac{1}{3x} = \dfrac{1}{3}$$

$$3x\left(\dfrac{1}{x}+\dfrac{1}{3x}\right) = 3x\left(\dfrac{1}{3}\right)$$

$$3+1 = x$$

$$4 = x$$

Numbers are 4 and $3(4) = 12$.

7. Integers : x and $x+1$

$$\frac{1}{x}+\frac{1}{x+1}=15\cdot\frac{1}{x(x+1)}$$

$$x(x+1)\left[\frac{1}{x}+\frac{1}{x+1}\right]=x(x+1)\cdot\frac{15}{x(x+1)}$$

$$x+1+x=15$$
$$2x=14$$
$$x=7$$

Integers are 7 and 8.

9. Numbers: x and $3x$

$$\frac{135}{x}=\frac{135}{3x}+10$$

$$3x\left(\frac{135}{x}\right)=3x\left(\frac{135}{3x}+10\right)$$

$$405=135+30x$$
$$270=30x$$
$$9=x$$

The numbers are 9 and 27.

13. x: lesser number $3x+1$: greater number

$$\frac{3x+1}{x}=3+\frac{1}{x}$$

$$x\left(\frac{3x+1}{x}\right)=x\left(3+\frac{1}{x}\right)$$

$$3x+1=3x+1$$
$$1=1\quad\{x\mid x\in\Re\}$$

Any pair of numbers such that the second number is 1 more than 3 times the first number will satisfy the conditions of this problem.

15. Numbers: x and $34-x$

$$\frac{34-x}{x}=4+\frac{4}{x}$$

$$x\left(\frac{34-x}{x}\right)=x\left(4+\frac{4}{x}\right)$$

$$34-x=4x+4$$
$$30=5x$$
$$6=x$$

The numbers are 6 and $34-6=28$.

17. Numerator: $3x$ Denominator: $4x$

$$\text{Fraction: }\frac{3x}{4x}$$

$$\frac{3x-2}{4x+4}=\frac{1}{2}$$
$$2(3x-2)=4x+4$$
$$6x-4=4x+4$$
$$2x=8$$
$$x=4$$

$$\text{Original Fraction : }\frac{3x}{4x}=\frac{3(4)}{4(4)}=\frac{12}{16}$$

19. x: Number of seagulls

$$x+x+\frac{1}{2}x+\frac{1}{4}x+1=100$$

$$4\left(2x+\frac{1}{2}x+\frac{1}{4}x+1\right)=4(100)$$

$$8x+2x+x+4=400$$
$$11x=396$$
$$x=36$$

36 seagulls

21. $c_1 = 3c_2 \quad c_3 = 5c_2$

$$\frac{1}{c_1} + \frac{1}{c_2} + \frac{1}{c_3} = \frac{1}{c}$$

$$\frac{1}{3c_2} + \frac{1}{c_2} + \frac{1}{5c_2} = \frac{1}{\frac{45}{23}}$$

$$\frac{1}{3c_2} + \frac{1}{c_2} + \frac{1}{5c_2} = \frac{23}{45}$$

$$45c_2\left(\frac{1}{3c_2} + \frac{1}{c_2} + \frac{1}{5c_2}\right) = 45c_2\left(\frac{23}{45}\right)$$

$$15 + 45 + 9 = 23c_2$$

$$69 = 23c_2$$

$$\frac{69}{23} = c_2$$

$$c_2 = 3\mu\,F$$

23. x: speed of stream's current

$18 + x$: rate of boat with the current

$18 - x$: rate of boat against the current

$$T = \frac{D}{R}$$

Time traveling 68 miles against current =
Time traveling 85 miles with current

$$\frac{68}{18-x} = \frac{85}{18+x}$$

$$(18-x)(18+x)\cdot\frac{68}{18-x} =$$

$$(18-x)(18+x)\cdot\frac{85}{18+x}$$

$$68(18+x) = 85(18-x)$$

$$1224 + 68x = 1530 - 85x$$

$$153x = 306$$

$$x = 2$$

Speed of current: 2 m/h

25. Speed of train B: x

Speed of train A: $x + 40$

Time it takes train A to travel 280 km =
Time it takes train B to travel 420 km

$$\left(T = \frac{D}{R}\right)$$

$$\frac{280}{x} = \frac{420}{x+40}; \; x = 80$$

Train B: 80 km/h

Train A: $80 + 40 = 120$ km/h

27 Walking Rate: x

Driving Rate: 12x

Time spent walking + Time spent driving

$$= 2\tfrac{1}{2} \text{ hours.}\left(T = \frac{D}{R}\right)$$

$$\frac{4}{x} + \frac{72}{12x} = \frac{5}{2}$$

$$\frac{4}{x} + \frac{6}{x} = \frac{5}{2}; \; x = 4$$

Walking Rate: 4 m/h

29. x: distance from home to the city

Time spent traveling to the city +
Time spent returning home = 5

$$\frac{x}{60} + \frac{x}{40} = 5$$

$$120\left(\frac{x}{60} + \frac{x}{40}\right) = 120(5)$$

$$2x + 3x = 600$$

$$5x = 600$$

$$x = 120$$

120 miles

31. Jogger's Rate: x   Cyclist's Rate: 2x

Time it takes jogger − time it takes

cyclist = 4 hours

$$\frac{40}{x} - \frac{40}{2x} = 4$$

$$\frac{40}{x} - \frac{20}{x} = 4; \; x = 5$$

Cyclist's Rate: 2(5) = 10 m/h

33. x: time it takes working together

Part of job done by Jean +

Part of job done by Jean's father = 1

$$\frac{x}{15} + \frac{x}{10} = 1$$

$$150\left(\frac{x}{15} + \frac{x}{10}\right) = 150(1)$$

$$10x + 15x = 150 \; ; \; x = 6$$

6 hours

37. x: time it takes for helper working alone

Part of the job done by the mason in

6 hours + part of the job done by the helper

in 6 hours = 1

$$\frac{6}{8} + \frac{6}{x} = 1$$

$$4x\left(\frac{3}{4} + \frac{6}{x}\right) = 4x(1)$$

$$3x + 24 = 4x$$

$$24 = x$$

24 hours

39. x: time it takes working together

Part of the job done by Sheila + part of the

job done by Bob = 1

$$\frac{x}{15} + \frac{x}{30} = 1 \; ; \; x = 10 \qquad \text{10 minutes}$$

41. x: number of hours needed to fill the tank

Fraction of tank filled by first pipe −

Fraction of the tank emptied by second

pipe = 1

$$\frac{x}{8} - \frac{x}{12} = 1 \; ; \; x = 24 \qquad \text{24 hours}$$

47. Area of first rectangle − Area of second

rectangle = Area of third rectangle

$$9 \cdot \frac{1}{x+5} - 1 \cdot \frac{1}{x-5} = 3x \cdot \frac{1}{x^2-25}$$

$$\frac{9}{x+5} - \frac{1}{x-5} = \frac{3x}{x^2-25}$$

$$(x+5)(x-5)\left[\frac{9}{x+5} - \frac{1}{x-5}\right]$$

$$= (x+5)(x-5)\left[\frac{3x}{x^2-25}\right]$$

$$9(x-5) - (x+5) = 3x$$

$$9x - 45 - x - 5 = 3x$$

$$8x - 50 = 3x$$

$$-50 = -5x$$

$$10 = x$$

49. x: cost of the boat

Cost per share for 3 people − Cost per share

for 4 people = 600

$$\frac{x}{3} - \frac{x}{4} = 600$$

$$12\left(\frac{x}{3} - \frac{x}{4}\right) = 12(600)$$

$$4x - 3x = 7200$$

$$x = 7200 \qquad \$7200$$

50.

$$\begin{array}{r|rrrr} -5 & 2 & -4 & 8 & 3 \\ & & -10 & 70 & -390 \\ \hline & 2 & -14 & 78 & -387 \\ & & & & \uparrow \end{array}$$

$$P(-5) = -387$$

51. $x^2 - 4xy + 4y^2 - 13x + 26y + 12$

$= (x - 2y)^2 - 13(x - 2y) + 12$ (let $A = x - 2$:

$= A^2 - 13A + 12$

$= (A - 12)(A - 1)$

$= (x - 2y - 12)(x - 2y - 1)$

52. $\qquad (y + 1)^2 - (y - 1)^2 = 2(y - 3)$

$y^2 + 2y + 1 - \left(y^2 - 2y + 1\right) = 2y - 6$

$y^2 + 2y + 1 - y^2 + 2y - 1 = 2y - 6$

$\qquad\qquad\qquad\qquad 4y = 2y - 6$

$\qquad\qquad\qquad\qquad 2y = -6$

$\qquad\qquad\qquad\qquad y = -3 \quad \{-3\}$

Problem set 4.7

1. $c = kd;\ 2\pi = k \cdot 2;\ \pi = k;\ c = \pi d;$

$c = \pi \cdot 16;\ c = 16\pi$ ft

3. $d = kr^2;\ 200 = k \cdot 60^2;\ \dfrac{200}{3600} = k;$

$k = \dfrac{1}{18};\ d = \dfrac{1}{18}r^2;\ d = \dfrac{1}{18}(100)^2$

$= \dfrac{10000}{18} \approx 555.6$

approximately 555.6 ft

7. $I = \dfrac{k}{d^2};\ 25 = \dfrac{k}{4^2};\ k = (25)(16) = 400$

$I = \dfrac{400}{d^2};\ I = \dfrac{400}{6^2} = \dfrac{400}{36} \approx 11.1$

approximately 11.1 foot candles

9. $V = \dfrac{KT}{P};\ 20 = \dfrac{K(100)}{15};\ 300 = K100;\ K = 3$

$V = \dfrac{3T}{P};\ V = \dfrac{3(150)}{30} = 15$

15 cubic meters

13. $F = \dfrac{KM_1 M_2}{D^2};\ 16 = \dfrac{K(4)(2)}{3^2};\ 16 = \dfrac{8K}{9}$

$144 = 8K;\ K = 18$

$F = \dfrac{18M_1 M_2}{D^2};\ F = \dfrac{18(5)(3)}{2^2} = \dfrac{270}{4}$

$= 67.5$

67.5 units of force

15. $H = \dfrac{KT}{P};\ 4 = \dfrac{K \cdot 6}{2};\ K = \dfrac{4}{3}$

$H = \dfrac{\left(\frac{4}{3}\right)T}{P}$

$8 = \dfrac{\left(\frac{4}{3}\right)(18)}{P};\ 8 = \dfrac{24}{P};\ 8P = 24;\ P = 3$

3 people

17. $R = \dfrac{KL}{D^2};\ \dfrac{3}{2} = \dfrac{L(720)}{\left(\frac{1}{4}\right)^2};\ \dfrac{3}{2} = \dfrac{720K}{\frac{1}{16}}$

$\dfrac{3}{2} = 11520K;\ 3 = 23040K;\ K = \dfrac{3}{23040}$

$= \dfrac{1}{7680}$

$R = \dfrac{\left(\frac{1}{7680}\right)(960)}{\left(\frac{1}{2}\right)^2} = \dfrac{\frac{960}{7680}}{\frac{1}{4}}$

$= \dfrac{960}{7680} \cdot \dfrac{4}{1} = 0.5$

Resistance is 0.5 ohms

21. No ; Consider, for example, $y = \dfrac{-3}{x}$

$\left(y = \dfrac{K}{x},\ K = -3\right)$. If $x = 1$, $y = -3$.

If $x = 3$, $y = -1$. As x increases, y is also

increasing. In general if $y = \dfrac{K}{x}$ and $K < 0$,

then as x increases, y will also increase.

25. Parts: $x, 92 - x$

$$\dfrac{92 - x}{x} = 1 + \dfrac{18}{x}$$

$$x\left(\dfrac{92 - x}{x}\right) = x\left(1 + \dfrac{18}{x}\right)$$

$$92 - x = x + 18$$

$$74 = 2x$$

$$37 = x$$

Lesser part: 37   Greater part: $92 - 37 = 55$

26. $(nr + R)I = (nr + R)\dfrac{nE}{nr + R}$

$$nrI + RI = nE$$

$$RI = nE - nrI$$

$$RI = n(E - rI)$$

$$\dfrac{RI}{E - rI} = n$$

27. x: time working together

$$\dfrac{x}{6} + \dfrac{x}{8} = 1$$

$$24\left(\dfrac{x}{6} + \dfrac{x}{8}\right) = 24(1)$$

$$4x + 3x = 24$$

$$7x = 24$$

$$x = \dfrac{24}{7} = 3\tfrac{3}{7}$$

$3\tfrac{3}{7}$ hours

Review Problems: Chapter 4

1. $\dfrac{7x}{9(x - 2)}$ $\{x \mid x \neq 2\}$

2. $\{x \mid x \neq 1 \text{ and } x \neq -5\}$

3. $\dfrac{7y + 14}{(2y - 1)(y + 3)}$ $\left\{y \,\middle|\, y \neq \dfrac{1}{2} \text{ and } y \neq -3\right\}$

4. Since the denominator, $x^2 + 4$, can never equal 0 for all real numbers x, the domain is the set of all real numbers. $\{x \mid x \in \mathfrak{R}\}$

5. $\dfrac{2x - 3xy}{9y^2 - 4} = \dfrac{x(2 - 3y)}{(3y + 2)(3y - 2)} = \dfrac{-x}{3y + 2}$

6. $\dfrac{x^2 + 6x - 7}{x^2 - 49} = \dfrac{(x + 7)(x - 1)}{(x + 7)(x - 7)} = \dfrac{x - 1}{x - 7}$

7. $\dfrac{6m^2 + 7m + 2}{2m^2 - 9m - 5} = \dfrac{(3m + 2)(2m + 1)}{(2m + 1)(m - 5)} = \dfrac{3m + 2}{m - 5}$

8. $\dfrac{x^{2n} - x^n - 2}{x^{2n} + 3x^n + 2} = \dfrac{\left(x^n - 2\right)\left(x^n + 1\right)}{\left(x^n + 1\right)\left(x^n + 2\right)} = \dfrac{x^n - 2}{x^n + 2}$

9. $\dfrac{y^3 - 8}{y^2 - 4} = \dfrac{(y - 2)\left(y^2 + 2y + 4\right)}{(y + 2)(y - 2)} = \dfrac{y^2 + 2y + 4}{y + 2}$

In 10 – 13, numerators and denominators are immediately expressed in factored form.

10. $\dfrac{(x - 2)(x - 7)}{x^2(x + 2)} \cdot \dfrac{(x + 2)(x - 2)}{(x - 2)^2} = \dfrac{x - 7}{x^2}$

11. $\dfrac{5y(x-2y)}{(x-2y)(x-y)} \cdot \dfrac{x+2y}{x^2} \cdot \dfrac{3x(x-y)}{y(x+2y)} = \dfrac{15}{x}$

12. $\dfrac{\left(y^2+9\right)(y+3)(y-3)}{y^2+9} \cdot \dfrac{4(y-5)}{(y-5)(y-3)}$

$= 4(y+3)$

13. $\dfrac{\left(y^n+4\right)\left(y^n-3\right)}{\left(y^n+1\right)\left(y^n-1\right)} \cdot \dfrac{\left(y^n+1\right)\left(y^n+2\right)}{\left(y^n-3\right)\left(y^n+2\right)}$

$= \dfrac{y^n+4}{y^n-1}$

In 14−16, the division is immediately
expressed as the multiplication by the reciprocal
of the divisor.

14. $\dfrac{x^2+16x+64}{2x^2-128} \cdot \dfrac{x^2-6x-16}{3x^2+30x+48}$

$= \dfrac{(x+8)^2}{2(x+8)(x-8)} \cdot \dfrac{(x-8)(x+2)}{3(x+8)(x+2)} = \dfrac{1}{6}$

15. $\dfrac{a^3-27}{a^2+3a+9} \cdot \dfrac{1}{ab+ac-3b-3c}$

$\dfrac{(a-3)\left(a^2+3a+9\right)}{a^2+3a+9} \cdot \dfrac{1}{(b+c)(a-3)} = \dfrac{1}{b+c}$

16. $\dfrac{y^3-8}{y^4-16} \cdot \dfrac{y^2+4}{y^2+2y+4}$

$= \dfrac{(y-2)\left(y^2+2y+4\right)}{\left(y^2+4\right)(y+2)(y-2)} \cdot \dfrac{y^2+4}{y^2+2y+4}$

$= \dfrac{1}{y+2}$

17. $\dfrac{2x-7}{x^2-9} - \dfrac{x-4}{x^2-9} = \dfrac{2x-7-(x-4)}{x^2-9}$

$= \dfrac{x-3}{x^2-9} = \dfrac{x-3}{(x-3)(x+3)} = \dfrac{1}{x+3}$

18. $\dfrac{1}{x} + \dfrac{2}{x-5}$　　$\left[\text{LCD is } x(x-5)\right]$

$= \dfrac{1}{x} \cdot \dfrac{x-5}{x-5} + \dfrac{2}{x-5} \cdot \dfrac{x}{x}$

$= \dfrac{x-5+2x}{x(x-5)} = \dfrac{3x-5}{x(x-5)}$

19. $\dfrac{3a^2}{(3a+4b)(3a-4b)} - \dfrac{a}{3a+4b}$

$\left[\text{LCD is } (3a+4b)(3a-4b)\right]$

$= \dfrac{3a^2}{(3a+4b)(3a-4b)} - \dfrac{a}{3a+4b} \cdot \dfrac{3a-4b}{3a-4b}$

$= \dfrac{3a^2-a(3a-4b)}{(3a+4b)(3a-4b)} = \dfrac{3a^2-3a^2+4ab}{(3a+4b)(3a-4b)}$

$= \dfrac{4ab}{(3a+4b)(3a-4b)}$

20. $\dfrac{7}{x^2y^3} - \dfrac{5}{xy^3} + \dfrac{4}{x^2y}$　　$\left[\text{LCD is } x^2y^3\right]$

$= \dfrac{7}{x^2y^3} - \dfrac{5}{xy^3} \cdot \dfrac{x}{x} + \dfrac{4}{x^2y} \cdot \dfrac{y^2}{y^2}$

$= \dfrac{7-5x+4y^2}{x^2y^3}$

21. $\dfrac{x}{x+3} + \dfrac{x}{x-3} - \dfrac{9}{(x+3)(x-3)}$

$\left[\text{LCD is } (x+3)(x-3)\right]$

$= \dfrac{x}{x+3} \cdot \dfrac{x-3}{x-3} + \dfrac{x}{x-3} \cdot \dfrac{x+3}{x+3} - \dfrac{9}{(x+3)(x-3)}$

$= \dfrac{x(x-3)+x(x+3)-9}{(x+3)(x-3)}$

$= \dfrac{x^2-3x+x^2+3x-9}{(x+3)(x-3)} = \dfrac{2x^2-9}{(x+3)(x-3)}$

22. $\dfrac{4}{(a+2)(a-1)} - \dfrac{2}{(a+2)(a-2)} + \dfrac{3}{(a-2)^2}$

$\qquad \left[\text{LCD is } (a-2)^2(a+2)(a-1)\right]$

$= \dfrac{4}{(a+2)(a-1)} \cdot \dfrac{(a-2)^2}{(a-2)^2}$

$\quad - \dfrac{2}{(a+2)(a-2)} \cdot \dfrac{(a-2)(a-1)}{(a-2)(a-1)}$

$\quad + \dfrac{3}{(a-2)^2} \cdot \dfrac{(a+2)(a-1)}{(a+2)(a-1)}$

$= \dfrac{4(a-2)^2 - 2(a-2)(a-1) + 3(a+2)(a-1)}{(a-2)^2(a+2)(a-1)}$

$= \dfrac{4a^2 - 16a + 16 - 2a^2 + 6a - 4 + 3a^2 + 3a - 6}{(a-2)^2(a+2)(a-1)}$

$= \dfrac{5a^2 - 7a + 6}{(a-2)^2(a+2)(a-1)}$

23. $\dfrac{3b^2}{(b+3)(b+2)} \cdot \dfrac{b+3}{b} + \dfrac{b-9}{6b}$

$= \dfrac{3b}{b+2} + \dfrac{b-9}{6b} \qquad \left[\text{LCD is } 6b(b+2)\right]$

$= \dfrac{3b}{b+2} \cdot \dfrac{6b}{6b} + \dfrac{b-9}{6b} \cdot \dfrac{b+2}{b+2}$

$= \dfrac{(3b)(6b) + (b-9)(b+2)}{6b(b+2)}$

$= \dfrac{18b^2 + b^2 - 7b - 18}{6b(b+2)} = \dfrac{19b^2 - 7b - 18}{6b(b+2)}$

24. $\left(\dfrac{1}{x+2} + \dfrac{1}{x+4}\right) \cdot \left(x^2 + 6x + 8\right)$

$= \left(\dfrac{1}{x+2} \cdot \dfrac{x+4}{x+4} + \dfrac{1}{x+4} \cdot \dfrac{x+2}{x+2}\right)$

$\qquad\qquad\qquad\qquad \cdot \left(x^2 + 6x + 8\right)$

$= \dfrac{2x+6}{(x+2)(x+4)} \cdot (x+2)(x+4) = 2x+6$

25. $\dfrac{\dfrac{b}{x} - \dfrac{b}{y}}{\dfrac{b}{x} + \dfrac{b}{y}} \cdot \dfrac{xy}{xy} = \dfrac{by - bx}{by + bx} = \dfrac{b(y-x)}{b(y+x)} = \dfrac{y-x}{y+x}$

26. $\dfrac{3 - \dfrac{1}{x+3}}{3 + \dfrac{1}{x+3}} \cdot \dfrac{x+3}{x+3} = \dfrac{3(x+3)-1}{3(x+3)+1} = \dfrac{3x+9-1}{3x+9+1}$

$= \dfrac{3x+8}{3x+10}$

27. $\dfrac{\dfrac{1}{y+5} + 1}{\dfrac{6}{(y+5)(y-3)} - \dfrac{1}{y-3}} \cdot \dfrac{(y+5)(y-3)}{(y+5)(y-3)}$

$= \dfrac{y-3 + (y+5)(y-3)}{6 - (y+5)} = \dfrac{y-3 + y^2 + 2y - 15}{6 - y - 5}$

$= \dfrac{y^2 + 3y - 18}{1 - y}$

28. $\dfrac{4 - \dfrac{1}{y^2}}{4 + \dfrac{4}{y} + \dfrac{1}{y^2}} \cdot \dfrac{y^2}{y^2} = \dfrac{4y^2 - 1}{4y^2 + 4y + 1}$

$= \dfrac{(2y+1)(2y-1)}{(2y+1)^2} = \dfrac{2y-1}{2y+1}$

29. $1 + \dfrac{y}{1 + \dfrac{1}{y}} \cdot \dfrac{y}{y} = 1 + \dfrac{y^2}{y+1}$

$= 1 \cdot \dfrac{y+1}{y+1} + \dfrac{y^2}{y+1} = \dfrac{y^2 + y + 1}{y+1}$

30. $(y-2)\left[\dfrac{2y}{y-2}-3\right]=(y-2)\dfrac{4}{y-2}$

$\qquad 2y-3(y-2)=4$

$\qquad 2y-3y+6=4$

$\qquad -y+6=4$

$\qquad\qquad 2=y$

Since 2 causes a denominator to be 0 in the original equation, 2 is not a solution. The equation has no solution. $\varnothing$

31. $\dfrac{1}{y-5}-\dfrac{3}{y+5}=\dfrac{6}{(y+5)(y-5)}$

$(y+5)(y-5)\left[\dfrac{1}{y-5}-\dfrac{3}{y+5}\right]$

$\qquad = (y+5)(y-5)\cdot\dfrac{6}{(y+5)(y-5)}$

$\qquad y+5-3(y-5)=6$

$\qquad y+5-3y+15=6$

$\qquad\qquad -2y+20=6$

$\qquad\qquad\quad -2y=-14$

$\qquad\qquad\qquad y=7\qquad\qquad \{7\}$

32. $\dfrac{x+5}{x+1}-\dfrac{x}{x+2}=\dfrac{4x+1}{(x+1)(x+2)}$

$(x+1)(x+2)\left[\dfrac{x+5}{x+1}-\dfrac{x}{x+2}\right]$

$\qquad = (x+1)(x+2)\cdot\dfrac{4x+1}{(x+1)(x+2)}$

$(x+5)(x+2)-x(x+1)=4x+1$

$\quad x^2+7x+10-x^2-x=4x+1$

$\qquad\qquad 6x+10=4x+1$

$\qquad\qquad\qquad 2x=-9$

$\qquad\qquad\qquad x=-\dfrac{9}{2}\quad\left\{-\dfrac{9}{2}\right\}$

33. $P = 29$ (since P is in thousands)

$\qquad P=30-\dfrac{9}{t+1}$

$\qquad 29=30-\dfrac{9}{t+1}$

$(t+1)29=(t+1)\left[30-\dfrac{9}{t+1}\right]$

$\qquad 29t+29=30t+30-9$

$\qquad 29t+29=30t+21$

$\qquad\qquad 8=t$

The population of 29,000 will occur 8 years from 1985, in 1993.

34. $nP=n\left(\dfrac{R-C}{n}\right)$

$\qquad Pn=R-C$

$\qquad C=R-Pn$

35. $\qquad T=\dfrac{A-p}{pr}$

$\qquad prT=pr\left(\dfrac{A-p}{pr}\right)$

$\qquad prT=A-p$

$\qquad prT+p=A$

$\qquad p(rT+1)=A$

$\qquad\qquad p=\dfrac{A}{rT+1}$

36. $\qquad A=\dfrac{rs}{r+s}$

$(r+s)A=(r+s)\dfrac{rs}{r+s}$

$\qquad Ar+As=rs$

$\qquad As=rs-Ar$

$\qquad As=r(s-A)$

$\qquad \dfrac{As}{s-A}=r$

37. Lesser number: x   Greater number: $6x - 4$

$$\frac{6x-4}{x} = 5 + \frac{1}{x}$$

$$x\left(\frac{6x-4}{x}\right) = x\left(5 + \frac{1}{x}\right)$$

$$6x - 4 = 5x + 1$$

$$x = 5$$

Lesser number: 5

Greater number: $6(5) - 4 = 26$

38. Consecutive integers: $x, \ x + 1$

$$\frac{1}{x} + \frac{1}{x+1} = 11 \cdot \frac{1}{x(x+1)}$$

$$x(x+1) \cdot \left[\frac{1}{x} + \frac{1}{x+1}\right] = x(x+1) \cdot \frac{11}{x(x+1)}$$

$$x + 1 + x = 11$$

$$2x = 10$$

$$x = 5$$

Integers are 5 and 6.

39. x: number

$$\frac{1}{x} - 2 = 5$$

$$x\left(\frac{1}{x} - 2\right) = x \cdot 5$$

$$1 - 2x = 5x$$

$$1 = 7x$$

$$\frac{1}{7} = x$$

The number is $\frac{1}{7}$.

40. x: wind speed

$320 + x$: speed of plane with wind

$320 - x$: speed of plane against the wind

Time plane travels 1400 miles with wind

= Time plane travels 1160 against the wind

$$\left[RT = D, \text{ so } T = \frac{D}{R}\right]$$

$$\frac{1400}{320+x} = \frac{1160}{320-x}$$

$$(320 + x)(320 - x) \cdot \frac{1400}{320+x}$$

$$= (320 + x)(320 - x) \cdot \frac{1160}{320-x}$$

$$1400(320 - x) = 1160(320 + x)$$

$$448,000 - 1400x = 371,200 + 1160x$$

$$76,800 = 2560x$$

$$30 = x$$

Wind speed is 30 m/h.

41. x : walking rate   3x: cycling rate

Time spent cycling + time spent walking

$$= 7 \text{ hours. } \quad \left(T = \frac{D}{R}\right)$$

$$\frac{60}{3x} + \frac{8}{x} = 7$$

$$\frac{20}{x} + \frac{8}{x} = 7$$

$$x\left(\frac{20}{x} + \frac{8}{x}\right) = x \cdot 7$$

$$20 + 8 = 7x$$

$$28 = 7x$$

$$4 = x$$

Cycling rate: $3x = 3(4) = 12$ m/h

42. x: Time needed by helper working alone

The fraction of the job done by the mason in 8 hours + The fraction of the job done by the helper in 8 hours = 1

$$\frac{8}{10} + \frac{8}{x} = 1$$

$$5x\left(\frac{4}{5} + \frac{8}{x}\right) = 5x(1)$$

$$4x + 40 = 5x$$

$$40 = x$$

40 hours

43. x: time spent working together

Part of job done by small pipe + Part of job done by large pipe = 1

$$\frac{x}{12} + \frac{x}{10} = 1$$

$$120\left(\frac{x}{12} + \frac{x}{10}\right) = 120(1)$$

$$10x + 12x = 120$$

$$22x = 120$$

$$x = \frac{120}{22} = 5\frac{5}{11}$$

$5\frac{5}{11}$ minutes

44. $D = KT^2$

$$144 = K \cdot 3^2$$

$$144 = 9K$$

$$16 = K$$

$$D = 16T^2$$

$$D = 16(7)^2$$

$$D = 784;\ 784\ \text{ft}$$

45. $F = \dfrac{K}{D^2}$

$$100 = \frac{K}{5^2}$$

$$100 = \frac{K}{25}$$

$$2500 = K$$

$$F = \frac{2500}{D^2}$$

$$F = \frac{2500}{10^2} = \frac{2500}{100}$$

$$F = 25\ ;\ 25\ \text{pounds}$$

46. $D = \dfrac{KC}{P}$

$$10 = \frac{K \cdot 100}{7}$$

$$70 = 100K$$

$$\frac{7}{10} = K$$

$$D = \frac{\frac{7}{10}C}{P}$$

$$D = \frac{\frac{7}{10} \cdot 400}{10} = \frac{280}{10} = 28;\ 28\ \text{days}$$

Problem set 5.1.1

1. $\left(5x^3y^4\right)^2\left(-3x^7y^{11}\right)$

$= 25x^6y^8\left(-3x^7y^{11}\right) = -75x^{13}y^{19}$

3. $\left(4ab^3\right)^3\left(-3a^{-5}b^8\right)$

$= 64a^3b^9\left(-3a^{-5}b^8\right) = -192a^{-2}b^{17}$

$= \dfrac{-192b^{17}}{a^2}$

5. $\left(54r^3s^9\right)\left(-3r^2s^{-4}\right)^{-3}$

$= \left(54r^3s^9\right)\left[(-3)^{-3}r^{-6}s^{12}\right]$

$= 54(-3)^{-3}r^{-3}s^{21} = \dfrac{54s^{21}}{(-3)^3r^3}$

$= \dfrac{54s^{21}}{-27r^3} = \dfrac{-2s^{21}}{r^3}$

7. $\left(-a^{-2}b^3c\right)\left(ab^{-1}c^{-4}\right)^{-3}$

$= \left(-a^{-2}b^3c\right)\left(a^{-3}b^3c^{12}\right)$

$= -a^{-5}b^6c^{13} = \dfrac{-b^6c^{13}}{a^5}$

9. $\left(3x^{-3}y^{-4}z\right)^3\left(3xy^{-5}z\right)^2 \cdot$

$\qquad\qquad\qquad \left(-3x^{-4}z^{12}\right)$

$= \left(27x^{-9}y^{-12}z^3\right) \cdot$

$\qquad\qquad \left(9x^2y^{-10}z^2\right)\left(-3x^{-4}z^{12}\right)$

$= (27)(9)(-3)x^{-11}y^{-22}z^{17}$

$= \dfrac{-729z^{17}}{x^{11}y^{22}}$

11. $\dfrac{-27x^{-8}y^4z}{15x^4y^4z^{-4}}$

$= -\dfrac{9}{5}x^{-8-4}y^{4-4}z^{1-(-4)}$

$= -\dfrac{9}{5}x^{-12}y^0z^5 = \dfrac{-9z^5}{5x^{12}}$

15. $\dfrac{\left(-4xy^{-3}z^2\right)^{-3}}{\left(2xy^{-3}z^2\right)^{-2}}$

$= \dfrac{(-4)^{-3}x^{-3}y^9z^{-6}}{2^{-2}x^{-2}y^6z^{-4}}$

$= \dfrac{(-4)^{-3}}{2^{-2}}x^{-3-(-2)}y^{9-6}z^{-6-(-4)}$

$= \dfrac{2^2}{(-4)^3}x^{-1}y^3z^{-2}$

$= \dfrac{4}{-64}x^{-1}y^3z^{-2} = \dfrac{-y^3}{16xz^2}$

21. $\left(\dfrac{-25a^3b^4c}{-75ab^6c^{-3}}\right)^{-2} = \left(\dfrac{1}{3}a^2b^{-2}c^4\right)^{-2}$

$= \left(\dfrac{1}{3}\right)^{-2}a^{-4}b^4c^{-8}$

$= \dfrac{1}{\left(\frac{1}{3}\right)^2}a^{-4}b^4c^{-8} = \dfrac{9b^4}{a^4c^8}$

23. $\left(\dfrac{-75a^3b^4c}{25a^{-2}b^4c^5}\right) = \left(-3a^5c^{-4}\right)^3$

$= (-3)^3a^{15}c^{-12} = \dfrac{-27a^{15}}{c^{12}}$

25. $\left(\dfrac{-30x^3y^2}{6xy^7}\right) \cdot \left(\dfrac{48xy^{-5}}{16xy^{-8}}\right)$

$= \left(-5x^2y^{-5}\right)\left(3x^0y^3\right)$

$= -15x^2y^{-2} = \dfrac{-15x^2}{y^2}$

27. $\left(\dfrac{54a^4b^3c^5}{-3a^4b^{-6}c}\right)\cdot\left(\dfrac{5a^{-5}c}{-45b^3c^{-7}}\right)$

$=\left(-18a^0b^9c^4\right)\left(-\dfrac{1}{9}a^{-5}b^{-3}c^8\right)$

$=2a^{-5}b^6c^{12}=\dfrac{2b^6c^{12}}{a^5}$

31. $\left(y^{-5n}\cdot y^{3n}\right)^{-4}=\left(y^{-5n+3n}\right)^{-4}$

$=\left(y^{-2n}\right)^{-4}=y^{(-2n)(-4)}=y^{8n}$

33. $\left(4x^{3n+1}y^{2n}\right)^3\left(-2x^{2n}y^{4n-3}\right)^{-2}$

$=64x^{9n+3}y^{6n}(-2)^{-2}x^{-4n}y^{-8n+6}$

$=\dfrac{64}{(-2)^2}x^{9n+3-4n}y^{6n-8n+6}$

$=\dfrac{64}{4}x^{5n+3}y^{6-2n}=16x^{5n+3}y^{6-2n}$

35. $\left(x^{-2}y^5\right)^{-3n}=x^{(-2)(-3n)}y^{(5)(-3n)}$

$=x^{6n}y^{-15n}=\dfrac{x^{6n}}{y^{15n}}$

37. $\left(y^{2n-5}\cdot y^{4n+1}\right)^{2n}$

$=\left(y^{6n-4}\right)^{2n}=y^{(6n-4)(2n)}$

$=y^{12n^2-8n}$

39. $\left(\dfrac{y^{1-n}}{y^{4-n}}\right)^{-2}=\left(y^{1-n-(4-n)}\right)^{-2}$

$=\left(y^{-3}\right)^{-2}=y^6$

41. $\left(\dfrac{a^{3n}}{a^{2n-2}}\right)^{-3}=\left(a^{3n-(2n-2)}\right)^{-3}$

$=\left(a^{n+2}\right)^{-3}=a^{-3n-6}$

Equivalently : $\left(a^{n+2}\right)^{-3}=\dfrac{1}{\left(a^{n+2}\right)^3}$

$=\dfrac{1}{a^{3n+6}}$

45. $\left(\dfrac{x^ny^{3n+2}}{x^{n-1}y^{3n-1}}\right)^{-2}$

$=\left(x^{n-(n-1)}y^{3n+2-(3n-1)}\right)^{-2}$

$=\left(xy^3\right)^{-2}=x^{-2}y^{-6}=\dfrac{1}{x^2y^6}$

49. $RT=D;\ T=\dfrac{D}{R}=\dfrac{4.58\times10^9}{3\times10^5}$

$=1.52667\times10^4=15266.7$ seconds

15266.7 seconds $\times\dfrac{1\text{minute}}{60\text{ seconds}}\times\dfrac{1\text{ hour}}{60\text{ minutes}}$

≈4.24 hours

51.

$$F = \frac{KM_1 M_2}{D^2}$$

$$6 \times 10^{-8} = \frac{K(30)(30)}{1^2}$$

$$6 \times 10^{-8} = 900K$$

$$\frac{6}{900} \times 10^{-8} = K$$

$$K = \frac{6 \times 10^{-8}}{9 \times 10^2} = \frac{2}{3} \times 10^{-10}$$

$$F = \frac{\frac{2}{3} \times 10^{-10} M_1 M_2}{D^2} \quad \text{Mass of earth: } M_2$$

$$9.8 = \frac{\frac{2}{3} \times 10^{-10}(1)M_2}{\left(6 \times 10^6\right)^2}$$

$$9.8 = \frac{\frac{2}{3} \times 10^{-10} M_2}{36 \times 10^{12}}$$

$$9.8 = \frac{1}{54} \times 10^{-22} M_2$$

$$M_2 = \frac{9.8}{\frac{1}{54} \times 10^{-22}} = 529.2 \times 10^{22}$$

Mass of Earth is approximately :

529.2×10^{22} kg.

Equivalently : $5.292 \times 10^2 \times 10^{22}$

$= 5.292 \times 10^{24}$ kg.

57. False; True statement: $\dfrac{1}{(-2)^3} = (-2)^{-3}$

Notice that the exponent changes sign when a base is moved from denominator to numerator and vice-versa. The sign of the base does not change.

59. False; True statement: $10^2 \cdot 10^3$

$= 10^{2+3} = 10^5$

61. False; $(-2)^4 = (-2)(-2)(-2)(-2) = 16$

$-2^4 = -(2 \cdot 2 \cdot 2 \cdot 2) = -16$

63. $\dfrac{(a+b)^3 (a+b)^{-2}}{(a+b)^4} = \dfrac{(a+b)^{3-2}}{(a+b)^4}$

$= \dfrac{(a+b)^1}{(a+b)^4} = (a+b)^{1-4} = (a+b)^{-3}$

$= \dfrac{1}{(a+b)^3}$

65. $A = \pi r^2 = \pi \left(3x^4\right)^2 = \pi \left(9x^8\right)$

$= 9\pi x^8$ sq. cm.

69. $V = \pi r^2 h \quad$ diameter $= 4a^2$ cm

radius $= 2a^2$ cm

height $= 10m = 10m \times \dfrac{1000 \text{ cm}}{1m} = 1000$ cm

$V = \pi \left(2a^2\right)^2 (1000) = \pi \left(4a^4\right)(1000)$

$= 4000\pi a^4$ cm^3

71. $a^6 = \left(a^3\right)^2 = 3^2 = 9$

$b^6 = \left(b^2\right)^3 = 2^3 = 8$

$a^6 > b^6$ becomes $9 > 8$, which is true.

73. $a^9 = \left(a^3\right)^3 = 3^3 = 27$

$b^{-2} = \dfrac{1}{b^2} = \dfrac{1}{2}$

$c^8 = \left(c^2\right)^4 = 5^4 = 625$

$a^9 b^{-2} < c^8$ becomes $27 \cdot \dfrac{1}{2} < 625$,

which is true.

75. $c^{-4} = \dfrac{1}{c^4} = \dfrac{1}{\left(c^2\right)^2} = \dfrac{1}{5^2} = \dfrac{1}{25}$

$a^{-3} = \dfrac{1}{a^3} = \dfrac{1}{3}$

$b^{-2} = \dfrac{1}{b^2} = \dfrac{1}{2}$

$c^{-4} < 2a^{-3} - b^{-2}$ becomes

$\dfrac{1}{25} < 2 \cdot \dfrac{1}{3} - \dfrac{1}{2}$

$\dfrac{1}{25} < \dfrac{2}{3} - \dfrac{1}{2}$

$\dfrac{1}{25} < \dfrac{1}{6}$, which is true.

77. $c^4 = \left(c^2\right)^2 = 5^2 = 25$

$\left(\dfrac{2}{a^2}\right)^3 = \dfrac{8}{a^6} = \dfrac{8}{\left(a^3\right)^2} = \dfrac{8}{3^2} = \dfrac{8}{9}$

$\dfrac{1}{c^4 + 1} + \left(\dfrac{2}{a^2}\right)^3 > 1$ becomes

$\dfrac{1}{25 + 1} + \dfrac{8}{9} > 1$

$\dfrac{1}{26} + \dfrac{8}{9} > 1$

$\dfrac{9}{(26)(9)} + \dfrac{(8)(26)}{(26)(9)} > 1$

$\dfrac{217}{234} > 1$, which is false.

78. $\dfrac{2y}{(y+2)(y-3)} - \dfrac{6(y-1)}{(2y-3)(y-3)} \div \dfrac{(y+2)(y-1)}{2y-3}$

$= \dfrac{2y}{(y+2)(y-3)} - \dfrac{6(y-1)}{(2y-3)(y-3)}$

$\qquad\qquad\qquad \cdot \dfrac{2y-3}{(y+2)(y-1)}$

$= \dfrac{2y}{(y+2)(y-3)} - \dfrac{6}{(y-3)(y+2)}$

$= \dfrac{2y-6}{(y-3)(y+2)} = \dfrac{2(y-3)}{(y-3)(y+2)} = \dfrac{2}{y+2}$

79. $1 + \dfrac{x}{2 + \frac{1}{x}} \cdot \dfrac{x}{x} = 1 + \dfrac{x^2}{2x+1}$

$= 1 \cdot \dfrac{2x+1}{2x+1} + \dfrac{x^2}{2x+1} = \dfrac{x^2 + 2x + 1}{2x+1}$

$\left[\text{Equivalently}: \dfrac{(x+1)^2}{2x+1}\right]$

80. $125 - 8x^3 y^3 = 5^3 - (2xy)^3$

$= (5 - 2xy)\left[5^2 + 5(2xy) + (2xy)^2\right]$

$= (5 - 2xy)\left(25 + 10xy + 4x^2 y^2\right)$

Problem set 5.1.2

1. $\left(3.7 \times 10^{-4}\right) \times \left(8.3 \times 10^6\right)$

$= 30.71 \times 10^2$

$= (3.071 \times 10) \times 10^2 = 3.071 \times 10^3$

3. $\dfrac{4.2 \times 10^{12}}{1.4 \times 10^4} = 3 \times 10^8$

5. $\left(8.4 \times 10^{5}\right) \times \left(7.1 \times 10^{-15}\right)$

$= 59.64 \times 10^{-10}$

$= (5.964 \times 10) \times 10^{-10}$

$= 5.964 \times 10^{1-10} = 5.964 \times 10^{-9}$

11. $\dfrac{\left(9 \times 10^{4}\right) \times \left(4 \times 10^{-3}\right)}{\left(3 \times 10^{-4}\right) \times \left(1.2 \times 10^{2}\right)} = \dfrac{36 \times 10^{1}}{3.6 \times 10^{-2}}$

$= 10 \times 10^{3} = 10^{4} = 1 \times 10^{4}$

15. $\dfrac{\text{Length of foot}}{\text{Length of atom}} = \dfrac{2 \times 10^{2}}{3 \times 10^{-8}}$

$= 0.\overline{6} \times 10^{10} = 6.\overline{6} \times 10^{9}$

Human foot is $6.\overline{6} \times 10^{9}$ times as large as the hydrogen atom.

19. $P = \dfrac{8.1 \times 10^{7}}{(300+700)^{4}} = \dfrac{8.1 \times 10^{7}}{(1000)^{4}}$

$= \dfrac{8.1 \times 10^{7}}{\left(1 \times 10^{3}\right)^{4}} = \dfrac{8.1 \times 10^{7}}{1 \times 10^{12}} = 8.1 \times 10^{-5}$

Proportion discharged is 8.1×10^{-5}

23. $0.45125 = 0.45 + \dfrac{0.75}{x}$

$0.45125x = 0.45x + 0.75$

$0.00125x = 0.75$

$x = \dfrac{0.75}{0.00125} = 600 \text{ ; } 600 \text{ gallons}$

24. $F = \dfrac{K\sqrt{T}}{L}$

$40 = \dfrac{K\sqrt{25}}{3}$

$120 = 5K$

$24 = K$

$F = \dfrac{24\sqrt{49}}{5} = \dfrac{24(7)}{5} = 33.6$

33.6 vibrations/sec.

25. $\dfrac{\left(5x^{-2}y^{4}\right)^{-3}}{\left(-2x^{-5}y\right)^{-2}} = \dfrac{5^{-3}x^{6}y^{-12}}{(-2)^{-2}x^{10}y^{-2}}$

$= \dfrac{5^{-3}}{(-2)^{-2}}x^{-4}y^{-10} = \dfrac{(-2)^{2}}{5^{3}x^{4}y^{10}}$

$= \dfrac{4}{125x^{4}y^{10}}$

Problem set 5.2.1

7. $180,000,000 \text{ lb} \times \dfrac{1 \text{ ton}}{2000 \text{ lb}} = 90,000 \text{ tons}$

$= 9 \times 10^{4} \text{ tons}$

$C = 10x + 200\sqrt{x}$

$= 10\left(9 \times 10^{4}\right) + 200\sqrt{9 \times 10^{4}}$

$= 9 \times 10^{5} + 200\left(3 \times 10^{2}\right)$

$= 900,000 + 60,000$

$= 960,000$

Cost is \$960,000.

11. $H = \left(10.45 + \sqrt{100(4)} - 4\right)(33 - 0)$

$\qquad = \left(10.45 + \sqrt{400} - 4\right)(33)$

$\qquad = (10.45 + 20 - 4)(33)$

$\qquad = (26.45)(33) = 872.85$

Since $872.45 < 2000$, exposed flesh will not freeze under these conditions.

15. $2\sqrt{36} - \sqrt[3]{-8} + 4\sqrt[5]{1} = 2(6) - (-2) + 4(1)$

$\qquad = 12 - (-2) + 4 = 12 + 2 + 4 = 18$

17. $A = \sqrt[3]{-\dfrac{2}{2}} + \sqrt{\dfrac{2^2}{4} + \dfrac{(-3)^3}{27}}$

$\qquad = \sqrt[3]{-1} + \sqrt{1 - 1} = \sqrt[3]{-1} + \sqrt{0}$

$\qquad = \sqrt[3]{-1} = -1$

19. $\dfrac{-3 \pm \sqrt{(3)^2 - 4(5)(-8)}}{2(5)} = \dfrac{-3 \pm \sqrt{9 - (-160)}}{10}$

$\qquad = \dfrac{-3 \pm \sqrt{169}}{10} = \dfrac{-3 \pm 13}{10} \ ;$

$\qquad \dfrac{-3 + 13}{10} = \dfrac{10}{10} = 1$

$\qquad \dfrac{-3 - 13}{10} = \dfrac{-16}{10} = -\dfrac{8}{5}$

21. a. $\sqrt{\left(\dfrac{4}{3}\right)^2 + \dfrac{2}{9}} = \sqrt{\dfrac{16}{9} + \dfrac{2}{9}} = \sqrt{\dfrac{18}{9}} = \sqrt{2}$

$\quad$ **b.** $a + \dfrac{h}{2a} = \dfrac{4}{3} + \dfrac{\frac{2}{9}}{2\left(\frac{4}{3}\right)} = \dfrac{4}{3} + \dfrac{\frac{2}{9}}{\frac{8}{3}}$

$\qquad = \dfrac{4}{3} + \dfrac{2}{9} \cdot \dfrac{3}{8} = \dfrac{4}{3} + \dfrac{1}{12} = \dfrac{16}{12} + \dfrac{1}{12}$

$\qquad = \dfrac{17}{12} \approx 1.42$

$\quad$ Since $\sqrt{a^2 + h} \approx a + \dfrac{h}{2a}$, $\sqrt{2} \approx 1.42$

23. In terms of specific numbers, the line with the square brackets says $\left[\dfrac{1}{2}\right]^2 = \left[-\dfrac{1}{2}\right]^2$, which is correct. The error occurs in the next step, where the writer took the square root of both sides. If $\left[\dfrac{1}{2}\right]^2 = \left[-\dfrac{1}{2}\right]^2$, it is incorrect to take the square root and assert that $\dfrac{1}{2} = -\dfrac{1}{2}$. You cannot take the square root of both sides of an equation without first checking to see if sign problems might occur.

25. $R_a = R_f \sqrt{1 - \left(\dfrac{.5c}{c}\right)^2} = \sqrt{1 - (.5)^2}$

$\qquad = \sqrt{1 - .25} = \sqrt{.75} \approx .87$

For each year the friend ages, the astronaut moving at half the speed of light would age approximately 0.87 years.

If $v = c$ (moving at the speed of light) :

$R_a = R_f \sqrt{1 - \left(\dfrac{c}{c}\right)^2} = R_f \sqrt{1 - 1^2}$

$\qquad = R_f \sqrt{0} = 0$

Since $R_a = 0$, for each year the friend ages, the astronaut moving at the speed of light would not age at all ! People aboard a spacecraft traveling at the speed of light would remain their same age as long as they were in motion!

26. $\dfrac{x^3 + 64}{x^2 - 16} = \dfrac{(x+4)\left(x^2 - 4x + 16\right)}{(x+4)(x-4)}$

$\qquad = \dfrac{x^2 - 4x + 16}{x - 4}$

27. Exterior improvements: x
     Apartment: 3x
     Interior improvements: 5x
$$x + 3x + 5x = 180,000$$
$$9x = 180,000$$
$$x = 20,000$$
     Cost of apartment $= 3(20,000) = \$60,000.$

28. Commutative properties of addition
     and multiplication.

Problem set 5.2.2

3. $8^{\frac{1}{3}} = \sqrt[3]{8} = 2$

7. $64^{-\frac{1}{2}} = \dfrac{1}{64^{\frac{1}{2}}} = \dfrac{1}{\sqrt{64}} = \dfrac{1}{8}$

9. $\left(\frac{1}{64}\right)^{-\frac{1}{3}} = \dfrac{1}{\left(\frac{1}{64}\right)^{\frac{1}{3}}} = \dfrac{1}{\frac{1}{4}} = 4$

11. $(-64)^{-\frac{1}{3}} = \dfrac{1}{(-64)^{\frac{1}{3}}} = \dfrac{1}{\sqrt[3]{-64}} = \dfrac{1}{-4} = -\dfrac{1}{4}$

13. $\left(-\frac{27}{64}\right)^{-\frac{1}{3}} = \dfrac{1}{\left(-\frac{27}{64}\right)^{\frac{1}{3}}} = \dfrac{1}{\sqrt[3]{\frac{-27}{64}}}$

$= \dfrac{1}{-\frac{3}{4}} = -\dfrac{4}{3}$

15. $16^{\frac{5}{2}} = \left(\sqrt{16}\right)^5 = 4^5 = 1024$

17. $8^{\frac{2}{3}} = \left(\sqrt[3]{8}\right)^2 = 2^2 = 4$

19. $(-1)^{\frac{17}{3}} = \left(\sqrt[3]{-1}\right)^{17} = (-1)^{17} = -1$

21. $-25^{\frac{3}{2}} = -\left(\sqrt{25}\right)^3 = -\left(5^3\right) = -125$

23. $\left(\frac{27}{64}\right)^{\frac{2}{3}} = \left(\sqrt[3]{\frac{27}{64}}\right)^2 = \left(\frac{3}{4}\right)^2 = \frac{9}{16}$

25. $\left(\frac{1}{32}\right)^{-\frac{3}{5}} = \dfrac{1}{\left(\frac{1}{32}\right)^{\frac{3}{5}}} = \dfrac{1}{\left(\sqrt[5]{\frac{1}{32}}\right)^3} = \dfrac{1}{\left(\frac{1}{2}\right)^3}$

$= \dfrac{1}{\frac{1}{8}} = 8$

29. $(-1)^{\frac{17}{125}} = \left(\sqrt[125]{-1}\right)^{17} = (-1)^{17} = -1$

31. $\left[3 + 27^{\frac{2}{3}} + 32^{\frac{2}{5}}\right]^{\frac{3}{2}} - 9^{\frac{1}{2}}$

$= \left[3 + \left(\sqrt[3]{27}\right)^2 + \left(\sqrt[5]{32}\right)^2\right]^{\frac{3}{2}} - 9^{\frac{1}{2}}$

$= \left[3 + 3^2 + 2^2\right]^{\frac{3}{2}} - 9^{\frac{1}{2}}$

$= 16^{\frac{3}{2}} - 9^{\frac{1}{2}} = \left(\sqrt{16}\right)^3 - \sqrt{9}$

$= 4^3 - 3 = 64 - 3 = 61$

33. $A = \left(\dfrac{800}{500}\right)^{\frac{1}{5}} - 1 = (1.6)^{\frac{1}{5}} - 1 = 1.099 - 1$

$= .099 = 9.9\%$

$\left[\text{Calculator: } (1.6)^{\frac{1}{5}} \quad 1.6\boxed{y^x}5\boxed{\tfrac{1}{x}}\boxed{=}\right]$

Annual rate of return is 9.9%.

39. $T = 10\left(60 + 120 \cdot 6^{-1}\right)100^{-\frac{1}{2}}$

$= 10\left(60 + 120 \cdot \tfrac{1}{6}\right) \cdot \dfrac{1}{100^{\frac{1}{2}}}$

$= 10(60 + 20) \cdot \dfrac{1}{\sqrt{100}} = 10(80) \cdot \dfrac{1}{10}$

$= 80$; 80 minutes.

43. $t = 16$ since 7:00 PM is 16 hours after 3:00 AM.

$O = \left[0.26 + 0.02(16)\right] \cdot$

$\left[36 + 16(16) - 16^2\right]^{-\frac{1}{2}}$

$= (0.58)(36)^{-\frac{1}{2}} = (0.58) \cdot \dfrac{1}{36^{\frac{1}{2}}}$

$= (0.58)\dfrac{1}{\sqrt{36}} = \dfrac{0.58}{6} = 0.09\overline{6}$

Ozone level is approximately 0.097 parts per millon.

45. $3(4)^{\frac{1}{2}} - (2.4)^0 + 16\left(4^{-2}\right)$

$= 3(2) - 1 + 16 \cdot \dfrac{1}{16} = 6 - 1 + 1 = 6$

47. $\dfrac{2\left(4^{-\frac{1}{3}}\right)}{3\left(4^{\frac{5}{3}}\right)} = \dfrac{2}{3} \cdot \dfrac{1}{4^{\frac{1}{3}} \cdot 4^{\frac{5}{3}}} = \dfrac{2}{3} \cdot \dfrac{1}{4^{\frac{6}{3}}}$

$= \dfrac{2}{3} \cdot \dfrac{1}{4^2} = \dfrac{2}{3(16)} = \dfrac{1}{24}$

49. $81^{\frac{1}{4}} = \sqrt[4]{81} = 3$

$81^{\frac{1}{2}} = \sqrt{81} = 9 \qquad 81^{\frac{1}{2}} > 81^{\frac{1}{4}}$

In general, if b > 1 and n > m,

$b^{\frac{1}{m}} > b^{\frac{1}{n}}$.

51. Suppose that $a = 16$ and $b = 9$

$(a+b)^{\frac{1}{2}} \neq a^{\frac{1}{2}} + b^{\frac{1}{2}}$

$(16+9)^{\frac{1}{2}} \neq 16^{\frac{1}{2}} + 9^{\frac{1}{2}}$

$25^{\frac{1}{2}} \neq 16^{\frac{1}{2}} + 9^{\frac{1}{2}}$

$\sqrt{25} \neq \sqrt{16} + \sqrt{9}$

$5 \neq 4 + 3$

$5 \neq 7$

53. $x^2y - xy^2 - xy + y^2$

$xy(x-y) - y(x-y) = (x-y)(xy-y)$

$= (x-y)y(x-1) = y(x-y)(x-1)$

54. $(y+2)(y+5)\left[\dfrac{5}{y+2} - \dfrac{3}{y+5}\right]$

$\qquad = (y+2)(y+5)\left[\dfrac{9}{(y+2)(y+5)}\right]$

$\qquad 5(y+5) - 3(y+2) = 9$

$\qquad 5y + 25 - 3y - 6 = 9$

$\qquad\qquad 2y + 19 = 9$

$\qquad\qquad\qquad 2y = -10$

$\qquad\qquad\qquad y = -5$

Since -5 causes denominators to be zero in the original equation, -5 is not a solution. The equation has no solution. $\varnothing$

55. $\dfrac{4y+1}{8y^3 - 1} + \dfrac{2y}{4y^2 + 2y + 1}$

$\quad = \dfrac{4y+1}{(2y-1)\left(4y^2 + 2y + 1\right)}$

$\qquad\qquad + \dfrac{2y}{4y^2 + 2y + 1} \cdot \dfrac{2y-1}{2y-1}$

$\quad = \dfrac{4y+1+2y(2y-1)}{(2y-1)\left(4y^2 + 2y + 1\right)}$

$\quad = \dfrac{4y+1+4y^2 - 2y}{(2y-1)\left(4y^2 + 2y + 1\right)}$

$\quad = \dfrac{4y^2 + 2y + 1}{(2y-1)\left(4y^2 + 2y + 1\right)} = \dfrac{1}{2y-1}$

Problem set 5.2.3

1. $\left(7y^{\frac{1}{3}}\right)\left(2y^{\frac{1}{4}}\right) = 14y^{\frac{1}{3} + \frac{1}{4}} = 14y^{\frac{4}{12} + \frac{3}{12}}$

$\quad = 14y^{\frac{7}{12}}$

3. $\left(3x^{\frac{3}{4}}\right)\left(-5x^{-\frac{1}{2}}\right) = -15x^{\frac{3}{4} - \frac{1}{2}}$

$\qquad = -15x^{\frac{3}{4} - \frac{2}{4}} = -15x^{\frac{1}{4}}$

5. $\dfrac{20x^{\frac{1}{2}}}{5x^{\frac{1}{4}}} = 4x^{\frac{1}{2} - \frac{1}{4}} = 4x^{\frac{2}{4} - \frac{1}{4}} = 4x^{\frac{1}{4}}$

7. $\dfrac{80y^{\frac{1}{6}}}{10y^{\frac{1}{4}}} = 8y^{\frac{1}{6} - \frac{1}{4}} = 8y^{\frac{2}{12} - \frac{3}{12}} = 8y^{-\frac{1}{12}}$

$\quad = \dfrac{8}{y^{\frac{1}{12}}}$

13. $\left(16xy^{\frac{1}{4}}z^{\frac{2}{3}}\right)^{\frac{1}{4}} = 16^{\frac{1}{4}}x^{\frac{1}{4}}y^{\frac{1}{16}}z^{\frac{1}{6}}$

$\quad = 2x^{\frac{1}{4}}y^{\frac{1}{16}}z^{\frac{1}{6}}$

15. $\left(\dfrac{2x^{\frac{1}{4}}}{5y^{\frac{1}{3}}}\right)^3 = \dfrac{\left(2x^{\frac{1}{4}}\right)^3}{\left(5y^{\frac{1}{3}}\right)^3} = \dfrac{8x^{\frac{3}{4}}}{125y}$

17. $\left(\dfrac{x^3}{y^5}\right)^{-\frac{1}{2}} = \dfrac{x^{-\frac{3}{2}}}{y^{-\frac{5}{2}}} = \dfrac{y^{\frac{5}{2}}}{x^{\frac{3}{2}}}$

19. $\left(\dfrac{27a^{-3}}{64b^{-3}}\right)^{-\frac{1}{3}} = \dfrac{27^{-\frac{1}{3}}a}{64^{-\frac{1}{3}}b} = \dfrac{64^{\frac{1}{3}}a}{27^{\frac{1}{3}}b}$

$= \dfrac{4a}{3b}$

21. $\dfrac{8a^2b^{-\frac{2}{3}}}{6(ab)^{\frac{1}{2}}} = \dfrac{8a^2b^{-\frac{2}{3}}}{6a^{\frac{1}{2}}b^{\frac{1}{2}}}$

$= \dfrac{4}{3}a^{2-\frac{1}{2}}b^{-\frac{2}{3}-\frac{1}{2}} = \dfrac{4}{3}a^{\frac{3}{2}}b^{-\frac{7}{6}}$

$= \dfrac{4a^{\frac{3}{2}}}{3b^{\frac{7}{6}}}$

23. $\left(\dfrac{27a^3b^{-\frac{3}{2}}}{8a^{-1}b^3}\right)^{-\frac{2}{3}} = \left(\dfrac{27}{8}a^4b^{-\frac{9}{2}}\right)^{-\frac{2}{3}}$

$= \left(\dfrac{27}{8}\right)^{-\frac{2}{3}}a^{-\frac{8}{3}}b^3 = \dfrac{b^3}{\left(\dfrac{27}{8}\right)^{\frac{2}{3}}a^{\frac{8}{3}}}$

$= \dfrac{b^3}{\left(\dfrac{3}{2}\right)^2 a^{\frac{8}{3}}} = \dfrac{b^3}{\dfrac{9}{4}a^{\frac{8}{3}}} = \dfrac{4b^3}{9a^{\frac{8}{3}}}$

25. $\left(x^{\frac{1}{a}}\right)^{a^2} = x^{\frac{1}{a}\cdot a^2} = x^a$

29. $\dfrac{\left(r^n s^{\frac{1}{n}}\right)^n}{r^{n^2}s^n} = \dfrac{r^{n^2}s}{r^{n^2}s^n} = s^{1-n}$

$\left(\text{or } \dfrac{1}{s^{n-1}}\right)$

31. $\dfrac{x^{\frac{1}{a}}y^{\frac{5}{a}}}{x^{\frac{2}{a}}} = x^{\frac{1}{a}-\frac{2}{a}}y^{\frac{5}{a}} = x^{-\frac{1}{a}}y^{\frac{5}{a}} = \dfrac{y^{\frac{5}{a}}}{x^{\frac{1}{a}}}$

33. $\left(\dfrac{x^{4a}y^{6a}}{z^{8a}w^{10a}}\right)^{-\frac{3}{2a}} = \dfrac{\left(x^{4a}y^{6a}\right)^{-\frac{3}{2a}}}{\left(z^{8a}w^{10a}\right)^{-\frac{3}{2a}}}$

$= \dfrac{x^{4a\cdot\left(-\frac{3}{2a}\right)}y^{6a\cdot\left(-\frac{3}{2a}\right)}}{\left(z^{8a}\right)^{-\frac{3}{2a}}\left(w^{10a}\right)^{-\frac{3}{2a}}}$

$= \dfrac{x^{-6}y^{-9}}{z^{-12}w^{-15}} = \dfrac{z^{12}w^{15}}{x^6y^9}$

35. $\left(\dfrac{r^{-\frac{3}{4a}}}{s^{\frac{2}{3a}}}\right)^{-12a}\left(r^{6a^2}s^{-8a^2}\right)^{-\frac{5}{2a^2}}$

$= \dfrac{r^9}{s^{-8}}\left(r^{-15}s^{20}\right) = \dfrac{r^{-6}s^{20}}{s^{-8}}$

$= r^{-6}s^{28} = \dfrac{s^{28}}{r^6}$

37. $\sqrt[3]{3}\cdot\sqrt{3} = 3^{\frac{1}{3}}\cdot 3^{\frac{1}{2}} = 3^{\frac{5}{6}} = \sqrt[6]{3^5}$

$\left(\text{or } \sqrt[6]{243}\right)$

41. $\dfrac{\sqrt{3}}{\sqrt[3]{3}} = \dfrac{3^{\frac{1}{2}}}{3^{\frac{1}{3}}} = 3^{\frac{1}{2}-\frac{1}{3}} = 3^{\frac{1}{6}} = \sqrt[6]{3}$

45. $\dfrac{\sqrt{8}}{\sqrt[3]{2}} = \dfrac{8^{\frac{1}{2}}}{2^{\frac{1}{3}}} = \dfrac{\left(2^3\right)^{\frac{1}{2}}}{2^{\frac{1}{3}}} = \dfrac{2^{\frac{3}{2}}}{2^{\frac{1}{3}}} = 2^{\frac{3}{2}-\frac{1}{3}}$

$= 2^{\frac{7}{6}} = \sqrt[6]{2^7} \quad \left(\text{or } \sqrt[6]{128}\right)$

47. $\dfrac{\sqrt[3]{8}}{\sqrt[6]{4}} = \dfrac{2}{4^{\frac{1}{6}}} = \dfrac{2}{\left(2^2\right)^{\frac{1}{6}}} = \dfrac{2}{2^{\frac{1}{3}}} = 2^{1-\frac{1}{3}}$

$2^{\frac{2}{3}} = \sqrt[3]{2^2} = \sqrt[3]{4}$

49. $\sqrt[9]{a^3} = \left(a^3\right)^{\frac{1}{9}} = a^{\frac{1}{3}} = \sqrt[3]{a}$

51. $\sqrt[9]{27} = \sqrt[9]{3^3} = \left(3^3\right)^{\frac{1}{9}} = 3^{\frac{1}{3}} = \sqrt[3]{3}$

53. $\sqrt[12]{16} = \sqrt[12]{2^4} = \left(2^4\right)^{\frac{1}{12}} = 2^{\frac{1}{3}} = \sqrt[3]{2}$

57. $\sqrt[9]{2^3 x^3 y^6} = \left(2^3 x^3 y^6\right)^{\frac{1}{9}}$

$= 2^{\frac{1}{3}} x^{\frac{1}{3}} y^{\frac{2}{3}} = \left(2xy^2\right)^{\frac{1}{3}} = \sqrt[3]{2xy^2}$

59. $\sqrt[9]{27x^3 y^6} = \sqrt[9]{3^3 x^3 y^6} = \left(3^3 x^3 y^6\right)^{\frac{1}{9}}$

$= 3^{\frac{1}{3}} x^{\frac{1}{3}} y^{\frac{2}{3}} = \left(3xy^2\right)^{\frac{1}{3}} = \sqrt[3]{3xy^2}$

61. $\sqrt[6]{x^2 - 2xy + y^2} = \sqrt[6]{(x-y)^2}$

$= \left[(x-y)^2\right]^{\frac{1}{6}} = (x-y)^{\frac{1}{3}}$

$= \sqrt[3]{x-y}$

65. $\dfrac{8^{-\frac{4}{3}} + 2^{-2}}{16^{-\frac{3}{4}} + 2 - 1} = \dfrac{8^{\frac{4}{3}} + \frac{1}{2^2}}{\frac{1}{16^{\frac{3}{4}}} + \frac{1}{2}}$

$= \dfrac{\dfrac{1}{\left(\sqrt[3]{8}\right)^2} + \frac{1}{4}}{\dfrac{1}{\left(\sqrt[4]{16}\right)^3} + \frac{1}{2}} = \dfrac{\frac{1}{16} + \frac{1}{4}}{\frac{1}{8} + \frac{1}{2}} \cdot \dfrac{16}{16}$

$= \dfrac{1+4}{2+8} = \dfrac{5}{10} = \dfrac{1}{2}$

The birthday boy ate $\frac{1}{2}$ of the cake. Thus, $\frac{1}{2}$ was left over. The professor ate half of what was left over,

or $\frac{1}{2} \cdot \frac{1}{2} = \frac{1}{4}$ of the cake.

67. $\sqrt{\sqrt{b}} = \sqrt{b^{\frac{1}{2}}} = \left(b^{\frac{1}{2}}\right)^{\frac{1}{2}} = b^{\frac{1}{4}} = \sqrt[4]{b}$

69. $\frac{1}{10} = \frac{1}{40} + \frac{1}{20} + \frac{1}{R_3}$

$40R_3\left(\frac{1}{10}\right) = 40R_3\left(\frac{1}{40} + \frac{1}{20} + \frac{1}{R_3}\right)$

$4R_3 = R_3 + 2R_3 + 40$

$4R_3 = 3R_3 + 40$

$R_3 = 40$; 40 ohms contained in

the third resistor.

70. $\text{abf}\left(\frac{1}{f}\right) = \text{abf}\left(\frac{1}{a} + \frac{1}{b}\right)$

$ab = bf + af$

$ab - bf = af$

$b(a-f) = af$

$b = \dfrac{af}{a-f}$

71. x: speed of wind

525 + x: speed of plane with wind

525 − x: speed of plane against wind

Time required to fly 1335 miles against

wind = Time required to fly 1815 miles

with wind

$\dfrac{1335}{525-x} = \dfrac{1815}{525+x}$

$1335(525 + x) = 1815(525 - x)$

$700,875 + 1335x = 952,875 - 1815x$

$3150x = 252,000$

$x = 80$

Wind is blowing at 80 m/h.

Problem set 5.2.4

11. $\left(3x^{\frac{1}{2}} + 4y^{\frac{2}{3}}\right)\left(5x^{\frac{1}{2}} - 2y^{\frac{2}{3}}\right)$

$= \left(3x^{\frac{1}{2}}\right)\left(5x^{\frac{1}{2}}\right) + \left(3x^{\frac{1}{2}}\right)\left(-2y^{\frac{2}{3}}\right)$

$\quad + \left(4y^{\frac{2}{3}}\right)\left(5x^{\frac{1}{2}}\right) + \left(4y^{\frac{2}{3}}\right)\left(-2y^{\frac{2}{3}}\right)$

$= 15x - 6x^{\frac{1}{2}}y^{\frac{2}{3}} + 20x^{\frac{1}{2}}y^{\frac{2}{3}} - 8y^{\frac{4}{3}}$

$= 15x + 14x^{\frac{1}{2}}y^{\frac{2}{3}} - 8y^{\frac{4}{3}}$

13. $\left(x^{\frac{2}{3}} + 3\right)^2 = \left(x^{\frac{2}{3}}\right)^2 + 2x^{\frac{2}{3}} \cdot 3 + 3^2$

$= x^{\frac{4}{3}} + 6x^{\frac{2}{3}} + 9$

17. $\left(3x^{\frac{1}{2}} + 4y^{\frac{1}{2}}\right)^2$

$= \left(3x^{\frac{1}{2}}\right)^2 + 2\left(3x^{\frac{1}{2}}\right)\left(4y^{\frac{1}{2}}\right) + \left(4y^{\frac{1}{2}}\right)^2$

$= 9x + 24x^{\frac{1}{2}}y^{\frac{1}{2}} + 16y$

19. $\left(x^{\frac{1}{2}} + 5^{\frac{1}{2}}\right)\left(x^{\frac{1}{2}} - 5^{\frac{1}{2}}\right)$

$= \left(x^{\frac{1}{2}}\right)^2 - \left(5^{\frac{1}{2}}\right)^2 = x - 5$

23. $\left(3x^{\frac{3}{2}}+4^{\frac{1}{2}}\right)\left(3x^{\frac{3}{2}}-4^{\frac{1}{2}}\right)$

$$=\left(3x^{\frac{3}{2}}\right)^2-\left(4^{\frac{1}{2}}\right)^2=9x^3-4$$

25. $\left(x^{\frac{1}{3}}-y^{\frac{1}{3}}\right)\left(x^{\frac{2}{3}}+x^{\frac{1}{3}}y^{\frac{1}{3}}+y^{\frac{2}{3}}\right)$

$$=x+x^{\frac{2}{3}}y^{\frac{1}{3}}+x^{\frac{1}{3}}y^{\frac{2}{3}}$$
$$\underline{\quad-x^{\frac{2}{3}}y^{\frac{1}{3}}-x^{\frac{1}{3}}y^{\frac{2}{3}}-y\quad}$$
$$x\qquad\qquad\qquad -y$$

27. $\left(x^{\frac{1}{2}}+1\right)\left[\left(x^{\frac{1}{4}}+1\right)\left(x^{\frac{1}{4}}-1\right)\right]$

$$=\left(x^{\frac{1}{2}}+1\right)\left(x^{\frac{1}{2}}-1\right)=\left(x^{\frac{1}{2}}\right)^2-1^2$$
$$x-1$$

31. $\dfrac{27x^{\frac{8}{5}}y^{\frac{4}{5}}-36x^{\frac{3}{5}}y^{\frac{9}{5}}}{9x^{\frac{3}{5}}y^{\frac{4}{5}}}$

$$=\frac{27x^{\frac{8}{5}}y^{\frac{4}{5}}}{9x^{\frac{3}{5}}y^{\frac{4}{5}}}-\frac{36x^{\frac{3}{5}}y^{\frac{9}{5}}}{9x^{\frac{3}{5}}y^{\frac{4}{5}}}$$
$$=3x^1y^0-4x^0y^1=3x-4y$$

35. $7(x-1)^{\frac{12}{5}}-14(x-1)^{\frac{7}{5}}$

$$=7(x-1)^{\frac{7}{5}}\left[(x-1)^{\frac{5}{5}}-2\right]$$
$$=7(x-1)^{\frac{7}{5}}(x-1-2)=7(x-1)^{\frac{7}{5}}(x-3)$$

45. $9x^{\frac{2}{3}}-16=\left(3x^{\frac{1}{3}}\right)^2-4^2$

$$=\left(3x^{\frac{1}{3}}+4\right)\left(3x^{\frac{1}{3}}-4\right)$$

47. $\dfrac{3}{y^{\frac{1}{2}}}-y^{\frac{1}{2}}\qquad\left[\text{LCD is }y^{\frac{1}{2}}\right]$

$$=\frac{3}{y^{\frac{1}{2}}}-y^{\frac{1}{2}}\cdot\frac{y^{\frac{1}{2}}}{y^{\frac{1}{2}}}=\frac{3-y}{y^{\frac{1}{2}}}$$

49. $y^{\frac{3}{4}}-\dfrac{4}{y^{\frac{1}{4}}}\qquad\left[\text{LCD is }y^{\frac{1}{4}}\right]$

$$=y^{\frac{3}{4}}\cdot\frac{y^{\frac{1}{4}}}{y^{\frac{1}{4}}}-\frac{4}{y^{\frac{1}{4}}}=\frac{y-4}{y^{\frac{1}{4}}}$$

51. $\dfrac{4y^3}{\left(y^2-1\right)^{\frac{1}{2}}}+3y\left(y^2-1\right)^{\frac{1}{2}}$

$$\left[\text{LCD is }\left(y^2-1\right)^{\frac{1}{2}}\right]$$

$$=\dfrac{4y^3}{\left(y^2-1\right)^{\frac{1}{2}}}+3y\left(y^2-1\right)^{\frac{1}{2}}\cdot\dfrac{\left(y^2-1\right)^{\frac{1}{2}}}{\left(y^2-1\right)^{\frac{1}{2}}}$$

$$=\dfrac{4y^3+3y\left(y^2-1\right)}{\left(y^2-1\right)^{\frac{1}{2}}}$$

$$=\dfrac{4y^3+3y^3-3y}{\left(y^2-1\right)^{\frac{1}{2}}}=\dfrac{7y^3-3y}{\left(y^2-1\right)^{\frac{1}{2}}}$$

53. $\dfrac{y^5}{\left(y^2+1\right)^{\frac{1}{2}}}+3y^3\left(y^2+1\right)^{\frac{1}{2}}$

$$\left[\text{LCD is }\left(y^2+1\right)^{\frac{1}{2}}\right]$$

$$=\dfrac{y^5}{\left(y^2+1\right)^{\frac{1}{2}}}+$$

$$3y^3\left(y^2+1\right)^{\frac{1}{2}}\cdot\dfrac{\left(y^2+1\right)^{\frac{1}{2}}}{\left(y^2+1\right)^{\frac{1}{2}}}$$

$$=\dfrac{y^5+3y^3\left(y^2+1\right)}{\left(y^2+1\right)^{\frac{1}{2}}}$$

$$=\dfrac{y^5+3y^5+3y^3}{\left(y^2+1\right)^{\frac{1}{2}}}=\dfrac{4y^5+3y^3}{\left(y^2+1\right)^{\frac{1}{2}}}$$

55. Consecutive integers $x,\ x+1,\ x+2$

$$x+x+1+x+2=4x-6$$
$$3x+3=4x-6$$
$$9=x$$

Integers are 9,10, and 11.

56. $4\left[\dfrac{3}{4}(2y-5)-\dfrac{5}{4}(3y+9)\right]=4(-6)$

$$6y-15-15y-45=-24$$
$$-9y-60=-24$$
$$-9y=36$$
$$y=-4\qquad\{-4\}$$

57. $\dfrac{a+b}{(a-5b)^2}-\dfrac{a-b}{(a+5b)(a-5b)}$

$$\left[\text{LCD is }(a-5b)^2(a+5b)\right]$$

$$=\dfrac{a+b}{(a-5b)^2}\cdot\dfrac{(a+5b)}{(a+5b)}$$

$$-\dfrac{a-b}{(a+5b)(a-5b)}\cdot\dfrac{a-5b}{a-5b}$$

$$=\dfrac{(a+b)(a+5b)-(a-b)(a-5b)}{(a-5b)^2(a+5b)}$$

$$=\dfrac{a^2+6ab+5b^2-a^2+6ab-5b^2}{(a-5b)^2(a+5b)}$$

$$=\dfrac{12ab}{(a-5b)^2(a+5b)}$$

Problem set 5.3.1

1. $\sqrt{20}=\sqrt{4\cdot5}=\sqrt{2^2\cdot5}=2\sqrt{5}$

7. $7\sqrt{28}=7\sqrt{4\cdot7}=7\sqrt{2^2\cdot7}=7(2)\sqrt{7}$
$$=14\sqrt{7}$$

11. $\sqrt[3]{54} = \sqrt[3]{27 \cdot 2} = \sqrt[3]{3^3 \cdot 2} = 3\sqrt[3]{2}$

13. $\sqrt[5]{64} = \sqrt[5]{32 \cdot 2} = \sqrt[5]{2^5 \cdot 2} = 2\sqrt[5]{2}$

15. $6\sqrt[3]{16} = 6\sqrt[3]{8 \cdot 2} = 6\sqrt[3]{2^3 \cdot 2}$
　　　$= 6(2)\left(\sqrt[3]{2}\right) = 12\sqrt[3]{2}$

19. $\sqrt{12y} = \sqrt{4 \cdot 3y} = \sqrt{2^2 \cdot 3y} = 2\sqrt{3y}$

23. $\sqrt{48x^3} = \sqrt{16 \cdot 3x^3} = \sqrt{4^2 \cdot 3x^2 x}$
　　　$= 4x\sqrt{3x}$

25. $\sqrt{20xy^3} = \sqrt{4 \cdot 5xy^3} = \sqrt{2^2 \cdot 5xy^2 y}$
　　　$= 2y\sqrt{5xy}$

27. $\sqrt{75xy^2 z^5} = \sqrt{25 \cdot 3xy^2 z^5}$
　　　$= \sqrt{5^2 \cdot 3xy^2\left(z^2\right)^2 z} = 5yz^2\sqrt{3xz}$

29. $4\sqrt{50x^7} = 4\sqrt{25 \cdot 2\left(x^3\right)^2 x}$
　　　$= 4 \cdot 5x^3\sqrt{2x} = 20x^3\sqrt{2x}$

31. $\sqrt[3]{32x^3} = \sqrt[3]{8 \cdot 4x^3} = \sqrt[3]{2^3 \cdot 4x^3}$
　　　$= 2x\sqrt[3]{4}$

33. $\sqrt[3]{-32xy^5 z^6} = \sqrt[3]{-8 \cdot 4xy^5 z^6}$
　　　$= \sqrt[3]{(-2)^3 \cdot 4xy^3 y^2\left(z^2\right)^3}$
　　　$= -2yz^2\sqrt[3]{4xy^2}$

35. $-2x^2 y\left(\sqrt[3]{54x^3 y^7 z^2}\right)$
　　　$= -2x^2 y\left(\sqrt[3]{27 \cdot 2x^3\left(y^2\right)^3 yz^2}\right)$
　　　$= -2x^2 y\left(3xy^2\right)\sqrt[3]{2yz^2}$
　　　$= -6x^3 y^3 \sqrt[3]{2yz^2}$

37. $-5\sqrt[4]{48y^7} = -5\sqrt[4]{16 \cdot 3y^7}$
　　　$= -5\sqrt[4]{2^4 \cdot 3y^4 y^3} = -5(2)y\sqrt[4]{3y^3}$
　　　$= -10y\sqrt[4]{3y^3}$

39. $-3y\left(\sqrt[5]{64x^3 y^6}\right) = -3y\left(\sqrt[5]{2^5 \cdot 2x^3 y^5 y}\right)$
　　　$= -3y(2y)\sqrt[5]{2x^3 y} = -6y^2\sqrt[5]{2x^3 y}$

47. $-2\sqrt{\dfrac{16}{100}} = -2\dfrac{\sqrt{16}}{\sqrt{100}} = -2 \cdot \dfrac{4}{10} = -\dfrac{4}{5}$

49. $-\sqrt[5]{-\dfrac{1}{32}} = -\dfrac{\sqrt[5]{-1}}{\sqrt[5]{32}} = -\dfrac{(-1)}{2} = \dfrac{1}{2}$

51. $\sqrt{\dfrac{80}{25}} = \dfrac{\sqrt{80}}{\sqrt{25}} = \dfrac{\sqrt{16 \cdot 5}}{5} = \dfrac{4\sqrt{5}}{5}$

55. $-5\sqrt{\dfrac{28}{121}} = -5\dfrac{\sqrt{28}}{\sqrt{121}} = -5\dfrac{\sqrt{4 \cdot 7}}{11}$
　　　$= -5\dfrac{(2)\sqrt{7}}{11} = \dfrac{-10\sqrt{7}}{11}$

59. $\sqrt{\dfrac{1}{3}} = \dfrac{1}{\sqrt{3}} = \dfrac{1}{\sqrt{3}} \cdot \dfrac{\sqrt{3}}{\sqrt{3}} = \dfrac{\sqrt{3}}{\sqrt{3^2}} = \dfrac{\sqrt{3}}{3}$

61. $\sqrt{\dfrac{2}{5}} = \dfrac{\sqrt{2}}{\sqrt{5}} = \dfrac{\sqrt{2}}{\sqrt{5}} \cdot \dfrac{\sqrt{5}}{\sqrt{5}} = \dfrac{\sqrt{10}}{\sqrt{5^2}} = \dfrac{\sqrt{10}}{5}$

63. $\sqrt{\dfrac{3}{8}} = \dfrac{\sqrt{3}}{\sqrt{8}} = \dfrac{\sqrt{3}}{\sqrt{8}} \cdot \dfrac{\sqrt{2}}{\sqrt{2}} = \dfrac{\sqrt{6}}{\sqrt{16}} = \dfrac{\sqrt{6}}{4}$

65. $\sqrt{\dfrac{5}{18}} = \dfrac{\sqrt{5}}{\sqrt{18}} = \dfrac{\sqrt{5}}{\sqrt{18}} \cdot \dfrac{\sqrt{2}}{\sqrt{2}} = \dfrac{\sqrt{10}}{\sqrt{36}} = \dfrac{\sqrt{10}}{6}$

67. $\sqrt{\dfrac{35}{4} + \dfrac{7}{9}} = \sqrt{\dfrac{315}{36} + \dfrac{28}{36}} = \sqrt{\dfrac{343}{36}}$

$\qquad = \dfrac{\sqrt{343}}{\sqrt{36}} = \dfrac{\sqrt{343}}{6}$

69. $-6\sqrt{\dfrac{5}{18}} = -6\dfrac{\sqrt{5}}{\sqrt{18}} = \dfrac{-6\sqrt{5}}{\sqrt{18}} \cdot \dfrac{\sqrt{2}}{\sqrt{2}}$

$\qquad = \dfrac{-6\sqrt{10}}{\sqrt{36}} = \dfrac{-6\sqrt{10}}{6} = -\sqrt{10}$

71. $\dfrac{1}{5}\sqrt{\dfrac{75}{63}} = \dfrac{1}{5}\dfrac{\sqrt{75}}{\sqrt{63}} = \dfrac{\sqrt{25 \cdot 3}}{5\sqrt{9 \cdot 7}} = \dfrac{5\sqrt{3}}{5 \cdot 3\sqrt{7}}$

$\qquad = \dfrac{\sqrt{3}}{3\sqrt{7}} = \dfrac{\sqrt{3}}{3\sqrt{7}} \cdot \dfrac{\sqrt{7}}{\sqrt{7}} = \dfrac{\sqrt{21}}{3\sqrt{49}}$

$\qquad = \dfrac{\sqrt{21}}{3 \cdot 7} = \dfrac{\sqrt{21}}{21}$

73. $\sqrt[3]{\dfrac{1}{2}} = \dfrac{\sqrt[3]{1}}{\sqrt[3]{2}} = \dfrac{1}{\sqrt[3]{2}} \cdot \dfrac{\sqrt[3]{2^2}}{\sqrt[3]{2^2}} = \dfrac{\sqrt[3]{2^2}}{\sqrt[3]{2^3}} = \dfrac{\sqrt[3]{4}}{2}$

75. $\sqrt[3]{\dfrac{2}{3}} = \dfrac{\sqrt[3]{2}}{\sqrt[3]{3}} = \dfrac{\sqrt[3]{2}}{\sqrt[3]{3}} \cdot \dfrac{\sqrt[3]{3^2}}{\sqrt[3]{3^2}} = \dfrac{\sqrt[3]{2 \cdot 3^2}}{\sqrt[3]{3^3}}$

$\qquad = \dfrac{\sqrt[3]{18}}{3}$

77. $-6\sqrt[3]{\dfrac{3}{2}} = -6\dfrac{\sqrt[3]{3}}{\sqrt[3]{2}} = \dfrac{-6\sqrt[3]{3}}{\sqrt[3]{2}} \cdot \dfrac{\sqrt[3]{2^2}}{\sqrt[3]{2^2}}$

$\qquad = \dfrac{-6\sqrt[3]{3 \cdot 2^2}}{\sqrt[3]{2^3}} = \dfrac{-6\sqrt[3]{12}}{2} = -3\sqrt[3]{12}$

79. $\sqrt{2 - \dfrac{1}{3}} = \sqrt{\dfrac{6}{3} - \dfrac{1}{3}} = \sqrt{\dfrac{5}{3}} = \dfrac{\sqrt{5}}{\sqrt{3}}$

$\qquad = \dfrac{\sqrt{5}}{\sqrt{3}} \cdot \dfrac{\sqrt{3}}{\sqrt{3}} = \dfrac{\sqrt{15}}{3}$

81. $\sqrt{4 + \dfrac{3}{5}} = \sqrt{\dfrac{20}{5} + \dfrac{3}{5}} = \sqrt{\dfrac{23}{5}} = \dfrac{\sqrt{23}}{\sqrt{5}}$

$\qquad = \dfrac{\sqrt{23}}{\sqrt{5}} \cdot \dfrac{\sqrt{5}}{\sqrt{5}} = \dfrac{\sqrt{115}}{5}$

83. $\left(\sqrt{\dfrac{1}{2}}\right)^3 = \left(\sqrt{\dfrac{1}{2}}\right)^2 \left(\sqrt{\dfrac{1}{2}}\right) = \dfrac{1}{2}\dfrac{\sqrt{1}}{\sqrt{2}} = \dfrac{1}{2\sqrt{2}}$

$\qquad = \dfrac{1}{2\sqrt{2}} \cdot \dfrac{\sqrt{2}}{\sqrt{2}} = \dfrac{\sqrt{2}}{2\sqrt{4}} = \dfrac{\sqrt{2}}{2(2)} = \dfrac{\sqrt{2}}{4}$

85. c: hypotenuse

$$5^2 + 5^2 = c^2$$

$$c^2 = 25 + 25 = 50$$

$$c = \sqrt{50} = \sqrt{25 \cdot 2} = 5\sqrt{2}$$

$5\sqrt{2}$ meters.

87. Legs: 3x, 4x

$$(3x)^2 + (4x)^2 = \left(\sqrt{200}\right)^2$$

$$9x^2 + 16x^2 = 200$$

$$25x^2 = 200$$

$$x^2 = 8$$

$$x = \sqrt{8} = \sqrt{4 \cdot 2} = 2\sqrt{2}$$

Legs: $3x = 3\left(2\sqrt{2}\right) = 6\sqrt{2}$ cm

$$4x = 4\left(2\sqrt{2}\right) = 8\sqrt{2} \text{ cm}$$

99. $\left(7.4 \times 10^{10}\right)\left(9.6 \times 10^{-4}\right) = 71.04 \times 10^6$

$$= 7.104 \times 10 \times 10^6 = 7.104 \times 10^7$$

100. $\left(x^2 y^2 - 3xy + 7\right) + \left(-4x^2 y^2 - 2xy - 9\right)$

$$\qquad\qquad - \left(3x^2 y^2 - 5xy - 2\right)$$

$$= -3x^2 y^2 - 5xy - 2 - 3x^2 y^2 + 5xy + 2$$

$$= -6x^2 y^2$$

101. x: ounces of solution A

40 − x: ounces of solution B

$$.4x + .15(40 - x) = .25(40)$$

$$.4x + 6 - .15x = 10$$

$$.25x + 6 = 10$$

$$.25x = 4$$

$$x = 16$$

Solution A: 16 ounces

Solution B: 40 − 16 = 24 ounces

Problem set 5.3.2

1. $\sqrt{\dfrac{3x}{7y}} = \dfrac{\sqrt{3x}}{\sqrt{7y}} \cdot \dfrac{\sqrt{7y}}{\sqrt{7y}} = \dfrac{\sqrt{21xy}}{\sqrt{(7y)^2}} = \dfrac{\sqrt{21xy}}{7y}$

3. $\sqrt{\dfrac{3}{32y^2}} = \dfrac{\sqrt{3}}{\sqrt{16 \cdot 2y^2}} = \dfrac{\sqrt{3}}{4y\sqrt{2}}$

$$= \dfrac{\sqrt{3}}{4y\sqrt{2}} \cdot \dfrac{\sqrt{2}}{\sqrt{2}} = \dfrac{\sqrt{6}}{4y(2)} = \dfrac{\sqrt{6}}{8y}$$

7. $\sqrt{\dfrac{11}{20y^3}} = \dfrac{\sqrt{11}}{\sqrt{4 \cdot 5y^2 y}} = \dfrac{\sqrt{11}}{2y\sqrt{5y}}$

$$= \dfrac{\sqrt{11}}{2y\sqrt{5y}} \cdot \dfrac{\sqrt{5y}}{\sqrt{5y}} = \dfrac{\sqrt{55y}}{2y(5y)} = \dfrac{\sqrt{55y}}{10y^2}$$

13. $\sqrt{\dfrac{3x}{32y^3}} = \dfrac{\sqrt{3x}}{\sqrt{16 \cdot 2y^2 y}} = \dfrac{\sqrt{3x}}{4y\sqrt{2y}}$

$$= \dfrac{\sqrt{3x}}{4y\sqrt{2y}} \cdot \dfrac{\sqrt{2y}}{\sqrt{2y}} = \dfrac{\sqrt{6xy}}{4y\sqrt{(2y)^2}}$$

$$= \dfrac{\sqrt{6xy}}{(4y)(2y)} = \dfrac{\sqrt{6xy}}{8y^2}$$

15. $\dfrac{3x}{\sqrt{8x^3}} = \dfrac{3x}{\sqrt{4 \cdot 2x^2 x}} = \dfrac{3x}{2x\sqrt{2x}}$

$$= \dfrac{3}{2\sqrt{2x}} \cdot \dfrac{\sqrt{2x}}{\sqrt{2x}} = \dfrac{3\sqrt{2x}}{2(2x)} = \dfrac{3\sqrt{2x}}{4x}$$

17. $\sqrt{\dfrac{27xy^3}{32z^5}} = \dfrac{\sqrt{9 \cdot 3xy^2 y}}{\sqrt{16 \cdot 2(z^2)^2 z}} = \dfrac{3y\sqrt{3xy}}{4z^2\sqrt{2z}}$

$= \dfrac{3y\sqrt{3xy}}{4z^2\sqrt{2z}} \cdot \dfrac{\sqrt{2z}}{\sqrt{2z}} = \dfrac{3y\sqrt{6xyz}}{4z^2(2z)}$

$= \dfrac{3y\sqrt{6xyz}}{8z^3}$

19. $\dfrac{27xy^3}{\sqrt{16 \cdot 2(x^2)^2 x}} = \dfrac{27xy^3}{4x^2\sqrt{2x}} = \dfrac{27y^3}{4x\sqrt{2x}}$

$= \dfrac{27y^3}{4x\sqrt{2x}} \cdot \dfrac{\sqrt{2x}}{\sqrt{2x}} = \dfrac{27y^3\sqrt{2x}}{(4x)(2x)}$

$= \dfrac{27y^3\sqrt{2x}}{8x^2}$

23. $\dfrac{\sqrt{20x^2 y}}{\sqrt{5x^3 y^3}} = \dfrac{\sqrt{4 \cdot 5x^2 y}}{\sqrt{5x^2 xy^2 y}} = \dfrac{2x\sqrt{5y}}{xy\sqrt{5xy}}$

$= \dfrac{2\sqrt{5y}}{y\sqrt{5xy}} \cdot \dfrac{\sqrt{5xy}}{\sqrt{5xy}} = \dfrac{2\sqrt{25xy^2}}{y(5xy)}$

$= \dfrac{2(5y)\sqrt{x}}{5xy^2} = \dfrac{2\sqrt{x}}{xy}$

25. $\sqrt[3]{\dfrac{7}{2x}} = \dfrac{\sqrt[3]{7}}{\sqrt[3]{2x}} \cdot \dfrac{\sqrt[3]{(2x)^2}}{\sqrt[3]{(2x)^2}} = \dfrac{\sqrt[3]{7(2x)^2}}{\sqrt[3]{(2x)^3}}$

$= \dfrac{\sqrt[3]{28x^2}}{2x}$

27. $\dfrac{7}{\sqrt[3]{2x^2}} \cdot \dfrac{\sqrt[3]{2^2 x}}{\sqrt[3]{2^2 x}} = \dfrac{7\sqrt[3]{4x}}{\sqrt[3]{2^3 x^3}} = \dfrac{7\sqrt[3]{4x}}{2x}$

29. $\dfrac{7x^2}{\sqrt[3]{2x^5}} = \dfrac{7x^2}{\sqrt[3]{2x^3 x^2}} = \dfrac{7x^2}{x\sqrt[3]{2x^2}}$

$= \dfrac{7x}{\sqrt[3]{2x^2}} \cdot \dfrac{\sqrt[3]{2^2 x}}{\sqrt[3]{2^2 x}} = \dfrac{7x\sqrt[3]{4x}}{2x} = \dfrac{7\sqrt[3]{4x}}{2}$

31. $\dfrac{2}{\sqrt[3]{8 \cdot 2x^2 y}} = \dfrac{2}{2\sqrt[3]{2x^2 y}}$

$= \dfrac{1}{\sqrt[3]{2x^2 y}} \cdot \dfrac{\sqrt[3]{2^2 xy^2}}{\sqrt[3]{2^2 xy^2}} = \dfrac{\sqrt[3]{2^2 xy^2}}{\sqrt{2^3 x^3 y^3}}$

$= \dfrac{\sqrt[3]{4xy^2}}{2xy}$

33. $\sqrt[5]{\dfrac{3}{2^3 x^3}} = \dfrac{\sqrt[5]{3}}{\sqrt[5]{2^3 x^3}} \cdot \dfrac{\sqrt[5]{2^2 x^2}}{\sqrt[5]{2^2 x^2}}$

$= \dfrac{\sqrt[5]{3 \cdot 2^2 x^2}}{\sqrt[5]{2^5 x^5}} = \dfrac{\sqrt[5]{12x^2}}{2x}$

35. $\dfrac{7}{\sqrt[5]{2^5 x^5 x^2 y^4}} = \dfrac{7}{2x\sqrt[5]{x^2 y^4}} \cdot \dfrac{\sqrt[5]{x^3 y}}{\sqrt[5]{x^3 y}}$

$= \dfrac{7\sqrt[5]{x^3 y}}{2x\sqrt[5]{x^5 y^5}} = \dfrac{7\sqrt[5]{x^3 y}}{2x(xy)} = \dfrac{7\sqrt[5]{x^3 y}}{2x^2 y}$

37. $\sqrt{1 + \dfrac{1}{y}} = \sqrt{\dfrac{y}{y} + \dfrac{1}{y}} = \sqrt{\dfrac{y+1}{y}}$

$= \dfrac{\sqrt{y+1}}{\sqrt{y}} \cdot \dfrac{\sqrt{y}}{\sqrt{y}} = \dfrac{\sqrt{y^2+y}}{y}$

39. $\sqrt{y^2 + \dfrac{1}{y^3}} = \sqrt{y^2 \cdot \dfrac{y^3}{y^3} + \dfrac{1}{y^3}} = \sqrt{\dfrac{y^{5}+1}{y^3}}$

$= \dfrac{\sqrt{y^{5}+1}}{\sqrt{y^{2} \cdot y}} = \dfrac{\sqrt{y^{5}+1}}{y\sqrt{y}} \cdot \dfrac{\sqrt{y}}{\sqrt{y}} = \dfrac{\sqrt{y^{6}+y}}{y^2}$

43. $\sqrt{9x - 27y} = \sqrt{9(x-3y)} = 3\sqrt{x-3y}$

45. $\sqrt{x^2 - 18x + 81} = \sqrt{(x-9)^2} = x - 9$

49. $P = \dfrac{8000}{\sqrt[3]{(x+1)^2}} \cdot \dfrac{\sqrt[3]{(x+1)}}{\sqrt[3]{(x+1)}}$

$= \dfrac{8000\sqrt[3]{(x+1)}}{\sqrt[3]{(x+1)^3}} = \dfrac{8000\sqrt[3]{(x+1)}}{x+1}$

51. $S = \dfrac{2xyz}{\sqrt[5]{2x^4y^2}} \cdot \dfrac{\sqrt[5]{2^4xy^3}}{\sqrt[5]{2^4xy^3}}$

$= \dfrac{2xyz\sqrt[5]{16xy^3}}{\sqrt[5]{2^5x^5y^5}} = \dfrac{2xyz\sqrt[5]{16xy^3}}{2xy}$

$= z\sqrt[5]{16xy^3}$

53. $\dfrac{\sqrt[3]{3v}}{\sqrt[3]{4\pi}} \cdot \dfrac{\sqrt[3]{2\pi^2}}{\sqrt[3]{2\pi^2}} = \dfrac{\sqrt[3]{6v\pi^2}}{\sqrt[3]{8\pi^3}} = \dfrac{\sqrt[3]{6v\pi^2}}{2\pi}$

54. $\dfrac{8x^3y^2}{-24x^7y^{-5}} = -\dfrac{1}{3}x^{-4}y^7 = \dfrac{-y^7}{3x^4}$

55.

$$
\begin{array}{r}
2y^2 - 6y + 7 \\
2y-5\overline{\smash{\big)}\,4y^3 - 22y^2 + 44y - 35} \\
\underline{4y^3 - 10y^2} \\
-12y^2 + 44y \\
\underline{-12y^2 + 30y} \\
14y - 35 \\
\underline{14y - 35} \\
0
\end{array}
$$

56. $\dfrac{2y-1}{(y+2)(y-2)} = \dfrac{y+6}{(y+2)(y+1)} + \dfrac{y+3}{(y+1)(y-2)}$

$(y+2)(y-2)(y+1) \cdot \dfrac{2y-1}{(y+2)(y-2)}$

$= (y+2)(y-2)(y+1)$

$\qquad \cdot \left[\dfrac{y+6}{(y+2)(y+1)} + \dfrac{y+3}{(y+1)(y-2)}\right]$

$(y+1)(2y-1) = (y-2)(y+6)$

$\qquad\qquad\qquad + (y+2)(y+3)$

$2y^2 + y - 1 = y^2 + 4y - 12 + y^2 + 5y + 6$

$2y^2 + y - 1 = 2y^2 + 9y - 6$

$y - 1 = 9y - 6$

$-8y = -5$

$y = \dfrac{-5}{-8} = \dfrac{5}{8} \qquad \left\{\dfrac{5}{8}\right\}$

Problem set 5.3.3

3. $\sqrt{28}\sqrt{\dfrac{3}{7}} = \sqrt{(28)\left(\dfrac{3}{7}\right)} = \sqrt{12} = \sqrt{4 \cdot 3}$

$= 2\sqrt{3}$

9. $\left(-8\sqrt{12}\right)\left(\frac{3}{2}\sqrt{2}\right) = (-8)\left(\frac{3}{2}\right)\sqrt{12 \cdot 2}$

$= -12\sqrt{24} = -12\sqrt{4 \cdot 6} = -12(2)\sqrt{6}$

$= -24\sqrt{6}$

13. $\left(4\sqrt[3]{5}\right)\left(6\sqrt[3]{50}\right) = 24\sqrt[3]{250} = 24\sqrt[3]{125 \cdot 2}$

$= 24(5)\sqrt[3]{2} = 120\sqrt[3]{2}$

17. $\dfrac{\sqrt{54x^3}}{\sqrt{6x}} = \sqrt{\dfrac{54x^3}{6x}} = \sqrt{9x^2} = 3x$

23. $\dfrac{18\sqrt{20} + 16\sqrt{28}}{2\sqrt{2}} = \dfrac{18\sqrt{20}}{2\sqrt{2}} + \dfrac{16\sqrt{28}}{2\sqrt{2}}$

$= 9\sqrt{10} + 8\sqrt{14}$

27. $\sqrt{12x^2 y}\sqrt{5xy} = \sqrt{60x^3 y^2}$

$= \sqrt{4 \cdot 15x^2 \, xy^2} = 2xy\sqrt{15x}$

29. $\sqrt[3]{4x^2 y}\sqrt[3]{2xy} = \sqrt[3]{8x^3 y^2} = 2x\sqrt[3]{y^2}$

31. $\left(-2xy^2\sqrt{3x}\right)\left(xy\sqrt{6x}\right)$

$= -2x^2 y^3\sqrt{9 \cdot 2x^2} = -2x^2 y^3 (3x)\sqrt{2}$

$= -6x^3 y^3\sqrt{2}$

33. $\left(2x^2 y\sqrt[4]{8xy}\right)\left(-3xy^2\sqrt[4]{2x^2 y^3}\right)$

$= -6x^3 y^3\sqrt[4]{16x^3 y^4}$

$= -6x^3 y^3 (2y)\sqrt[4]{x^3} = -12x^3 y^4\sqrt[4]{x^3}$

35. $\dfrac{\sqrt{200x^3}}{\sqrt{10x^{-1}}} = \sqrt{\dfrac{200x^3}{10x^{-1}}} = \sqrt{20x^4}$

$= \sqrt{4 \cdot 5x^4} = 2x^2\sqrt{5}$

37. $\dfrac{\sqrt[3]{108x^4 y^5}}{\sqrt[3]{2xy^3}} = \sqrt[3]{\dfrac{108x^4 y^5}{2xy^3}} = \sqrt[3]{54x^3 y^2}$

$= \sqrt[3]{27 \cdot 2x^3 y^2} = 3x\sqrt[3]{2y^2}$

39. $\dfrac{\sqrt{98x^2 y}}{\sqrt{2x^{-3}y}} = \sqrt{\dfrac{98x^2 y}{2x^{-3}y}} = \sqrt{49x^5}$

$= \sqrt{49\left(x^2\right)^2 x} = 7x^2\sqrt{x}$

41. $\dfrac{\sqrt{x^2 - y^2}}{\sqrt{x-y}} = \sqrt{\dfrac{x^2 - y^2}{x-y}} = \sqrt{\dfrac{(x+y)(x-y)}{x-y}}$

$= \sqrt{x+y}$

43. $\dfrac{\sqrt{a^2 - 4b^2}}{\sqrt{a+2b}} = \sqrt{\dfrac{(a+2b)(a-2b)}{a+2b}} = \sqrt{a-2b}$

45. $\qquad H = \dfrac{62.4Ns}{33,000}$

$33,000H = 62.4Ns$

$\dfrac{33,000H}{62.4N} = s$

$s = \dfrac{33,000(468)}{62.4(1500)} = \dfrac{15,444,000}{93,600} = 165$

The dam is 165 feet high.

46. $\dfrac{\left(x^a y^{\frac{1}{a}}\right)^a}{\left(x^{\frac{1}{a}} y^a\right)^{a^2}} = \dfrac{x^{a^2} y}{x^a y^{a^3}} = x^{a^2-a} y^{1-a^3}$

47. $\dfrac{\dfrac{1}{y^2} - \dfrac{1}{100}}{\dfrac{1}{y} - \dfrac{1}{10}} \cdot \dfrac{100y^2}{100y^2} = \dfrac{100 - y^2}{100y - 10y^2}$

$\quad = \dfrac{(10+y)(10-y)}{10y(10-y)} = \dfrac{10+y}{10y}$

Problem set 5.3.4

7. $\sqrt{2} - \sqrt{11} + 6\sqrt{2} + 4\sqrt{11}$
$\quad = \sqrt{2} + 6\sqrt{2} - \sqrt{11} + 4\sqrt{11}$
$\quad = (1+6)\sqrt{2} + (-1+4)\sqrt{11}$
$\quad = 7\sqrt{2} + 3\sqrt{11}$

11. $\sqrt{50} + \sqrt{18} = \sqrt{25 \cdot 2} + \sqrt{9 \cdot 2}$
$\quad = 5\sqrt{2} + 3\sqrt{2} = 8\sqrt{2}$

15. $3\sqrt{8} - \sqrt{32} + 3\sqrt{72} - \sqrt{75}$
$\quad = 3\sqrt{4 \cdot 2} - \sqrt{16 \cdot 2} + 3\sqrt{36 \cdot 2} - \sqrt{25 \cdot 3}$
$\quad = 3(2)\sqrt{2} - 4\sqrt{2} + 3(6)\sqrt{2} - 5\sqrt{3}$
$\quad = 6\sqrt{2} - 4\sqrt{2} + 18\sqrt{2} - 5\sqrt{3}$
$\quad = 20\sqrt{2} - 5\sqrt{3}$

17. $8\sqrt{\dfrac{1}{2}} - \dfrac{1}{2}\sqrt{8} = 8 \cdot \dfrac{1}{\sqrt{2}} \cdot \dfrac{\sqrt{2}}{\sqrt{2}} - \dfrac{1}{2}\sqrt{4 \cdot 2}$
$\quad = \dfrac{8\sqrt{2}}{2} - \dfrac{1}{2}(2)\sqrt{2} = 4\sqrt{2} - \sqrt{2} = 3\sqrt{2}$

19. $\dfrac{\sqrt{63}}{3} + 7\sqrt{3} = \dfrac{\sqrt{9 \cdot 7}}{3} + 7\sqrt{3}$
$\quad = \dfrac{3\sqrt{7}}{3} + 7\sqrt{3} = \sqrt{7} + 7\sqrt{3}$

21. $\sqrt{25x} + \sqrt{16x} = 5\sqrt{x} + 4\sqrt{x} = 9\sqrt{x}$

23. $\sqrt[4]{32} + 3\sqrt[4]{1250}$
$\quad = \sqrt[4]{16 \cdot 2} + 3\sqrt[4]{625 \cdot 2}$
$\quad = 2\sqrt[4]{2} + 3 \cdot 5\sqrt[4]{2} = 2\sqrt[4]{2} + 15\sqrt[4]{2}$
$\quad = 17\sqrt[4]{2}$

27. $\dfrac{1}{4}\sqrt{2} + \dfrac{2}{3}\sqrt{8} = \dfrac{\sqrt{2}}{4} + \dfrac{2}{3}\sqrt{4 \cdot 2}$
$\quad = \dfrac{\sqrt{2}}{4} + \dfrac{2 \cdot 2\sqrt{2}}{3} = \dfrac{\sqrt{2}}{4} + \dfrac{4\sqrt{2}}{3}$
$\quad = \dfrac{3\sqrt{2}}{12} + \dfrac{16\sqrt{2}}{12} = \dfrac{19\sqrt{2}}{12}$

29. $\dfrac{\sqrt{45}}{4} - \sqrt{80} + \dfrac{\sqrt{20}}{3}$
$\quad = \dfrac{\sqrt{9 \cdot 5}}{4} - \sqrt{16 \cdot 5} + \dfrac{\sqrt{4 \cdot 5}}{3}$
$\quad = \dfrac{3\sqrt{5}}{4} - \dfrac{4\sqrt{5}}{1} + \dfrac{2\sqrt{5}}{3}$
$\quad = \dfrac{9\sqrt{5}}{12} - \dfrac{48\sqrt{5}}{12} + \dfrac{8\sqrt{5}}{12} = \dfrac{-31\sqrt{5}}{12}$

31. $7\sqrt[3]{2}+8\sqrt[3]{16}-2\sqrt[3]{54}$

$\quad = 7\sqrt[3]{2}+8\sqrt[3]{8\cdot 2}-2\sqrt[3]{27\cdot 2}$

$\quad = 7\sqrt[3]{2}+8(2)\sqrt[3]{2}-2(3)\sqrt[3]{2}$

$\quad = 7\sqrt[3]{2}+16\sqrt[3]{2}-6\sqrt[3]{2}=17\sqrt[3]{2}$

35. $5\sqrt{12x}-2\sqrt{3x}=5\sqrt{4\cdot 3x}-2\sqrt{3x}$

$\quad = 10\sqrt{3x}-2\sqrt{3x}=8\sqrt{3x}$

37. $\sqrt[3]{54x^5}+2x\sqrt[3]{16x^2}-7\sqrt[3]{2x^5}$

$\quad = \sqrt[3]{27\cdot 2x^3x^2}$

$\quad\quad\quad +2x\sqrt[3]{8\cdot 2x^2}-7\sqrt[3]{2x^3x^2}$

$\quad = 3x\sqrt[3]{2x^2}+4x\sqrt[3]{2x^2}-7x\sqrt[3]{2x^2}=0$

39. $16\sqrt{\dfrac{5}{8}}+6\sqrt{\dfrac{5}{2}}=\dfrac{16\sqrt{5}}{\sqrt{8}}\cdot\dfrac{\sqrt{2}}{\sqrt{2}}+\dfrac{6\sqrt{5}}{\sqrt{2}}\cdot\dfrac{\sqrt{2}}{\sqrt{2}}$

$\quad = \dfrac{16\sqrt{10}}{4}+\dfrac{6\sqrt{10}}{2}=4\sqrt{10}+3\sqrt{10}=7\sqrt{10}$

41. $12\sqrt{\dfrac{2}{3}}+24\sqrt{\dfrac{1}{6}}=\dfrac{12\sqrt{2}}{\sqrt{3}}+\dfrac{24}{\sqrt{6}}$

$\quad = \dfrac{12\sqrt{2}}{\sqrt{3}}\cdot\dfrac{\sqrt{3}}{\sqrt{3}}+\dfrac{24}{\sqrt{6}}\cdot\dfrac{\sqrt{6}}{\sqrt{6}}$

$\quad = \dfrac{12\sqrt{6}}{3}+\dfrac{24\sqrt{6}}{6}=4\sqrt{6}+4\sqrt{6}=8\sqrt{6}$

43. $2x\sqrt{\dfrac{y^3}{x}}-2y^2\sqrt{\dfrac{x}{y}}=\dfrac{2x\sqrt{y^2\cdot y}}{\sqrt{x}}-2y^2\sqrt{\dfrac{x}{y}}$

$\quad = 2xy\dfrac{\sqrt{y}}{\sqrt{x}}\cdot\dfrac{\sqrt{x}}{\sqrt{x}}-2y^2\dfrac{\sqrt{x}}{\sqrt{y}}\cdot\dfrac{\sqrt{y}}{\sqrt{y}}$

$\quad = \dfrac{2xy\sqrt{xy}}{x}-\dfrac{2y^2\sqrt{xy}}{y}$

$\quad = 2y\sqrt{xy}-2y\sqrt{xy}=0$

47. $\sqrt{\dfrac{5}{2}}-\sqrt{\dfrac{9}{10}}+15\sqrt{\dfrac{2}{5}}$

$\quad = \dfrac{\sqrt{5}}{\sqrt{2}}\cdot\dfrac{\sqrt{2}}{\sqrt{2}}-\dfrac{3}{\sqrt{10}}\cdot\dfrac{\sqrt{10}}{\sqrt{10}}+\dfrac{15\sqrt{2}}{\sqrt{5}}\cdot\dfrac{\sqrt{5}}{\sqrt{5}}$

$\quad = \dfrac{\sqrt{10}}{2}-\dfrac{3\sqrt{10}}{10}+\dfrac{15\sqrt{10}}{5}$

$\quad = \dfrac{5\sqrt{10}}{10}-\dfrac{3\sqrt{10}}{10}+\dfrac{30\sqrt{10}}{10}=\dfrac{32\sqrt{10}}{10}$

$\quad = \dfrac{16\sqrt{10}}{5}$

49. $7\sqrt{3}+\sqrt[4]{12^2}=7\sqrt{3}+\left(12^2\right)^{\frac{1}{4}}$

$\quad = 7\sqrt{3}+12^{\frac{1}{2}}=7\sqrt{3}+\sqrt{12}$

$\quad = 7\sqrt{3}+\sqrt{4\cdot 3}=7\sqrt{3}+2\sqrt{3}=9\sqrt{3}$

51. $8m\sqrt{m}-\sqrt[10]{m^{15}}=8m\sqrt{m}-\left(m^{15}\right)^{\frac{1}{10}}$

$\quad = 8m\sqrt{m}-m^{\frac{3}{2}}=8m\sqrt{m}-\sqrt{m^3}$

$\quad = 8m\sqrt{m}-\sqrt{m^2\cdot m}=8m\sqrt{m}-m\sqrt{m}$

$\quad = 7m\sqrt{m}$

53. $\sqrt{45}+\sqrt[4]{25}=\sqrt{9\cdot 5}+\left(5^2\right)^{\frac{1}{4}}$

$\quad = 3\sqrt{5}+5^{\frac{1}{2}}=3\sqrt{5}+\sqrt{5}=4\sqrt{5}$

55. x: number

$\quad x-\left(2\sqrt{18}-\sqrt{50}\right)=\sqrt{2}$

$\quad x-2\sqrt{18}+\sqrt{50}=\sqrt{2}$

$\quad x-2\sqrt{9\cdot 2}+\sqrt{25\cdot 2}=\sqrt{2}$

$\quad x-6\sqrt{2}+5\sqrt{2}=\sqrt{2}$

$\quad x-\sqrt{2}=\sqrt{2}$

$\quad x=\sqrt{2}+\sqrt{2}$

$\quad x=2\sqrt{2}$; The number is $2\sqrt{2}$.

57. $\left|\dfrac{3x+\sqrt{32}}{2}\right| = \sqrt{50}$

$\left|\dfrac{3x+4\sqrt{2}}{2}\right| = 5\sqrt{2}$

$\dfrac{3x+4\sqrt{2}}{2} = 5\sqrt{2}$ or $\dfrac{3x+4\sqrt{2}}{2} = -5\sqrt{2}$

$3x+4\sqrt{2} = 10\sqrt{2}$ $3x+4\sqrt{2} = -10\sqrt{2}$

$3x = 6\sqrt{2}$ $3x = -14\sqrt{2}$

$x = 2\sqrt{2}$ $x = \dfrac{-14}{3}\sqrt{2}$

$$\left\{2\sqrt{2}, \ \dfrac{-14}{3}\sqrt{2}\right\}$$

59. $P = 2L + 2W = 2 \cdot \dfrac{12}{\sqrt{6}} + 2 \cdot 7\sqrt{\dfrac{3}{2}}$

$= \dfrac{24}{\sqrt{6}} \cdot \dfrac{\sqrt{6}}{\sqrt{6}} + \dfrac{14\sqrt{3}}{\sqrt{2}} \cdot \dfrac{\sqrt{2}}{\sqrt{2}}$

$= \dfrac{24\sqrt{6}}{6} + \dfrac{14\sqrt{6}}{2} = 4\sqrt{6} + 7\sqrt{6} = 11\sqrt{6}$

60. $128a^3b - 50ab = 2ab\left(64a^2 - 25\right)$

$= 2ab(8a+5)(8a-5)$

61. $\dfrac{3}{y-1} - \dfrac{2}{y+1} + \dfrac{2y}{(y+1)(y-1)}$

      {LCD is $(y+1)(y-1)$}

$= \dfrac{3}{y-1} \cdot \dfrac{y+1}{y+1} - \dfrac{2}{y+1} \cdot \dfrac{y-1}{y-1} + \dfrac{2y}{(y+1)(y-1)}$

$= \dfrac{3(y+1) - 2(y-1) + 2y}{(y+1)(y-1)}$

$= \dfrac{3y+3-2y+2+2y}{(y+1)(y-1)} = \dfrac{3y+5}{(y+1)(y-1)}$

62. $\left(4x^5y^{10}\right)\left(-2x^3y^2\right)^{-3}$

$= \left(4x^5y^{10}\right)(-2)^{-3}x^{-9}y^{-6}$

$= (4)(-2)^{-3}x^{-4}y^4 = \dfrac{4y^4}{(-2)^3x^4}$

$= \dfrac{4y^4}{-8x^4} = -\dfrac{y^4}{2x^4}$

Problem set 5.3.5

5. $5\sqrt{6}\left(7\sqrt{8} - 2\sqrt{12}\right) = 35\sqrt{48} - 10\sqrt{72}$

$= 35\sqrt{16\cdot 3} - 10\sqrt{36\cdot 2}$

$= 35(4)\sqrt{3} - 10(6)\sqrt{2}$

$= 140\sqrt{3} - 60\sqrt{2}$

9. $\sqrt{2x}\left(\sqrt{6x} - 3\sqrt{x}\right) = \sqrt{12x^2} - 3\sqrt{2x^2}$

$= \sqrt{4\cdot 3x^2} - 3\sqrt{2x^2} = 2x\sqrt{3} - 3x\sqrt{2}$

11. $\sqrt{3x}\left(2y\sqrt{12xy} + 2\sqrt{12xy^3}\right)$

$= 2y\sqrt{36x^2y} + 2\sqrt{36x^2y^3}$

$= 12xy\sqrt{y} + 12xy\sqrt{y} = 24xy\sqrt{y}$

13. $\sqrt{3}\left(1 - \sqrt{5x}\right) + \sqrt{5}\left(\sqrt{3x} - 1\right)$

$= \sqrt{3} - \sqrt{15x} + \sqrt{15x} - \sqrt{5} = \sqrt{3} - \sqrt{5}$

17. $4\sqrt[3]{3}\left(9\sqrt[3]{9} + 7\sqrt[3]{7}\right) = 36\sqrt[3]{27} + 28\sqrt[3]{21}$

$= 36(3) + 28\sqrt[3]{21} = 108 + 28\sqrt[3]{21}$

19. $\left(\sqrt{2}+\sqrt{7}\right)\left(\sqrt{3}+\sqrt{5}\right)$

$= \sqrt{6} + \sqrt{10} + \sqrt{21} + \sqrt{35}$

25. $\left(3\sqrt{2}-2\sqrt{8}\right)\left(2\sqrt{3}-4\sqrt{5}\right)$

$= 6\sqrt{6} - 12\sqrt{10} - 4\sqrt{24} + 8\sqrt{40}$

$= 6\sqrt{6} - 12\sqrt{10} - 4\sqrt{4\cdot 6} + 8\sqrt{4\cdot 10}$

$= 6\sqrt{6} - 12\sqrt{10} - 4(2)\sqrt{6} + 8(2)\sqrt{10}$

$= 6\sqrt{6} - 12\sqrt{10} - 8\sqrt{6} + 16\sqrt{10}$

$= 4\sqrt{10} - 2\sqrt{6}$

27. $\left(\sqrt{5}+7\right)\left(\sqrt{5}-\sqrt{7}\right) = \sqrt{25} - 49$

$= 5 - 49 = -44$

29. $\left(2+5\sqrt{3}\right)\left(2-5\sqrt{3}\right) = 4 - 25\sqrt{9}$

$= 4 - 75 = -71$

31. $\left(\sqrt{7}+\sqrt{5}\right)\left(\sqrt{7}-\sqrt{5}\right) = \sqrt{49} - \sqrt{25}$

$= 7 - 5 = 2$

37. $\left(\sqrt{2x}+\sqrt{50}\right)^2 = \left(\sqrt{2x}+\sqrt{25\cdot 2}\right)^2$

$= \left(\sqrt{2x}+5\sqrt{2}\right)^2$

$= \left(\sqrt{2x}+5\sqrt{2}\right)\left(\sqrt{2x}+5\sqrt{2}\right)$

$= \left(\sqrt{2x}\right)^2 + 5\sqrt{4x} + 5\sqrt{4x} + 25\left(\sqrt{2}\right)^2$

$= 2x + 10\sqrt{4x} + 25\cdot 2$

$= 2x + 10\cdot 2\sqrt{x} + 50$

$= 2x + 20\sqrt{x} + 50$

39. $\left(\sqrt{2a+b}+\sqrt{c}\right)\left(\sqrt{2a+b}-\sqrt{c}\right)$

$= \left(\sqrt{2a+b}\right)^2 - \left(\sqrt{c}\right)^2 = 2a+b-c$

41. $\left(\sqrt{x-4}+\sqrt{x+4}\right)^2$

$= \left(\sqrt{x-4}+\sqrt{x+4}\right)\left(\sqrt{x-4}+\sqrt{x+4}\right)$

$= x-4+\sqrt{x-4}\sqrt{x+4}+\sqrt{x-4}\sqrt{x+4}$
$\qquad\qquad +x+4$

$= 2x + 2\sqrt{(x-4)(x+4)} = 2x + 2\sqrt{x^2-16}$

45. $\left(y-\sqrt{2y-5}\right)^2$

$= \left(y-\sqrt{2y-5}\right)\left(y-\sqrt{2y-5}\right)$

$= y^2 - y\sqrt{2y-5} - y\sqrt{2y-5} + 2y - 5$

$= y^2 + 2y - 5 - 2y\sqrt{2y-5}$

47. $\left(x+\sqrt{1-x^2}\right)^2$

$= \left(x+\sqrt{1-x^2}\right)\left(x+\sqrt{1-x^2}\right)$

$= x^2 + x\sqrt{1-x^2} + x\sqrt{1-x^2} + 1 - x^2$

$= 2x\sqrt{1-x^2} + 1$

49. $7\left[(2y-5)-(y+1)\right] = \left(\sqrt{7}-2\right)\left(\sqrt{7}+2\right)$

$7\left[2y-5-y-1\right] = \sqrt{49} - 4$

$7(y-6) = 7 - 4$

$7y - 42 = 3$

$7y = 45$

$y = \dfrac{45}{7} \qquad \left\{\dfrac{45}{7}\right\}$

51. Consecutive integers $x, x+1, x+2, x+3$

$x+(x+1)+(x+2)+(x+3)$

$$= \left(\sqrt{27}+1\right)\left(\sqrt{27}-1\right)$$

$$4x+6 = 27-1$$

$$4x+6 = 26$$

$$4x = 20$$

$$x = 5$$

Integers are $5, 6, 7,$ and 8.

53. Width: x Length: $x+4$

Perimeter $= 2W + 2L$

$$2x+2(x+4) = \left(\sqrt{37}+1\right)\left(\sqrt{37}-1\right)$$

$$2x+2x+8 = 37-1$$

$$4x+8 = 36$$

$$4x = 28$$

$$x = 7$$

Width: 7 inches

Length: $7+4 = 11$ inches

55. $\left|\dfrac{3x+3}{5}\right| > \left(\sqrt{5}+\sqrt{3}\right)\left(\sqrt{5}-\sqrt{3}\right)$

$\left|\dfrac{3x+3}{5}\right| > \sqrt{25}-\sqrt{9}$

$\left|\dfrac{3x+3}{5}\right| > 5-3$

$\left|\dfrac{3x+3}{5}\right| > 2$

$\dfrac{3x+3}{5} > 2$ or $\dfrac{3x+3}{5} < -2$

$3x+3 > 10$ $3x+3 < -10$

$3x > 7$ $3x < -13$

$x > \dfrac{7}{3}$ $x < \dfrac{-13}{3}$

$\left\{x \mid x > \dfrac{7}{3} \text{ or } x < \dfrac{-13}{3}\right\}$

57. $5x^2 - 8 = -4x$

$$5\left(\frac{-2+2\sqrt{11}}{5}\right)^2 - 8 \overset{?}{=} -4\left(\frac{-2+2\sqrt{11}}{5}\right)$$

$$5\frac{\left(-2+2\sqrt{11}\right)\left(-2+2\sqrt{11}\right)}{25} - 8$$

$$\overset{?}{=} \frac{8-8\sqrt{11}}{5}$$

$$\frac{5\left(4-8\sqrt{11}+4\cdot 11\right)}{25} - 8 \overset{?}{=} \frac{8-8\sqrt{11}}{5}$$

$$\frac{48-8\sqrt{11}}{5} - 8 \overset{?}{=} \frac{8-8\sqrt{11}}{5}$$

$$\frac{48-8\sqrt{11}}{5} - \frac{40}{5} \overset{?}{=} \frac{8-8\sqrt{11}}{5}$$

$$\frac{8-8\sqrt{11}}{5} = \frac{8-8\sqrt{11}}{5}$$

59. $\left(64x^3y^8\right)\left(-2x^2y^{-4}\right)^{-2}$

$$= \left(64x^3y^8\right)(-2)^{-2}x^{-4}y^8$$

$$= 64(-2)^{-2}x^{-1}y^{16}$$

$$= \frac{64y^{16}}{(-2)^2 x} = \frac{16y^{16}}{x}$$

60. $F = K\sqrt{T}$

$216 = K\sqrt{3}$

$\dfrac{216}{\sqrt{3}} = K$

$F = \dfrac{216}{\sqrt{3}}\sqrt{27} = 216\sqrt{\dfrac{27}{3}} = 216\sqrt{9}$

$F = 216(3) = 648$

648 times per second.

61. $\dfrac{2y+1}{9-y^2} = \dfrac{1-y}{y^2+4y+3} - \dfrac{y+2}{y^2-2y-3}$

$\dfrac{2y+1}{(3+y)(3-y)} = \dfrac{1-y}{(3+y)(y+1)} - \dfrac{y+2}{(y-3)(y+1)}$

$(y+3)(y-3)(y+1)\cdot \dfrac{2y+1}{(3+y)(3-y)}$

$= (y+3)(y-3)(y+1)$

$\cdot \left[\dfrac{1-y}{(3+y)(y+1)} - \dfrac{y+2}{(y-3)(y+1)} \right]$

$-(y+1)(2y+1) = (y-3)(1-y)$
$ -(y+3)(y+2)$

$-2y^2-3y-1 = 4y-y^2-3-y^2-5y-6$

$-2y^2-3y-1 = -2y^2-y-9$

$-3y-1 = -y-9$

$-2y = -8$

$y = 4 \qquad \{4\}$

Problem set 5.3.6

1. $\dfrac{8}{\sqrt{5}-2} \cdot \dfrac{\sqrt{5}+2}{\sqrt{5}+2} = \dfrac{8\left(\sqrt{5}+2\right)}{5-4} = 8\left(\sqrt{5}+2\right)$

$\left(\text{or } 8\sqrt{5}+16\right)$

5. $\dfrac{6}{\sqrt{5}+\sqrt{3}} \cdot \dfrac{\sqrt{5}-\sqrt{3}}{\sqrt{5}-\sqrt{3}} = \dfrac{6\left(\sqrt{5}-\sqrt{3}\right)}{5-3}$

$= \dfrac{6\left(\sqrt{5}-\sqrt{3}\right)}{2} = 3\left(\sqrt{5}-\sqrt{3}\right) = 3\sqrt{5}-3\sqrt{3}$

7. $\dfrac{11}{\sqrt{7}-\sqrt{3}} \cdot \dfrac{\sqrt{7}+\sqrt{3}}{\sqrt{7}+\sqrt{3}} = \dfrac{11\left(\sqrt{7}+\sqrt{3}\right)}{7-3}$

$= \dfrac{11\sqrt{7}+11\sqrt{3}}{4}$

---

9. $\dfrac{\sqrt{5}}{\sqrt{7}+\sqrt{3}} \cdot \dfrac{\sqrt{7}-\sqrt{3}}{\sqrt{7}-\sqrt{3}} = \dfrac{\sqrt{5}\left(\sqrt{7}-\sqrt{3}\right)}{7-3}$

$= \dfrac{\sqrt{35}-\sqrt{15}}{4}$

13. $\dfrac{8}{3+2\sqrt{2}} \cdot \dfrac{3-2\sqrt{2}}{3-2\sqrt{2}} = \dfrac{8\left(3-2\sqrt{2}\right)}{9-4\cdot 2}$

$= \dfrac{8\left(3-2\sqrt{2}\right)}{1} = 24-16\sqrt{2}$

15. $\dfrac{25}{5\sqrt{2}-3\sqrt{5}} \cdot \dfrac{5\sqrt{2}+3\sqrt{5}}{5\sqrt{2}+3\sqrt{5}} = \dfrac{25\left(5\sqrt{2}+3\sqrt{5}\right)}{25\cdot 2-9\cdot 5}$

$= \dfrac{25\left(5\sqrt{2}+3\sqrt{5}\right)}{5} = 5\left(5\sqrt{2}+3\sqrt{5}\right)$

$= 25\sqrt{2}+15\sqrt{5}$

17. $\dfrac{\sqrt{3}}{2\sqrt{3}+3\sqrt{5}} \cdot \dfrac{2\sqrt{3}-3\sqrt{5}}{2\sqrt{3}-3\sqrt{5}} = \dfrac{\sqrt{3}\left(2\sqrt{3}-3\sqrt{5}\right)}{4\cdot 3-9\cdot 5}$

$= \dfrac{2\cdot 3-3\sqrt{15}}{12-45} = \dfrac{6-3\sqrt{15}}{-33} = \dfrac{3\left(2-\sqrt{15}\right)}{-33}$

$= \dfrac{2-\sqrt{15}}{-11} \left[\text{or } -\dfrac{2-\sqrt{15}}{11} \text{ or } \dfrac{-2+\sqrt{15}}{11} \right]$

19. $\dfrac{7}{\sqrt{x}-5} \cdot \dfrac{\sqrt{x}+5}{\sqrt{x}+5} = \dfrac{7\left(\sqrt{x}+5\right)}{x-25} = \dfrac{7\sqrt{x}+35}{x-25}$

21. $\dfrac{\sqrt{y}}{\sqrt{y}+3} \cdot \dfrac{\sqrt{y}-3}{\sqrt{y}-3} = \dfrac{\sqrt{y}\left(\sqrt{y}-3\right)}{y-9} = \dfrac{y-3\sqrt{y}}{y-9}$

23. $\dfrac{2\sqrt{3}-1}{2\sqrt{3}+1} \cdot \dfrac{2\sqrt{3}-1}{2\sqrt{3}-1}$

$= \dfrac{4 \cdot 3 - 2\sqrt{3} - 2\sqrt{3} + 1}{4 \cdot 3 - 1} = \dfrac{13 - 4\sqrt{3}}{11}$

25. $\dfrac{\sqrt{5}+\sqrt{3}}{\sqrt{5}-\sqrt{3}} \cdot \dfrac{\sqrt{5}+\sqrt{3}}{\sqrt{5}+\sqrt{3}}$

$= \dfrac{5 + \sqrt{15} + \sqrt{15} + 3}{5 - 3} = \dfrac{8 + 2\sqrt{15}}{2}$

$= \dfrac{2\left(4 + \sqrt{15}\right)}{2} = 4 + \sqrt{15}$

27. $\dfrac{\sqrt{x}+1}{\sqrt{x}+3} \cdot \dfrac{\sqrt{x}-3}{\sqrt{x}-3} = \dfrac{x - 3\sqrt{x} + \sqrt{x} - 3}{x - 9}$

$= \dfrac{x - 2\sqrt{x} - 3}{x - 9}$

31. $\dfrac{3\sqrt{5}+2\sqrt{2}}{4\sqrt{5}+\sqrt{2}} \cdot \dfrac{4\sqrt{5}-\sqrt{2}}{4\sqrt{5}-\sqrt{2}}$

$= \dfrac{12 \cdot 5 - 3\sqrt{10} + 8\sqrt{10} - 2 \cdot 2}{16 \cdot 5 - 2} = \dfrac{56 + 5\sqrt{10}}{78}$

33. $\dfrac{2\sqrt{x}+\sqrt{y}}{4\sqrt{x}+3\sqrt{y}} \cdot \dfrac{4\sqrt{x}-3\sqrt{y}}{4\sqrt{x}-3\sqrt{y}}$

$= \dfrac{8x - 6\sqrt{xy} + 4\sqrt{xy} - 3y}{16x - 9y}$

$= \dfrac{8x - 2\sqrt{xy} - 3y}{16x - 9y}$

35. $\dfrac{x^3 - y^3}{\sqrt{x}+\sqrt{y}} \cdot \dfrac{\sqrt{x}-\sqrt{y}}{\sqrt{x}-\sqrt{y}}$

$= \dfrac{\left(x^3 - y^3\right)\left(\sqrt{x}-\sqrt{y}\right)}{x - y}$

$= \dfrac{(x-y)\left(x^2 + xy + y^2\right)\left(\sqrt{x}-\sqrt{y}\right)}{x - y}$

$= \left(x^2 + xy + y^2\right)\left(\sqrt{x}-\sqrt{y}\right)$

37. $\dfrac{x^2 - y^2}{\sqrt{x-y}} \cdot \dfrac{\sqrt{x-y}}{\sqrt{x-y}}$

$= \dfrac{(x+y)(x-y)\sqrt{x-y}}{x-y} = (x+y)\sqrt{x-y}$

39. $\dfrac{\sqrt{x+y}+\sqrt{x-y}}{\sqrt{x+y}-\sqrt{x-y}} \cdot \dfrac{\sqrt{x+y}+\sqrt{x-y}}{\sqrt{x+y}+\sqrt{x-y}}$

$= \dfrac{(x+y) + 2\sqrt{x+y}\sqrt{x-y} + (x-y)}{(x+y) - (x-y)}$

$= \dfrac{2x + 2\sqrt{x^2 - y^2}}{2y} = \dfrac{2\left(x + \sqrt{x^2 - y^2}\right)}{2y}$

$= \dfrac{x + \sqrt{x^2 - y^2}}{y}$

43. $\dfrac{\dfrac{1}{y-2} - \dfrac{1}{y+2}}{1 + \dfrac{1}{(y+2)(y-2)}} \cdot \dfrac{(y+2)(y-2)}{(y+2)(y-2)}$

$= \dfrac{y+2 - (y-2)}{(y+2)(y-2)+1} = \dfrac{4}{y^2 - 4 + 1} = \dfrac{4}{y^2 - 3}$

44. $\dfrac{x^3+y^3}{x^2-xy+y^2} \cdot \dfrac{x^2-2xy+y^2}{x^2-y^2}$

$= \dfrac{(x+y)\left(x^2-xy+y^2\right)}{x^2-xy+y^2} \cdot \dfrac{(x-y)^2}{(x+y)(x-y)}$

$= x-y$

45. x: time it takes both pipes working together
1 complete job = pool is completely emptied
Fraction of the pool emptied by the second
pipe in x minutes − Fraction of the pool
filled by the first pipe in x minutes = 1

$$\frac{x}{20}-\frac{x}{30}=1$$

$$60\left(\frac{x}{20}-\frac{x}{30}\right)=60(1)$$

$$3x-2x=60$$

$$x=60$$

It will take 60 minutes (or 1 hour) before
the pool has no water.

Problem set 5.4

1. $\sqrt{x+3}=1$ check: $\sqrt{-2+3}=1$

$\left(\sqrt{x+3}\right)^2=1^2$ $\sqrt{1}=1$

$x+3=1$ $1=1$

$x=-2$ $\{-2\}$

5. $\sqrt{3x}=3$ check: $\sqrt{3(3)}=3$

$\left(\sqrt{3x}\right)^2=3^2$ $\sqrt{9}=3$

$3x=9$ $3=3$

$x=3$ $\{3\}$

7. $\sqrt{5x}=-5$ check: $\sqrt{5(5)}=-5$

$\left(\sqrt{5x}\right)^2=(-5)^2$ $\sqrt{25}=-5$

$5x=25$ $5\ne-5$

$x=5$ 5 is an extraneous
                 solution. $\varnothing$

11. $2\sqrt{x}=7$ check: $2\sqrt{\dfrac{49}{4}}=7$

$\left(2\sqrt{x}\right)^2=7^2$ $2\left(\dfrac{7}{2}\right)=7$

$4x=49$ $7=7$

$x=\dfrac{49}{4}$ $\left\{\dfrac{49}{4}\right\}$

15. $\sqrt{2x+4}-6=0$

$\sqrt{2x+4}=6$

$\left(\sqrt{2x+4}\right)^2=6^2$

$2x+4=36$

$2x=32$

$x=16$

check: $\sqrt{2(16)+4}-6=0$

$\sqrt{32+4}-6=0$

$\sqrt{36}-6=0$

$6-6=0$

$0=0$ $\{16\}$

17. $\sqrt{3x-1}=-4$

$\left(\sqrt{3x-1}\right)^2=(-4)^2$

$3x-1=16$

$3x=17$

$x=\dfrac{17}{3}$

check: $\sqrt{3\left(\dfrac{17}{3}\right)-1}=-4$

$\sqrt{16}=-4$

$4\ne-4$ $\varnothing$

21. $-\sqrt{x-2} = 3$

$\sqrt{x-2} = -3$

$\left(\sqrt{x-2}\right)^2 = (-3)^2$

$x - 2 = 9$

$x = 11$

check: $-\sqrt{11-2} = 3$

$-\sqrt{9} = 3$

$-3 \neq 3$ $\varnothing$

25. $\sqrt{7x-10} = \sqrt{x+2}$

$\left(\sqrt{7x-10}\right)^2 = \left(\sqrt{x+2}\right)^2$

$7x - 10 = x + 2$

$6x = 12$

$x = 2$

check: $\sqrt{7(2)-10} = \sqrt{2+2}$

$\sqrt{4} = \sqrt{4}$

$2 = 2$ $\{2\}$

27. $\sqrt{2x+7} - \sqrt{3-2x} = 0$

$\sqrt{2x+7} = \sqrt{3-2x}$

$\left(\sqrt{2x+7}\right)^2 = \left(\sqrt{3-2x}\right)^2$

$2x + 7 = 3 - 2x$

$4x = -4$

$x = -1$

check: $\sqrt{2(-1)+7} - \sqrt{3-2(-1)} = 0$

$\sqrt{5} - \sqrt{5} = 0$

$0 = 0$ $\{-1\}$

33. $\sqrt{3y+4} = -\sqrt{2y+3}$

$\left(\sqrt{3y+4}\right)^2 = \left(-\sqrt{2y+3}\right)^2$

$3y + 4 = 2y + 3$

$y = -1$

check: $\sqrt{3(-1)+4} = -\sqrt{2(-1)+3}$

$1 \neq -1$ $\varnothing$

39. $2\sqrt{x} = \sqrt{5x-16}$

$\left(2\sqrt{x}\right)^2 = \left(\sqrt{5x-16}\right)^2$

$4x = 5x - 16$

$16 = x$

check: $2\sqrt{16} = \sqrt{5(16)-16}$

$2(4) = \sqrt{64}$

$8 = 8$ $\{16\}$

41. $3\sqrt{10y+25} = 5\sqrt{4y+1}$

$\left(3\sqrt{10y+25}\right)^2 = \left(5\sqrt{4y+1}\right)^2$

$9(10y+25) = 25(4y+1)$

$90y + 225 = 100y + 25$

$200 = 10y$

$20 = y$

check: $3\sqrt{10(20)+25} = 5\sqrt{4(20)+1}$

$3\sqrt{225} = 5\sqrt{81}$

$45 = 45$ $\{20\}$

45. $\sqrt[3]{3y-1}+4=0$

$$\sqrt[3]{3y-1}=-4$$

$$\left(\sqrt[3]{3y-1}\right)^3=(-4)^3$$

$$3y-1=-64$$

$$3y=-63$$

$$y=-21$$

check: $\sqrt[3]{3(-21)-1}+4=0$

$$\sqrt[3]{-64}+4=0$$

$$-4+4=0$$

$$0=0 \quad \{-21\}$$

49. $\sqrt[3]{y^2+4y}=\sqrt[3]{y^2+5y+1}$

$$\left(\sqrt[3]{y^2+4y}\right)^3=\left(\sqrt[3]{y^2+5y+1}\right)^3$$

$$y^2+4y=y^2+5y+1$$

$$4y=5y+1$$

$$-1=y$$

check:

$$\sqrt[3]{(-1)^2+4(-1)}=\sqrt[3]{(-1)^2+5(-1)+1}$$

$$\sqrt[3]{-3}=\sqrt[3]{-3} \quad \{-1\}$$

51. $\sqrt[4]{x+8}=\sqrt[4]{2x}$

$$\left(\sqrt[4]{x+8}\right)^4=\left(\sqrt[4]{2x}\right)^4$$

$$x+8=2x$$

$$8=x$$

check: $\sqrt[4]{8+8}=\sqrt[4]{2(8)}$

$$\sqrt[4]{16}=\sqrt[4]{16} \quad \{8\}$$

53. $\sqrt{y-7}=7-\sqrt{y}$

$$\left(\sqrt{y-7}\right)^2=\left(7-\sqrt{y}\right)^2$$

$$y-7=\left(7-\sqrt{y}\right)\left(7-\sqrt{y}\right)$$

$$y-7=49-14\sqrt{y}+y$$

$$-56=-14\sqrt{y}$$

$$4=\sqrt{y}$$

$$(4)^2=\left(\sqrt{y}\right)^2$$

$$16=y$$

check: $\sqrt{16-7}=7-\sqrt{16}$

$$\sqrt{9}=7-\sqrt{16}$$

$$3=7-4$$

$$3=3 \quad \{16\}$$

57. $\sqrt{y+8}=\sqrt{y-4}+2$

$$\left(\sqrt{y+8}\right)^2=\left(\sqrt{y-4}+2\right)^2$$

$$y+8=\left(\sqrt{y-4}+2\right)\left(\sqrt{y-4}+2\right)$$

$$y+8=y-4+4\sqrt{y-4}+4$$

$$y+8=y+4\sqrt{y-4}$$

$$8=4\sqrt{y-4}$$

$$2=\sqrt{y-4}$$

$$2^2=\left(\sqrt{y-4}\right)^2$$

$$4=y-4$$

$$8=y$$

check: $\sqrt{8+8}=\sqrt{8-4}+2$

$$\sqrt{16}=\sqrt{4}+2$$

$$4=4 \quad \{8\}$$

59. $\sqrt{p-5}-\sqrt{p-8}=3$

$\quad\quad \sqrt{p-5}=3+\sqrt{p-8}$

$\quad\quad \left(\sqrt{p-5}\right)^2=\left(3+\sqrt{p-8}\right)^2$

$\quad\quad p-5=9+6\sqrt{p-8}+p-8$

$\quad\quad p-5=1+p+6\sqrt{p-8}$

$\quad\quad -6=6\sqrt{p-8}$

$\quad\quad -1=\sqrt{p-8}$

$\quad\quad (-1)^2=\left(\sqrt{p-8}\right)^2$

$\quad\quad 1=p-8$

$\quad\quad 9=p$

check: $\sqrt{9-5}-\sqrt{9-8}=3$

$\quad\quad \sqrt{4}-\sqrt{1}=3$

$\quad\quad 2-1=3$

$\quad\quad 1\neq 3\quad\quad \varnothing$

63. $\quad \sqrt{x+6}=2-\sqrt{x-2}$

$\quad \left(\sqrt{x+6}\right)^2=\left(2-\sqrt{x-2}\right)^2$

$\quad x+6=\left(2-\sqrt{x-2}\right)\left(2-\sqrt{x-2}\right)$

$\quad x+6=4-4\sqrt{x-2}+x-2$

$\quad x+6=x+2-4\sqrt{x-2}$

$\quad 4=-4\sqrt{x-2}$

$\quad -1=\sqrt{x-2}$

$\quad (-1)^2=\left(\sqrt{x-2}\right)^2$

$\quad 1=x-2$

$\quad 3=x$

check: $\sqrt{3+6}=2-\sqrt{3-2}$

$\quad 3=2-1$

$\quad 3\neq 1\quad\quad \varnothing$

67. $(y+5)^{\frac{1}{2}}+y^{\frac{1}{2}}=5$

$\quad \sqrt{y+5}+\sqrt{y}=5$

$\quad \sqrt{y+5}=5-\sqrt{y}$

$\quad \left(\sqrt{y+5}\right)^2=\left(5-\sqrt{y}\right)^2$

$\quad y+5=25-10\sqrt{y}+y$

$\quad -20=-10\sqrt{y}$

$\quad 2=\sqrt{y}$

$\quad (2)^2=\left(\sqrt{y}\right)^2$

$\quad 4=y$

check: $(4+5)^{\frac{1}{2}}+4^{\frac{1}{2}}=5$

$\quad 3+2=5$

$\quad 5=5\quad \{4\}$

69. $(y+5)^{\frac{1}{2}}+(y-3)^{\frac{1}{2}}=4$

$\quad \sqrt{y+5}+\sqrt{y-3}=4$

$\quad \sqrt{y+5}=4-\sqrt{y-3}$

$\quad \left(\sqrt{y+5}\right)^2=\left(4-\sqrt{y-3}\right)^2$

$\quad y+5=16-8\sqrt{y-3}+y-3$

$\quad y+5=y+13-8\sqrt{y-3}$

$\quad -8=-8\sqrt{y-3}$

$\quad 1=\sqrt{y-3}$

$\quad 1^2=\left(\sqrt{y-3}\right)^2$

$\quad 1=y-3$

$\quad 4=y$

check: $(4+5)^{\frac{1}{2}}+(4-3)^{\frac{1}{2}}=4$

$\quad \sqrt{9}+\sqrt{1}=4$

$\quad 4=4\quad \{4\}$

73. $\sqrt{y^2-4y-8}=y$

$\left(\sqrt{y^2-4y-8}\right)^2=(y)^2$

$y^2-4y-8=y^2$

$-4y-8=0$

$-4y=8$

$y=-2$

check: $\sqrt{(-2)^2-4(-2)-8}=-2$

$\sqrt{4}=-2$

$2\neq -2 \qquad \varnothing$

75. $p+3=\sqrt{p^2+2p+9}$

$(p+3)^2=\left(\sqrt{p^2+2p+9}\right)^2$

$p^2+6p+9=p^2+2p+9$

$6p+9=2p+9$

$4p=0$

$p=0$

check: $0+3=\sqrt{0^2+2(0)+9}$

$3=\sqrt{9}$

$3=3 \qquad \{0\}$

79. $\sqrt{y^2+12y-4}+4-y=0$

$\sqrt{y^2+12y-4}=y-4$

$\left(\sqrt{y^2+12y-4}\right)^2=(y-4)^2$

$y^2+12y-4=y^2-8y+16$

$12y-4=-8y+16$

$20y=20$

$y=1$

check: $\sqrt{1^2+12(1)-4}+4-1=0$

$\sqrt{9}+4-1=0$

$3+4-1=0$

$6\neq 0 \qquad \varnothing$

85. x: number

$\sqrt{\sqrt{\sqrt{x+1}+1}+1}=2$

$\left(\sqrt{\sqrt{\sqrt{x+1}+1}+1}\right)^2=(2)^2$

$\sqrt{\sqrt{x+1}+1}+1=4$

$\sqrt{\sqrt{x+1}+1}=3$

$\left(\sqrt{\sqrt{x+1}+1}\right)^2=(3)^2$

$\sqrt{x+1}+1=9$

$\sqrt{x+1}=8$

$x+1=64$

$x=63$

The number is 63.

86. $\left(4x^{-4}y^{\frac{1}{2}}\right)^{-\frac{3}{2}}$

$=4^{-\frac{3}{2}}\left(x^{-4}\right)^{-\frac{3}{2}}\left(y^{\frac{1}{2}}\right)^{-\frac{3}{2}}$

$=\frac{1}{4^{\frac{3}{2}}}x^{6}y^{-\frac{3}{4}}=\frac{x^6}{\left(\sqrt{4}\right)^3 y^{\frac{3}{4}}}$

$=\frac{x^6}{8y^{\frac{3}{4}}}$

87. $\dfrac{\sqrt[3]{5}}{\sqrt[4]{5}}=\dfrac{5^{\frac{1}{3}}}{5^{\frac{1}{4}}}=5^{\frac{1}{3}-\frac{1}{4}}=5^{\frac{4}{12}-\frac{3}{12}}$

$=5^{\frac{1}{12}}=\sqrt[12]{5}$

88. $\dfrac{2}{y-3} - \dfrac{y}{y^2-y-6} \cdot \dfrac{y^2-2y-3}{y^2+y}$

$= \dfrac{2}{y-3} - \dfrac{y}{(y-3)(y+2)} \cdot \dfrac{(y-3)(y+1)}{y(y+1)}$

$= \dfrac{2}{y-3} - \dfrac{1}{y+2}$

$= \dfrac{2}{y-3} \cdot \dfrac{y+2}{y+2} - \dfrac{1}{y+2} \cdot \dfrac{y-3}{y-3}$

$= \dfrac{2(y+2)-(y-3)}{(y-3)(y+2)} = \dfrac{2y+4-y+3}{(y-3)(y+2)}$

$= \dfrac{y+7}{(y-3)(y+2)}$

Problem set 5.5.1

1. $\sqrt{-4} = \sqrt{4}\,i = 2i$

5. $\sqrt{-28} = \sqrt{28}\,i = \sqrt{4 \cdot 7}\,i = 2\sqrt{7}\,i$

11. $5\sqrt{-12} = 5\sqrt{4 \cdot 3}\,i = 5(2)\sqrt{3}\,i = 10\sqrt{3}\,i$

13. $\sqrt{-49} + \sqrt{-100} = \sqrt{49}\,i + \sqrt{100}\,i$
$= 7i + 10i = 17i$

17. $3\sqrt{-49} + 5\sqrt{-100} = 3(7i\ ) + 5(10i\ )$
$= 21i + 50i = 71i$

19. $\sqrt{-72} + \sqrt{-50} = \sqrt{36 \cdot 2}\,i + \sqrt{25 \cdot 2}\,i$
$= 6\sqrt{2}\,i + 5\sqrt{2}\,i = 11\sqrt{2}\,i$

21. $5\sqrt{-8} - 3\sqrt{-18} = 5\sqrt{4 \cdot 2}\,i - 3\sqrt{9 \cdot 2}\,i$
$= 10\sqrt{2}\,i - 9\sqrt{2}\,i = \sqrt{2}\,i$

23. $\frac{3}{5}\sqrt{-50} + \frac{1}{2}\sqrt{-32}$

$= \frac{3}{5}\sqrt{25 \cdot 2}\,i + \frac{1}{2}\sqrt{16 \cdot 2}\,i$

$= \frac{3}{5}(5)\sqrt{2}\,i + \frac{1}{2}(4)\sqrt{2}\,i$

$= 3\sqrt{2}\,i + 2\sqrt{2}\,i = 5\sqrt{2}\,i$

25. $\sqrt{-7}\sqrt{-2} = \left(\sqrt{7}\,i\right)\left(\sqrt{2}\,i\right) = \sqrt{14}\,i^2$

$= \sqrt{14}(-1) = -\sqrt{14}$

29. $\sqrt{-7}\sqrt{-25} = \left(\sqrt{7}\,i\right)(5i) = 5\sqrt{7}\,i^2$

$= 5\sqrt{7}(-1) = -5\sqrt{7}$

33. $\left(2\sqrt{-8}\right)\left(3\sqrt{-6}\right) = \left(2\sqrt{8}\,i\right)\left(3\sqrt{6}\,i\right)$

$= 6\sqrt{48}\,i^2 = 6\sqrt{16 \cdot 3}(-1) = 6(4)\sqrt{3}(-1)$

$= -24\sqrt{3}$

35. $\left(3\sqrt{-5}\right)\left(-4\sqrt{-12}\right) = \left(3\sqrt{5}\,i\right)\left(-4\sqrt{12}\,i\right)$

$= -12\sqrt{60}\,i^2 = -12\sqrt{4 \cdot 15}(-1)$

$= -12(2)\sqrt{15}(-1) = 24\sqrt{15}$

37. $\dfrac{\sqrt{-125}}{\sqrt{5}} = \sqrt{\dfrac{125}{5}}\,i = \sqrt{25}\,i = 5i$

39. $\dfrac{\sqrt{-24}}{\sqrt{-6}} = \sqrt{\dfrac{24}{6}}\,\dfrac{i}{i} = \sqrt{4} = 2$

43. $\sqrt{\frac{-1}{2}} + \sqrt{\frac{-9}{2}} = \sqrt{\frac{1}{2}}i + \sqrt{\frac{9}{2}}i$

$= \frac{1}{\sqrt{2}}i + \frac{3}{\sqrt{2}}i$

$= \frac{1}{\sqrt{2}} \cdot \frac{\sqrt{2}}{\sqrt{2}}i + \frac{3}{\sqrt{2}} \cdot \frac{\sqrt{2}}{\sqrt{2}}i = \frac{\sqrt{2}}{2}i + \frac{3\sqrt{2}}{2}i$

$= \frac{4\sqrt{2}}{2}i = 2\sqrt{2}i$

45. $\sqrt{-36}\sqrt{-4} = (6i)(2i) = 12i^2 = 12(-1)$

$= -12$

Integers: $x, x+1, x+2$

$x + x + 1 + x + 2 = -12$

$3x + 3 = -12$

$3x = -15$

$x = -5$

Integers are $-5, -4,$ and -3.

47. Here's the correct solution to each part. Contrast these with the disasters in the text.

a. $-\sqrt{-25} = -\sqrt{25}i = -5i$

b. $\left(\sqrt{-4}\right)^2 = \sqrt{-4}\sqrt{-4} = (2i)(2i)$

$= 4i^2 = 4(-1) = -4$

c. $\sqrt{-9} + \sqrt{-36} = \sqrt{9}i + \sqrt{36}i$

$= 3i + 6i = 9i$

d. $\frac{\sqrt{-20}}{\sqrt{-5}} = \frac{\sqrt{20}i}{\sqrt{5}i} = \sqrt{\frac{20}{5}} = \sqrt{4} = 2$

e. The other square root of negative 1 is $-i$

$(-i)^2 = (-i)(-i) = -i^2 = -(-1) = 1$

51. a. $i^{17} = i^{16}i = \left(i^2\right)^8 i = (-1)^8 i$

$= 1i = i$

b. $i^{98} = \left(i^2\right)^{49} = (-1)^{49} = -1$

c. $i^{75} = i^{74}i = \left(i^2\right)^{37}i$

$= (-1)^{37}i = (-1)i = -i$

d. $i^{1000} = \left(i^2\right)^{500} = (-1)^{500} = 1$

52. $y^5 - 4y^3 - 8y^2 + 32$

$= y^3\left(y^2 - 4\right) - 8\left(y^2 - 4\right)$

$= \left(y^3 - 8\right)\left(y^2 - 4\right)$

$= (y-2)\left(y^2 + 2y + 4\right)(y+2)(y-2)$

$= (y+2)(y-2)^2\left(y^2 + 2y + 4\right)$

53. inverses of addition

54. x: distance from the reef to the shore

Time spent traveling to the reef + time spent returning to the shore $= 5$

$$\left(RT = D \; ; \;\; T = \frac{D}{R}\right)$$

$\frac{x}{20} + \frac{x}{12} = 5$

$60\left(\frac{x}{20} + \frac{x}{12}\right) = 60(5)$

$3x + 5x = 300$

$8x = 300$

$x = \frac{300}{8} = 37.5$

The reef is 37.5 miles from the shore.

Problem set 5.5.2

1. $8x = 40$ and $3y = -6$

$\quad$ $x = 5$ and $y = -2$

5. $3x - 5 = 10$ and $-4 = -20y$

$\qquad 3x = 15 \qquad \dfrac{-4}{-20} = y$

$\qquad x = 5 \qquad\quad y = \dfrac{1}{5}$

17. $\left(5 + 2\sqrt{32}i\right) + \left(11 - 5\sqrt{8}i\right)$

$\quad = (5 + 11) + \left(2\sqrt{32} - 5\sqrt{8}\right)i$

$\quad = 16 + \left(2\sqrt{16 \cdot 2} - 5\sqrt{4 \cdot 2}\right)i$

$\quad = 16 + \left(8\sqrt{2} - 10\sqrt{2}\right)i = 16 - 2\sqrt{2}i$

19. $\left(5 + 2\sqrt{32}i\right) - \left(11 - 5\sqrt{8}i\right)$

$\quad = \left(5 + 2\sqrt{32}i\right) + \left(-11 + 5\sqrt{8}i\right)$

$\quad = (5 - 11) + \left(2\sqrt{32} + 5\sqrt{8}\right)i$

$\quad = -6 + \left(2\sqrt{16 \cdot 2} + 5\sqrt{4 \cdot 2}\right)i$

$\quad = -6 + \left(8\sqrt{2} + 10\sqrt{2}\right)i = -6 + 18\sqrt{2}i$

21. $\left(5\sqrt{20} + 3\sqrt{63}i\right) - \left(\sqrt{80} - 2\sqrt{28}i\right)$

$\quad = \left(5\sqrt{20} + 3\sqrt{63}i\right) + \left(-\sqrt{80} + 2\sqrt{28}i\right)$

$\quad = \left(5\sqrt{20} - \sqrt{80}\right) + \left(3\sqrt{63} + 2\sqrt{28}\right)i$

$\quad = \left(5\sqrt{4 \cdot 5} - \sqrt{16 \cdot 5}\right)$

$\qquad\qquad + \left(3\sqrt{9 \cdot 7} + 2\sqrt{4 \cdot 7}\right)i$

$\quad = \left(10\sqrt{5} - 4\sqrt{5}\right) + \left(9\sqrt{7} + 4\sqrt{7}\right)i$

$\quad = 6\sqrt{5} + 13\sqrt{7}i$

23. $\left(\tfrac{1}{3} - \tfrac{2}{5}i\right) - \left(\tfrac{1}{2} - \tfrac{3}{4}i\right)$

$\quad = \left(\tfrac{1}{3} - \tfrac{1}{2}\right) + \left(-\tfrac{2}{5} + \tfrac{3}{4}\right)i$

$\quad = \left(\tfrac{2}{6} - \tfrac{3}{6}\right) + \left(\tfrac{-8}{20} + \tfrac{15}{20}\right)i = -\tfrac{1}{6} + \tfrac{7}{20}i$

25. $(7 + 3i)(5 + 2i) = 35 + 14i + 15i + 6i^2$

$\quad = 35 + 14i + 15i + (-6) = 29 + 29i$

31. $(7 - 5i)(-2 - 3i)$

$\quad = -14 - 21i + 10i + 15i^2$

$\quad = -14 - 21i + 10i + (-15) = -29 - 11i$

33. $(3 + 5i)(3 - 5i)$

$\quad = 9 - 15i + 15i - 25i^2$

$\quad = 9 - 25(-1) = 9 + 25 = 34$

37. $(2 + 3i)^2 = (2 + 3i)(2 + 3i)$

$\quad = 4 + 12i + 9i^2 = 4 + 12i + 9(-1)$

$\quad = -5 + 12i$

39. $(1+i)^3 = (1+i)[(1+i)(1+i)]$

$= (1+i)(1+2i+i^2)$

$= (1+i)(1+2i-1) = (1+i)2i$

$= 2i + 2i^2 = 2i + 2(-1) = -2 + 2i$

43. $(8+9i)(2-i)-(1-i)(1+i)$

$= (16 + 10i - 9i^2) - (1 - i^2)$

$= (25 + 10i) - (2) = 23 + 10i$

45. $(2+i)^2 - (3-i)^2$

$= (2+i)(2+i) - (3-i)(3-i)$

$= (4 + 4i + i^2) - (9 - 6i + i^2)$

$= (3 + 4i) - (8 - 6i)$

$= (3 + 4i) + (-8 + 6i) = -5 + 10i$

49. $\left(-\dfrac{\sqrt{6}}{2} + \dfrac{\sqrt{2}}{2}i\right)\left(-\dfrac{\sqrt{6}}{2} + \dfrac{\sqrt{2}}{2}i\right)$

$= \dfrac{6}{4} - \dfrac{\sqrt{12}}{4}i - \dfrac{\sqrt{12}}{4}i + \dfrac{2}{4}i^2$

$= \dfrac{3}{2} - \dfrac{2\sqrt{12}}{4}i + \dfrac{1}{2}(-1)$

$= \dfrac{2}{2} - \dfrac{\sqrt{12}}{2}i = 1 - \dfrac{\sqrt{4\cdot3}}{2}i = 1 - \dfrac{2\sqrt{3}}{2}i$

$= 1 - \sqrt{3}i$

$\left(\dfrac{\sqrt{6}}{2} - \dfrac{\sqrt{2}}{2}i\right)\left(\dfrac{\sqrt{6}}{2} - \dfrac{\sqrt{2}}{2}i\right)$

$= \dfrac{6}{4} - \dfrac{\sqrt{12}}{4}i - \dfrac{\sqrt{12}}{4}i + \dfrac{2}{4}i^2$

$= \dfrac{3}{2} - \dfrac{2\sqrt{12}}{4}i + \dfrac{1}{2}(-1)$

$= \dfrac{2}{2} - \dfrac{\sqrt{12}}{2}i = 1 - \dfrac{\sqrt{4\cdot3}}{2}i = 1 - \dfrac{2\sqrt{3}}{2}i$

$= 1 - \sqrt{3}i$

51. $\dfrac{2i}{3-i} \cdot \dfrac{3+i}{3+i} = \dfrac{2i(3+i)}{9-i^2} = \dfrac{2i(3+i)}{10}$

$= \dfrac{i(3+i)}{5} = \dfrac{3i+i^2}{5} = \dfrac{3i-1}{5}$

$= -\dfrac{1}{5} + \dfrac{3}{5}i$

53. $\dfrac{3-i}{2i} \cdot \dfrac{i}{i} = \dfrac{3i-i^2}{2i^2} = \dfrac{3i-(-1)}{2(-1)} = \dfrac{3i+1}{-2}$

$= -\dfrac{1}{2} - \dfrac{3}{2}i$

55. $\dfrac{1+i}{1-i} \cdot \dfrac{1+i}{1+i} = \dfrac{1+2i+i^2}{1-i^2}$

$= \dfrac{1+2i+(-1)}{1-(-1)} = \dfrac{2i}{2} = i$

57. $\dfrac{2+3i}{3-i} \cdot \dfrac{3+i}{3+i} = \dfrac{6+11i+3i^2}{9-i^2}$

$= \dfrac{3+11i}{10} = \dfrac{3}{10} + \dfrac{11}{10}i$

61. $\dfrac{-4+7i}{-2-5i} \cdot \dfrac{-2+5i}{-2+5i} = \dfrac{8-34i+35i^2}{4-25i^2}$

$= \dfrac{-27-34i}{29} = -\dfrac{27}{29} - \dfrac{34}{29}i$

63. $\dfrac{-4+7i}{-5i} \cdot \dfrac{i}{i} = \dfrac{-4i+7i^2}{-5i^2}$

$= \dfrac{-4i-7}{5} = -\dfrac{7}{5} - \dfrac{4}{5}i$

65. False; The conjugate of $7+3i$ is $7-3i$.

69. $\dfrac{3+x}{2-x} = \dfrac{3+(2-3i)}{2-(2-3i)}$

$= \dfrac{5-3i}{3i} \cdot \dfrac{i}{i} = \dfrac{5i-3i^2}{3i^2} = \dfrac{5i-3(-1)}{3(-1)}$

$= \dfrac{5i+3}{-3} = \dfrac{5i}{-3} + \dfrac{3}{-3} = -\dfrac{5}{3}i - 1$

Consequently, the statement is true.

73. $(1-2i)^2 - 2(1-2i) + 5$

$= (1-2i)(1-2i) - 2(1-2i) + 5$

$= 1 - 4i + 4i^2 - 2 + 4i + 5$

$= 1 - 4i - 4 - 2 + 4i + 5 = 0$

75. Show that the product of $2+3i$ and

$\dfrac{2}{13} - \dfrac{3}{13}i$ is 1.

$(2+3i)\left(\dfrac{2}{13} - \dfrac{3}{13}i\right)$

$= \dfrac{4}{13} - \dfrac{6}{13}i + \dfrac{6}{13}i - \dfrac{9}{13}i^2$

$= \dfrac{4}{13} - \dfrac{9}{13}(-1) = \dfrac{4}{13} + \dfrac{9}{13} = \dfrac{13}{13} = 1$

77. $\dfrac{(1+i)(-1+2i)+(2-i)}{2-3i}$

$= \dfrac{-1+i+2i^2+2-i}{2-3i}$

$= \dfrac{-1}{2-3i} \cdot \dfrac{2+3i}{2+3i} = \dfrac{-2-3i}{4-9i^2}$

$= \dfrac{-2-3i}{13} = -\dfrac{2}{13} - \dfrac{3}{13}i$

78. $4y^2 - 2y\left[3 - y\left(2 - y + y^2\right)\right]$

$= 4y^2 - 2y\left[3 - 2y + y^2 - y^3\right]$

$= 4y^2 - 6y + 4y^2 - 2y^3 + 2y^4$

$= 2y^4 - 2y^3 + 8y^2 - 6y$

79. $a^6b^3 + a^3 - a^3b^3 - 1$

$= a^6b^3 - a^3b^3 + a^3 - 1$

$= a^3b^3\left(a^3 - 1\right) + a^3 - 1$

$= \left(a^3 - 1\right)\left(a^3b^3 + 1\right)$

$= \left(a^3 - 1\right)\left[(ab)^3 + 1^3\right]$

$= (a-1)\left(a^2+a+1\right)(ab+1)\cdot$

$\qquad \left[(ab)^2 - (ab)\cdot 1 + 1^2\right]$

$= (a-1)\left(a^2+a+1\right)(ab+1)\cdot$

$\qquad\qquad \left(a^2b^2 - ab + 1\right)$

80. $r^2F = r^2\left(\dfrac{Gm_1m_2}{r^2}\right)$

$\qquad r^2F = Gm_1m_2$

$\qquad \dfrac{r^2F}{Gm_2} = m_1$

Review Problems: Chapter 5

1. $7^{-3} = \dfrac{1}{7^3} = \dfrac{1}{343}$

2. $\left(\dfrac{3}{4}\right)^{-3} = \dfrac{3^{-3}}{4^{-3}} = \dfrac{4^3}{3^3} = \dfrac{64}{27}$

3. $\left(4^5 \cdot 4^{-7}\right)^{-2} = \left(4^{-2}\right)^{-2} = 4^4 = 256$

4. $\sqrt[3]{-27} = \sqrt[3]{(-3)^3} = -3$

5. $\sqrt[3]{\frac{27}{64}} = \sqrt[3]{\left(\frac{3}{4}\right)^3} = \frac{3}{4}$

6. $16^{\frac{3}{2}} = \left(\sqrt{16}\right)^3 = 4^3 = 64$

7. $8^{-\frac{2}{3}} = \frac{1}{8^{\frac{2}{3}}} = \frac{1}{\left(\sqrt[3]{8}\right)^2} = \frac{1}{2^2} = \frac{1}{4}$

8. $\left(\frac{1}{32}\right)^{\frac{4}{5}} = \left(\sqrt[5]{\frac{1}{32}}\right)^4 = \left(\frac{1}{2}\right)^4 = \frac{1}{16}$

9. $\frac{3^3}{3^{-2}} = 3^{3-(-2)} = 3^5 = 243$

10. $\left(5^{-3} \cdot 5^3\right)^{-4} = \left(5^0\right)^{-4} = 1^{-4} = 1$

11. $\left(\frac{2^{-1}}{2^{-3}}\right)^{-4} = \left(2^{-1-(-3)}\right)^{-4}$

$= \left(2^2\right)^{-4} = 2^{-8} = \frac{1}{2^8} = \frac{1}{256}$

12. $A = (2000)(16)^{-1} - 70(16)^{\frac{3}{4}} + 3000$

$= \frac{2000}{16} - 70\left(\sqrt[4]{16}\right)^3 + 3000$

$= 125 - 70(2)^3 + 3000$

$= 125 - 560 + 3000 = 2565$

2,565 clocks can be sold.

13. $\left(54x^3y^7\right)\left(-3x^2y^{-5}\right)^{-3}$

$= \left(54x^3y^7\right)(-3)^{-3}x^{-6}y^{15}$

$= (54)(-3)^{-3}x^{-3}y^{22}$

$= \frac{54y^{22}}{(-3)^3x^3} = \frac{54y^{22}}{-27x^3} = \frac{-2y^{22}}{x^3}$

14. $\frac{15x^4y^2}{-45x^7y^{-3}} = -\frac{1}{3}x^{4-7}y^{2-(-3)}$

$= -\frac{1}{3}x^{-3}y^5 = \frac{-y^5}{3x^3}$

15. $\left(\frac{2x^{-2}}{3y^3}\right)^{-4} = \frac{\left(2x^{-2}\right)^{-4}}{\left(3y^3\right)^{-4}}$

$= \frac{2^{-4}x^8}{3^{-4}y^{-12}} = \frac{3^4x^8y^{12}}{2^4} = \frac{81x^8y^{12}}{16}$

16. $\left(\frac{-10a^2bc^2}{-30ab^{-2}c^6}\right)^{-2} = \left(\frac{1}{3}ab^3c^{-4}\right)^{-2}$

$= \frac{a^{-2}b^{-6}c^8}{3^{-2}} = \frac{3^2c^8}{a^2b^6} = \frac{9c^8}{a^2b^6}$

17. $\left(7x^{\frac{1}{3}}\right)\left(4x^{\frac{1}{4}}\right) = 28x^{\frac{1}{3}+\frac{1}{4}}$

$= 28x^{\frac{4}{12}+\frac{3}{12}} = 28x^{\frac{7}{12}}$

18. $\dfrac{80y^{\frac{3}{4}}}{-20y^{\frac{1}{5}}} = -4y^{\frac{3}{4}-\frac{1}{5}}$

$= -4y^{\frac{15}{20}-\frac{4}{20}} = -4y^{\frac{11}{20}}$

19. $\left(\dfrac{32x^5}{y^{-4}}\right)^{-\frac{1}{5}} = \dfrac{(32)^{-\frac{1}{5}}\left(x^5\right)^{-\frac{1}{5}}}{\left(y^{-4}\right)^{-\frac{1}{5}}}$

$= \dfrac{(32)^{-\frac{1}{5}}x^{-1}}{y^{\frac{4}{5}}} = \dfrac{1}{(32)^{\frac{1}{5}}xy^{\frac{4}{5}}}$

$= \dfrac{1}{\sqrt[5]{32}\,xy^{\frac{4}{5}}} = \dfrac{1}{2xy^{\frac{4}{5}}}$

20. $\left(9x^{-2}y^{\frac{1}{2}}\right)^{-\frac{3}{2}}$

$= (9)^{-\frac{3}{2}}\left(x^{-2}\right)^{-\frac{3}{2}}\left(y^{\frac{1}{2}}\right)^{-\frac{3}{2}}$

$= 9^{-\frac{3}{2}}x^3y^{-\frac{3}{4}} = \dfrac{x^3}{9^{\frac{3}{2}}y^{\frac{3}{4}}}$

$= \dfrac{x^3}{\left(\sqrt{9}\right)^3 y^{\frac{3}{4}}} = \dfrac{x^3}{27y^{\frac{3}{4}}}$

21. $\left(x^{2n+5}\cdot x^{3n-2}\right)^{4n}$

$= \left(x^{2n+5+3n-2}\right)^{4n} = \left(x^{5n+3}\right)^{4n}$

$= x^{4n(5n+3)} = x^{20n^2+12n}$

22. $\left(\dfrac{x^{1-n}}{x^{7-n}}\right)^3 = \left(x^{1-n-(7-n)}\right)^3$

$= \left(x^{-6}\right)^3 = x^{-18} = \dfrac{1}{x^{18}}$

23. $\left(\dfrac{x^{3n+2}y^{2n}}{x^{3n}}\right)^4 = \left(x^{3n+2-3n}y^{2n}\right)^4$

$= \left(x^2y^{2n}\right)^4 = \left(x^2\right)^4\left(y^{2n}\right)^4$

$= x^8y^{8n}$

24. $\left(x^{b^2}y^b\right)^{\frac{1}{b}} = \left(x^{b^2}\right)^{\frac{1}{b}}\left(y^b\right)^{\frac{1}{b}} = x^by$

25. $\dfrac{\left(x^ay^{\frac{1}{a}}\right)^a}{x^{a^2}y^a} = \dfrac{\left(x^a\right)^a\left(y^{\frac{1}{a}}\right)^a}{x^{a^2}y^a} = \dfrac{x^{a^2}y}{x^{a^2}y^a}$

$= x^{a^2-a^2}y^{1-a} = y^{1-a}\left(\text{or }\dfrac{1}{y^{a-1}}\right)$

26. $x^{\frac{3}{4}}\left(x^{\frac{1}{4}}-x^{\frac{1}{2}}\right) = x^{\frac{3}{4}+\frac{1}{4}}-x^{\frac{3}{4}+\frac{1}{2}}$

$= x - x^{\frac{5}{4}}$

27. $\left(3x^{\frac{2}{3}} - 2y^{\frac{1}{2}}\right)\left(5x^{\frac{2}{3}} + 3y^{\frac{1}{2}}\right)$

$$= 15x^{\frac{4}{3}} + 9x^{\frac{2}{3}}y^{\frac{1}{2}} - 10x^{\frac{2}{3}}y^{\frac{1}{2}} - 6y$$

$$= 15x^{\frac{4}{3}} - x^{\frac{2}{3}}y^{\frac{1}{2}} - 6y$$

28. $\left(x^{\frac{2}{3}} - 5\right)^2 = \left(x^{\frac{2}{3}} - 5\right)\left(x^{\frac{2}{3}} - 5\right)$

$$= x^{\frac{4}{3}} - 5x^{\frac{2}{3}} - 5x^{\frac{2}{3}} + 25$$

$$= x^{\frac{4}{3}} - 10x^{\frac{2}{3}} + 25$$

29. $\dfrac{30x^{\frac{4}{3}}y^{\frac{1}{3}} - 20x^{\frac{1}{3}}y^{\frac{4}{3}}}{10x^{\frac{1}{3}}y^{\frac{1}{3}}}$

$$= \dfrac{30x^{\frac{4}{3}}y^{\frac{1}{3}}}{10x^{\frac{1}{3}}y^{\frac{1}{3}}} - \dfrac{20x^{\frac{1}{3}}y^{\frac{4}{3}}}{10x^{\frac{1}{3}}y^{\frac{1}{3}}}$$

$$= 3x^{\frac{4}{3}-\frac{1}{3}}y^{\frac{1}{3}-\frac{1}{3}} - 2x^{\frac{1}{3}-\frac{1}{3}}y^{\frac{4}{3}-\frac{1}{3}}$$

$$= 3x - 2y$$

30. $3x^{\frac{2}{3}} - 5x^{\frac{1}{3}} + 2$

$$= \left(3x^{\frac{1}{3}} - 2\right)\left(x^{\frac{1}{3}} - 1\right)$$

31. $4x^{\frac{2}{3}} - 25 = \left(2x^{\frac{1}{3}}\right)^2 - 5^2$

$$= \left(2x^{\frac{1}{3}} + 5\right)\left(2x^{\frac{1}{3}} - 5\right)$$

32. $30(x+4)^{\frac{4}{3}} - 20(x+4)^{\frac{1}{3}}$

$$= 10(x+4)^{\frac{1}{3}}\left[3(x+4) - 2\right]$$

$$= 10(x+4)^{\frac{1}{3}}(3x + 10)$$

33. $\dfrac{5}{y^{\frac{1}{2}}} + y^{\frac{1}{2}} \cdot \dfrac{y^{\frac{1}{2}}}{y^{\frac{1}{2}}} = \dfrac{5+y}{y^{\frac{1}{2}}}$

34. $\dfrac{y^2}{\left(y^2+4\right)^{\frac{1}{2}}} - \left(y^2+4\right)^{\frac{1}{2}} \cdot \dfrac{\left(y^2+4\right)^{\frac{1}{2}}}{\left(y^2+4\right)^{\frac{1}{2}}}$

$$= \dfrac{y^2 - \left(y^2+4\right)}{\left(y^2+4\right)^{\frac{1}{2}}} = \dfrac{-4}{\left(y^2+4\right)^{\frac{1}{2}}}$$

35. $\sqrt{20xy^3} = \sqrt{4 \cdot 5xy^2 y} = 2y\sqrt{5xy}$

36. $\sqrt{\dfrac{3}{7}} = \dfrac{\sqrt{3}}{\sqrt{7}} \cdot \dfrac{\sqrt{7}}{\sqrt{7}} = \dfrac{\sqrt{21}}{\sqrt{49}} = \dfrac{\sqrt{21}}{7}$

37. $\sqrt{\dfrac{3x}{7y}} = \dfrac{\sqrt{3x}}{\sqrt{7y}} \cdot \dfrac{\sqrt{7y}}{\sqrt{7y}} = \dfrac{\sqrt{21xy}}{\sqrt{49y^2}} = \dfrac{\sqrt{21xy}}{7y}$

38. $\sqrt{\dfrac{5}{8x^3}} = \dfrac{\sqrt{5}}{\sqrt{8x^3}} = \dfrac{\sqrt{5}}{\sqrt{4 \cdot 2x^2 x}} = \dfrac{\sqrt{5}}{2x\sqrt{2x}}$

$= \dfrac{\sqrt{5}}{2x\sqrt{2x}} \cdot \dfrac{\sqrt{2x}}{\sqrt{2x}} = \dfrac{\sqrt{10x}}{2x\sqrt{4x^2}} = \dfrac{\sqrt{10x}}{2x(2x)}$

$= \dfrac{\sqrt{10x}}{4x^2}$

39. $\sqrt[5]{64x^3 y^{12} z^6}$

$= \sqrt[5]{32 \cdot 2x^3 \left(y^2\right)^5 y^2 z^5 z}$

$= 2y^2 z \sqrt[5]{2x^3 y^2 z}$

40. $\dfrac{3y}{\sqrt{8y^3}} = \dfrac{3y}{\sqrt{4 \cdot 2y^2 y}} = \dfrac{3y}{2y\sqrt{2y}}$

$= \dfrac{3}{2\sqrt{2y}} \cdot \dfrac{\sqrt{2y}}{\sqrt{2y}} = \dfrac{3\sqrt{2y}}{2(2y)} = \dfrac{3\sqrt{2y}}{4y}$

41. $\dfrac{5}{\sqrt[5]{32x^4 y}} = \dfrac{5}{2\sqrt[5]{x^4 y}} \cdot \dfrac{\sqrt[5]{xy^4}}{\sqrt[5]{xy^4}}$

$= \dfrac{5\sqrt[5]{xy^4}}{2\sqrt[5]{x^5 y^5}} = \dfrac{5\sqrt[5]{xy^4}}{2xy}$

42. $\left(9.2 \times 10^{-4}\right)\left(7.4 \times 10^{12}\right)$

$= 68.08 \times 10^{-4+12}$

$= 68.08 \times 10^8 = 6.808 \times 10 \times 10^8$

$= 6.808 \times 10^9$

43. $\dfrac{3.3 \times 10^{-13}}{6.6 \times 10^{-4}} = 0.5 \times 10^{-13-(-4)}$

$= 0.5 \times 10^{-9} = 5 \times 10^{-1} \times 10^{-9}$

$= 5 \times 10^{-10}$

44. $V = \dfrac{4}{3}\pi \left(1.3 \times 10^{-5}\right)^3$

$= \dfrac{4}{3}\pi \left(2.197 \times 10^{-15}\right)$

$\approx 2.93\pi \times 10^{-15} \text{ cm}^3$

45. $T = \dfrac{D}{R} = \dfrac{6 \times 10^{17}}{20,000} = \dfrac{6 \times 10^{17}}{2 \times 10^4} = 3 \times 10^{13}$

3×10^{13} hours

46. $5\sqrt{18} + 3\sqrt{8} - \sqrt{2}$

$= 5\sqrt{9 \cdot 2} + 3\sqrt{4 \cdot 2} - \sqrt{2}$

$= 5(3)\sqrt{2} + 3(2)\sqrt{2} - \sqrt{2}$

$= 15\sqrt{2} + 6\sqrt{2} - \sqrt{2} = 20\sqrt{2}$

47. $12\sqrt{\dfrac{1}{3}} + 4\sqrt{\dfrac{1}{12}} = \dfrac{12\sqrt{1}}{\sqrt{3}} + \dfrac{4\sqrt{1}}{\sqrt{12}}$

$= \dfrac{12}{\sqrt{3}} + \dfrac{4}{\sqrt{12}} = \dfrac{12}{\sqrt{3}} + \dfrac{4}{\sqrt{4 \cdot 3}}$

$= \dfrac{12}{\sqrt{3}} + \dfrac{4}{2\sqrt{3}} = \dfrac{12}{\sqrt{3}} + \dfrac{2}{\sqrt{3}}$

$= \dfrac{12}{\sqrt{3}} \cdot \dfrac{\sqrt{3}}{\sqrt{3}} + \dfrac{2}{\sqrt{3}} \cdot \dfrac{\sqrt{3}}{\sqrt{3}} = \dfrac{12\sqrt{3}}{3} + \dfrac{2\sqrt{3}}{3}$

$= \dfrac{14\sqrt{3}}{3}$

48. $3y^2\sqrt{\dfrac{x}{y}} + \dfrac{5y}{x}\sqrt{x^3 y} = 3y^2\dfrac{\sqrt{x}}{\sqrt{y}} + \dfrac{5y}{x}\sqrt{x^2 xy}$

$= 3y^2\dfrac{\sqrt{x}}{\sqrt{y}} \cdot \dfrac{\sqrt{y}}{\sqrt{y}} + \dfrac{5y}{x}x\sqrt{xy}$

$= \dfrac{3y^2\sqrt{xy}}{y} + 5y\sqrt{xy}$

$= 3y\sqrt{xy} + 5y\sqrt{xy} = 8y\sqrt{xy}$

49. $\left(7\sqrt{10}\right)\left(-3\sqrt{5}\right) = -21\sqrt{50}$

$= -21\sqrt{25 \cdot 2} = -21(5)\sqrt{2} = -105\sqrt{2}$

50. $\left(7\sqrt[3]{2}\right)\left(-\frac{2}{7}\sqrt[3]{4}\right) = (7)\left(-\frac{2}{7}\right)\sqrt[3]{4 \cdot 2}$

$= -2\sqrt[3]{8} = -2(2) = -4$

51. $\sqrt[3]{4xy^2}\,\sqrt[3]{8xy^5} = \sqrt[3]{32x^2y^7}$

$= \sqrt[3]{8 \cdot 4x^2\left(y^2\right)^3 y} = 2y^2\sqrt[3]{4x^2y}$

52. $\dfrac{\sqrt{98x^3}}{\sqrt{2x^{-2}}} = \sqrt{\dfrac{98}{2}x^{3-(-2)}} = \sqrt{49x^5}$

$= \sqrt{49\left(x^2\right)^2 x} = 7x^2\sqrt{x}$

53. $5\sqrt{3}\left(2\sqrt{6}+4\sqrt{15}\right) = 10\sqrt{18}+20\sqrt{45}$

$= 10\sqrt{9 \cdot 2}+20\sqrt{9 \cdot 5}$

$= 10(3)\sqrt{2}+20(3)\sqrt{5}$

$= 30\sqrt{2}+60\sqrt{5}$

54. $\left(5\sqrt{2}-4\sqrt{3}\right)\left(7\sqrt{2}+3\sqrt{3}\right)$

$= 35(2)+15\sqrt{6}-28\sqrt{6}-12(3)$

$= 70+15\sqrt{6}-28\sqrt{6}-36$

$= 34-13\sqrt{6}$

55. $\sqrt{2x}\left(\sqrt{6x}+3\sqrt{x}\right) = \sqrt{12x^2}+3\sqrt{2x^2}$

$= \sqrt{4 \cdot 3x^2}+3\sqrt{2x^2} = 2x\sqrt{3}+3x\sqrt{2}$

56. $\dfrac{\sqrt[4]{64x^7}}{\sqrt[4]{2x^2}} = \sqrt[4]{\dfrac{64x^7}{2x^2}} = \sqrt[4]{32x^5}$

$= \sqrt[4]{16 \cdot 2x^4x} = 2x\sqrt[4]{2x}$

57. $\left(\sqrt{7x}+\sqrt{3y}\right)\left(\sqrt{7x}-\sqrt{3y}\right) = 7x-3y$

58. $11\sqrt{3}-7\sqrt[6]{3^3} = 11\sqrt{3}-7\left(3^3\right)^{\frac{1}{6}}$

$= 11\sqrt{3}-7 \cdot 3^{\frac{1}{2}} = 11\sqrt{3}-7\sqrt{3} = 4\sqrt{3}$

59. $\dfrac{6}{\sqrt{3}-1} \cdot \dfrac{\sqrt{3}+1}{\sqrt{3}+1} = \dfrac{6\left(\sqrt{3}+1\right)}{3-1} = \dfrac{6\left(\sqrt{3}+1\right)}{2}$

$= 3\left(\sqrt{3}+1\right) \quad \left[\text{or } 3\sqrt{3}+3\right]$

60. $\dfrac{\sqrt{7}}{\sqrt{5}+\sqrt{3}} \cdot \dfrac{\sqrt{5}-\sqrt{3}}{\sqrt{5}-\sqrt{3}} = \dfrac{\sqrt{35}-\sqrt{21}}{5-3}$

$= \dfrac{\sqrt{35}-\sqrt{21}}{2}$

61. $\dfrac{7}{2\sqrt{5}-3\sqrt{7}} \cdot \dfrac{2\sqrt{5}+3\sqrt{7}}{2\sqrt{5}+3\sqrt{7}} = \dfrac{7\left(2\sqrt{5}+3\sqrt{7}\right)}{4(5)-9(7)}$

$= \dfrac{7\left(2\sqrt{5}+3\sqrt{7}\right)}{-43} = \dfrac{14\sqrt{5}+21\sqrt{7}}{-43}$

$\left[\text{or } -\dfrac{14\sqrt{5}+21\sqrt{7}}{43} = \dfrac{-14\sqrt{5}-21\sqrt{7}}{43}\right]$

62. $\dfrac{\sqrt{y}+5}{\sqrt{y}-3} \cdot \dfrac{\sqrt{y}+3}{\sqrt{y}+3} = \dfrac{y+3\sqrt{y}+5\sqrt{y}+15}{y-9}$

$= \dfrac{y+8\sqrt{y}+15}{y-9}$

63. $\dfrac{\sqrt{7}+\sqrt{3}}{\sqrt{7}-\sqrt{3}}\cdot\dfrac{\sqrt{7}+\sqrt{3}}{\sqrt{7}+\sqrt{3}}=\dfrac{7+\sqrt{21}+\sqrt{21}+3}{7-3}$

$=\dfrac{10+2\sqrt{21}}{4}=\dfrac{2\left(5+\sqrt{21}\right)}{4}=\dfrac{5+\sqrt{21}}{2}$

64. $\dfrac{2\sqrt{x}}{\sqrt{x}+\sqrt{y}}\cdot\dfrac{\sqrt{x}-\sqrt{y}}{\sqrt{x}-\sqrt{y}}=\dfrac{2x-2\sqrt{xy}}{x-y}$

65. $\sqrt[3]{5}\sqrt{5}=5^{\frac{1}{3}}\cdot 5^{\frac{1}{2}}=5^{\frac{1}{3}+\frac{1}{2}}$

$=5^{\frac{2}{6}+\frac{3}{6}}=5^{\frac{5}{6}}=\sqrt[6]{5^{5}}\left(\text{or } \sqrt[6]{3125}\right)$

66. $\dfrac{\sqrt[3]{3}}{\sqrt[4]{3}}=\dfrac{3^{\frac{1}{3}}}{3^{\frac{1}{4}}}=3^{\frac{1}{3}-\frac{1}{4}}=3^{\frac{4}{12}-\frac{3}{12}}$

$=3^{\frac{1}{12}}=\sqrt[12]{3}$

67. $\sqrt[12]{x^{4}}=\left(x^{4}\right)^{\frac{1}{12}}=x^{\frac{1}{3}}=\sqrt[3]{x}$

68. $\sqrt{2x+4}=6$ check: $\sqrt{2(16)+4}=6$

$\left(\sqrt{2x+4}\right)^{2}=6^{2}$ $\sqrt{36}=6$

$2x+4=36$ $6=6$

$2x=32$ $\{16\}$

$x=16$

69. $\sqrt[3]{2y-1}-3=0$

$\sqrt[3]{2y-1}=3$

$\left(\sqrt[3]{2y-1}\right)^{3}=3^{3}$

$2y-1=27$

$2y=28$

$y=14$

check: $\sqrt[3]{2(14)-1}-3=0$

$\sqrt[3]{27}-3=0$

$3-3=0$ $\{14\}$

70. $\sqrt{5y-5}=\sqrt{4y-3}$

$\left(\sqrt{5y-5}\right)^{2}=\left(\sqrt{4y-3}\right)^{2}$

$5y-5=4y-3$

$y-5=-3$

$y=2$

check: $\sqrt{5(2)-5}=\sqrt{4(2)-3}$

$\sqrt{5}=\sqrt{5}$ $\{2\}$

71. $\sqrt{4x}+4=0$

$\sqrt{4x}=-4$

$\left(\sqrt{4x}\right)^{2}=(-4)^{2}$

$4x=16$

$x=4$

check: $\sqrt{4(4)}+4=0$

$\sqrt{16}+4=0$

$4+4=0$

$8\neq 0$ $\varnothing$

72.　$\sqrt{y-3} = 1 - \sqrt{y+2}$

$\left(\sqrt{y-3}\right)^2 = \left(1 - \sqrt{y+2}\right)^2$

$y - 3 = \left(1 - \sqrt{y+2}\right)\left(1 - \sqrt{y+2}\right)$

$y - 3 = 1 - 2\sqrt{y+2} + y + 2$

$y - 3 = y + 3 - 2\sqrt{y+2}$

$-6 = -2\sqrt{y+2}$

$3 = \sqrt{y+2}$

$3^2 = \left(\sqrt{y+2}\right)^2$

$9 = y + 2$

$7 = y$

check: $\sqrt{7-3} = 1 - \sqrt{7+2}$

$\sqrt{4} = 1 - \sqrt{9}$

$2 = 1 - 3$

$2 \neq -2$　　$\varnothing$

73.　$1010 = 200\sqrt{x} + 10$

$1000 = 200\sqrt{x}$

$5 = \sqrt{x}$

$5^2 = \left(\sqrt{x}\right)^2$

$25 = x$; 25 tons

74.　$\sqrt{-20} = \sqrt{20}i = \sqrt{4 \cdot 5}i = 2\sqrt{5}i$

75.　$2\sqrt{-100} + 3\sqrt{-36} = 2\sqrt{100}i + 3\sqrt{36}i$

$= 2(10)i + 3(6)i = 20i + 18i = 38i$

76.　$\sqrt{-5}\sqrt{-9} = \sqrt{5}i\sqrt{9}i = \sqrt{45}i^2$

$\sqrt{9 \cdot 5}(-1) = -3\sqrt{5}$

77.　$\dfrac{10\sqrt{-24}}{-2\sqrt{6}} = \dfrac{10\sqrt{6 \cdot 4}i}{-2\sqrt{6}} = \dfrac{20\sqrt{6}i}{-2\sqrt{6}} = -10i$

78.　$7x - 5 = 9$　and　$-3 = 3y - 2$

$7x = 14$　　　　　$-1 = 3y$

$x = 2$　and　$-\dfrac{1}{3} = y$

79.　$(7 + 12i) + (5 - 10i) = 12 + 2i$

80.　$(7 - 12i) - (-3 - 7i)$

$= (7 - 12i) + (3 + 7i) = 10 - 5i$

81.　$(7 - 5i)(2 + 3i) = 14 + 21i - 10i - 15i^2$

$= 14 + 21i - 10i - 15(-1) = 29 + 11i$

82.　$(2 + 5i)^2 = (2 + 5i)(2 + 5i)$

$= 4 + 10i + 10i + 25i^2$

$= 4 + 20i + 25(-1) = -21 + 20i$

83.　$\dfrac{3i}{5+i} \cdot \dfrac{5-i}{5-i} = \dfrac{15i - 3i^2}{25 - i^2}$

$= \dfrac{15i - 3(-1)}{25 - (-1)} = \dfrac{3 + 15i}{26}$

$= \dfrac{3}{26} + \dfrac{15i}{26}$

84.　$\dfrac{3-4i}{4+2i} \cdot \dfrac{4-2i}{4-2i} = \dfrac{12 - 6i - 16i + 8i^2}{16 - 4i^2}$

$= \dfrac{12 - 22i + 8(-1)}{16 - 4(-1)} = \dfrac{4 - 22i}{20}$

$= \dfrac{4}{20} - \dfrac{22}{20}i = \dfrac{1}{5} - \dfrac{11}{10}i$

85. $\dfrac{5+i}{3i} \cdot \dfrac{i}{i} = \dfrac{5i+i^2}{3i^2} = \dfrac{5i+(-1)}{3(-1)}$

$\quad = \dfrac{-1+5i}{-3} = \dfrac{-1}{-3} + \dfrac{5i}{-3} = \dfrac{1}{3} - \dfrac{5i}{3}$

86. $i^{23} = i^{22}i = \left(i^2\right)^{11} i$

$\quad = (-1)^{11} i = (-1)i = -i$

Problem set 6.1.1

3. $(2x+3)(5x-1)=0$

$2x+3=0$ or $5x-1=0$

$2x=-3$ $\qquad$ $5x=1$

$x=-\dfrac{3}{2}$ $\qquad$ $x=\dfrac{1}{5}$ $\quad \left\{-\dfrac{3}{2},\dfrac{1}{5}\right\}$

5. $3x^2+10x-8=0$

$(3x-2)(x+4)=0$

$3x-2=0$ or $x+4=0$

$3x=2$ $\qquad$ $x=-4$

$x=\dfrac{2}{3}$ $\qquad \left\{-4,\dfrac{2}{3}\right\}$

11. $5x^2+26x+5=0$

$(5x+1)(x+5)=0$

$5x+1=0$ or $x+5=0$

$5x=-1$ $\qquad$ $x=-5$

$x=-\dfrac{1}{5}$ $\qquad \left\{-5,-\dfrac{1}{5}\right\}$

17. $5x^2-8x-21=0$

$(5x+7)(x-3)=0$

$5x+7=0$ or $x-3=0$

$5x=-7$ $\qquad$ $x=3$

$x=-\dfrac{7}{5}$

check $-\dfrac{7}{5}$:

$$5\left(-\dfrac{7}{5}\right)^2-8\left(-\dfrac{7}{5}\right)-21=0$$

$$5\left(\dfrac{49}{25}\right)+\dfrac{56}{5}-21=0$$

$$\dfrac{49}{5}+\dfrac{56}{5}-21=0$$

$$\dfrac{105}{5}-21=0$$

$$21-21=0$$

$$0=0$$

check 3 :

$$5(3)^2-8(3)-21=0$$

$$45-24-21=0$$

$$21-21=0$$

$$0=0$$

$$\left\{3,-\dfrac{7}{5}\right\}$$

19. $\qquad x^2-x=2$

$x^2-x-2=0$

$(x-2)(x+1)=0$

$x-2=0$ or $x+1=0$

$x=2$ $\qquad$ $x=-1$ $\quad \{-1,2\}$

21. $3x^2 - 17x = -10$

$3x^2 - 17x + 10 = 0$

$(3x - 2)(x - 5) = 0$

$3x - 2 = 0$ or $x - 5 = 0$

$x = \frac{2}{3}$ $x = 5$ $\left\{\frac{2}{3},\, 5\right\}$

23. $x(x - 3) = 54$

$x^2 - 3x - 54 = 0$

$(x + 6)(x - 9) = 0$

$x + 6 = 0$ or $x - 9 = 0$

$x = -6$ $x = 9$

check -6 : $-6(-6 - 3) = 54$

$-6(-9) = 54$

$54 = 54$

check 9 : $9(9 - 3) = 54$

$9(6) = 54$

$54 = 54$

$\{-6,\, 9\}$

27. $x^2 = \frac{5}{6}x + \frac{2}{3}$

$6x^2 = 6\left(\frac{5}{6}x + \frac{2}{3}\right)$

$6x^2 = 5x + 4$

$6x^2 - 5x - 4 = 0$

$(3x - 4)(2x + 1) = 0$

$3x - 4 = 0$ or $2x + 1 = 0$

$3x = 4$ $2x = -1$

$x = \frac{4}{3}$ $x = -\frac{1}{2}$ $\left\{-\frac{1}{2},\, \frac{4}{3}\right\}$

29. $(x + 1)^2 - 5(x + 2) = 3x + 7$

$x^2 + 2x + 1 - 5x - 10 = 3x + 7$

$x^2 - 3x - 9 = 3x + 7$

$x^2 - 6x - 16 = 0$

$(x - 8)(x + 2) = 0$

$x - 8 = 0$ or $x + 2 = 0$

$x = 8$ $x = -2$ $\{-2,\, 8\}$

31. $\frac{1}{6}x^2 + x - \frac{1}{2} = -2$

$6\left(\frac{1}{6}x^2 + x - \frac{1}{2}\right) = 6(-2)$

$x^2 + 6x - 3 = -12$

$x^2 + 6x + 9 = 0$

$(x + 3)(x + 3) = 0$

$x + 3 = 0$

$x = -3$ $\{-3\}$

35. $3\left(x^2 - 4x - 1\right) = 2(x + 1)$

$3x^2 - 12x - 3 = 2x + 2$

$3x^2 - 14x - 5 = 0$

$(3x + 1)(x - 5) = 0$

$3x + 1 = 0$ or $x - 5 = 0$

$x = -\frac{1}{3}$ $x = 5$ $\left\{-\frac{1}{3},\, 5\right\}$

37. $9x^2 + 6x = -1$

$9x^2 + 6x + 1 = 0$

$(3x + 1)(3x + 1) = 0$

$3x + 1 = 0$

$x = -\frac{1}{3}$ $\left\{-\frac{1}{3}\right\}$

41. $(x + 2)(x - 5) = 8$

$x^2 - 3x - 10 = 8$

$x^2 - 3x - 18 = 0$

$(x - 6)(x + 3) = 0$

$x - 6 = 0$ or $x + 3 = 0$

$x = 6$ $x = -3$ $\{-3,\, 6\}$

43. $2x - [(x+2)(x-3)+8] = 0$

$2x - [x^2 - x - 6 + 8] = 0$

$2x - [x^2 - x + 2] = 0$

$2x - x^2 + x - 2 = 0$

$-x^2 + 3x - 2 = 0$

$(-1)(-x^2 + 3x - 2) = (-1)0$

$x^2 - 3x + 2 = 0$

$(x-1)(x-2) = 0$

$x - 1 = 0 \quad \text{or} \quad x - 2 = 0$

$x = 1 \qquad\qquad x = 2 \qquad \{1, 2\}$

45. $3\left[(x+2)^2 - 4x\right] = 15$

$3\left[x^2 + 4x + 4 - 4x\right] = 15$

$3\left[x^2 + 4\right] = 15$

$3x^2 + 12 = 15$

$3x^2 - 3 = 0$

$\frac{1}{3}(3x^2 - 3) = \frac{1}{3}(0)$

$x^2 - 1 = 0$

$(x+1)(x-1) = 0$

$x + 1 = 0 \quad \text{or} \quad x - 1 = 0$

$x = -1 \qquad\qquad x = 1 \qquad \{-1, 1\}$

47. $x^2 - 4c^2 = 0$

$(x - 2c)(x + 2c) = 0$

$x - 2c = 0 \quad \text{or} \quad x + 2c = 0$

$x = 2c \qquad\qquad x = -2c \qquad \{2c, -2c\}$

49. $x^2 - 4cx + 4c^2 = 0$

$(x - 2c)(x - 2c) = 0$

$x = 2c$

check : $(2c)^2 - 4c(2c) + 4c^2 = 0$

$4c^2 - 8c^2 + 4c^2 = 0$

$8c^2 - 8c^2 = 0$

$0 = 0 \qquad \{2c\}$

51. a. When the ball hits the ground, h = 0.

$-16t^2 + 64t + 80 = 0$

$-\frac{1}{16}(-16t^2 + 64t + 80) = -\frac{1}{16}(0)$

$t^2 - 4t - 5 = 0$

$(t - 5)(t + 1) = 0$

$t - 5 = 0 \quad \text{or} \quad t + 1 = 0$

$t = 5 \qquad\qquad t = -1$ (reject; A negative value for time is meaningless)

The ball reaches the ground in 5 seconds.

b. Since the building is 80 feet tall, the ball passes the edge of the top when h = 80.

$-16t^2 + 64t + 80 = 80$

$-16t^2 + 64t = 0$

$-\frac{1}{16}(-16t^2 + 64t) = -\frac{1}{16}(0)$

$t^2 - 4t = 0$

$t(t - 4) = 0$

$t = 0 \quad \text{or} \quad t - 4 = 0$

$t = 0 \qquad\qquad t = 4$

This occurs after 4 seconds. (The value t = 0 is the value of time when the ball is first thrown upward.)

53. $h = vt - 16t^2 + h_o$

$h = 624 - 224 = 400$

$400 = 80t - 16t^2 + 624$

$16t^2 - 80t - 224 = 0$

$\frac{1}{16}\left(16t^2 - 80t - 224\right) = \frac{1}{16}(0)$

$t^2 - 5t - 14 = 0$

$(t - 7)(t + 2) = 0$

$t - 7 = 0 \qquad t + 2 = 0$

$t = 7 \qquad t = -2 \,(\text{reject})$

The rocket will be 224 feet below the cliff after 7 seconds.

55. $D = \frac{7}{10}x^2 + \frac{3}{4}x$

$295 = \frac{7}{10}x^2 + \frac{3}{4}x$

$20(295) = 20\left(\frac{7}{10}x^2 + \frac{3}{4}x\right)$

$5900 = 14x^2 + 15x$

$0 = 14x^2 + 15x - 5900$

$0 = (x - 20)(14x + 295)$

$x - 20 = 0 \qquad 14x + 295 = 0$

$x = 20 \qquad\qquad x = -\frac{295}{14}\,(\text{reject})$

Keep the speed at 20 m/h.

57. $P = 3500 + 475t - 10t^2$

$7500 = 3500 + 475t - 10t^2$

$10t^2 - 475t + 3750 = 0$

$\frac{1}{5}\left(10t^2 - 475t + 3750\right) = \frac{1}{5}(0)$

$2t^2 - 95t + 750 = 0$

$(2t - 75)(t - 10) = 0$

$2t - 75 = 0 \qquad t - 10 = 0$

$t = \frac{75}{2} \qquad\qquad t = 10$

Since $0 \le t \le 20$, it will take 10 years.

63. $3a^3b^3 + 6a^2b^2 - 24ab$

$= 3ab\left(a^2b^2 + 2ab - 8\right)$

$= 3ab(ab + 4)(ab - 2)$

63. $\sqrt[3]{\dfrac{16x^4}{y^7}} = \sqrt[3]{\dfrac{8 \cdot 2x^3x}{\left(y^2\right)^3 y}} = \dfrac{2x\sqrt[3]{2x}}{y^2\sqrt[3]{y}} \cdot \dfrac{\sqrt[3]{y^2}}{\sqrt[3]{y^2}}$

$= \dfrac{2x\sqrt[3]{2xy^2}}{y^2\sqrt[3]{y^3}} = \dfrac{2x\sqrt[3]{2xy^2}}{y^3}$

64. $x - 1 \,\big|\, x^5 + 0x^4 + 0x^3 + 0x^2 + 0x + 1$

$$\begin{array}{r|rrrrrr} 1 & 1 & 0 & 0 & 0 & 0 & 1 \\ & & 1 & 1 & 1 & 1 & 1 \\ \hline & 1 & 1 & 1 & 1 & 1 & 2 \end{array}$$

Quotient:

$$x^4 + x^3 + x^2 + x + 1 + \frac{2}{x - 1}$$

Problem set 6.1.2

3. $x(x+4)\left(\dfrac{2}{x}+\dfrac{9}{x+4}\right)=x(x+4)\cdot 1$

$2(x+4)+9x=x(x+4)$

$2x+8+9x=x^2+4x$

$11x+8=x^2+4x$

$0=x^2-7x+8$

$0=(x+1)(x-8)$

$x=-1\qquad x=8\qquad\{-1,\,8\}$

5. $4(x-1)(x-4)\left[\dfrac{1}{x-1}+\dfrac{1}{x-4}\right]$

$=4(x-1)(x-4)\cdot\dfrac{5}{4}$

$4(x-4)+4(x-1)=5(x-1)(x-4)$

$4x-16+4x-4=5x^2-25x+20$

$8x-20=5x^2-25x+20$

$0=5x^2-33x+40$

$0=(5x-8)(x-5)$

$5x-8=0\qquad x-5=0$

$x=\dfrac{8}{5}\qquad x=5\qquad\left\{\dfrac{8}{5},\,5\right\}$

9. $(y+5)(y-1)(y-6)\left[\dfrac{7}{y+5}-\dfrac{3}{y-1}\right]$

$=(y+5)(y-1)(y-6)\dfrac{8}{y-6}$

$7(y-1)(y-6)-3(y+5)(y-6)$

$=8(y+5)(y-1)$

$7y^2-49y+42-3y^2+3y+90$

$=8y^2+32y-40$

$4y^2-46y+132=8y^2+32y-40$

$0=4y^2+78y-172$

$\dfrac{1}{2}\cdot 0=\dfrac{1}{2}\left(4y^2+78y-172\right)$

$0=2y^2+39y-86$

$0=(2y+43)(y-2)$

$2y+43=0\qquad y-2=0$

$y=-\dfrac{43}{2}\qquad y=2\qquad\left\{-\dfrac{43}{2},\,2\right\}$

13. $\dfrac{x}{x-5}+\dfrac{17}{25-x^2}=\dfrac{1}{x+5}$

$(x-5)(x+5)\left[\dfrac{x}{x-5}+\dfrac{17}{25-x^2}\right]$

$=(x-5)(x+5)\left[\dfrac{1}{x+5}\right]$

$x(x+5)+(-1)17=x-5$

$x^2+5x-17=x-5$

$x^2+4x-12=0$

$(x+6)(x-2)=0$

$x+6=0\qquad x-2=0$

$x=-6\qquad x=2\qquad\{-6,\,2\}$

15. $\dfrac{5}{y-3} = \dfrac{30}{y^2-9} + 1$

$(y+3)(y-3)\left[\dfrac{5}{y-3}\right]$

$\qquad = (y+3)(y-3)\left[\dfrac{30}{y^2-9} + 1\right]$

$5(y+3) = 30 + (y+3)(y-3)$

$5y + 15 = 30 + y^2 - 9$

$0 = y^2 - 5y + 6$

$0 = (y-3)(y-2)$

$y - 3 = 0 \qquad y - 2 = 0$

$y = 3 \qquad y = 2$

Reject 3 (it causes division by zero in the original equation). $\{2\}$

21. $\dfrac{x+2}{x^2-x} - \dfrac{6}{x^2-1} = 0$

$x(x-1)(x+1)\left[\dfrac{x+2}{x(x-1)} - \dfrac{6}{(x-1)(x+1)}\right]$

$\qquad = x(x-1)(x+1)\cdot 0$

$(x+1)(x+2) - 6x = 0$

$x^2 + 3x + 2 - 6x = 0$

$x^2 - 3x + 2 = 0$

$(x-1)(x-2) = 0$

$x - 1 = 0 \qquad x - 2 = 0$

$x = 1$ (reject ; causes division by 0)

$x = 2 \qquad \{2\}$

23. Even though this equation appears "complicated," it is not quadratic.

$\dfrac{1}{x^3-8} - \dfrac{2}{x^2+2x+4} = \dfrac{3}{(2-x)\left(x^2+2x+4\right)}$

$(x-2)\left(x^2+2x+4\right)\cdot$

$\left[\dfrac{1}{(x-2)\left(x^2+2x+4\right)} - \dfrac{2}{x^2+2x+4}\right]$

$\qquad = (x-2)\left(x^2+2x+4\right)\cdot$

$\left[\dfrac{3}{(2-x)\left(x^2+2x+4\right)}\right]$

$1 - 2(x-2) = (-1)3$

$1 - 2x + 4 = -3$

$-2x = -8$

$x = 4 \qquad \{4\}$

25. $5y^{-2} + 1 = 6y^{-1}$

$\dfrac{5}{y^2} + 1 = \dfrac{6}{y}$

$y^2\left(\dfrac{5}{y^2} + 1\right) = y^2\left(\dfrac{6}{y}\right)$

$5 + y^2 = 6y$

$y^2 - 6y + 5 = 0$

$(y-5)(y-1) = 0$

$y = 5 \qquad y = 1 \qquad \{1, 5\}$

27. $\sqrt{10y} = y$

$$\left(\sqrt{10y}\right)^2 = (y)^2$$

$$10y = y^2$$

$$0 = y^2 - 10y$$

$$0 = y(y - 10)$$

$$y = 0 \qquad y = 10$$

check 0: $\sqrt{10(0)} = 0$

$$0 = 0$$

check 10: $\sqrt{10(10)} = 10$

$$10 = 10 \qquad \{0, \ 10\}$$

29. $5 = x - \sqrt{x-3}$

$$\sqrt{x-3} = x - 5$$

$$\left(\sqrt{x-3}\right)^2 = (x-5)^2$$

$$x - 3 = x^2 - 10x + 25$$

$$0 = x^2 - 11x + 28$$

$$0 = (x-7)(x-4)$$

$$x = 7 \qquad x = 4$$

check 7 : $5 = 7 - \sqrt{7-3}$

$$5 = 7 - \sqrt{4}$$

$$5 = 7 - 2$$

$$5 = 5$$

check 4 : $5 = 4 - \sqrt{4-3}$

$$5 = 4 - \sqrt{1}$$

$$5 \neq 3 \qquad \text{4 is extraneous.}$$

$$\{7\}$$

33. $\sqrt{2x+6} - 3x = -11$

$$\sqrt{2x+6} = 3x - 11$$

$$\left(\sqrt{2x+6}\right)^2 = (3x-11)^2$$

$$2x + 6 = 9x^2 - 66x + 121$$

$$0 = 9x^2 - 68x + 115$$

$$0 = (x-5)(9x-23)$$

$$x = 5 \qquad x = \frac{23}{9}$$

check 5 :

$$\sqrt{2(5)+6} - 3(5) = -11$$

$$4 - 15 = -11$$

$$-11 = -11$$

check $\frac{23}{9}$:

$$\sqrt{2\left(\frac{23}{9}\right)+6} - 3\left(\frac{23}{9}\right) = -11$$

$$\sqrt{\frac{46}{9} + \frac{54}{9}} - \frac{23}{3} = -11$$

$$\sqrt{\frac{100}{9}} - \frac{23}{3} = -11$$

$$\frac{10}{3} - \frac{23}{3} \neq -11$$

$\frac{23}{9}$ is extraneous. $\{5\}$

37. $\left(\sqrt{y}+1\right)^2 = \left(\sqrt{5y-1}\right)^2$

$\left(\sqrt{y}+1\right)\left(\sqrt{y}+1\right) = 5y-1$

$y+2\sqrt{y}+1 = 5y-1$

$2\sqrt{y} = 4y-2$

$\frac{1}{2}\left(2\sqrt{y}\right) = \frac{1}{2}(4y-2)$

$\sqrt{y} = 2y-1$

$\left(\sqrt{y}\right)^2 = (2y-1)^2$

$y = 4y^2-4y+1$

$0 = 4y^2-5y+1$

$0 = (4y-1)(y-1)$

$y = \frac{1}{4} \qquad y = 1$

check $\frac{1}{4}$:

$\sqrt{\frac{1}{4}}+1 = \sqrt{5\left(\frac{1}{4}\right)-1}$

$\frac{1}{2}+1 = \sqrt{\frac{5}{4}-1}$

$\frac{1}{2}+1 = \sqrt{\frac{1}{4}}$

$1\frac{1}{2} \neq \frac{1}{2} \qquad \frac{1}{4}$ is extraneous.

check 1 :

$\sqrt{1}+1 = \sqrt{5(1)-1}$

$1+1 = \sqrt{4}$

$2 = 2 \qquad \{1\}$

39. $\sqrt{4z-3} - \sqrt{8z+1} + 2 = 0$

$\sqrt{4z-3} = \sqrt{8z+1} - 2$

$\left(\sqrt{4z-3}\right)^2 = \left(\sqrt{8z+1} - 2\right)^2$

$4z-3 = \left(\sqrt{8z+1} - 2\right)\left(\sqrt{8z+1} - 2\right)$

$4z-3 = 8z+1 - 4\sqrt{8z+1} + 4$

$4z-3 = 8z+5 - 4\sqrt{8z+1}$

$-4z-8 = -4\sqrt{8z+1}$

$\frac{1}{4}(-4z-8) = \frac{1}{4}\left(-4\sqrt{8z+1}\right)$

$z+2 = \sqrt{8z+1}$

$(z+2)^2 = \left(\sqrt{8z+1}\right)^2$

$z^2+4z+4 = 8z+1$

$z^2-4z+3 = 0$

$(z-1)(z-3) = 0$

$z = 1 \qquad z = 3 \qquad$ (Both check.)

$\{1, 3\}$

41. $2\sqrt{3x-2}+\sqrt{2x-3}=5$

$$\left(2\sqrt{3x-2}\right)^2=\left(5-\sqrt{2x-3}\right)^2$$

$$4(3x-2)=25-10\sqrt{2x-3}+2x-3$$

$$12x-8=2x+22-10\sqrt{2x-3}$$

$$10x-30=-10\sqrt{2x-3}$$

$$x-3=-\sqrt{2x-3}$$

$$(x-3)^2=\left(-\sqrt{2x-3}\right)^2$$

$$x^2-6x+9=2x-3$$

$$x^2-8x+12=0$$

$$(x-6)(x-2)=0$$

$$x=6 \qquad x=2$$

check 6 : $2\sqrt{3(6)-2}+\sqrt{2(6)-3}=5$

$$2(4)+3=5$$

$$11\neq5$$

6 is extraneous.

check 2 : $2\sqrt{3(2)-2}+\sqrt{2(2)-3}=5$

$$2(2)+1=5$$

$$5=5$$

$$\{2\}$$

47. $\left(\sqrt{6y-2}\right)^2=\left(\sqrt{2y+3}-\sqrt{4y-1}\right)^2$

$$6y-2=2y+3-2\sqrt{2y+3}\sqrt{4y-1}+4y-1$$

$$6y-2=6y+2-2\sqrt{2y+3}\sqrt{4y-1}$$

$$-4=-2\sqrt{2y+3}\sqrt{4y-1}$$

$$2=\sqrt{2y+3}\sqrt{4y-1}$$

$$2^2=\left(\sqrt{2y+3}\sqrt{4y-1}\right)^2$$

$$4=(2y+3)(4y-1)$$

$$4=8y^2+10y-3$$

$$0=8y^2+10y-7$$

$$0=(4y+7)(2y-1)$$

$$4y+7=0 \qquad 2y-1=0$$

$$y=-\frac{7}{4} \qquad y=\frac{1}{2}$$

check $-\frac{7}{4}$:

$$\sqrt{6\left(-\frac{7}{4}\right)-2}=\sqrt{2\left(-\frac{7}{4}\right)+3}-\sqrt{4\left(-\frac{7}{4}\right)-1}$$

$$\sqrt{\frac{-21}{2}-\frac{4}{2}}=\sqrt{-\frac{7}{2}+\frac{6}{2}}-\sqrt{-7-1}$$

$$\sqrt{-\frac{25}{2}}=\sqrt{\frac{-1}{2}}-\sqrt{-8}$$

$$\frac{5i}{\sqrt{2}}=\frac{i}{\sqrt{2}}-2\sqrt{2}i\cdot\frac{\sqrt{2}}{\sqrt{2}}$$

$$\frac{5i}{\sqrt{2}}=\frac{i}{\sqrt{2}}-\frac{4i}{\sqrt{2}}$$

$$\frac{5i}{\sqrt{2}}\neq-\frac{3i}{\sqrt{2}} \qquad -\frac{7}{4}\text{ is extraneous.}$$

check $\frac{1}{2}$:

$$\sqrt{6\left(\frac{1}{2}\right)-2}=\sqrt{2\left(\frac{1}{2}\right)+3}-\sqrt{4\left(\frac{1}{2}\right)-1}$$

$$\sqrt{1}=\sqrt{4}-\sqrt{1}$$

$$1=1$$

$$\left\{\frac{1}{2}\right\}$$

51. $\left(\sqrt{y+1+\sqrt{7y+4}}\right)^2 = 3^2$

$y+1+\sqrt{7y+4} = 9$

$\sqrt{7y+4} = 8-y$

$\left(\sqrt{7y+4}\right)^2 = (8-y)^2$

$7y+4 = 64-16y+y^2$

$0 = y^2 - 23y + 60$

$0 = (y-20)(y-3)$

$y-20 = 0 \qquad y-3 = 0$

$y = 20 \qquad y = 3$

check 20 :

$\sqrt{20+1+\sqrt{7(20)+4}} - 3 = 0$

$\sqrt{21+\sqrt{144}} - 3 = 0$

$\sqrt{21+12} - 3 = 0$

$\sqrt{33} - 3 \neq 0 \qquad$ 20 is extraneous.

check 3 :

$\sqrt{3+1\sqrt{7(3)+4}} - 3 = 0$

$\sqrt{4+\sqrt{25}} - 3 = 0$

$\sqrt{4+5} - 3 = 0$

$\sqrt{9} - 3 = 0$

$0 = 0 \qquad \{3\}$

53. $\sqrt[3]{z^2-1} = 2$

$\left(\sqrt[3]{z^2-1}\right)^3 = 2^3$

$z^2 - 1 = 8$

$z^2 - 9 = 0$

$(z+3)(z-3) = 0$

$z = -3 \qquad z = 3$

$\{-3, 3\}$ (both check)

57. x : number

$\sqrt{2x-3} = \sqrt{x+2} + 1$

$\left(\sqrt{2x-3}\right)^2 = \left(\sqrt{x+2}+1\right)^2$

$2x-3 = x+2+2\sqrt{x+2}+1$

$x-6 = 2\sqrt{x+2}$

$(x-6)^2 = \left(2\sqrt{x+2}\right)^2$

$x^2 - 12x + 36 = 4(x+2)$

$x^2 - 12x + 36 = 4x + 8$

$x^2 - 16x + 28 = 0$

$(x-14)(x-2) = 0$

$x = 14 \qquad x = 2$ (extraneous)

The number is 14.

59. If there is no demand, no calculators sell, so N = 0

$\sqrt{100-x^2} - 6 = 0$

$\sqrt{100-x^2} = 6$

$100 - x^2 = 36$

$0 = x^2 - 64$

$0 = (x-8)(x+8)$

$x = 8 \qquad x = -8$ (meaningless)

The smallest price is $8.

61. $y^{\frac{2}{3}} - y^{\frac{1}{3}} - 6 = 0$

$$\text{let } t = y^{\frac{1}{3}}$$

$$t^2 - t - 6 = 0$$

$$(t-3)(t+2) = 0$$

$$t = 3 \quad \text{or} \quad t = -2$$

$$y^{\frac{1}{3}} = t$$

$$y^{\frac{1}{3}} = 3 \qquad\qquad y^{\frac{1}{3}} = -2$$

$$\left(y^{\frac{1}{3}}\right)^3 = 3^3 \qquad \left(y^{\frac{1}{3}}\right)^3 = (-2)^3$$

$$y = 27 \qquad\qquad y = -8$$

$$\{-8,\ 27\}$$

65. $2y - 3y^{\frac{1}{2}} + 1 = 0$

$$\text{let } t = y^{\frac{1}{2}}$$

$$2t^2 - 3t + 1 = 0$$

$$(t-1)(2t-1) = 0$$

$$t = 1 \qquad\qquad t = \frac{1}{2}$$

$$y^{\frac{1}{2}} = t$$

$$y^{\frac{1}{2}} = 1 \qquad\qquad y^{\frac{1}{2}} = \frac{1}{2}$$

$$\left(y^{\frac{1}{2}}\right)^2 = 1^2 \qquad \left(y^{\frac{1}{2}}\right)^2 = \left(\frac{1}{2}\right)^2$$

(Note: check solutions, since both sides are raised to an even power.)

$$y = 1 \qquad\qquad y = \frac{1}{4}$$

check 1: $2(1) - 3(1)^{\frac{1}{2}} + 1 = 0$

$$2 - 3\sqrt{1} + 1 = 0$$

$$2 - 3 + 1 = 0$$

$$-1 + 1 = 0$$

$$0 = 0$$

check $\frac{1}{4}$: $2\left(\frac{1}{4}\right) - 3\left(\frac{1}{4}\right)^{\frac{1}{2}} + 1 = 0$

$$\frac{1}{2} - 3\sqrt{\frac{1}{4}} + 1 = 0$$

$$\frac{1}{2} - 3\left(\frac{1}{2}\right) + 1 = 0$$

$$\frac{1}{2} - \frac{3}{2} + 1 = 0$$

$$0 = 0$$

$$\left\{\frac{1}{4},\ 1\right\}$$

67. $x + 2x^{\frac{1}{2}} - 24 = 0$

$$\text{Let } t = x^{\frac{1}{2}}$$
$$t^2 + 2t - 24 = 0$$
$$(t + 6)(t - 4) = 0$$
$$t = -6 \qquad t = 4$$
$$x^{\frac{1}{2}} = t$$
$$x^{\frac{1}{2}} = -6 \qquad\qquad x^{\frac{1}{2}} = 4$$
$$\left(x^{\frac{1}{2}}\right)^2 = (-6)^2 \qquad \left(x^{\frac{1}{2}}\right)^2 = 4^2$$
$$x = 36 \qquad\qquad x = 16$$

check 36 :
$$36 + 2(36)^{\frac{1}{2}} - 24 = 0$$
$$36 + 2\sqrt{36} - 24 = 0$$
$$36 + 12 - 24 \neq 0$$

36 is extraneous.

check 16 :
$$16 + 2(16)^{\frac{1}{2}} - 24 = 0$$
$$16 + 2\sqrt{16} - 24 = 0$$
$$16 + 8 - 24 = 0$$
$$0 = 0 \qquad \{16\}$$

69. $y^4 - 13y^2 + 36 = 0$

$$\text{Let } t = y^2$$
$$t^2 - 13t + 36 = 0$$
$$(t - 4)(t - 9) = 0$$
$$t = 4 \qquad t = 9$$
$$y^2 = t$$
$$y^2 = 4 \qquad\qquad y^2 = 9$$
$$y^2 - 4 = 0 \qquad\qquad y^2 - 9 = 0$$
$$(y + 2)(y - 2) \qquad (y - 3)(y + 3) = 0$$
$$y = -2 \quad y = 2 \qquad y = 3 \quad y = -3$$
$$\{-3, \ -2, \ 2, \ 3\}$$

73. $z - 2\sqrt{z} - 15 = 0$

$$\text{Let } t = \sqrt{z}$$
$$t^2 - 2t - 15 = 0$$
$$(t + 3)(t - 5) = 0$$
$$t = -3 \qquad t = 5$$
$$\sqrt{z} = t$$
$$\sqrt{z} = -3 \qquad\qquad \sqrt{z} = 5$$
$$\left(\sqrt{z}\right)^2 = (-3)^2 \qquad \left(\sqrt{z}\right)^2 = 5^2$$
$$z = 9 \qquad\qquad z = 25$$

check 9 :
$$9 - 2\sqrt{9} - 15 = 0$$
$$9 - 6 - 15 = 0$$
$$-12 \neq 0 \quad \text{9 is extraneous.}$$

check 25 :
$$25 - 2\sqrt{25} - 15 = 0$$
$$25 - 2(5) - 15 = 0$$
$$25 - 10 - 15 = 0$$
$$0 = 0 \qquad \{25\}$$

75. $(y - 2)^{\frac{2}{3}} - 3(y - 2)^{\frac{1}{3}} + 2 = 0$

$$\text{Let } t = (y - 2)^{\frac{1}{3}}$$
$$t^2 - 3t + 2 = 0$$
$$(t - 1)(t - 2) = 0$$
$$t = 1 \qquad t = 2$$
$$(y - 2)^{\frac{1}{3}} = 1 \qquad\qquad (y - 2)^{\frac{1}{3}} = 2$$
$$\left[(y - 2)^{\frac{1}{3}}\right]^3 = 1^3 \qquad \left[(y - 2)^{\frac{1}{3}}\right]^3 = 2^3$$
$$y - 2 = 1 \qquad\qquad y - 2 = 8$$
$$y = 3 \qquad\qquad y = 10$$
$$\{3, \ 10\}$$

77. $(z-3)^2 + (z-3) - 20 = 0$

Let $t = z - 3$

$t^2 + t - 20 = 0$

$(t+5)(t-4) = 0$

$t = -5 \qquad t = 4$

$z - 3 = -5 \quad z - 3 = 4$

$z = -2 \qquad z = 7$

$\{-2,\ 7\}$

81. $(x-1)^{-2} + 2(x-1)^{-1} - 3 = 0$

Let $t = (x-1)^{-1}$

$t^2 + 2t - 3 = 0$

$(t+3)(t-1) = 0$

$t = -3 \quad t = 1$

$(x-1)^{-1} = -3 \qquad (x-1)^{-1} = 1$

$\dfrac{1}{(x-1)} = \dfrac{-3}{1} \qquad \dfrac{1}{(x-1)} = \dfrac{1}{1}$

using the principle for proportions:

$-3(x-1) = 1 \qquad x - 1 = 1$

$-3x + 3 = 1 \qquad\quad x = 2$

$-3x = -2$

$x = \dfrac{2}{3} \qquad \left\{\dfrac{2}{3},\ 2\right\}$

83. $(y-2)^{\frac{1}{2}} = 11(y-2)^{\frac{1}{4}} - 18$

$(y-2)^{\frac{1}{2}} - 11(y-2)^{\frac{1}{4}} + 18 = 0$

Let $t = (y-2)^{\frac{1}{4}}$

$t^2 - 11t + 18 = 0$

$(t-9)(t-2) = 0$

$t = 9 \qquad t = 2$

$(y-2)^{\frac{1}{4}} = 9 \qquad\qquad (y-2)^{\frac{1}{4}} = 2$

$\left[(y-2)^{\frac{1}{4}}\right]^4 = 9^4 \qquad \left[(y-2)^{\frac{1}{4}}\right]^4 = 2^4$

$y - 2 = 6561 \qquad\qquad y - 2 = 16$

$y = 6563 \qquad\qquad y = 18$

Since we raised both sides of the equations

containing $(y-2)^{\frac{1}{4}}$ to an even power, we

must check for extraneous solutions.

check 6563 :

$(6563-2)^{\frac{1}{2}} = 11(6563-2)^{\frac{1}{4}} - 18$

$\sqrt{6561} = 11\left(\sqrt[4]{6561}\right) - 18$

$81 = 11(9) - 18$

$81 = 81$

check 18 :

$(18-2)^{\frac{1}{2}} = 11(18-2)^{\frac{1}{4}} - 18$

$\sqrt{16} = 11\left(\sqrt[4]{16}\right) - 18$

$4 = 11(2) - 18$

$4 = 4$

$\{18,\ 6563\}$

85. $2\left(\sqrt{z}+1\right)^2 + 12 = 11\left(\sqrt{z}+1\right)$

$\quad 2\left(\sqrt{z}+1\right)^2 - 11\left(\sqrt{z}+1\right) + 12 = 0$

$\quad\quad$ Let $t = \sqrt{z}+1$

$\quad 2t^2 - 11t + 12 = 0$

$\quad (t-4)(2t-3) = 0$

$\quad\quad t = 4 \quad\quad\quad\quad t = \frac{3}{2}$

$\quad \sqrt{z}+1 = 4 \quad\quad \sqrt{z}+1 = \frac{3}{2}$

$\quad\quad \sqrt{z} = 3 \quad\quad\quad \sqrt{z} = \frac{1}{2}$

$\quad \left(\sqrt{z}\right)^2 = 3^2 \quad\quad \left(\sqrt{z}\right)^2 = \left(\frac{1}{2}\right)^2$

$\quad\quad z = 9 \quad\quad\quad\quad z = \frac{1}{4}$

$\quad\quad \left\{9, \frac{1}{4}\right\} \quad$ (Both answers check.)

87. $\left(\frac{y^2+5y}{6}\right)^2 = 7\left(\frac{y^2+5y}{6}\right) - 6$

$\quad \left(\frac{y^2+5y}{6}\right) - 7\left(\frac{y^2+5y}{6}\right) + 6 = 0$

$\quad\quad$ Let $t = \frac{y^2+5y}{6}$

$\quad t^2 - 7t + 6 = 0$

$\quad (t-1)(t-6) = 0$

$\quad\quad\quad t = 1 \quad\quad\quad\quad\quad t = 6$

$\quad \frac{y^2+5y}{6} = 1 \quad\quad \frac{y^2+5y}{6} = 6$

$\quad y^2 + 5y = 6 \quad\quad y^2 + 5y = 36$

$\quad y^2 + 5y - 6 = 0 \quad y^2 + 5y - 36 = 0$

$\quad (y+6)(y-1) = 0 \quad (y+9)(y-4) = 0$

$\quad y = -6 \quad y = 1 \quad\quad y = -9 \quad y = 4$

$\quad\quad\quad \{-9, -6, 1, 4\}$

88. $\sqrt[3]{\frac{81a^5}{b^6}} = \sqrt[3]{\frac{27 \cdot 3a^3 a^2}{\left(b^2\right)^3}} = \frac{3a\sqrt[3]{3a^2}}{b^2}$

89. $\dfrac{\frac{1}{b^2} - \frac{1}{ab} - \frac{2}{a^2}}{\frac{1}{b^2} - \frac{3}{ab} + \frac{2}{a^2}} \cdot \dfrac{a^2b^2}{a^2b^2}$

$\quad = \dfrac{a^2 - ab - 2b^2}{a^2 - 3ab + 2b^2} = \dfrac{(a+b)(a-2b)}{(a-b)(a-2b)}$

$\quad = \dfrac{a+b}{a-b}$

90. $\dfrac{\left(x^{\frac{1}{n}}y^n\right)^n}{x^n y^{n^2}} = \dfrac{xy^{n^2}}{x^n y^{n^2}} = x^{1-n}$

$\quad\quad \left(\text{or } \dfrac{1}{x^{n-1}}\right)$

Problem set 6.2.1

3. x : number(s)

$\quad\quad 2x^2 - 5x = 3$

$\quad\quad 2x^2 - 5x - 3 = 0$

$\quad\quad (2x+1)(x-3) = 0$

$\quad\quad 2x+1 = 0 \quad\quad x-3 = 0$

$\quad\quad\quad x = -\frac{1}{2} \quad\quad\quad x = 3$

$\quad$ Numbers : $-\frac{1}{2}, 3$

5. Numbers : x and $6x - 11$

$$x(6x - 11) = 7$$
$$6x^2 - 11x - 7 = 0$$
$$(3x - 7)(2x + 1) = 0$$
$$x = \frac{7}{3} \qquad x = -\frac{1}{2}$$

Numbers : x and $6x - 11$

$$\frac{7}{3} \text{ and } 6\left(\frac{7}{3}\right) - 11 = 3$$

Numbers : x and $6x - 11$

$$-\frac{1}{2} \text{ and } 6\left(-\frac{1}{2}\right) - 11 = -14$$

Thus, the numbers are $\frac{7}{3}$ and 3

or $-\frac{1}{2}$ and -14.

9. Consecutive positive integers:

$$x, \ x + 1, \ x + 2$$
$$(x + 1)^2 - 5(x + 2) = 3x + 7$$
$$x^2 + 2x + 1 - 5x - 10 = 3x + 7$$
$$x^2 - 3x - 9 = 3x + 7$$
$$x^2 - 6x - 16 = 0$$
$$(x - 8)(x + 2) = 0$$
$$x = 8 \qquad x = -2 \text{ (reject ; not positive)}$$

The integers are 8, 9 and 10.

13. Number(s) : x

$$\frac{1}{x} + x = \frac{61}{30}$$
$$30x\left(\frac{1}{x} + x\right) = 30x\left(\frac{61}{30}\right)$$
$$30 + 30x^2 = 61x$$
$$30x^2 - 61x + 30 = 0$$
$$(5x - 6)(6x - 5) = 0$$
$$5x - 6 = 0 \qquad 6x - 5 = 0$$
$$x = \frac{6}{5} \qquad\qquad x = \frac{5}{6}$$

Numbers : $\frac{6}{5}, \frac{5}{6}$

15. Consecutive positive integers:

$$x, \ x + 1$$
$$\frac{1}{x} + \frac{1}{x + 1} = \frac{5}{6}$$
$$6x(x + 1)\left[\frac{1}{x} + \frac{1}{x + 1}\right] = 6x(x + 1) \cdot \frac{5}{6}$$
$$6x + 6 + 6x = 5x^2 + 5x$$
$$0 = 5x^2 - 7x - 6$$
$$0 = (5x + 3)(x - 2)$$
$$5x + 3 = 0 \qquad x - 2 = 0$$
$$x = -\frac{3}{5} \text{ (reject ; not positive)} \quad x = 2$$

Numbers: 2 and 3

18. $|6 - 4x| - 4 = 3$

$$|6 - 4x| = 7$$
$$6 - 4x = 7 \qquad 6 - 4x = -7$$
$$-4x = 1 \qquad\qquad -4x = -13$$
$$x = -\frac{1}{4} \qquad\qquad x = \frac{13}{4}$$
$$\left\{ -\frac{1}{4}, \ \frac{13}{4} \right\}$$

19. $\dfrac{4y+1}{(2y-1)\left(4y^2+2y+1\right)}$

$$+\dfrac{2y}{4y^2+2y+1}\cdot\dfrac{2y-1}{2y-1}$$

$$=\dfrac{4y+1+4y^2-2y}{(2y-1)\left(4y^2+2y+1\right)}$$

$$=\dfrac{4y^2+2y+1}{(2y-1)\left(4y^2+2y+1\right)}=\dfrac{1}{2y-1}$$

20. $3\sqrt[3]{a^5b^7}-8ab\sqrt[3]{a^2b^4}$

$$=3\sqrt[3]{a^3a^2\left(b^2\right)^3b}-8ab\sqrt[3]{a^2b^3b}$$

$$=3ab^2\sqrt[3]{a^2b}-8ab^2\sqrt[3]{a^2b}$$

$$=-5ab^2\sqrt[3]{a^2b}$$

Problem set 6.2.2

3.

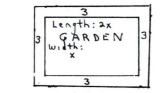

Area of path $=\dfrac{1}{2}$ Area of garden $-\,4$

(Notice that the area of path = area of large rectangle − area of small rectangle)

$$(x+6)(2x+6)-x(2x)=\tfrac{1}{2}\left(2x^2\right)-4$$

$$2x^2+18x+36-2x^2=x^2-4$$

$$18x+36=x^2-4$$

$$0=x^2-18x-40$$

$$(x-20)(x+2)=0$$

$$x=20\qquad x=-2\ (\text{reject})$$

width : $x=20$ ft

7. Height : x   Shorter Base : 2x

Longer Base : 3x + 1

$A=\tfrac{1}{2}H(B_1+B_2)$

$93=\tfrac{1}{2}x(2x+3x+1)$

$186=x(5x+1)$

$0=5x^2+x-186$

$0=(5x+31)(x-6)$

$5x+31=0\qquad\qquad x-6=0$

$x=-\dfrac{31}{5}\ (\text{reject})\qquad x=6$

Height : 6m   Shorter Base : $2(6)=12$m

Longer Base : $3(6)+1=19$m

13. width : x   Height : 4x + 1

V = LWH

$$54=3x(4x+1)$$

$$54=12x^2+3x$$

$$0=12x^2+3x-54$$

$$0=4x^2+x-18$$

$$(4x+9)(x-2)=0$$

$$x=-\dfrac{9}{4}\ (\text{reject})\qquad x=2$$

Width : 2yd   Height : $4(2)+1=9$yd

15. Height : x Length : 2x Width : 4x − 5

$2HL+2LW+2HW=\text{S.A.}$

$2x(2x)+2(2x)(4x-5)+2x(4x-5)=162$

$4x^2+16x^2-20x+8x^2-10x=162$

$28x^2-30x-162=0$

$14x^2-15x-81=0$

$(14x+27)(x-3)=0$

$x=-\dfrac{27}{14}\ (\text{reject})\qquad x=3$

Height : 3yd Length : $2(3)=6$yd

Width : $4(3)-5=7$yd

17. One Leg : x

Other Leg : Perimeter –

$\qquad$ (Hypotenuse + First Leg)

$\qquad = 56 - (25 + x)$

$\qquad = 31 - x$

Pythagorean Theorem:

$(\text{One Leg})^2 + (\text{Other Leg})^2$

$\qquad\qquad\qquad = (\text{Hypotenuse})^2$

$\qquad x^2 + (31 - x)^2 = 25^2$

$\qquad x^2 + 961 - 62x + x^2 = 625$

$\qquad 2x^2 - 62x + 336 = 0$

$\qquad x^2 - 31x + 168 = 0$

$\qquad (x - 24)(x - 7) = 0$

$\qquad x = 24 \qquad x = 7$

Legs : 24 in and 7 in

21. $V = LWH$

$\qquad 168 = (2x - 6)(x - 6)3$

$\qquad 56 = 2x^2 - 18x + 36$

$\qquad 0 = 2x^2 - 18x - 20$

$\qquad 0 = x^2 - 9x - 10$

$\qquad 0 = (x - 10)(x + 1)$

$\qquad x = 10 \qquad x = -1 \text{ (reject)}$

Cardboard: x by $2x$ = 10 in by 20 in

23. 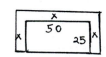 x: Width of path

Area of path = 196

Area of large rectangle – Area of small rectangle = 196

$\qquad (20 + 2x)(25 + 2x) - (20)(25) = 196$

$\qquad 500 + 90x + 4x^2 - 500 = 196$

$\qquad 4x^2 + 90x - 196 = 0$

$\qquad 2x^2 + 45x - 98 = 0$

$\qquad (2x + 49)(x - 2) = 0$

$\qquad x = -\dfrac{49}{2} \text{ (reject)} \qquad x = 2$

Width of path : 2m

25. x: Width of path

Area of pool plus path = Area of pool + 318

$\qquad (50 + 2x)(25 + x) = (50)(25) + 318$

$\qquad 1250 + 100x + 2x^2 = 1568$

$\qquad 2x^2 + 100x - 318 = 0$

$\qquad x^2 + 50x - 159 = 0$

$\qquad (x - 3)(x + 53) = 0$

$\qquad x = 3 \qquad x = -53 \text{ (reject)}$

Width of path : 3 ft.

29. Height of rectangle : x

Length of rectangle : 4x

Altitude of triangle: $x + 1$

Area of rectangle + Area of triangle = 60

$$x(4x) + \frac{1}{2} \cdot 4x(x+1) = 60$$

$$4x^2 + 2x^2 + 2x = 60$$

$$6x^2 + 2x - 60 = 0$$

$$3x^2 + x - 30 = 0$$

$$(x-3)(3x+10) = 0$$

$$x = 3 \qquad x = -\frac{10}{3} \text{ (reject)}$$

Height of rectangle: 3yd

31. x: radius of hollow center

$x + 3$: radius of ball

Volume of ball – Volume of hollow

center $= 684\pi$

$$\frac{4}{3}\pi(x+3)^3 - \frac{4}{3}\pi x^3 = 684\pi$$

$$\frac{4}{3}\pi\left(x^3 + 9x^2 + 27x + 27\right) - \frac{4}{3}\pi x^3 = 684\pi$$

$$\frac{4}{3}\pi x^3 + 12\pi x^2 + 36\pi x + 36\pi$$

$$- \frac{4}{3}\pi x^3 = 684\pi$$

$$12\pi x^2 + 36\pi x + 36\pi = 684\pi$$

$$\frac{1}{12\pi}\left(12\pi x^2 + 36\pi x + 36\pi\right)$$

$$= \frac{1}{12\pi}(684\pi)$$

$$x^2 + 3x + 3 = 57$$

$$x^2 + 3x - 54 = 0$$

$$(x+9)(x-6) = 0$$

$$x = -9 \qquad x = 6$$

radius of ball : $6 + 3 = 9$ in

32. $\dfrac{\sqrt{2}-\sqrt{3}}{\sqrt{2}+\sqrt{3}} \cdot \dfrac{\sqrt{2}-\sqrt{3}}{\sqrt{2}-\sqrt{3}} = \dfrac{2-2\sqrt{6}+3}{2-3}$

$$= \frac{5-2\sqrt{6}}{-1} = -5 + 2\sqrt{6}$$

33. $\left(\dfrac{2}{3} - \dfrac{1}{3}i\right)\left(\dfrac{6}{5} + \dfrac{3}{5}i\right)$

$$= \frac{12}{15} + \frac{6}{15}i - \frac{6}{15}i - \frac{3}{15}i^2$$

$$= \frac{12}{15} - \frac{3}{15}(-1) = \frac{15}{15} = 1$$

34. $\dfrac{2y^2}{y^4-1} - \dfrac{1}{y^2-1} + \dfrac{1}{y^2+1}$

$$= \frac{2y^2}{\left(y^2-1\right)\left(y^2+1\right)} - \frac{1}{y^2-1} \cdot \frac{y^2+1}{y^2+1}$$

$$+ \frac{1}{y^2+1} \cdot \frac{y^2-1}{y^2-1}$$

$$= \frac{2y^2 - y^2 - 1 + y^2 - 1}{\left(y^2-1\right)\left(y^2+1\right)}$$

$$= \frac{2y^2 - 2}{\left(y^2-1\right)\left(y^2+1\right)} = \frac{2\left(y^2-1\right)}{\left(y^2-1\right)\left(y^2+1\right)}$$

$$= \frac{2}{y^2+1}$$

Problem set 6.2.3

1. x: speed on outgoing trip
    x + 30: speed on return trip
    Time on outgoing trip + time on return
    trip = 2

$$\left[RT = D \; ; \; T = \frac{D}{R} \right]$$

$$\frac{40}{x} + \frac{40}{x+30} = 2$$

$$x(x+30)\left[\frac{40}{x} + \frac{40}{x+30} \right] = x(x+30) \cdot 2$$

$$40x + 1200 + 40x = 2x^2 + 60x$$

$$0 = 2x^2 - 20x - 1200$$

$$0 = x^2 - 10x - 600$$

$$0 = (x-30)(x+20)$$

$$x = 30 \qquad x = -20 \text{ (reject)}$$

Speed on outgoing trip: 30 m/h

3. x: speed of first engine
    x + 5: speed of second engine
    Time of first engine + time of second
    engine = 9

$$\left[T = \frac{D}{R} \right]$$

$$\frac{140}{x} + \frac{200}{x+5} = 9$$

$$x(x+5)\left[\frac{140}{x} + \frac{200}{x+5} \right] = x(x+5) \cdot 9$$

$$140x + 700 + 200x = 9x^2 + 45x$$

$$0 = 9x^2 - 295x - 700$$

$$(x-35)(9x+20) = 0$$

$$x = 35 \qquad x = -\frac{20}{9} \text{ (reject)}$$

speed of first engine: 35 m/h
speed of second engine: 40 m/h

5. Time for Mrs. Lovett working alone: x
    Time for Mr. Todd working alone: x + 4
    Part of job done by Mrs. Lovett in 2
    days + Part of job done by Mr. Todd in
    9 days = One complete job

$$\frac{2}{x} + \frac{9}{x+4} = 1$$

$$x(x+4)\left[\frac{2}{x} + \frac{9}{x+4} \right] = x(x+4) \cdot 1$$

$$2x + 8 + 9x = x^2 + 4x$$

$$0 = x^2 - 7x - 8$$

$$0 = (x-8)(x+1)$$

$$x = 8 \qquad x = -1 \text{ (reject)}$$

Time for Mrs. Lovett working alone:
8 days.

9. x: number of people on faculty
    x − 5: number of people who contributed
    Contribution of each person with 5
    fewer people = Contribution of each
    person in the original faculty + 0.40

$$\frac{60}{x-5} = \frac{60}{x} + 0.4$$

$$\frac{60}{x-5} = \frac{60}{x} + \frac{2}{5}$$

$$5x(x-5)\left[\frac{60}{x-5} \right] = 5x(x-5)\left[\frac{60}{x} + \frac{2}{5} \right]$$

$$300x = 300x - 1500 + 2x^2 - 10x$$

$$0 = 2x^2 - 10x - 1500$$

$$(2x+50)(x-30) = 0$$

$$x = -25 \text{ (reject)} \qquad x = 30$$

Number of people who contributed:
x − 5 = 30 − 5 = 25

11. Bob's present age: x

Aunt's present age: $(x-1)^2$

Aunt's age in 2 years = 6 times Bob's age in 2 years

$$(x-1)^2 + 2 = 6(x+2)$$
$$x^2 - 2x + 3 = 6x + 12$$
$$x^2 - 8x - 9 = 0$$
$$(x-9)(x+1) = 0$$
$$x = 9 \qquad x = -1 \text{ (reject)}$$

Bob's present age: 9

Aunt's present age: $(9-1)^2 = 64$

15. (Number of calculators sold)

$\qquad$ · (price per calculator) = 4000

p: price per calculator

$$(1300 - 100p)p = 4000$$
$$0 = 100p^2 - 1300p + 4000$$
$$0 = p^2 - 13p + 40$$
$$(p-8)(p-5) = 0$$
$$p = 8 \qquad p = 5$$

price per calculator: $5 or $8.

17. x: train's speed on old schedule

x + 2: train's speed on new schedule

Time on new schedule = time on old schedule $- \frac{1}{4}$

$$\left[RT = D \; ; \; T = \frac{D}{R} \right]$$

$$\frac{351}{x+2} = \frac{351}{x} - \frac{1}{4}$$

$$4x(x+2)\left[\frac{351}{x+2} \right] = 4x(x+2)\left[\frac{351}{x} - \frac{1}{4} \right]$$

$$1404x = 1404x + 2808 - x^2 - 2x$$
$$x^2 + 2x - 2808 = 0$$
$$(x+54)(x-52) = 0$$
$$x = -54 \text{ (reject)} \qquad x = 52$$

Train's speed on new schedule:

$52 + 2 = 54$ m/h

21.

```
-2 |  1    5    7    2
         -2   -6   -2
      _____
       1    3    1    0  : x² + 3x + 1
```

or

$$
\begin{array}{r}
x^2 + 3x + 1 \\
x+2 \overline{\smash)x^3 + 5x^2 + 7x + 2} \\
\underline{x^3 + 2x^2} \\
3x^2 + 7x \\
\underline{3x^2 + 6x} \\
x + 2 \\
\underline{x + 2} \\
0
\end{array}
$$

Using either approach, $x + 2$ is a factor.

22. $\dfrac{y^{\frac{1}{4}}y^{-\frac{1}{2}}}{y^{\frac{2}{3}}} = \dfrac{y^{-\frac{1}{4}}}{y^{\frac{2}{3}}} = y^{-\frac{1}{4}-\frac{2}{3}}$

$= y^{-\frac{3}{12}-\frac{8}{12}} = y^{-\frac{11}{12}} = \dfrac{1}{y^{\frac{11}{12}}}$

23. $\sqrt{128y^2} - 2y\sqrt{2} + 3\sqrt{32y^2}$

$= \sqrt{64 \cdot 2y^2} - 2y\sqrt{2} + 3\sqrt{16 \cdot 2y^2}$

$= 8y\sqrt{2} - 2y\sqrt{2} + 12y\sqrt{2}$

$= 18y\sqrt{2}$

Problem set 6.3

5. $x^2 = 75$

$x = \pm\sqrt{75} = \pm\sqrt{25 \cdot 3} = \pm 5\sqrt{3}$

$\{-5\sqrt{3},\ 5\sqrt{3}\}$

7. $z^2 = -4$

$z = \pm\sqrt{-4} = \pm\sqrt{4}i$

$z = \pm 2i$

$\{-2i,\ 2i\}$

11. $3x^2 = 25$

$x^2 = \dfrac{25}{3}$

$x = \pm\sqrt{\dfrac{25}{3}} = \pm\dfrac{5}{\sqrt{3}} \cdot \dfrac{\sqrt{3}}{\sqrt{3}} = \pm\dfrac{5\sqrt{3}}{3}$

$\left\{-\dfrac{5\sqrt{3}}{3},\ \dfrac{5\sqrt{3}}{3}\right\}$

17. $4\left(x^2+2x\right)+7 = 3x^2+8x+2$

$4x^2+8x+7 = 3x^2+8x+2$

$x^2 = -5$

$x = \pm\sqrt{-5} = \pm\sqrt{5}i$

$\{-\sqrt{5}i,\ \sqrt{5}i\}$

19. $(x+4)(x+1) = 5x-71$

$x^2+5x+4 = 5x-71$

$x^2 = -75$

$x = \pm\sqrt{-75} = \pm\sqrt{(25)(3)(-1)}$

$x = \pm 5\sqrt{3}i$

$\{-5\sqrt{3}i,\ 5\sqrt{3}i\}$

21. $3y^4 - 2y^2 - 5 = 0$

Let $t = y^2$

$3t^2 - 2t - 5 = 0$

$(3t-5)(t+1) = 0$

$t = \dfrac{5}{3} \qquad t = -1$

$y^2 = \dfrac{5}{3} \qquad\qquad y^2 = -1$

$y = \pm\dfrac{\sqrt{5}}{\sqrt{3}} \cdot \dfrac{\sqrt{3}}{\sqrt{3}} \qquad y = \pm\sqrt{-1}$

$y = \pm\dfrac{\sqrt{15}}{3} \qquad\qquad y = \pm i$

$\left\{-i,\ i,\ -\dfrac{\sqrt{15}}{3},\ \dfrac{\sqrt{15}}{3}\right\}$

25. $3(x^2-1)+\sqrt{x^2-1}=2$

Let $t=\sqrt{x^2-1}$

$3t^2+t-2=0$

$(3t-2)(t+1)=0$

$t=\frac{2}{3}\qquad t=-1$

$\sqrt{x^2-1}=\frac{2}{3}\qquad\qquad \sqrt{x^2-1}=-1$

$\left(\sqrt{x^2-1}\right)^2=\left(\frac{2}{3}\right)^2\quad \left(\sqrt{x^2-1}\right)^2=(-1)^2$

$x^2-1=\frac{4}{9}\qquad\qquad x^2-1=1$

$x^2=\frac{13}{9}\qquad\qquad x^2=2$

$x=\pm\sqrt{\frac{13}{9}}=\pm\frac{\sqrt{13}}{3}\qquad\qquad x=\pm\sqrt{2}$

Since both sides were squared in the solution process, all potential answers must be checked.

check $\sqrt{2}$:

$3(2-1)+\sqrt{2-1}=2$

$3+1\ne2$

Similarly, $-\sqrt{2}$ is extraneous.

check $\frac{\sqrt{13}}{3}$:

$3\left(\frac{13}{9}-1\right)+\sqrt{\frac{13}{9}-1}=2$

$3\left(\frac{4}{9}\right)+\sqrt{\frac{4}{9}}=2$

$\frac{4}{3}+\frac{2}{3}=2$

$2=2$

Similarly, $-\frac{\sqrt{13}}{3}$ checks.

$$\left\{-\frac{\sqrt{13}}{3},\ \frac{\sqrt{13}}{3}\right\}$$

27. $\left(\sqrt{y^2+7}\right)^2=\left(1+\sqrt{y^2+2}\right)^2$

$y^2+7=1+2\sqrt{y^2+2}+y^2+2$

$y^2+7=y^2+3+2\sqrt{y^2+2}$

$4=2\sqrt{y^2+2}$

$2=\sqrt{y^2+2}$

$4=y^2+2$

$y^2=2$

$y=\pm\sqrt{2}$ (Both check)

$\{-\sqrt{2},\ \sqrt{2}\}$

31.
$$\sqrt{x^4 - 2} = x$$
$$\left(\sqrt{x^4 - 2}\right)^2 = x^2$$
$$x^4 - 2 = x^2$$
$$x^4 - x^2 - 2 = 0$$
Let $t = x^2$
$$t^2 - t - 2 = 0$$
$$(t - 2)(t + 1) = 0$$
$$t = 2 \qquad t = -1$$
$$x^2 = 2 \qquad x^2 = -1$$
$$x = \pm\sqrt{2} \quad x = \pm i$$
check $\sqrt{2}$:
$$\sqrt{\left(\sqrt{2}\right)^4 - 2} = \sqrt{2}$$
$$\sqrt{4 - 2} = \sqrt{2}$$
$$\sqrt{2} = \sqrt{2}$$
check $-\sqrt{2}$:
$$\sqrt{\left(-\sqrt{2}\right)^4 - 2} = -\sqrt{2}$$
$$\sqrt{4 - 2} = -\sqrt{2}$$
$$\sqrt{2} \neq -\sqrt{2}$$
check i :
$$\sqrt{i^4 - 2} = i \qquad i^4 = \left(i^2\right)^2 = (-1)^2 = 1$$
$$\sqrt{1 - 2} = i$$
$$\sqrt{-1} = i$$
$$i = i$$
check $-i$:
$$\sqrt{(-i)^4 - 2} = -i$$
$$\sqrt{1 - 2} = -i$$
$$i \neq -i$$
$\{\sqrt{2}, \, i\}$

37.
$$10^5\left(1 + 2t^2\right) = 33 \times 10^5$$
$$\frac{1}{10^5}\left[10^5\left(1 + 2t^2\right)\right] = \left(33 \times 10^5\right)\left(\frac{1}{10^5}\right)$$
$$1 + 2t^2 = 33$$
$$2t^2 = 32$$
$$t^2 = 16$$
$$t = 4 \qquad t = -4 \text{ (reject)}$$
After 4 hours.

39. width: x
Length: $x + 8$
$V = LWH$
$$256 = (x - 4)(x + 8 - 4)2$$
$$256 = (x - 4)(x + 4)2$$
$$128 = (x - 4)(x + 4)$$
$$128 = x^2 - 16$$
$$144 = x^2$$
$$\pm\sqrt{144} = x$$
$$x = 12 \qquad x = -12 \text{ (reject)}$$
width: 12 in   Length: $12 + 8 = 20$ in

43. $4\pi r^2 = A$
$$r^2 = \frac{A}{4\pi}$$
$$r = \sqrt{\frac{A}{4\pi}} \qquad \left(\text{reject } -\sqrt{\frac{A}{4\pi}}\right)$$
$$r = \frac{\sqrt{A}}{2\sqrt{\pi}} \cdot \frac{\sqrt{\pi}}{\sqrt{\pi}}$$
$$r = \frac{\sqrt{A\pi}}{2\pi}$$

47.
$$y = \frac{b}{a}\sqrt{a^2 - x^2}$$
$$y^2 = \frac{b^2}{a^2}\left(a^2 - x^2\right)$$
$$a^2y^2 = a^2\left[\frac{b^2}{a^2}\left(a^2 - x^2\right)\right]$$
$$a^2y^2 = b^2a^2 - b^2x^2$$
$$b^2x^2 = b^2a^2 - a^2y^2$$
$$x^2 = \frac{b^2a^2 - a^2y^2}{b^2}$$
$$x = \pm\sqrt{\frac{b^2a^2 - a^2y^2}{b^2}} = \pm\sqrt{\frac{a^2\left(b^2 - y^2\right)}{b^2}}$$
$$x = \pm\frac{a}{b}\sqrt{b^2 - y^2}$$

49.
$$m_v = \frac{m_o}{\sqrt{1 - \frac{v^2}{c^2}}}$$
$$m_v^2 = \frac{m_o^2}{1 - \frac{v^2}{c^2}} \cdot \frac{c^2}{c^2} = \frac{m_o^2 c^2}{c^2 - v^2}$$
$$\left(c^2 - v^2\right)\left(m_v^2\right) = \left(c^2 - v^2\right)\frac{m_o^2 c^2}{c^2 - v^2}$$
$$m_v^2 c^2 - m_v^2 v^2 = m_o^2 c^2$$
$$m_v^2 c^2 - m_o^2 c^2 = m_v^2 v^2$$
$$\frac{m_v^2 c^2 - m_o^2 c^2}{m_v^2} = v^2$$
$$v = \sqrt{\frac{m_v^2 c^2 - m_o^2 c^2}{m_v^2}} = \sqrt{\frac{c^2\left(m_v^2 - m_o^2\right)}{m_v^2}}$$
$$v = \frac{c\sqrt{m_v^2 - m_o^2}}{m_v}$$

50.
$$\sqrt[4]{36x^2y^4}\,\sqrt[4]{12x^5y^3}$$
$$= \sqrt[4]{432x^7y^7}$$
$$= \sqrt[4]{16 \cdot 27x^4 3y^4 y^3} = 2xy\sqrt[4]{27x^3y^3}$$

51.
$$\left(3.2\times10^{12}\right)\left(6.5\times10^{-6}\right)$$
$$= 20.8\times10^6$$
$$= 2.08\times10\times10^6$$
$$= 2.08\times10^7$$

52.
$$3 - \frac{3}{3 - \frac{3}{3-y}} \cdot \frac{3-y}{3-y}$$
$$= 3 - \frac{3(3-y)}{3(3-y)-3}$$
$$= 3 - \frac{9-3y}{6-3y}$$
$$= 3 - \frac{3(3-y)}{3(2-y)}$$
$$= 3 - \frac{3-y}{2-y}$$
$$= 3 \cdot \frac{2-y}{2-y} - \frac{3-y}{2-y}$$
$$= \frac{3(2-y)-(3-y)}{2-y} = \frac{6-3y-3+y}{2-y}$$
$$= \frac{3-2y}{2-y}$$

Problem set 6.4

1. $x^2 - 4x = 21$

$x^2 - 4x + 4 = 21 + 4$

$(x - 2)^2 = 25$

$x - 2 = \pm 5$

$x = 2 \pm 5$

$x = 7 \qquad x = -3$

$\{-3, 7\}$

5. $2y^2 - 5y = 3$

$y^2 - \frac{5}{2}y = \frac{3}{2}$

Note: $\left(\frac{1}{2}\right)\left(-\frac{5}{2}\right) = -\frac{5}{4}$ and $\left(-\frac{5}{4}\right)^2 = \frac{25}{16}$

$y^2 - \frac{5}{2}y + \frac{25}{16} = \frac{3}{2} + \frac{25}{16}$

$\left(y - \frac{5}{4}\right)^2 = \frac{24}{16} + \frac{25}{16}$

$\left(y - \frac{5}{4}\right)^2 = \frac{49}{16}$

$y - \frac{5}{4} = \pm\frac{7}{4}$

$y = \frac{5}{4} \pm \frac{7}{4}$

$y = \frac{12}{4} = 3 \qquad y = -\frac{2}{4} = -\frac{1}{2}$

$\left\{-\frac{1}{2}, 3\right\}$

9. $y^2 - 6y + 2 = 0$

$y^2 - 6y = -2$

$\left(\frac{1}{2}\right)(-6) = -3$ and $(-3)^2 = 9$

$y^2 - 6y + 9 = -2 + 9$

$(y - 3)^2 = 7$

$y - 3 = \pm\sqrt{7}$

$y = 3 \pm \sqrt{7}$

$\{3 + \sqrt{7}, \ 3 - \sqrt{7}\}$

13. $2z^2 + z = 5$

$z^2 + \frac{1}{2}z = \frac{5}{2}$

Note: $\frac{1}{2}\left(\frac{1}{2}\right) = \frac{1}{4}$ and $\left(\frac{1}{4}\right)^2 = \frac{1}{16}$

$z^2 + \frac{1}{2}z + \frac{1}{16} = \frac{5}{2} + \frac{1}{16}$

$\left(z + \frac{1}{4}\right)^2 = \frac{41}{16}$

$z + \frac{1}{4} = \pm\sqrt{\frac{41}{16}}$

$z + \frac{1}{4} = \pm\frac{\sqrt{41}}{4}$

$z = -\frac{1}{4} \pm \frac{\sqrt{41}}{4}$

$z = \frac{-1 \pm \sqrt{41}}{4}$

$\left\{\frac{-1 + \sqrt{41}}{4}, \ \frac{-1 - \sqrt{41}}{4}\right\}$

17. $y^2 + 2y + 2 = 0$

$y^2 + 2y + 1 = -2 + 1$

$(y + 1)^2 = -1$

$y + 1 = \pm\sqrt{-1}$

$y + 1 = \pm i$

$y = -1 \pm i$

(For practice) check $-1 - i$:

$(-1 - i)^2 + 2(-1 - i) + 2 = 0$

$1 + 2i + i^2 - 2 - 2i + 2 = 0$

$1 + i^2 = 0$

$1 + (-1) = 0$

$0 = 0$

$\{-1 + i, \ -1 - i\}$

21. $8z^2 - 4z = -1$

$$z^2 - \frac{1}{2}z = -\frac{1}{8} \qquad \left[\text{Note: } \frac{1}{2}\left(-\frac{1}{2}\right) = -\frac{1}{4}; \quad \left(-\frac{1}{4}\right)^2 = \frac{1}{16}\right]$$

$$z^2 - \frac{1}{2}z + \frac{1}{16} = -\frac{1}{8} + \frac{1}{16}$$

$$\left(z - \frac{1}{4}\right)^2 = -\frac{1}{16}$$

$$z - \frac{1}{4} = \pm\sqrt{-\frac{1}{16}}$$

$$z - \frac{1}{4} = \pm\frac{i}{4}$$

$$z = \frac{1}{4} \pm \frac{i}{4} = \frac{1 \pm i}{4}$$

$$\left\{\frac{1+i}{4}, \frac{1-i}{4}\right\}$$

23. $3y^2 + 2y + 4 = 0$

$$y^2 + \frac{2}{3}y + \frac{4}{3} = 0$$

$$y^2 + \frac{2}{3}y = -\frac{4}{3} \qquad \left[\text{Note: } \frac{1}{2}\left(\frac{2}{3}\right) = \frac{1}{3}; \quad \left(\frac{1}{3}\right)^2 = \frac{1}{9}\right]$$

$$y^2 + \frac{2}{3}y + \frac{1}{9} = -\frac{4}{3} + \frac{1}{9}$$

$$\left(y + \frac{1}{3}\right)^2 = -\frac{12}{9} + \frac{1}{9}$$

$$\left(y + \frac{1}{3}\right)^2 = -\frac{11}{9}$$

$$y + \frac{1}{3} = \pm\sqrt{-\frac{11}{9}} = \pm\frac{\sqrt{11}i}{3}$$

$$y = -\frac{1}{3} \pm \frac{\sqrt{11}i}{3} = \frac{-1 \pm \sqrt{11}i}{3}$$

$$\left\{\frac{-1 + \sqrt{11}i}{3}, \frac{-1 - \sqrt{11}i}{3}\right\}$$

25. x: side of square

Area $= $ side $+ 3\frac{3}{4}$

$$x^2 = x + \frac{15}{4}$$

$$x^2 - x + \frac{1}{4} = \frac{15}{4} + \frac{1}{4}$$

$$\left(x - \frac{1}{2}\right)^2 = 4$$

$$x - \frac{1}{2} = \pm 2$$

$$x = \frac{1}{2} \pm 2$$

$x = \frac{5}{2}$ $x = -\frac{3}{2}$ (reject ; the square's side cannot be negative)

side of square : $\frac{5}{2}$

26. $6\sqrt{\frac{1}{2}} + 24\sqrt{\frac{1}{8}} - 3\sqrt{2}$

$$= \frac{6}{\sqrt{2}} \cdot \frac{\sqrt{2}}{\sqrt{2}} + \frac{24}{2\sqrt{2}} \cdot \frac{\sqrt{2}}{\sqrt{2}} - 3\sqrt{2}$$

$$= \frac{6\sqrt{2}}{2} + \frac{24\sqrt{2}}{4} - 3\sqrt{2}$$

$$= 3\sqrt{2} + 6\sqrt{2} - 3\sqrt{2}$$

$$= 6\sqrt{2}$$

27. $\dfrac{3y+2}{y+5} + \dfrac{32y-9}{(y+5)(2y-3)} + \dfrac{y-2}{3-2y}$

$$= \frac{3y+2}{y+5} \cdot \frac{2y-3}{2y-3} + \frac{32y-9}{(y+5)(2y-3)}$$

$$+ \frac{y-2}{3-2y} \cdot \frac{-1}{-1} \cdot \frac{y+5}{y+5}$$

$$= \frac{(3y+2)(2y-3) + 32y-9 - (y-2)(y+5)}{(y+5)(2y-3)}$$

$$= \frac{6y^2 - 5y - 6 + 32y - 9 - y^2 - 3y + 10}{(y+5)(2y-3)}$$

$$= \frac{5y^2 + 24y - 5}{(y+5)(2y-3)} = \frac{(5y-1)(y+5)}{(y+5)(2y-3)}$$

$$= \frac{5y-1}{2y-3}$$

28. $30\left(\dfrac{y-5}{6}-\dfrac{1}{3}\right) < 30\left(\dfrac{y+2}{5}\right)$

$\qquad 5(y-5)-10 < 6(y+2)$

$\qquad 5y-25-10 < 6y+12$

$\qquad\quad 5y-35 < 6y+12$

$\qquad\qquad\quad -y < 47$

$\qquad\qquad\quad\ \ y > -47 \qquad \{y \mid y > -47\}$

Problem set 6.5.1

3. $\qquad 3+\dfrac{7}{x}=\dfrac{6}{x^2}$

$\quad x^2\left(3+\dfrac{7}{x}\right)=x^2\left(\dfrac{6}{x^2}\right)$

$\quad 3x^2+7x-6=0$

$\quad a=3, b=7, c=-6$

$\quad x=\dfrac{-b\pm\sqrt{b^2-4ac}}{2a}=\dfrac{-7\pm\sqrt{49-4(3)(-6)}}{6}$

$\quad =\dfrac{-7\pm\sqrt{49-(-72)}}{6}=\dfrac{-7\pm\sqrt{121}}{6}=\dfrac{-7\pm11}{6}$

$\quad x=\dfrac{-7+11}{6}=\dfrac{4}{6}=\dfrac{2}{3}$

$\quad x=\dfrac{-7-11}{6}=\dfrac{-18}{6}=-3$

$\qquad\qquad\left\{-3,\ \dfrac{2}{3}\right\}$

9. $\qquad \dfrac{5}{x^2}=3+\dfrac{8}{x}$

$\quad x^2\left(\dfrac{5}{x^2}\right)=x^2\left(3+\dfrac{8}{x}\right)$

$\quad 5=3x^2+8x$

$\quad 0=3x^2+8x-5$

$\quad a=3, b=8, c=-5$

$\quad x=\dfrac{-b\pm\sqrt{b^2-4ac}}{2a}=\dfrac{-8\pm\sqrt{64-4(3)(-5)}}{6}$

$\quad =\dfrac{-8\pm\sqrt{64-(-60)}}{6}=\dfrac{-8\pm\sqrt{124}}{6}$

$\quad =\dfrac{-8\pm\sqrt{4\cdot 31}}{6}=\dfrac{-8\pm2\sqrt{31}}{6}$

$\quad =\dfrac{2\left(-4\pm\sqrt{31}\right)}{6}=\dfrac{-4\pm\sqrt{31}}{3}$

$\qquad\left\{\dfrac{-4+\sqrt{31}}{3},\ \dfrac{-4-\sqrt{31}}{3}\right\}$

11. $2z^2=2z-1$

$\quad 2z^2-2z+1=0$

$\quad a=2, b=-2, c=1$

$\quad z=\dfrac{-b\pm\sqrt{b^2-4ac}}{2a}=\dfrac{-(-2)\pm\sqrt{4-4(2)(1)}}{4}$

$\quad =\dfrac{2\pm\sqrt{-4}}{4}=\dfrac{2\pm2i}{4}=\dfrac{2(1\pm i)}{4}=\dfrac{1\pm i}{2}$

$\qquad\left\{\dfrac{1+i}{2},\ \dfrac{1-i}{2}\right\}$

17.

$$\frac{3}{z} = \frac{2z}{z-1}$$

$$3(z-1) = z(2z)$$

$$3z-3 = 2z^2$$

$$0 = 2z^2 - 3z + 3$$

$$a = 2, b = -3, c = 3$$

$$z = \frac{-b \pm \sqrt{b^2-4ac}}{2a} = \frac{-(-3) \pm \sqrt{9-4(2)(3)}}{4}$$

$$= \frac{3 \pm \sqrt{-15}}{4} = \frac{3 \pm \sqrt{15}i}{4}$$

$$\left\{ \frac{3+\sqrt{15}i}{4}, \frac{3-\sqrt{15}i}{4} \right\}$$

19. $8y^3 - 1 = 0$

$$(2y)^3 - 1^3 = 0$$

$$(2y-1)\left[(2y)^2 + (2y)(1) + 1^2\right] = 0$$

$$(2y-1)\left(4y^2 + 2y + 1\right) = 0$$

$$2y-1 = 0 \qquad 4y^2 + 2y + 1 = 0$$

$$y = \frac{1}{2} \qquad y = \frac{-2 \pm \sqrt{4-4(4)(1)}}{8}$$

$$y = \frac{-2 \pm \sqrt{-12}}{8}$$

$$y = \frac{-2 \pm \sqrt{4(3)(-1)}}{8}$$

$$y = \frac{-2 \pm 2\sqrt{3}i}{8} = \frac{-1 \pm \sqrt{3}i}{4}$$

$$\left\{ \frac{-1+\sqrt{3}i}{4}, \frac{-1-\sqrt{3}i}{4}, \frac{1}{2} \right\}$$

23.

$$x = \frac{25-5x}{3x+3}$$

$$x(3x+3) = 25 - 5x$$

$$3x^2 + 3x = 25 - 5x$$

$$3x^2 + 8x - 25 = 0$$

$$a = 3, b = 8, c = -25$$

$$x = \frac{-b \pm \sqrt{b^2-4ac}}{2a} = \frac{-8 \pm \sqrt{64-4(3)(-25)}}{6}$$

$$= \frac{-8 \pm \sqrt{64-(-300)}}{6} = \frac{-8 \pm \sqrt{364}}{6}$$

$$= \frac{-8 \pm \sqrt{4 \cdot 91}}{6} = \frac{-8 \pm 2\sqrt{91}}{6} = \frac{-4 \pm \sqrt{91}}{3}$$

$$\left\{ \frac{-4+\sqrt{91}}{3}, \frac{-4-\sqrt{91}}{3} \right\}$$

25.

$$\frac{1}{y^2-3y+2} = \frac{1}{y+2} + \frac{5}{y^2-4}$$

$$\frac{1}{(y-2)(y-1)} = \frac{1}{y+2} + \frac{5}{(y+2)(y-2)}$$

$$(y+2)(y-2)(y-1)\left[\frac{1}{(y-2)(y-1)}\right]$$

$$= (y+2)(y-2)(y-1)$$

$$\cdot \left[\frac{1}{y+2} + \frac{5}{(y+2)(y-2)}\right]$$

$$y+2 = (y-2)(y-1) + 5(y-1)$$

$$y+2 = y^2 - 3y + 2 + 5y - 5$$

$$y+2 = y^2 + 2y - 3$$

$$0 = y^2 + y - 5$$

$$a = 1, b = 1, c = -5$$

$$y = \frac{-b \pm \sqrt{b^2-4ac}}{2a} = \frac{-1 \pm \sqrt{1-4(1)(-5)}}{2}$$

$$= \frac{-1 \pm \sqrt{21}}{2}$$

$$\left\{ \frac{-1+\sqrt{21}}{2}, \frac{-1-\sqrt{21}}{2} \right\}$$

27. $\sqrt{2}x^2 + 3x - 2\sqrt{2} = 0$

$a = \sqrt{2}, b = 3, c = -2\sqrt{2}$

$x = \dfrac{-3 \pm \sqrt{9 - 4(\sqrt{2})(-2\sqrt{2})}}{2\sqrt{2}}$

$= \dfrac{-3 \pm \sqrt{9 - 4(-2\sqrt{4})}}{2\sqrt{2}}$

$= \dfrac{-3 \pm \sqrt{9 - 4(-4)}}{2\sqrt{2}} = \dfrac{-3 \pm \sqrt{25}}{2\sqrt{2}}$

$= \dfrac{-3 \pm 5}{2\sqrt{2}}$

$x = \dfrac{2}{2\sqrt{2}} = \dfrac{1}{\sqrt{2}} \cdot \dfrac{\sqrt{2}}{\sqrt{2}} = \dfrac{\sqrt{2}}{2}$

$x = \dfrac{-8}{2\sqrt{2}} = \dfrac{-4}{\sqrt{2}} \cdot \dfrac{\sqrt{2}}{\sqrt{2}} = \dfrac{-4\sqrt{2}}{2} = -2\sqrt{2}$

$\left\{ -2\sqrt{2}, \dfrac{\sqrt{2}}{2} \right\}$

31. $ix^2 - 5x + 2i = 0$

$a = i, b = -5, c = 2i$

$x = \dfrac{-(-5) \pm \sqrt{25 - 4(i)(2i)}}{2i}$

$= \dfrac{5 \pm \sqrt{25 - 8i^2}}{2i} = \dfrac{5 \pm \sqrt{25 - 8(-1)}}{2i}$

$= \dfrac{5 \pm \sqrt{33}}{2i} = \dfrac{5 \pm \sqrt{33}}{2i} \cdot \dfrac{i}{i}$

$= \dfrac{5i \pm \sqrt{33}i}{2i^2} = \dfrac{5i \pm \sqrt{33}i}{2(-1)} = \dfrac{5i \pm \sqrt{33}i}{-2}$

$\left\{ \dfrac{5i + \sqrt{33}i}{-2}, \dfrac{5i - \sqrt{33}i}{-2} \right\}$

33. $|y^2 + 2y| = 3$

$y^2 + 2y = 3$

$y^2 + 2y - 3 = 0$

$y = \dfrac{-2 \pm \sqrt{4 - 4(1)(-3)}}{2}$

$= \dfrac{-2 \pm \sqrt{16}}{2} = \dfrac{-2 \pm 4}{2}$

$= 1, \ -3$

or

$y^2 + 2y = -3$

$y^2 + 2y + 3 = 0$

$y = \dfrac{-2 \pm \sqrt{4 - 4(1)(3)}}{2}$

$= \dfrac{-2 \pm \sqrt{-8}}{2} = \dfrac{-2 \pm 2\sqrt{2}i}{2} = -1 \pm \sqrt{2}i$

$\{-3, \ 1, \ -1 + \sqrt{2}i, \ -1 - \sqrt{2}i\}$

39. $x^2 + xy + 1 = 0$

$a = 1, b = y, c = 1$

$x = \dfrac{-y \pm \sqrt{y^2 - 4(1)(1)}}{2}$

$= \dfrac{-y \pm \sqrt{y^2 - 4}}{2}$

$\left\{ \dfrac{-y + \sqrt{y^2 - 4}}{2}, \dfrac{-y - \sqrt{y^2 - 4}}{2} \right\}$

41. $x^2 + 3xy - y^2 = 0$

$a = 1, b = 3y, c = -y^2$

$x = \dfrac{-3y \pm \sqrt{(3y)^2 - 4(1)(-y^2)}}{2}$

$= \dfrac{-3y \pm \sqrt{9y^2 + 4y^2}}{2} = \dfrac{-3y \pm \sqrt{13y^2}}{2}$

$= \dfrac{-3y \pm \sqrt{13}y}{2}$

$\left\{ \dfrac{-3y + \sqrt{13}y}{2}, \dfrac{-3y - \sqrt{13}y}{2} \right\}$

43. $P = -5I^2 + 80I$

$5I^2 - 80I + P = 0$

$a = 5, b = -80, c = P$

$I = \dfrac{80 \pm \sqrt{6400 - 4(5)P}}{10}$

$= \dfrac{80 \pm \sqrt{4(1600 - 5P)}}{10} = \dfrac{80 \pm 2\sqrt{1600 - 5P}}{10}$

$= \dfrac{40 \pm \sqrt{1600 - 5P}}{5}$

45. $\left(x^2 + 2x - 3\right)^2 + 6\left(x^2 + 2x - 3\right) + 8 = 0$

Let $t = x^2 + 2x - 3$

$t^2 + 6t + 8 = 0$

$t = \dfrac{-6 \pm \sqrt{36 - 4(1)(8)}}{2} = \dfrac{-6 \pm \sqrt{4}}{2} = \dfrac{-6 \pm 2}{2}$

$t = -2$ or $t = -4$. First let $t = -2$.

$x^2 + 2x - 3 = -2$

$x^2 + 2x - 1 = 0$

$x = \dfrac{-2 \pm \sqrt{4 - 4(1)(-1)}}{2}$

$= \dfrac{-2 \pm \sqrt{8}}{2} = \dfrac{-2 \pm 2\sqrt{2}}{2} = -1 \pm \sqrt{2}$

$t = -4$

$x^2 + 2x - 3 = -4$

$x^2 + 2x + 1 = 0$

$x = \dfrac{-2 \pm \sqrt{4 - 4(1)(1)}}{2}$

$= \dfrac{-2 \pm \sqrt{0}}{2} = -1$

$\{-1,\ -1 + \sqrt{2},\ -1 - \sqrt{2}\}$

47. $\left(y - \dfrac{3}{y}\right)^2 - \left(y - \dfrac{3}{y}\right) - 2 = 0$

Let $t = y - \dfrac{3}{y}$

$t^2 - t - 2 = 0$

$(t - 2)(t + 1) = 0; \qquad t = 2 \qquad t = -1$

$t = 2$

$y - \dfrac{3}{y} = 2$

$y^2 - 3 = 2y$

$y^2 - 2y - 3 = 0$

$(y - 3)(y + 1) = 0$

$y = 3 \qquad y = -1$

$t = -1$

$y - \dfrac{3}{y} = -1$

$y^2 - 3 = -y$

$y^2 + y - 3 = 0$

$y = \dfrac{-1 \pm \sqrt{1 - 4(1)(-3)}}{2} = \dfrac{-1 \pm \sqrt{13}}{2}$

$\left\{-1,\ 3,\ \dfrac{-1 + \sqrt{13}}{2},\ \dfrac{-1 - \sqrt{13}}{2}\right\}$

49. $y^{-2} - 4y^{-1} - 3 = 0$

$\quad$ Let $t = y^{-1}$

$\quad t^2 - 4t - 3 = 0$

$\quad t = \dfrac{4 \pm \sqrt{16 - 4(1)(-3)}}{2} = \dfrac{4 \pm \sqrt{28}}{2} = \dfrac{4 \pm 2\sqrt{7}}{2}$

$\quad = 2 \pm \sqrt{7}$

$\quad y^{-1} = t$

$\quad y^{-1} = 2 \pm \sqrt{7}$

$\quad \dfrac{1}{y} = 2 \pm \sqrt{7}$

$\quad 1 = \left(2 \pm \sqrt{7}\right)y$

$\quad y = \dfrac{1}{2 \pm \sqrt{7}}$

$\quad y = \dfrac{1}{2 + \sqrt{7}} \cdot \dfrac{2 - \sqrt{7}}{2 - \sqrt{7}} = \dfrac{2 - \sqrt{7}}{4 - 7} = \dfrac{2 - \sqrt{7}}{-3}$

$\quad y = \dfrac{1}{2 - \sqrt{7}} \cdot \dfrac{2 + \sqrt{7}}{2 + \sqrt{7}} = \dfrac{2 + \sqrt{7}}{4 - 7} = \dfrac{2 + \sqrt{7}}{-3}$

$\quad \left\{ \dfrac{2 - \sqrt{7}}{-3}, \ \dfrac{2 + \sqrt{7}}{-3} \right\}$

53. $6y^5 = -4y^4 - 2y^3$

$\quad 6y^5 + 4y^4 + 2y^3 = 0$

$\quad 2y^3\left(3y^2 + 2y + 1\right) = 0$

$\quad 2y^3 = 0$

$\quad y = 0$

$\quad$ or

$\quad 3y^2 + 2y + 1 = 0$

$\quad y = \dfrac{-2 \pm \sqrt{4 - 4(3)(1)}}{6} = \dfrac{-2 \pm \sqrt{-8}}{6}$

$\quad = \dfrac{-2 \pm \sqrt{4 \cdot 2}i}{6} = \dfrac{-2 \pm 2\sqrt{2}i}{6} = \dfrac{-1 \pm \sqrt{2}i}{3}$

$\quad \left\{ 0, \ \dfrac{-1 + \sqrt{2}i}{3}, \ \dfrac{-1 - \sqrt{2}i}{3} \right\}$

55. $\sqrt{2y + 3} = y - 1$

$\quad 2y + 3 = y^2 - 2y + 1$

$\quad 0 = y^2 - 4y - 2$

$\quad y = \dfrac{4 \pm \sqrt{16 - 4(1)(-2)}}{2} = \dfrac{4 \pm \sqrt{24}}{2}$

$\quad = \dfrac{4 \pm 2\sqrt{6}}{2} = 2 \pm \sqrt{6}$

check $2 + \sqrt{6}$ (using a calculator) :

$\quad \sqrt{2\left(2 + \sqrt{6}\right) + 3} = 2 + \sqrt{6} - 1$

$\quad\quad \sqrt{7 + 2\sqrt{6}} = 1 + \sqrt{6}$

$\quad\quad\quad 3.449 = 3.449$

check $2 - \sqrt{6}$:

$\quad \sqrt{2\left(2 - \sqrt{6}\right) + 3} = 2 - \sqrt{6} - 1$

$\quad\quad \sqrt{7 - 2\sqrt{6}} = 1 - \sqrt{6}$

$\quad\quad\quad 1.449 \neq -1.449$

$\quad\quad\quad\quad\quad 2 - \sqrt{6}$ is extraneous

$\quad\quad\quad \{2 + \sqrt{6}\}$

61. x^2 + (sum of roots with sign changed)x
$\qquad$ + product of roots = 0

b. $-2 + 4 = 2$ $\qquad$ $(-2)(4) = -8$

$\qquad x^2 - 2x - 8 = 0$

d. $-\frac{1}{4} + 6 = \frac{23}{4}$ $\qquad$ $\left(-\frac{1}{4}\right)(6) = -\frac{3}{2}$

$\qquad x^2 - \frac{23}{4}x - \frac{3}{2} = 0$

f. $\left(2 + \sqrt{3}\right) + \left(2 - \sqrt{3}\right) = 4$

$\qquad \left(2 + \sqrt{3}\right)\left(2 - \sqrt{3}\right) = 4 - 3 = 1$

$\qquad x^2 - 4x + 1 = 0$

h. $(3 + 2i) + (3 - 2i) = 6$

$\qquad (3 + 2i)(3 - 2i) = 9 - 4i^2 = 9 - 4(-1) = 13$

$\qquad x^2 - 6x + 13 = 0$

i. $\left(\frac{1}{2} + \frac{\sqrt{3}}{2}i\right) + \left(\frac{1}{2} - \frac{\sqrt{3}}{2}i\right) = 1$

$\qquad \left(\frac{1}{2} + \frac{\sqrt{3}}{2}i\right)\left(\frac{1}{2} - \frac{\sqrt{3}}{2}i\right) = \frac{1}{4} - \frac{3}{4}i^2$

$\qquad = \frac{1}{4} - \frac{3}{4}(-1) = 1$

$\qquad x^2 - x + 1 = 0$

j. $\sqrt{7} - \sqrt{7} = 0$

$\qquad \left(\sqrt{7}\right)\left(-\sqrt{7}\right) = -\sqrt{49} = -7$

$\qquad x^2 + 0x - 7 = 0$

$\qquad x^2 - 7 = 0$

62. $f = \dfrac{1}{2\pi \sqrt{\left(1.2 \times 10^{-2}\right)\left(50 \times 10^{-6}\right)}}$

$\qquad = \dfrac{1}{2\pi \sqrt{60 \times 10^{-8}}}$

$\qquad = \dfrac{1}{2\pi \sqrt{(4)(15)\left(10^{-4}\right)^2}}$

$\qquad = \dfrac{1}{2\pi (2)\left(10^{-4}\right)\sqrt{15}} = \dfrac{10^4}{4\pi \sqrt{15}} \cdot \dfrac{\sqrt{15}}{\sqrt{15}}$

$\qquad = \dfrac{10^4 \sqrt{15}}{60\pi}$ Hertz (approximately 202)

63. $\dfrac{x - 9}{\sqrt{x+3}} \cdot \dfrac{\sqrt{x-3}}{\sqrt{x-3}} = \dfrac{(x-9)\left(\sqrt{x-3}\right)}{x-9} = \sqrt{x-3}$

64. Consecutive odd integers: x, $x + 2$, $x + 4$

$\qquad x^2 + (x+2)^2 + (x+4)^2 = 11$

$\qquad x^2 + x^2 + 4x + 4 + x^2 + 8x + 16 = 11$

$\qquad 3x^2 + 12x + 20 = 11$

$\qquad 3x^2 + 12x + 9 = 0$

$\qquad x^2 + 4x + 3 = 0$

$\qquad (x + 3)(x + 1) = 0$

$\qquad x = -3 \qquad x = -1$

$\qquad$ Integers: -3, -1, 1

$\qquad\qquad$ or -1, 1, 3

Problem set 6.5.2

3. $x^2 - 2x = -5$

$x^2 - 2x + 5 = 0$

$x = \dfrac{2 \pm \sqrt{4 - 4(1)(5)}}{2} = \dfrac{2 \pm \sqrt{-16}}{2}$

$= \dfrac{2 \pm 4i}{2} = 1 \pm 2i$

x is not a real number.

5. x: width of frame

Area of frame $= \frac{1}{4}$ (area of picture)

$(30 + 2x)(24 + 2x) - 30(24) = \frac{1}{4}(30)(24)$

$720 + 108x + 4x^2 - 720 = 180$

$4x^2 + 108x - 180 = 0$

$x^2 + 27x - 45 = 0$

$x = \dfrac{-27 \pm \sqrt{(27)^2 - 4(1)(-45)}}{2}$

$= \dfrac{-27 \pm \sqrt{729 + 180}}{2} = \dfrac{-27 \pm \sqrt{909}}{2}$

$= \dfrac{-27 \pm \sqrt{9(101)}}{2} = \dfrac{-27 \pm 3\sqrt{101}}{2}$

width of frame $= \dfrac{-27 + 3\sqrt{101}}{2}$ in

≈ 1.57 in.

7. x: one leg

400 − x: other leg

$x^2 + (400 - x)^2 = 300^2$

$x^2 + 160,000 - 800x + x^2 = 90,000$

$2x^2 - 800x + 70,000 = 0$

$x^2 - 400x + 35,000 = 0$

$x = \dfrac{400 \pm \sqrt{160,000 - 4(1)(35,000)}}{2}$

$= \dfrac{400 \pm \sqrt{20,000}}{2} = \dfrac{400 \pm \sqrt{10,000(2)}}{2}$

$= \dfrac{400 \pm 100\sqrt{2}}{2} = 200 \pm 50\sqrt{2}$

One leg : $200 + 50\sqrt{2}$ ft ≈ 270.7 ft

Other leg : $400 - \left(200 + 50\sqrt{2}\right)$

$= 200 - 50\sqrt{2}$ ft ≈ 129.3 ft

13. Time for faster person working alone: x

Time for slower person working alone:

x + 3

Portion of the job done by the faster person in 5 hours + portion of the job done by the slower person in 5 hours = One whole job

$\dfrac{5}{x} + \dfrac{5}{x+3} = 1$

$x(x+3)\left[\dfrac{5}{x} + \dfrac{5}{x+3}\right] = x(x+3) \cdot 1$

$5x + 15 + 5x = x^2 + 3x$

$0 = x^2 - 7x - 15$

$x = \dfrac{7 \pm \sqrt{49 - 4(1)(-15)}}{2} = \dfrac{7 \pm \sqrt{109}}{2}$

Faster person : $\dfrac{7 \pm \sqrt{109}}{2}$ hours

Slower person : $\dfrac{7 \pm \sqrt{109}}{2} + 3$

$= \dfrac{13 \pm \sqrt{109}}{2}$ hours

15. x: speed of each boat in still water

Time spent by the boat going downstream (with current) + 1 hour = Time spent by the boat going upstream (against current).

$$\left[RT = D \; ; \; T = \frac{D}{R} \right]$$

$$\frac{75}{x+5} + 1 = \frac{44}{x-5}$$

$$(x+5)(x-5)\left[\frac{75}{x+5} + 1 \right]$$

$$= (x+5)(x-5)\left[\frac{44}{x-5} \right]$$

$$75x - 375 + x^2 - 25 = 44x + 220$$

$$x^2 + 75x - 400 = 44x + 220$$

$$x^2 + 31x - 620 = 0$$

$$x = \frac{-31 \pm \sqrt{961 - 4(1)(-620)}}{2}$$

$$= \frac{-31 \pm \sqrt{3441}}{2}$$

Speed of each boat: $\dfrac{-31 \pm \sqrt{3441}}{2}$ m/h

≈ 13.8 m/h

17. x: length of A tiles

x + 4: length of B tiles

$$1440x^2 = 160(x+4)^2$$

$$1440x^2 = 160x^2 + 1280x + 2560$$

$$1280x^2 - 1280x - 2560 = 0$$

$$x^2 - x - 2 = 0$$

$$x = \frac{1 \pm \sqrt{1 - 4(1)(-2)}}{2} = \frac{1 \pm \sqrt{9}}{2} = \frac{1 \pm 3}{2}$$

$$x = -1 \quad \text{or} \quad x = 2$$

The length of A tiles: 2 in

19. Area of triangle − Area of rectangle
= shaded region

$$\tfrac{1}{2}\left[(y+5+y+1+3)(2y) \right] - y(y+1) = 10$$

$$\tfrac{1}{2}(2y+9)(2y) - y(y+1) = 10$$

$$2y^2 + 9y - y^2 - y = 10$$

$$y^2 + 8y - 10 = 0$$

$$y = \frac{-8 \pm \sqrt{64 - 4(1)(-10)}}{2} = \frac{-8 \pm \sqrt{104}}{2}$$

$$= \frac{-8 \pm 2\sqrt{26}}{2} = -4 \pm \sqrt{26}$$

y: $-4 + \sqrt{26}$ yd

21. One side of rectangle: x

Adjacent side (call it y) : $2y + 2x = 36$

$$2y = 36 - 2x$$

$$y = 18 - x$$

So, adjacent side: $18 - x$

Area is 36

$$x(18 - x) = 36$$

$$18x - x^2 = 36$$

$$0 = x^2 - 18x + 36$$

$$x = \frac{18 \pm \sqrt{324 - 4(1)(36)}}{2} = \frac{18 \pm \sqrt{180}}{2}$$

$$= \frac{18 \pm \sqrt{36 \cdot 5}}{2} = \frac{18 \pm 6\sqrt{5}}{2} = 9 \pm 3\sqrt{5}$$

One side: $9 + 3\sqrt{5}$ or $9 - 3\sqrt{5}$ ft

Other side: $18 - \left(9 + 3\sqrt{5} \right) = 9 - 3\sqrt{5}$

or $18 - \left(9 - 3\sqrt{5} \right) = 9 + 3\sqrt{5}$ ft

Dimensions: $9 + 3\sqrt{5}$ ft by $9 - 3\sqrt{5}$ ft.

25. edge of open cube: x

edge of closed cube: $x - 1$

Surface area of open cube = Surface area of closed cube

$$5x^2 = 6(x-1)^2$$

$$5x^2 = 6x^2 - 12x + 6$$

$$0 = x^2 - 12x + 6$$

$$x = \frac{12 \pm \sqrt{144 - 4(1)(6)}}{2} = \frac{12 \pm \sqrt{120}}{2}$$

$$= \frac{12 \pm 2\sqrt{30}}{2} = 6 \pm \sqrt{30}$$

edge of closed cube: $6 + \sqrt{30} - 1$

$$= 5 + \sqrt{30} \text{ in}$$

26. $\sigma = \sqrt{npq}$

$$8 = \sqrt{400p\left(\frac{4}{5}\right)}$$

$$8 = \sqrt{320p}$$

$$64 = 320p$$

$$p = \frac{64}{320} = \frac{1}{5}$$

27. $v = \dfrac{k}{\sqrt{d}}$

$$16,000 \text{ km} = 16,000 \text{ km} \times \frac{1000m}{1 \text{ km}}$$

$$= 1.6 \times 10^7 m$$

$$1.58 \times 10^5 = \frac{k}{\sqrt{1.6 \times 10^7}}$$

$$1.58 \times 10^5 = \frac{k}{4000}$$

$$k = 6320 \times 10^5 = 6.32 \times 10^3 \times 10^5$$

$$= 6.32 \times 10^8$$

$$400,000 \text{ km} = 400,000 \text{ km} \times \frac{1000m}{1 \text{ km}}$$

$$= 4 \times 10^8 m$$

$$v = \frac{6.32 \times 10^8}{\sqrt{4 \times 10^8}} = \frac{6.32 \times 10^8}{2 \times 10^4} = 3.16 \times 10^4$$

speed of the moon: 3.16×10^4 m/s

28. $\left(\dfrac{y+3}{1-3y} - y\right) \div \left(\dfrac{y^2+3y}{1-3y} + 1\right)$

$$= \left(\frac{y+3}{1-3y} - y \cdot \frac{1-3y}{1-3y}\right)$$

$$\div \left(\frac{y^2+3y}{1-3y} + 1 \cdot \frac{1-3y}{1-3y}\right)$$

$$= \left(\frac{y+3-y+3y^2}{1-3y}\right) \div \left(\frac{y^2+3y+1-3y}{1-3y}\right)$$

$$= \left(\frac{3y^2+3}{1-3y}\right) \div \left(\frac{y^2+1}{1-3y}\right)$$

$$= \frac{3y^2+3}{1-3y} \cdot \frac{1-3y}{y^2+1} = \frac{3\left(y^2+1\right)(1-3y)}{(1-3y)\left(y^2+1\right)} = 3$$

Problem set 6.5.3

3. $a = 2, b = -11, c = 3$

$b^2 - 4ac = (-11)^2 - 4(2)(3)$

$= 121 - 24 = 97$; irrational solutions

9. $a = 4, b = 2, c = 5$

$b^2 - 4ac = 2^2 - 4(4)(5)$

$= 4 - 80 = -76$; solutions not real numbers

11. set $b^2 - 4ac = 0$

$h^2 - 4(5)(3) = 0$

$h^2 - 60 = 0$

$h^2 = 60$

$h = \pm\sqrt{60} = \pm\sqrt{4 \cdot 15}$

$h = \pm 2\sqrt{15}$

13. set $b^2 - 4ac < 0$

$4 - 4a \cdot 3 < 0$

$4 - 12a < 0$

$-12a < -4$

$a > \dfrac{1}{3}$

15. $x =$

$$\sqrt[3]{-\frac{q}{2} + \sqrt{\frac{q^2}{4} + \frac{p^3}{27}}} + \sqrt[3]{-\frac{q}{2} - \sqrt{\frac{q^2}{4} + \frac{p^3}{27}}}$$

$(p = -15, \; q = -4)$

$$= \sqrt[3]{-\frac{(-4)}{2} + \sqrt{\frac{16}{4} - \frac{3375}{27}}}$$

$$+ \sqrt[3]{-\frac{(-4)}{2} - \sqrt{\frac{16}{4} - \frac{3375}{27}}}$$

$= \sqrt[3]{2 + \sqrt{4 - 125}} + \sqrt[3]{2 - \sqrt{4 - 125}}$

$= \sqrt[3]{2 + \sqrt{-121}} + \sqrt[3]{2 - \sqrt{-121}}$

$= \sqrt[3]{2 + 11i} + \sqrt[3]{2 - 11i}$

$= \sqrt[3]{(2 + i)^3} + \sqrt[3]{(2 - i)^3}$

$= 2 + i + 2 - i = 4$

A solution is 4.

check: $x^3 - 15x - 4 = 0$

$4^3 - 15 \cdot 4 - 4 = 0$

$64 - 60 - 4 = 0$

$0 = 0$

16. $8(x - 5)^{\frac{4}{3}} - 2(x - 5)^{\frac{1}{3}}$

$= 2(x - 5)^{\frac{1}{3}}\left[4(x - 5)^1 - 1 \right]$

$= 2(x - 5)^{\frac{1}{3}}(4x - 20 - 1)$

$= 2(x - 5)^{\frac{1}{3}}(4x - 21)$

17. $25x^{3b}y^a + 30x^by^{4a} - 20x^{2b}y^{2a}$

$= 5x^by^a\left(5x^{2b} + 6y^{3a} - 4x^by^a \right)$

18. $\dfrac{3 - 5i}{2 + 4i} \cdot \dfrac{2 - 4i}{2 - 4i} = \dfrac{6 - 22i + 20i^2}{4 - 16i^2}$

$= \dfrac{6 - 22i + 20(-1)}{4 - 16(-1)} = \dfrac{-14 - 22i}{20}$

$= \dfrac{-14}{20} - \dfrac{22i}{20} = -\dfrac{7}{10} - \dfrac{11i}{10}$

Problem set 6.6.1

1. $x^2 - 5x + 4 > 0$

$x^2 - 5x + 4 = 0$

$(x - 4)(x - 1) = 0$

$x = 4 \qquad x = 1$

T   F   T

←———+———+———→
    1   4

Test 0: $0^2 - 5 \cdot 0 + 4 > 0$

$\qquad\qquad\qquad 4 > 0$ True

Test 2: $2^2 - 5 \cdot 2 + 4 > 0$

$\qquad\qquad\qquad -2 > 0$ False

Test 5: $5^2 - 5 \cdot 5 + 4 > 0$

$\qquad\qquad\qquad 4 > 0$ True

$\{x \mid x < 1 \ \text{ or } \ x > 4\}$

5. $x^2 - 6x + 9 < 0$

$(x - 3)^2 < 0$

F   F

←———+———→
     3

$\varnothing$

7. $x^2 - 6x + 8 \le 0$

$(x - 2)(x - 4) \le 0$

$x = 2 \qquad x = 4$

F   T   F

←———+———+———→
    2   4

Test 0: $0^2 - 6 \cdot 0 + 8 \le 0$

$\qquad\qquad\qquad 8 \le 0$ False

Test 3: $3^2 - 6 \cdot 3 + 8 \le 0$

$\qquad\qquad\qquad -1 \le 0$ True

Test 5: $5^2 - 6 \cdot 5 + 8 \le 0$

$\qquad\qquad\qquad 3 \le 0$ False

$\{x \mid 2 \le x \le 4\}$

11. $\qquad 2x^2 + x < 15$

$2x^2 + x - 15 < 0$

$(2x - 5)(x + 3) < 0$

F   T   F

←———+———+———→
  -3 $\frac{5}{2}$

$\left\{ x \mid -3 < x < \frac{5}{2} \right\}$

15. $\qquad\qquad 5x \le 2 - 3x^2$

$3x^2 + 5x - 2 \le 0$

$(3x - 1)(x + 2) \le 0$

F   T   F

←———+———+———→
  -2 $\frac{1}{3}$

Testing 0 (middle interval) :

$\qquad 5 \cdot 0 \le 2 - 3 \cdot 0^2$

$\qquad\qquad 0 \le 2,$ True

$\left\{ x \mid -2 \le x \le \frac{1}{3} \right\}$

17. $x^2 - 4x \ge 0$

$x(x - 4) \ge 0$

T   F   T

←———+———+———→
   0   4

$\{x \mid x \le 0 \text{ or } x \ge 4\}$

21. $-x^2 + x \ge 0$

$x^2 - x \le 0$

$x(x - 1) \le 0$

F   T   F

←———+———+———→
   0   1

$\{x \mid 0 \le x \le 1\}$

23. $-100x^2 + 800x + 500 < 1700$

$-100x^2 + 800x - 1200 < 0$

$x^2 - 8x + 12 > 0$

$(x-2)(x-6) > 0$

```
    T   F   T
  ←─┼───┼──→
    2   6
```

Manufacture fewer than 2 items or more than 6 items.

25. width: x

Length: $2x + 5$

Area ≥ 33

$x(2x+5) \geq 33$

$2x^2 + 5x - 33 \geq 0$

$(2x+11)(x-3) \geq 0$

```
     T    F   T
  ←──┼────┼──→
    -11   3
    ──
     2
```

The width must be greater than or equal to 3 meters.

27. $b^2 - 4ac < 0$

$(2k)^2 - 4(1)(9) < 0$

$4k^2 - 36 < 0$

$k^2 - 9 < 0$

$(k+3)(k-3) < 0$

```
    F   T   F
  ←─┼───┼──→
   -3   3
```

$x^2 + 2kx + 9 = 0$ has no real solutions for $-3 < k < 3$.

29. $y = \sqrt{x^2 + 3x - 10}$

y is a real number when

$x^2 + 3x - 10 \geq 0$

$(x+5)(x-2) \geq 0$

```
     T   F   T
  ←──┼───┼──→
    -5   2
```

y is a real number when $x \leq -5$ or $x \geq 2$.

33. $4x^4 - 7x^2 - 36 = 0$

Let $t = x^2$

$4t^2 - 7t - 36 = 0$

$(4t+9)(t-4) = 0$

$t = -\dfrac{9}{4}$ $t = 4$

$x^2 = -\dfrac{9}{4}$ $x^2 = 4$

$x = \pm\sqrt{-\dfrac{9}{4}}$ $x = \pm\sqrt{4}$

$x = \pm\dfrac{3}{2}i$ $x = \pm 2$

$\left\{\dfrac{3}{2}i, -\dfrac{3}{2}i, -2, 2\right\}$

34. $\sqrt{2y+7}+\sqrt{y+3}-1=0$

$$\left(\sqrt{2y+7}\right)^2=\left(1-\sqrt{y+3}\right)^2$$

$$2y+7=1-2\sqrt{y+3}+y+3$$

$$2y+7=y+4-2\sqrt{y+3}$$

$$y+3=-2\sqrt{y+3}$$

$$(y+3)^2=\left(-2\sqrt{y+3}\right)^2$$

$$y^2+6y+9=4(y+3)$$

$$y^2+6y+9=4y+12$$

$$y^2+2y-3=0$$

$$(y+3)(y-1)=0$$

$$y=-3 \qquad y=1$$

check -3 :

$$\sqrt{2(-3)+7}+\sqrt{-3+3}-1=0$$

$$\sqrt{1}+\sqrt{0}-1=0$$

$$0=0$$

check 1:

$$\sqrt{2(1)+7}+\sqrt{1+3}-1=0$$

$$3+2-1=0$$

$$4\neq 0 \quad \text{1 is extraneous}$$

$$\{-3\}$$

35. Consecutive even integers: $x,\ x+2$

$$\frac{1}{x}+\frac{1}{x+2}=\frac{9}{40}$$

$$40x(x+2)\left[\frac{1}{x}+\frac{1}{x+2}\right]=40x(x+2)\left(\frac{9}{40}\right)$$

$$40x+80+40x=9x^2+18x$$

$$0=9x^2-62x-80$$

$$0=(x-8)(9x+10)$$

$$x-8=0 \qquad 9x+10=0$$

$$x=8 \qquad\quad x=-\frac{10}{9} \text{ (reject ; not an even}$$
$$\text{integer)}$$

Consecutive even integers : 8, 10

Problem set 6.6.2

1. $\dfrac{x-4}{x+3}>0$

$$x-4=0 \qquad x+3=0$$

$$x=4 \qquad\quad x=-3$$

$$\begin{array}{ccc} \text{T} & \text{F} & \text{T} \\ \hline & & \end{array}$$
$$\quad -3 \quad\; 4$$

Test -4 : $\dfrac{-4-4}{-4+3}>0$

$$\dfrac{-8}{-1}>0$$

$$8>0 \text{ True}$$

Test 0 : $\dfrac{0-4}{0+3}>0$

$$\dfrac{-4}{3}>0 \text{ False}$$

Test 5 : $\dfrac{5-4}{5+3}>0$

$$\dfrac{1}{8}>0 \text{ True}$$

$$\{x\,|\,x<-3 \;\text{ or }\; x>4\}$$

5. $\dfrac{-x+2}{x-4}\geq 0$

$$-x+2=0 \qquad x-4=0$$

$$x=2 \qquad\quad x=4$$

$$\begin{array}{ccc} \text{F} & \text{T} & \text{F} \\ \hline & & \end{array}$$
$$\quad 2 \quad\; 4$$

Test 0 : $\dfrac{0+2}{0-4}\geq 0$

$$-\tfrac{1}{2}\geq 0 \text{ False}$$

Test 3 : $\dfrac{-3+2}{3-4}\geq 0$

$$1\geq 0 \text{ True}$$

Test 5 : $\dfrac{-5+2}{5-4}\geq 0$

$$-3\geq 0 \text{ False}$$

$$\{x\,|\,2\leq x<4\}$$

7.
$$\frac{x+1}{x+3} < 2$$

$$\frac{x+1}{x+3} - 2 < 0$$

$$\frac{x+1}{x+3} - 2 \cdot \frac{x+3}{x+3} < 0$$

$$\frac{x+1-2x-6}{x+3} < 0$$

$$\frac{-x-5}{x+3} < 0$$

$$-x-5 = 0 \qquad x+3 = 0$$

$$x = -5 \qquad\quad x = -3$$

$$\begin{array}{ccc} \text{T} & \text{F} & \text{T} \\ \hline & -5 & -3 \end{array}$$

Test -6 : $\dfrac{-6+1}{-6+3} < 2$

$$\frac{-5}{-3} < 2$$

$$\frac{5}{3} < 2 \text{ True}$$

Test -4 : $\dfrac{-4+1}{-4+3} < 2$

$$\frac{-3}{-1} < 2$$

$$3 < 2 \text{ False}$$

Test 0 : $\dfrac{0+1}{0+3} < 2$

$$\frac{1}{3} < 2 \text{ True}$$

$$\{x \mid x < -5 \text{ or } x > -3\}$$

11.
$$\frac{x-2}{x+2} \le 2$$

$$\frac{x-2}{x+2} - 2 \cdot \frac{x+2}{x+2} \le 0$$

$$\frac{x-2-2x-4}{x+2} \le 0$$

$$\frac{-x-6}{x+2} \le 0$$

$$-x-6 = 0 \qquad x+2 = 0$$

$$x = -6 \qquad\quad x = -2$$

$$\begin{array}{ccc} \text{T} & \text{F} & \text{T} \\ \hline & -6 & -2 \end{array}$$

$$\{x \mid x \le -6 \text{ or } x > -2\}$$

13.
$$\frac{x}{x+5} \le 1$$

$$\frac{x}{x+5} - 1 \cdot \frac{x+5}{x+5} \le 0$$

$$\frac{x-x-5}{x+5} \le 0$$

$$\frac{-5}{x+5} \le 0$$

$$x = -5$$

$$\begin{array}{cc} \text{F} & \text{T} \\ \hline & -5 \end{array}$$

Test -6 : $\dfrac{-6}{-6+5} \le 1$

$$6 \le 1 \text{ False}$$

Test -4 : $\dfrac{-4}{-4+5} \le 1$

$$-4 \le 1 \text{ True}$$

$$\{x \mid x > -5\}$$

15. $\dfrac{x^2-x-2}{x^2-4x+3} > 0$

$$x^2-x-2=0 \qquad\qquad x^2-4x+3=0$$
$$(x-2)(x+1)=0 \qquad (x-3)(x-1)=0$$
$$x=2 \qquad x=-1 \qquad\qquad x=3 \qquad x=1$$

$$
\begin{array}{ccccc}
\text{T} & \text{F} & \text{T} & \text{F} & \text{T}
\end{array}
$$

$-1 \quad 1 \quad 2 \quad 3$

Test -2 : $\dfrac{(-2)^2-(-2)-2}{(-2)^2-4(-2)+3} > 0$

$$\dfrac{4}{15} > 0 \text{ True}$$

Test 0 : $\dfrac{(0)^2-(0)-2}{(0)^2-4(0)+3} > 0$

$$-\dfrac{2}{3} > 0 \text{ False}$$

Test 1.5 : $\dfrac{(1.5)^2-(1.5)-2}{(1.5)^2-4(1.5)+3} > 0$

$$\dfrac{-1.25}{-0.75} > 0 \text{ True}$$

Test 2.5 : $\dfrac{(2.5)^2-(2.5)-2}{(2.5)^2-4(2.5)+3} > 0$

$$\dfrac{1.75}{-0.75} > 0 \text{ False}$$

Test 4 : $\dfrac{(4)^2-(4)-2}{(4)^2-4(4)+3} > 0$

$$\dfrac{10}{3} > 0 \text{ True}$$

$$\{x \mid x < -1 \text{ or } 1 < x < 2 \text{ or } x > 3\}$$

16. $\dfrac{1}{x(x-z)-y(x-z)} + \dfrac{1}{y(z-y)-x(z-y)}$

$$- \dfrac{1}{x(y-z)-z(y-z)}$$

$$= \dfrac{1}{(x-z)(x-y)} + \dfrac{1}{(z-y)(y-x)} - \dfrac{1}{(y-z)(x-z)}$$

LCD is $(x-z)(x-y)(y-z)$

$$= \dfrac{1}{(x-z)(x-y)} \cdot \dfrac{(y-z)}{(y-z)}$$

$$+ \dfrac{1}{(z-y)(y-x)} \cdot \dfrac{-1}{-1} \cdot \dfrac{-1}{-1} \cdot \dfrac{(x-z)}{(x-z)}$$

$$- \dfrac{1}{(y-z)(x-z)} \cdot \dfrac{(x-y)}{(x-y)}$$

$$= \dfrac{(y-z)+(x-z)-(x-y)}{(x-z)(x-y)(y-z)}$$

$$= \dfrac{y-z+x-z-x+y}{(x-z)(x-y)(y-z)} = \dfrac{2y-2z}{(x-z)(x-y)(y-z)}$$

$$= \dfrac{2(y-z)}{(x-z)(x-y)(y-z)} = \dfrac{2}{(x-z)(x-y)}$$

17. $8\sqrt[3]{81} + 3\sqrt[3]{24} - 2\sqrt[3]{\dfrac{3}{27}}$

$$= 8\sqrt[3]{27 \cdot 3} + 3\sqrt[3]{8 \cdot 3} - 2\dfrac{\sqrt[3]{3}}{3}$$

$$= 24\sqrt[3]{3} + 6\sqrt[3]{3} - 2\dfrac{\sqrt[3]{3}}{3}$$

$$= 30\sqrt[3]{3} - 2\dfrac{\sqrt[3]{3}}{3}$$

$$= 30\sqrt[3]{3} \cdot \dfrac{3}{3} - 2\dfrac{\sqrt[3]{3}}{3}$$

$$= \dfrac{90\sqrt[3]{3}}{3} - 2\dfrac{\sqrt[3]{3}}{3} = \dfrac{88\sqrt[3]{3}}{3}$$

18. number of quarters : x

number of dimes : $4x$

number of nickles : $5(4x) = 20x$

$$25x + 10(4x) + 5(20x) = 1155$$
$$25x + 40x + 100x = 1155$$
$$165x = 1155$$
$$x = 7$$

quarters : 7 dimes : $4(7) = 28$

nickles : $20(7) = 140$

Review Problems : Chapter 6

1. $x(12x+31) = 15$

$12x^2 + 31x - 15 = 0$

$(x+3)(12x-5) = 0$

$x+3 = 0 \qquad 12x-5 = 0$

$x = -3 \qquad\quad 12x = 5$

$x = \dfrac{5}{12}$

$\left\{-3, \ \dfrac{5}{12}\right\}$

2. $y^2 + \dfrac{3}{2}y + \dfrac{1}{2} = 0$

$2\left(y^2 + \dfrac{3}{2}y + \dfrac{1}{2}\right) = 2(0)$

$2y^2 + 3y + 1 = 0$

$(2y+1)(y+1) = 0$

$2y+1 = 0 \qquad y+1 = 0$

$2y = -1 \qquad\quad y = -1$

$y = -\dfrac{1}{2}$

$\left\{-\dfrac{1}{2}, \ -1\right\}$

3. $(2y+6)(y+6) - 2y^2 = y^2 - 4$

$2y^2 + 18y + 36 - 2y^2 = y^2 - 4$

$18y + 36 = y^2 - 4$

$0 = y^2 - 18y - 40$

$0 = (y+2)(y-20)$

$y+2 = 0 \qquad y-20 = 0$

$y = -2 \qquad\quad y = 20$

$\{-2, \ 20\}$

4. $\dfrac{2y-10}{y} = \dfrac{15}{2} + \dfrac{2y}{y-5}$

$2y(y-5)\left[\dfrac{2y-10}{y}\right] = 2y(y-5)\left[\dfrac{15}{2} + \dfrac{2y}{y-5}\right]$

$2(y-5)(2y-10) = 15y(y-5) + (2y)(2y)$

$4y^2 - 40y + 100 = 15y^2 - 75y + 4y^2$

$4y^2 - 40y + 100 = 19y^2 - 75y$

$0 = 15y^2 - 35y - 100$

$0 = 3y^2 - 7y - 20$

$0 = (y-4)(3y+5)$

$y-4 = 0 \qquad 3y+5 = 0$

$y = 4 \qquad\quad 3y = -5$

$y = -\dfrac{5}{3}$

$\left\{-\dfrac{5}{3}, \ 4\right\}$

5. $\dfrac{2y}{y+2} = \dfrac{y}{y+1} + \dfrac{1}{y^2+3y+2}$

$(y+2)(y+1)\left[\dfrac{2y}{y+2}\right]$

$\qquad = (y+2)(y+1)\left[\dfrac{y}{y+1} + \dfrac{1}{(y+2)(y+1)}\right]$

$2y(y+1) = y(y+2) + 1$

$2y^2 + 2y = y^2 + 2y + 1$

$y^2 = 1$

$y = \pm\sqrt{1}$

$y = \pm 1$

Reject -1 since it causes 0 in the original equation's denominator.

$\{1\}$

6. $\dfrac{y^2-7}{y^2-4y}=2-\dfrac{y^2-1}{y^2+4y}$

$\dfrac{y^2-7}{y(y-4)}=2-\dfrac{y^2-1}{y(y+4)}$

$y(y+4)(y-4)\left[\dfrac{y^2-7}{y(y-4)}\right]$

$\qquad =y(y+4)(y-4)\left[2-\dfrac{y^2-1}{y(y+4)}\right]$

$(y+4)\left(y^2-7\right)$

$\qquad =2y(y+4)(y-4)-\left(y^2-1\right)(y-4)$

$y^3-7y+4y^2-28$

$\qquad =2y^3-32y-\left(y^3-4y^2-y+4\right)$

$y^3-7y+4y^2-28=y^3+4y^2-31y-4$

$\qquad\qquad -7y-28=-31y-4$

$\qquad\qquad\qquad 24y=24$

$\qquad\qquad\qquad\quad y=1 \qquad \{1\}$

7. $2x^2-3=0$

$2x^2=3$

$x^2=\dfrac{3}{2}$

$x=\pm\sqrt{\dfrac{3}{2}}=\pm\dfrac{\sqrt{3}}{\sqrt{2}}\cdot\dfrac{\sqrt{2}}{\sqrt{2}}=\pm\dfrac{\sqrt{6}}{2}$

$\left\{-\dfrac{\sqrt{6}}{2},\dfrac{\sqrt{6}}{2}\right\}$

8. $\dfrac{2}{4y^2+1}=\dfrac{3}{5y^2-1}$

$2\left(5y^2-1\right)=3\left(4y^2+1\right)$

$10y^2-2=12y^2+3$

$-5=2y^2$

$-\dfrac{5}{2}=y^2$

$y=\pm\sqrt{-\dfrac{5}{2}}=\pm\dfrac{\sqrt{5}i}{\sqrt{2}}\cdot\dfrac{\sqrt{2}}{\sqrt{2}}$

$y=\pm\dfrac{\sqrt{10}i}{2}$

$\left\{-\dfrac{\sqrt{10}i}{2},\dfrac{\sqrt{10}i}{2}\right\}$

9. $3x^2+2x=4$

$3x^2+2x-4=0$

$a=3,b=2,c=-4$

$x=\dfrac{-b\pm\sqrt{b^2-4ac}}{2a}=\dfrac{-2\pm\sqrt{4-4(3)(-4)}}{2(3)}$

$\qquad =\dfrac{-2\pm\sqrt{52}}{6}=\dfrac{-2\pm\sqrt{4\cdot 13}}{6}=\dfrac{-2\pm2\sqrt{13}}{6}$

$\qquad =\dfrac{-1\pm\sqrt{13}}{3}$

$\left\{\dfrac{-1+\sqrt{13}}{3},\dfrac{-1-\sqrt{13}}{3}\right\}$

10. $\dfrac{5}{x+1} + \dfrac{x-1}{4} = 2$

$4(x+1)\left[\dfrac{5}{x+1} + \dfrac{x-1}{4}\right] = 4(x+1)(2)$

$20 + (x+1)(x-1) = 8(x+1)$

$20 + x^2 - 1 = 8x + 8$

$x^2 - 8x + 11 = 0$

$a = 1, b = -8, c = 11$

$x = \dfrac{-b \pm \sqrt{b^2 - 4ac}}{2a} = \dfrac{8 \pm \sqrt{64 - 4(1)(11)}}{2}$

$= \dfrac{8 \pm \sqrt{20}}{2} = \dfrac{8 \pm \sqrt{4 \cdot 5}}{2} = \dfrac{8 \pm 2\sqrt{5}}{2}$

$= 4 \pm \sqrt{5}$

$\{4 + \sqrt{5},\ 4 - \sqrt{5}\}$

11. $x(x-2) = -5$

$x^2 - 2x + 5 = 0$

$a = 1, b = -2, c = 5$

$x = \dfrac{-b \pm \sqrt{b^2 - 4ac}}{2a} = \dfrac{2 \pm \sqrt{4 - 4(1)(5)}}{2}$

$= \dfrac{2 \pm \sqrt{-16}}{2}$

$= \dfrac{2 \pm 4i}{2} = 1 \pm 2i$

$\{1 + 2i,\ 1 - 2i\}$

12. $\sqrt{2y+1} - y + 1 = 0$

$\sqrt{2y+1} = y - 1$

$\left(\sqrt{2y+1}\right)^2 = (y-1)^2$

$2y + 1 = y^2 - 2y + 1$

$0 = y^2 - 4y$

$0 = y(y-4)$

$y = 0 \qquad y - 4 = 0$

$\qquad\qquad y = 4$

check 0 : $\sqrt{2(0)+1} - 0 + 1 = 0$

$1 - 0 + 1 = 0$

$2 \neq 0$

0 is extraneous

check 4: $\sqrt{2(4)+1} - 4 + 1 = 0$

$3 - 4 + 1 = 0$

$0 = 0$

$\{4\}$

13. $2 - \sqrt{3y+1} + \sqrt{y-1} = 0$

$$\sqrt{y-1} = \sqrt{3y+1} - 2$$

$$\left(\sqrt{y-1}\right)^2 = \left(\sqrt{3y+1} - 2\right)^2$$

$$y - 1 = 3y + 1 - 4\sqrt{3y+1} + 4$$

$$y - 1 = 3y + 5 - 4\sqrt{3y+1}$$

$$-2y - 6 = -4\sqrt{3y+1}$$

$$y + 3 = 2\sqrt{3y+1}$$

$$(y+3)^2 = \left(2\sqrt{3y+1}\right)^2$$

$$y^2 + 6y + 9 = 4(3y+1)$$

$$y^2 + 6y + 9 = 12y + 4$$

$$y^2 - 6y + 5 = 0$$

$$(y-1)(y-5) = 0$$

$$y - 1 = 0 \qquad y - 5 = 0$$

$$y = 1 \qquad\quad y = 5$$

check 1: $2 - \sqrt{3(1)+1} + \sqrt{1-1} = 0$

$$2 - 2 + 0 = 0$$

$$0 = 0$$

check 5: $2 - \sqrt{3(5)+1} + \sqrt{5-1} = 0$

$$2 - 4 + 2 = 0$$

$$0 = 0$$

$\{1, 5\}$

14. $\sqrt[3]{3(x+3)^2} - 3 = 0$

$$\sqrt[3]{3(x+3)^2} = 3$$

$$\left[\sqrt[3]{3(x+3)^2}\right]^3 = 3^3$$

$$3(x+3)^2 = 3^3$$

$$x^2 + 6x + 9 = 9$$

$$x^2 + 6x = 0$$

$$x(x+6) = 0$$

$$x = 0 \qquad x + 6 = 0$$

$$x = -6$$

$\{-6, 0\}$

15. $2x^2 + 5x - 3 < 0$

$$(2x-1)(x+3) < 0$$

$$x = \tfrac{1}{2} \qquad x = -3$$

$$\begin{array}{ccc} F & T & F \\ \xleftarrow{\qquad} \!+\!\!-\!\!-\!\!+\! \xrightarrow{\qquad} \\ -3 & \tfrac{1}{2} & \end{array}$$

Test -4 : $2(-4)^2 + 5(-4) - 3 < 0$

$$9 < 0 \text{ False}$$

Test 0 : $2(0)^2 + 5(0) - 3 < 0$

$$-3 < 0 \text{ True}$$

Test 1 : $2(1)^2 + 5(1) - 3 < 0$

$$4 < 0 \text{ False}$$

$$\left\{x \mid -3 < x < \tfrac{1}{2}\right\}$$

16. $2x^2 + 9x + 4 \geq 0$

$$(2x+1)(x+4) \geq 0$$

$$x = -\tfrac{1}{2} \qquad x = -4$$

$$\begin{array}{ccc} T & F & T \\ \xleftarrow{\qquad} \!+\!\!-\!\!-\!\!+\! \xrightarrow{\qquad} \\ -4 & -\tfrac{1}{2} & \end{array}$$

Test -5 : $2(-5)^2 + 9(-5) + 4 \geq 0$

$$9 \geq 0 \text{ True}$$

Test -3 : $2(-3)^2 + 9(-3) + 4 \geq 0$

$$-5 \geq 0 \text{ False}$$

Test 0 : $2(0)^2 + 9(0) + 4 \geq 0$

$$4 \geq 0 \text{ True}$$

$$\left\{x \mid x \leq -4 \text{ or } x \geq -\tfrac{1}{2}\right\}$$

17. $\dfrac{x+7}{x-3} > 0$

$x+7 = 0 \qquad x-3 = 0$

$x = -7 \qquad x = 3$

$$\begin{array}{ccc} T & F & T \end{array}$$

$\longleftarrow \!\!\! \underset{-7}{|} \,\, \underset{3}{|} \!\!\! \longrightarrow$

Test $-8 : \dfrac{-8+7}{-8-3} > 0$

$ \dfrac{1}{11} > 0$ True

Test $0 : \dfrac{0+7}{0-3} > 0$

$ -\dfrac{7}{3} > 0$ False

Test $4 : \dfrac{4+7}{4-3} > 0$

$ 11 > 0$ True

$\{x \mid x < -7 \ \text{or} \ x > 3\}$

18. $ \dfrac{x}{x+3} \le 1$

$ \dfrac{x}{x+3} - 1 \le 0$

$\dfrac{x}{x+3} - 1 \cdot \dfrac{x+3}{x+3} \le 0$

$ \dfrac{-3}{x+3} \le 0$

$ x + 3 = 0$

$ x = -3$

$$\begin{array}{cc} F & T \end{array}$$

$\longleftarrow \!\!\! \underset{-3}{|} \!\!\! \longrightarrow$

Test $-4 : \dfrac{-4}{-4+3} \le 1$

$ 4 \le 1$ False

Test $0 : \dfrac{0}{0+3} \le 1$

$ 0 \le 1$ True

$\{x \mid x > -3\}$

19. $x^4 - 5x^2 + 4 = 0$

Let $t = x^2$

$t^2 - 5t + 4 = 0$

$(t-4)(t-1) = 0$

$t = 4 \qquad t = 1$

$x^2 = 4 \qquad x^2 = 1$

$x = \pm\sqrt{4} \quad x = \pm\sqrt{1}$

$x = \pm 2 \qquad x = \pm 1$

$\{-2, \ -1, \ 1, \ 2\}$

20. $\left(x^2 + 2x\right)^2 - 14\left(x^2 + 2x\right) = 15$

$\left(x^2 + 2x\right)^2 - 14\left(x^2 + 2x\right) - 15 = 0$

Let $t = x^2 + 2x$

$t^2 - 14t - 15 = 0$

$(t - 15)(t + 1) = 0$

$ t = 15 \qquad\qquad t = -1$

$ x^2 + 2x = 15 \qquad x^2 + 2x = -1$

$ x^2 + 2x - 15 = 0 \quad x^2 + 2x + 1 = 0$

$ (x+5)(x-3) = 0 \qquad (x+1)^2 = 0$

$ x = -5 \quad x = 3 \qquad\qquad x = -1$

$\{-5, \ -1, \ 3\}$

21. $x^{\frac{2}{3}} - x^{\frac{1}{3}} - 12 = 0$

Let $t = x^{\frac{1}{3}}$

$t^2 - t - 12 = 0$

$(t-4)(t+3) = 0$

$ t = 4 \qquad\qquad\qquad t = -3$

$ x^{\frac{1}{3}} = 4 \qquad\qquad\quad x^{\frac{1}{3}} = -3$

$\left(x^{\frac{1}{3}}\right)^3 = 4^3 \qquad\quad \left(x^{\frac{1}{3}}\right)^3 = (-3)^3$

$ x = 64 \qquad\qquad\quad x = -27$

$\{-27, \ 64\}$

22. $-16t^2 + 80t = 0$

$\qquad -16t(t-5) = 0$

$\qquad -16t = 0 \qquad t-5 = 0$

$\qquad\qquad t = 0 \qquad\qquad t = 5$

Object strikes ground after 5 seconds.

23. $100 + 25x - 5x^2 = 120$

$\qquad 0 = 5x^2 - 25x + 20$

$\qquad 0 = x^2 - 5x + 4$

$\qquad (x-4)(x-1) = 0$

$\qquad x = 4 \qquad x = 1$

4 milligrams (at the most)

24. $\quad V = \frac{1}{3}\pi r^2 h$

$\qquad 3V = \pi r^2 h$

$\qquad \dfrac{3V}{\pi h} = r^2$

$\qquad r = \sqrt{\dfrac{3V}{\pi h}} = \dfrac{\sqrt{3V}}{\sqrt{\pi h}} \cdot \dfrac{\sqrt{\pi h}}{\sqrt{\pi h}}$

$\qquad r = \dfrac{\sqrt{3V\pi h}}{\pi h}$

25. $A = \pi r^2 + \pi rh$

$\qquad \pi r^2 + \pi rh - A = 0$

$\qquad a = \pi, b = \pi h, c = -A$

$\qquad r = \dfrac{-b \pm \sqrt{b^2 - 4ac}}{2a}$

$\qquad = \dfrac{-\pi h \pm \sqrt{\pi^2 h^2 - 4\pi(-A)}}{2\pi}$

$\qquad = \dfrac{-\pi h \pm \sqrt{\pi^2 h^2 + 4\pi A}}{2\pi}$

Since r is positive,

$\qquad r = \dfrac{-\pi h + \sqrt{\pi^2 h^2 + 4\pi A}}{2\pi}$

26. $4x^2 + 8x = 5$

$\qquad 4x^2 + 8x - 5 = 0$

$\qquad a = 4, b = 8, c - 5$

$\qquad b^2 - 4ac = 8^2 - 4(4)(-5) = 64 + 80 = 144$

$\qquad$ solutions are rational

27. $y(y-2) + 4 = 0$

$\qquad y^2 - 2y + 4 = 0$

$\qquad a = 1, b = -2, c = 4$

$\qquad b^2 - 4ac = (-2)^2 - 4(1)(4) = 4 - 16 = -12$

$\qquad$ solutions are imaginary

28. $\qquad \dfrac{z^2}{2} - \dfrac{4}{5}z = \dfrac{3}{10}$

$\qquad 10\left(\dfrac{z^2}{2} - \dfrac{4}{5}z\right) = 10\left(\dfrac{3}{10}\right)$

$\qquad\qquad 5z^2 - 8z = 3$

$\qquad\qquad 5z^2 - 8z - 3 = 0$

$\qquad\qquad a = 5, b = -8, c = -3$

$\qquad b^2 - 4ac = 64 - 4(5)(-3) = 64 + 60 = 124$

$\qquad\qquad$ solutions are irrational

29. Consecutive even integers : $x, x+2$

$\qquad (x+2)^2 - 7x = 44$

$\qquad x^2 + 4x + 4 - 7x = 44$

$\qquad x^2 - 3x - 40 = 0$

$\qquad (x-8)(x+5) = 0$

$\qquad x = 8 \qquad x = -5$ (reject ; not even)

Consecutive even integers: 8, 10

30. x : even number

$$x(3x-16) = 12$$
$$3x^2 - 16x - 12 = 0$$
$$(x-6)(3x+2) = 0$$
$$x - 6 = 0 \qquad 3x + 2 = 0$$
$$x = 6 \qquad x = -\frac{2}{3} \text{ (reject ; not even)}$$

Even number: 6

31. x: negative number

$$2x - \frac{3}{x} = 1$$
$$x\left(2x - \frac{3}{x}\right) = x(1)$$
$$2x^2 - 3 = x$$
$$2x^2 - x - 3 = 0$$
$$(2x-3)(x+1) = 0$$
$$2x - 3 = 0 \qquad x + 1 = 0$$
$$x = \frac{3}{2} \text{ (reject)} \qquad x = -1$$

Negative number : -1

32. Base: x

Altitude: x − 5

$$\frac{1}{2}Bh = A$$
$$\frac{1}{2}x(x-5) = 72$$
$$x^2 - 5x = 144$$
$$x^2 - 5x - 144 = 0$$
$$a = 1, b = -5, c = -144$$
$$x = \frac{-b \pm \sqrt{b^2 - 4ac}}{2a}$$
$$= \frac{5 \pm \sqrt{25 - 4(1)(-144)}}{2}$$
$$= \frac{5 \pm \sqrt{25 - (-576)}}{2} = \frac{5 \pm \sqrt{601}}{2}$$

Base: $\frac{5 + \sqrt{601}}{2}$ m ≈ 14.8 m

33. Shorter leg: x

Longer leg: x + 7

Pythagorean Theorem :

(shorter leg)^2 + (Longer leg)^2

$\qquad\qquad$ = (hypotenuse)^2

$$x^2 + (x+7)^2 = 13^2$$
$$x^2 + x^2 + 14x + 49 = 169$$
$$2x^2 + 14x - 120 = 0$$
$$x^2 + 7x - 60 = 0$$
$$(x+12)(x-5) = 0$$
$$x + 12 = 0 \qquad\qquad x - 5 = 0$$
$$x = -12 \text{ (reject)} \qquad x = 5$$

Longer leg: $5 + 7 = 12$ m

34.
$$A = \pi r^2$$
$$49\pi = \pi r^2$$
$$49 = r^2$$
$$r = \sqrt{49} = 7 \qquad \text{(reject } -7)$$
$$C = 2\pi r = 2\pi (7) = 14\pi$$

Circumference: 14π cm.

35. width : x

length : 3x

$$(x-1)(3x+3) = 72$$
$$3x^2 - 3 = 72$$
$$3x^2 = 75$$
$$x^2 = 25$$
$$x = \pm\sqrt{25}$$
$$x = 5 \qquad x = -5 \text{ (reject)}$$

Width : 5 yd $\qquad$ Length : $3(5) = 15$ yd

36. x: width of border

Area of border = 2(area of picture)
Area of large rectangle − area of small
inner rectangle = 2(area of picture)

$(1+2x)(2+2x)-2=2(2)$

$$2+6x+4x^2-2=4$$

$$6x+4x^2=4$$

$$4x^2+6x-4=0$$

$$2x^2+3x-2=0$$

$$(2x-1)(x+2)=0$$

$$2x-1=0 \qquad x+2=0$$

$$x=\frac{1}{2} \qquad x=-2 \text{ (reject)}$$

width of border: $\frac{1}{2}$ ft.

37. x − 10 : rate for first 15 miles
x : rate for last 60 miles
Time for entire trip = 2 hours
Time for first 15 miles + Time for last
60 miles = 2

$$\left[RT=D \; ; \; T=\frac{D}{R} \right]$$

$$\frac{15}{x-10}+\frac{60}{x}=2$$

$$x(x-10)\left[\frac{15}{x-10}+\frac{60}{x} \right]=x(x-10)2$$

$$15x+60(x-10)=2x(x-10)$$

$$15x+60x-600=2x^2-20x$$

$$75x-600=2x^2-20x$$

$$0=2x^2-95x-600$$

$$0=(2x-15)(x-40)$$

$$2x-15=0 \qquad x-40=0$$

$$x=\frac{15}{2} \text{ (reject; leads to a negative rate for}$$

first 15 miles)

$$x=40$$

Rate for first 15 miles: 40-10=30 m/h

38. x: Time for Wilhelmina working alone
x − 10 : Time for Jerry working alone
Portion of the job done by Jerry in 15
minutes + portion of the job done by
Wilhelmina in 20 minutes = One whole
job

$$\frac{15}{x-10} + \frac{20}{x} = 1$$

$$x(x-10)\left[\frac{15}{x-10} + \frac{20}{x}\right] = x(x-10)1$$

$$15x + 20(x-10) = x^2 - 10x$$

$$15x + 20x - 200 = x^2 - 10x$$

$$35x - 200 = x^2 - 10x$$

$$0 = x^2 - 45x + 200$$

$$0 = (x-5)(x-40)$$

$$x - 5 = 0 \qquad x - 40 = 0$$

$$x = 5 \qquad\quad x = 40$$

(Reject 5. It leads to a negative working
time for Jerry.)
Time for Wilhelmina working alone:
40 minutes.

39. Rate of second person: x
Rate of first person: x + 4
(Distance covered by first person in
2 hours)^2 + (Distance covered by second
person in 2 hours)^2 = 40^2

$$[2(x+4)]^2 + (2x)^2 = 40^2$$

$$(2x+8)^2 + (2x)^2 = 40^2$$

$$4x^2 + 32x + 64 + 4x^2 = 1600$$

$$8x^2 + 32x + 64 = 1600$$

$$8x^2 + 32x - 1536 = 0$$

$$x^2 + 4x - 192 = 0$$

$$(x+16)(x-12) = 0$$

$$x = -16 \text{ (reject)} \qquad x = 12$$

Rate of slower person: 12 m/h

Problem set 7.1.1

1.

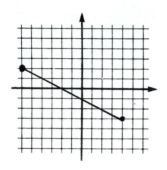

$$d = \sqrt{(-6-4)^2 + (2+3)^2}$$
$$= \sqrt{(-10)^2 + (5)^2} = \sqrt{100+25} = \sqrt{125}$$
$$= \sqrt{25(5)} = 5\sqrt{5}$$

11.

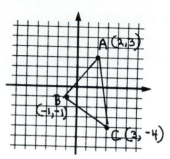

$$AB = \sqrt{(1-5)^2 + (10-7)^2} = \sqrt{16+9}$$
$$= \sqrt{25} = 5$$
$$CB = \sqrt{(1+3)^2 + (10+8)^2} = \sqrt{16+324}$$
$$= \sqrt{340} = \sqrt{4 \cdot 85} = 2\sqrt{85}$$
$$CA = \sqrt{(5+3)^2 + (7+8)^2} = \sqrt{64+225}$$
$$= \sqrt{289} = 17$$
$$\text{Perimeter} = 5 + 2\sqrt{85} + 17 = 22 + 2\sqrt{85}$$

13.

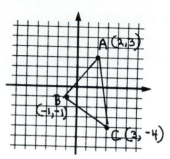

$$AB = \sqrt{(2+1)^2 + (3+1)^2} = \sqrt{9+16}$$
$$= \sqrt{25} = 5$$
$$AC = \sqrt{(2-3)^2 + (3+4)^2} = \sqrt{1+49}$$
$$= \sqrt{50} = 5\sqrt{2}$$
$$BC = \sqrt{(-1-3)^2 + (-1+4)^2} = \sqrt{16+9}$$
$$= \sqrt{25} = 5$$

Since two sides have equal measure
$(AB = BC)$, the triangle is isosceles.

15.
$$\sqrt{(5-5)^2 + (y_1-1)^2} = 8$$
$$\sqrt{(y_1-1)^2} = 8$$
$$(y_1-1)^2 = 64$$
$$y_1 - 1 = \pm\sqrt{64}$$

$y_1 - 1 = 8$	$y_1 - 1 = -8$
$y_1 = 9$	$y_1 = -7$

23.

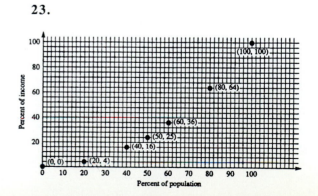

25. $AB = \sqrt{(1+4)^2 + (0+6)^2} = \sqrt{25+36}$

$\qquad = \sqrt{61}$

$BC = \sqrt{(11-1)^2 + (12-0)^2}$

$\qquad = \sqrt{100+144} = \sqrt{244} = \sqrt{4 \cdot 61} = 2\sqrt{61}$

$AC = \sqrt{(11+4)^2 + (12+6)^2}$

$\qquad = \sqrt{225+324} = \sqrt{549} = \sqrt{9 \cdot 61} = 3\sqrt{61}$

$AB + BC = \sqrt{61} + 2\sqrt{61} = 3\sqrt{61} = AC$

Thus A, B, and C are collinear.

27. $\sqrt{(x_1+1)^2 + (y_1-0)^2}$

$\qquad\qquad = \sqrt{(x_1+1)^2 + (y_1-5)^2}$

$(x_1+1)^2 + y_1^2 = (x_1+1)^2 + (y_1-5)^2$

$\qquad\qquad y_1^2 = (y_1-5)^2$

$\qquad\qquad y_1^2 = y_1^2 - 10y_1 + 25$

$\qquad\qquad 0 = -10y_1 + 25$

$\qquad\qquad y_1 = \frac{25}{10} = \frac{5}{2}$

31.

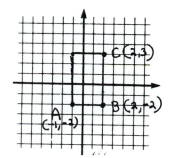

$AB = \sqrt{(2+1)^2 + (-2+2)^2} = \sqrt{9} = 3$

$BC = \sqrt{(2-2)^2 + (3+2)^2} = \sqrt{25} = 5$

Area $= (3)(5) = 15$ square units

32. $\frac{2}{\sqrt{3}+1} \cdot \frac{\sqrt{3}-1}{\sqrt{3}-1} = \frac{2(\sqrt{3}-1)}{3-1} = \frac{2(\sqrt{3}-1)}{2}$

$\qquad = \sqrt{3}-1$

33. width : x

Length : x + 4

$\qquad x(x+4) = 96$

$\qquad x^2 + 4x - 96 = 0$

$\qquad (x+12)(x-8) = 0$

$\qquad x = -12 \text{ (reject)} \qquad x = 8$

width: 8 yd   Length: $8 + 4 = 12$yd

34. $\sqrt[5]{32x^6y^2} - \sqrt[5]{x^6y^2}$

$\qquad = \sqrt[5]{32x^5xy^2} - \sqrt[5]{x^5xy^2}$

$\qquad = 2x\sqrt[5]{xy^2} - x\sqrt[5]{xy^2}$

$\qquad = x\sqrt[5]{xy^2}$

Problem set 7.1.2

5. $2x + 3y = 6$

$\qquad 3y = 6 - 2x$

$\qquad y = \frac{6-2x}{3}$

x	−1	0	1	2
y	$\frac{8}{3}$	2	$\frac{4}{3}$	$\frac{2}{3}$

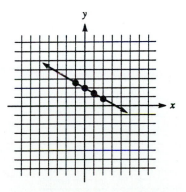

9. $2x - y = 4$
$2x - 4 = y$

x	−1	0	1	2
y	−6	−4	−2	0

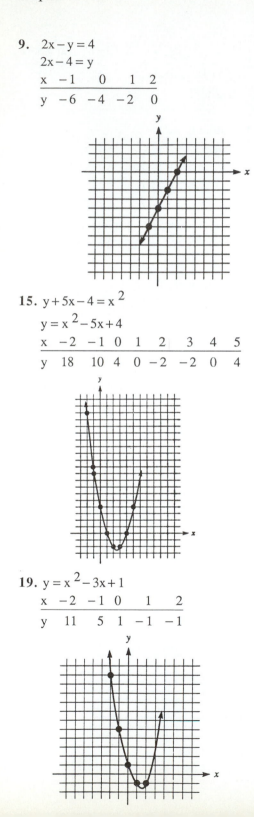

15. $y + 5x - 4 = x^2$
$y = x^2 - 5x + 4$

x	−2	−1	0	1	2	3	4	5
y	18	10	4	0	−2	−2	0	4

19. $y = x^2 - 3x + 1$

x	−2	−1	0	1	2
y	11	5	1	−1	−1

27. $y = -x^2$

x	−2	−1	0	1	2
y	−4	−1	0	−1	−4

Note: If $x = -2$, $y = -(-2)^2 = -4$

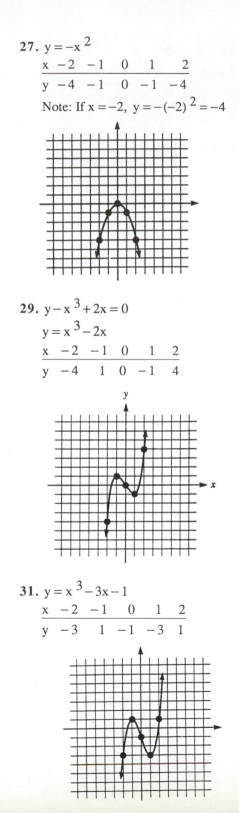

29. $y - x^3 + 2x = 0$
$y = x^3 - 2x$

x	−2	−1	0	1	2
y	−4	1	0	−1	4

31. $y = x^3 - 3x - 1$

x	−2	−1	0	1	2
y	−3	1	−1	−3	1

33. $C = \frac{5}{9}(F - 32)$

F	14	32	50	59	68
C	−10	0	10	15	20

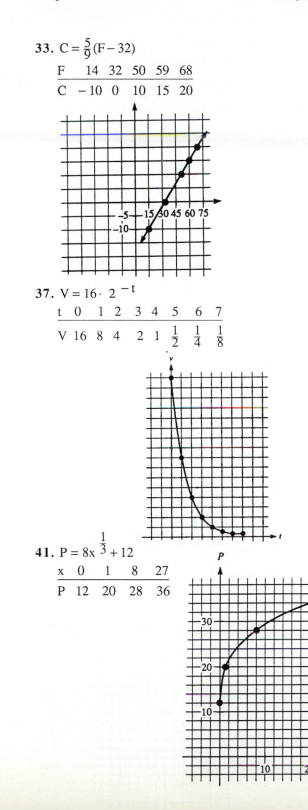

37. $V = 16 \cdot 2^{-t}$

t	0	1	2	3	4	5	6	7
V	16	8	4	2	1	$\frac{1}{2}$	$\frac{1}{4}$	$\frac{1}{8}$

41. $P = 8x^{\frac{1}{3}} + 12$

x	0	1	8	27
P	12	20	28	36

47.

t	0	1	2	3
h_{moon}	6	51.3	91.2	125.7
h_{earth}	6	38	38	6

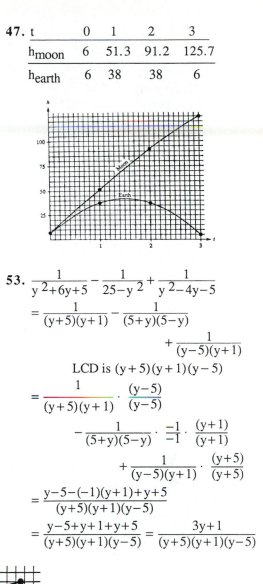

53.

$$\frac{1}{y^2+6y+5} - \frac{1}{25-y^2} + \frac{1}{y^2-4y-5}$$

$$= \frac{1}{(y+5)(y+1)} - \frac{1}{(5+y)(5-y)}$$

$$+ \frac{1}{(y-5)(y+1)}$$

LCD is $(y+5)(y+1)(y-5)$

$$= \frac{1}{(y+5)(y+1)} \cdot \frac{(y-5)}{(y-5)}$$

$$- \frac{1}{(5+y)(5-y)} \cdot \frac{-1}{-1} \cdot \frac{(y+1)}{(y+1)}$$

$$+ \frac{1}{(y-5)(y+1)} \cdot \frac{(y+5)}{(y+5)}$$

$$= \frac{y-5-(-1)(y+1)+y+5}{(y+5)(y+1)(y-5)}$$

$$= \frac{y-5+y+1+y+5}{(y+5)(y+1)(y-5)} = \frac{3y+1}{(y+5)(y+1)(y-5)}$$

54. $\left(y^2-2y\right)^2 - 11\left(y^2-2y\right)+24=0$

　　　　Let $t=y^2-2y$

$t^2-11t+24=0$

$(t-8)(t-3)=0$

　　　　$t=8$　　　　　　　　$t=3$

　　　$y^2-2y=8$　　　　　$y^2-2y=3$

　$y^2-2y-8=0$　　　$y^2-2y-3=0$

　$(y-4)(y+2)=0$　　$(y-3)(y+1)=0$

　$y=4$　　$y=-2$　　　$y=3$　　$y=-1$

　　　　　$\{-2,\ -1,\ 3,\ 4\}$

55. Consecutive even integers : x and $x+2$

　　　$x(x+2)+4=52$

　　　$x^2+2x-48=0$

　　　$(x-6)(x+8)=0$

　　　$x=6$　　　$x=-8$

　　　Consecutive even integers: 6 and 8

　　　or -8 and -6

Problem set 7.2.1

1. $x-2y=6$

　　　If $x=0$, $-2y=6$ and $y=-3$. $(0,\ -3)$

　　　If $y=0$, $x=6$. $(6,\ 0)$

　　　If $x=4$, $4-2y=6$ and $y=-1$.

　　　$(4,\ -1)$

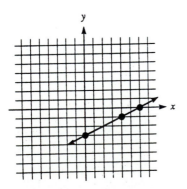

11. $y=-2x+1$

　　　If $x=0$, $y=-2(0)+1=1$. $(0,\ 1)$

　　　If $y=0$, $0=-2x+1$ and $x=\frac{1}{2}$. $\left(\frac{1}{2},\ 0\right)$

　　　If $x=1$, $y=-2(1)+1=-1$. $(1,\ -1)$

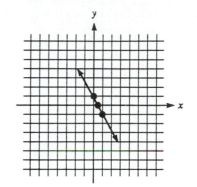

19. $x-y=0$

　　　If $x=0$, $-y=0$ and $y=0$. $(0,\ 0)$

　　　If $x=1$, $1-y=0$ and $y=1$. $(1,\ 1)$

　　　If $x=3$, $3-y=0$ and $y=3$. $(3,\ 3)$

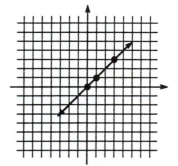

21. $y-3=0$

　　　$y=3$

　　　If $x=0$, $y=3$. $(0,\ 3)$

　　　If $x=1$, $y=3$. $(1,\ 3)$

　　　(For all x, $y=3$)

25. $y = 0$

The graph of $y = 0$ is the x-axis.

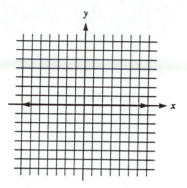

27. $y + 2 = -3$

$\quad y = -5$

$\quad (0, -5), (1, -5), (2, -5)$

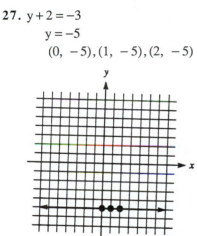

33. The graph of $x = 0$ is the y-axis.

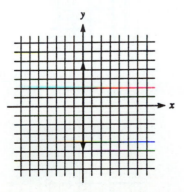

35. $A - 30P = -15$

If $A = 0, -30P = -15$ and $P = \frac{1}{2}$. $\left(0, \frac{1}{2}\right)$

If $P = 0$, $A = -15$. $(-15, 0)$

If $A = 30$, $30 - 30P = -15$ and $P = 1\frac{1}{2}$.

$\left(30, 1\frac{1}{2}\right)$

If $A = 60$, $60 - 30P = -15$ and $P = 2\frac{1}{2}$.

$\left(60, 2\frac{1}{2}\right)$

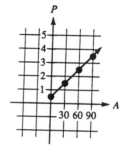

37. $33p + 15d = 495$

p	0	15	5	10	20
d	33	0	22	11	-11

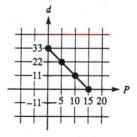

39. $E = 20I$ $0 \le I \le 1$

I	0	$\frac{1}{4}$	$\frac{1}{2}$	$\frac{3}{4}$	1
E	0	5	10	15	20

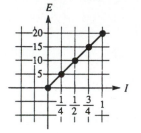

41. $\sqrt[3]{5y^2 - 16} - 4 = 0$

$$\sqrt[3]{5y^2 - 16} = 4$$

$$\left(\sqrt[3]{5y^2 - 16}\right)^3 = 4^3$$

$$5y^2 - 16 = 64$$

$$5y^2 = 80$$

$$y^2 = 16$$

$$y = \pm\sqrt{16}$$

$$y = \pm 4$$

$$\{-4, 4\}$$

42. x: width

x + 7 : length

By Pythagorean Theorem:

$$x^2 + (x+7)^2 = 13^2$$

$$x^2 + x^2 + 14x + 49 = 169$$

$$2x^2 + 14x - 120 = 0$$

$$x^2 + 7x - 60 = 0$$

$$(x - 5)(x + 12) = 0$$

$$x = 5 \qquad x = -12 \text{ (reject)}$$

width : 5 m

length : 5 + 7 = 12 m

43. $5\sqrt{\frac{1}{3}} - 7\sqrt{\frac{1}{27}} = \frac{5}{\sqrt{3}} \cdot \frac{\sqrt{3}}{\sqrt{3}} - \frac{7}{3\sqrt{3}} \cdot \frac{\sqrt{3}}{\sqrt{3}}$

$= \frac{5\sqrt{3}}{3} - \frac{7\sqrt{3}}{9} = \frac{15\sqrt{3}}{9} - \frac{7\sqrt{3}}{9} = \frac{8\sqrt{3}}{9}$

Problem set 7.2.2

23. $2x + 3y > 6$

Consider $2x + 3y = 6$

If x = 0, 3y = 6 and y = 2.

If y = 0, 2x = 6 and x = 3.

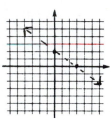

Test point : (0, 0)

$2(0) + 3(0) > 6$; $0 > 6$ False

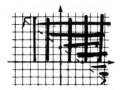

The graph of $y \ge 0$ consists of the x-axis and the half-plane above the x-axis.

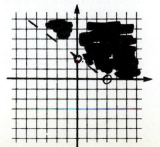

$2x + 3y > 6$ and $y \ge 0$ is the intersection of the two regions. The final graph is:

27. $x - y \geq 4$

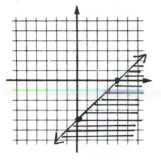

Graph $x + y \leq 6$ on the same axes.

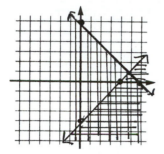

The graph of $x - y \geq 4$ and $x + y \leq 6$ is the intersection of the two regions, shown by:

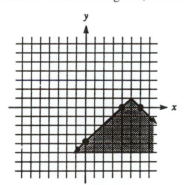

31. Graph $60x + 120y \leq 6000$ in Quadrant I.

$60x + 120y = 6000$ If $x = 0$, $y = 50$.
                       If $y = 0$, $x = 100$.

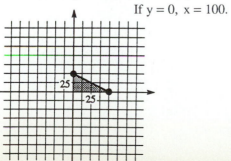

33. $|x| \leq 2$
    Equivalently: $-2 \leq x \leq 2$

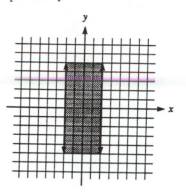

37. $|2x + 1| \leq 3$
    Equivalently: $-3 \leq 2x + 1 \leq 3$
                 $-4 \leq 2x \leq 2$
                 $-2 \leq x \leq 1$

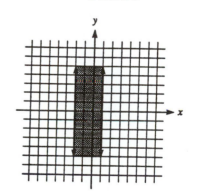

45. $|2x + 1| > 3$
    Equivalently: $2x + 1 > 3$ or $2x + 1 < -3$
                 $2x > 2$ or $2x < -4$
                 $x > 1$ or $x < -2$

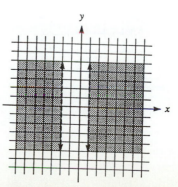

49. $|2y + 1| \geq 3$

Equivalently: $2y + 1 \geq 3$ or $2y + 1 \leq -3$

$y \geq 1$ or $y \leq -2$

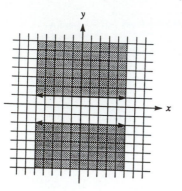

51. $|x| \leq 2$ and $|y| \leq 3$

$-2 \leq x \leq 2$ and $-3 \leq y \leq 3$

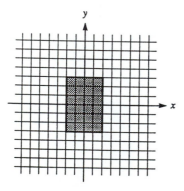

53. $|x| \geq 2$ and $|y| \geq 3$

$x \geq 2$ or $x \leq -2$ and $y \geq 3$ or $y \leq -3$

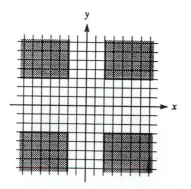

55. $|2x + 1| \leq 3$ and $|y| \leq 3$

$-3 \leq 2x + 1 \leq 3$ and $-3 \leq y \leq 3$

$-2 \leq x \leq 1$ and $-3 \leq y \leq 3$

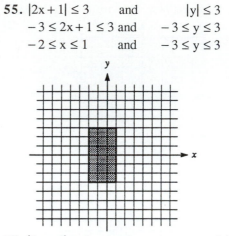

57. $|2x + 4| \geq 6$ and $|y| \leq 2$

$2x + 4 \geq 6$ or $2x + 4 \leq -6$ and $-2 \leq y \leq 2$

$x \geq 1$ or $x \leq -5$ and $-2 \leq y \leq 2$

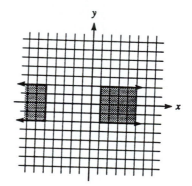

59. x: number of A pots (throw: 1h Glaze: 2h)

y: number of B pots (throw: 2h Glaze: 2h)

Time throwing is not to exceed 6 hours.

$$x + 2y \le 6$$

Glazing time is not to exceed 8 hours.

$$2x + 2y \le 8$$

Equivalently $x + y \le 4$

Thus: $x + 2y \le 6$

$$x + y \le 4$$

$$x \ge 0 \text{ and } y \ge 0$$

The graph consists of the intersections of $x + 2y \le 6$ and $x + y \le 4$ in Quadrant I.

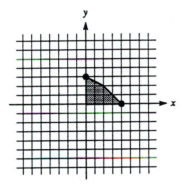

61. The maximum income must occur at
(0, 3) or (0, 6) or (3, 0) or (5, 4),
one of the vertices of the graphed region.

$$I = 25x + 15y$$

At (0, 3) : $I = 25(0) + 15(3) = 45$

At (0, 6) : $I = 25(0) + 15(6) = 90$

At (3, 0) : $I = 25(3) + 15(0) = 75$

At (5, 4) : $I = 25(5) + 45(4) = 185$

The maximum income is $185.

63. a. x: number of $10 calculators

y: number of $25 calculators

$$P = 6x + 20y$$

b. $x + y \le 200$

$$10x + 25y \le 5000$$

Equivalently: $2x + 5y \le 1000$

c.

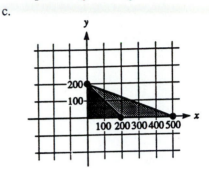

d. $P = 6x + 20y$ is maximum at (0, 0) or
(0, 200) or (200, 0).

At (0, 0) : $P = 6(0) + 20(0) = 0$

At (200, 0) : $P = 6(200) + 20(0) = 1200$

At (0, 200) : $P = 6(0) + 20(200) = 4000$

maximum weekly profit: $4000.

64. $\sqrt[5]{64x^{11}y^{10}}$

$$= \sqrt[5]{32 \cdot 2\left(x^2\right)^5 x\left(y^2\right)^5}$$

$$= 2x^2 y^2 \sqrt[5]{2x}$$

65. $fny\left(\dfrac{1}{y} + \dfrac{1}{ny}\right) = fny\left(\dfrac{1}{f}\right)$

$$fn + f = ny$$

$$\dfrac{fn+f}{n} = y$$

66. $\dfrac{y^6+8}{y^3-3y^2+2y-6}=\dfrac{\left(y^2\right)^3+2^3}{y^2(y-3)+2(y-3)}$

$=\dfrac{\left(y^2+2\right)\left(y^4-2y^2+4\right)}{(y-3)\left(y^2+2\right)}=\dfrac{y^4-2y^2+4}{y-3}$

Problem set 7.3.1

1. $m=\dfrac{y_2-y_1}{x_2-x_1}=\dfrac{2-(-2)}{4+2}=\dfrac{4}{2}=2$

7. $m=\dfrac{y_2-y_1}{x_2-x_1}=\dfrac{7-(-1)}{2-4}=\dfrac{7+1}{2-4}=\dfrac{8}{-2}=4$

9. $m=\dfrac{y_2-y_1}{x_2-x_1}=\dfrac{5-5}{4-3}=\dfrac{0}{1}=0$

11. $m=\dfrac{y_2-y_1}{x_2-x_1}=\dfrac{-9-2}{-6-(-6)}=\dfrac{-11}{-6+6}=\dfrac{-11}{0}$

Slope is undefined.

13. $\dfrac{6-y}{-3-2}=\dfrac{1}{2}$

$\dfrac{6-y}{-5}=\dfrac{1}{2}$

$2(6-y)=-5$

$12-2y=-5$

$-2y=-17$

$y=\dfrac{17}{2}$

15. $\dfrac{a^2+3a}{2-1}=-2$

$a^2+3a=-2$

$a^2+3a+2=0$

$(a+1)(a+2)=0$

$a=-1 \qquad a=-2$

17. $\dfrac{a-2-5}{1-(a+3)}=4$

$\dfrac{a-7}{-a-2}=\dfrac{4}{1}$

$a-7=4(-a-2)$

$a-7=-4a-8$

$5a=-1$

$a=-\dfrac{1}{5}$

33.

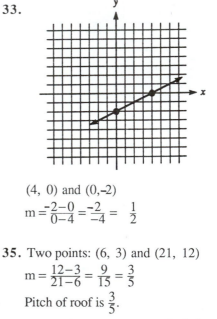

(4, 0) and (0,–2)

$m=\dfrac{-2-0}{0-4}=\dfrac{-2}{-4}=\dfrac{1}{2}$

35. Two points: (6, 3) and (21, 12)

$m=\dfrac{12-3}{21-6}=\dfrac{9}{15}=\dfrac{3}{5}$

Pitch of roof is $\dfrac{3}{5}$.

Another method: $m=\dfrac{\text{vertical change}}{\text{horizontal change}}$

$=\dfrac{9}{15}=\dfrac{3}{5}$

37. $m = \dfrac{\text{vertical change}}{\text{horizontal change}}$

h: height of pyramid

horizontal change $= \dfrac{1}{2}(650) = 325$

$\dfrac{h}{325} = \dfrac{13}{11}$

$11h = (325)(13)$

$11h = 4225$

$h \approx 384$

Pyramid is 384 ft tall.

41. slope of line through $(-3, -1)$ and $(1, 4)$:

$m = \dfrac{4+1}{1+3} = \dfrac{5}{4}$

Slope of any line parallel to this line is

also $\dfrac{5}{4}$ (parallel lines have the same slope).

43. slope of line through $(2, -3)$ and $(5, 2)$:

$m = \dfrac{2+3}{5-2} = \dfrac{5}{3}$

Slope of perpendicular line is the negative

reciprocal of $\dfrac{5}{3}$, which is $-\dfrac{3}{5}$.

45. $\dfrac{\text{vertical change}}{\text{horizontal change}} = m$

$\dfrac{v}{16} = \dfrac{3}{4}$

$4v = 48$

$v = 12$

Vertical change is 12.

47. slope of line passing through $(x, 2)$ and $(1, 0) =$ slope of line joining $(2, 3)$ and $(-2, 1)$

$\dfrac{2}{x-1} = \dfrac{3-1}{2+2}$

$\dfrac{2}{x-1} = \dfrac{1}{2}$

$x - 1 = 4$

$x = 5$

49. $\dfrac{n^2 + 3n}{2-1} = \dfrac{3-1}{4-5}$

$n^2 + 3n = -2$

$n^2 + 3n + 2 = 0$

$(n+2)(n+1) = 0$

$n = -2 \qquad n = -1$

51. Line through $(-2, 4)$ and $(19, 3)$:

$m = \dfrac{3-4}{19+2} = \dfrac{-1}{21}$

$\dfrac{c^2 + 4c}{3-2} = 21$

$\left(\text{the negative reciprocal of } -\dfrac{1}{21}\right)$

$c^2 + 4c - 21 = 0$

$(c+7)(c-3) = 0$

$c = -7 \qquad c = 3$

55.

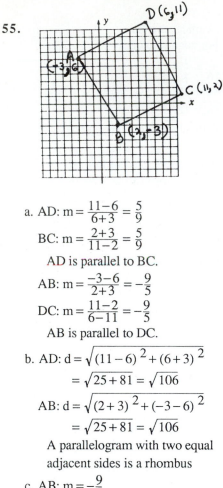

a. AD: $m = \dfrac{11-6}{6+3} = \dfrac{5}{9}$

   BC: $m = \dfrac{2+3}{11-2} = \dfrac{5}{9}$

   AD is parallel to BC.

   AB: $m = \dfrac{-3-6}{2+3} = -\dfrac{9}{5}$

   DC: $m = \dfrac{11-2}{6-11} = -\dfrac{9}{5}$

   AB is parallel to DC.

b. AD: $d = \sqrt{(11-6)^2 + (6+3)^2}$

        $= \sqrt{25+81} = \sqrt{106}$

   AB: $d = \sqrt{(2+3)^2 + (-3-6)^2}$

        $= \sqrt{25+81} = \sqrt{106}$

   A parallelogram with two equal
   adjacent sides is a rhombus

c. AB: $m = -\dfrac{9}{5}$

   BC : $m = \dfrac{5}{9}$

Since the slopes are negative reciprocals,
AB and BC are perpendicular.

57. a. AC: $m = \dfrac{b+c-b}{a+c-a} = \dfrac{c}{c} = 1$

    BD: $m = \dfrac{b+c-b}{a-(a+c)} = -\dfrac{c}{c} = -1$

    Since the slopes are negative
    reciprocals, diagonals AC and BD
    are perpendicular.

59 $\dfrac{(y-1)(y-1)}{(3y-5)(y+4)} \cdot \dfrac{(3y-5)(y+1)}{(y+4)(y-1)}$

                $\cdot \dfrac{(y+4)}{(y-3)(y-1)}$

    $= \dfrac{y+1}{(y+4)(y-3)}$

60. $\sqrt[3]{4y^2} \cdot \sqrt[3]{10x^4 y} = \sqrt[3]{40x^4 y^3}$

     $= \sqrt[3]{8 \cdot 5x^3 xy^3} = 2xy \sqrt[3]{5x}$

61. x: speed of current

    Time to cover 2 miles against the current

     $=$ time to cover 2 miles with the current $+ \dfrac{1}{6}$

     $\left(10 \text{ minutes} = \dfrac{1}{6} \text{ of an hour} \right)$

     $\left[RT = D \,;\, T = \dfrac{D}{R} \right]$

     $\dfrac{2}{5-x} = \dfrac{2}{5+x} + \dfrac{1}{6}$

     $6(5-x)(5+x) \left[\dfrac{2}{5-x} \right]$

         $= 6(5-x)(5+x) \left[\dfrac{2}{5+x} + \dfrac{1}{6} \right]$

     $12(5+x) = 12(5-x) + (5-x)(5+x)$

     $60 + 12x = 60 - 12x + 25 - x^2$

     $60 + 12x = 85 - 12x - x^2$

     $x^2 + 24x - 25 = 0$

     $(x+25)(x-1) = 0$

     $x = -25$ (reject) $x = 1$

    speed of current : 1 m/h

Problem set 7.3.2

1. $y - y_1 = m(x - x_1)$

    $y - 2 = 2(x + 3)$

      $y = 2x + 8$

5. $y - y_1 = m(x - x_1)$

$\qquad y - 6 = -\frac{2}{3}(x + 8)$

$\qquad\qquad y = -\frac{2}{3}x - \frac{16}{3} + 6$

$\qquad\qquad y = -\frac{2}{3}x + \frac{2}{3}$

7. $m = \frac{16 - 8}{3 - 5} = \frac{8}{-2} = -4$

$\qquad y - y_1 = m(x - x_1)$

$\qquad y - 8 = -4(x - 5) \left[\text{or } y - 16 = -4(x - 3)\right]$

$\qquad\qquad y = -4x + 28$

9. x-intercept $= 11$, so point $= (11, 0)$

Line passes through $(5, 6)$ and $(11, 0)$.

$m = \frac{0 - 6}{11 - 5} = \frac{-6}{6} = -1$

$y - y_1 = m(x - x_1)$

$\qquad y - 6 = -1(x - 5) \left[\text{or } y - 0 = -1(x - 11)\right]$

$\qquad\qquad y = -x + 11$

11. $m = \frac{125 - 115}{30 - 10} = \frac{10}{20} = \frac{1}{2}$

$\qquad y - y_1 = m(x - x_1)$

$\qquad y - 115 = \frac{1}{2}(x - 10)$

$\qquad\qquad y = \frac{1}{2}x + 110$

If $x = 80$, $y = \frac{1}{2}(80) + 110 = 150$

Blood pressure is 150.

17. $m = \frac{8 - 5}{120 - 100} = \frac{3}{20}$

$\qquad y - y_1 = m(x - x_1)$

$\qquad y - 8 = \frac{3}{20}(x - 120)$

$\qquad y - 8 = \frac{3}{20}x - 18$

$\qquad\qquad y = \frac{3}{20}x - 10$

a. If $x = 140$, $y = \frac{3}{20}(140) - 10 = 21 - 10$

$\qquad = 11$. Grade point average: 11

If $x = 110$, $y = \frac{3}{20}(110) - 10 = 16.5 - 10$

$\qquad = 6.5$. Grade point average: 6.5

21. $m = \frac{1.58 - 1.34}{8 - 2} = \frac{0.24}{6} = 0.04$

$\qquad y - 1.34 = 0.04(x - 2)$

$\qquad\qquad y = 0.04x + 1.26$

Year 2000 is 27 years after 1973.

If $x = 27$, $y = 0.04(27) + 1.26 = 2.34$

In the year 2000, 2.34 million college degrees will be conferred.

25.
$$z = \frac{k\sqrt{x}}{\sqrt{y}}$$

$$30 = \frac{k\sqrt{9}}{\sqrt{2}}$$

$$30\sqrt{2} = 3k$$

$$10\sqrt{2} = k$$

$$z = \frac{10\sqrt{2}\sqrt{x}}{\sqrt{y}}$$

$$50 = \frac{10\sqrt{2}\sqrt{4}}{\sqrt{y}}$$

$$50\sqrt{y} = 20\sqrt{2}$$

$$5\sqrt{y} = 2\sqrt{2}$$

$$\left(5\sqrt{y}\right)^2 = \left(2\sqrt{2}\right)^2$$

$$25y = 8$$

$$y = \frac{8}{25}$$

26. $y^{\frac{1}{3}} - 4y^{\frac{1}{6}} + 3 = 0$

Let $t = y^{\frac{1}{6}}$

$$t^2 - 4t + 3 = 0$$

$$(t-3)(t-1) = 0$$

$$t = 3 \qquad\qquad t = 1$$

$$y^{\frac{1}{6}} = 3 \qquad\qquad y^{\frac{1}{6}} = 1$$

$$\left(y^{\frac{1}{6}}\right)^6 = 3^6 \qquad \left(y^{\frac{1}{6}}\right)^6 = 1^6$$

$$y = 729 \qquad\qquad y = 1$$

check 729:

$$(729)^{\frac{1}{3}} - 4(729)^{\frac{1}{6}} + 3 = 0$$

$$\sqrt[3]{729} - 4\sqrt[6]{726} + 3 = 0$$

$$9 - 4(3) + 3 = 0$$

$$0 = 0$$

check 1:

$$(1)^{\frac{1}{3}} - 4(1)^{\frac{1}{6}} + 3 = 0$$

$$1 - 4 + 3 = 0$$

$$0 = 0$$

$$\{1,\ 729\}$$

27. $\left(9^{-2} + 3^{-3}\right)^{-\frac{1}{2}}$

$$= \left(\frac{1}{9^2} + \frac{1}{3^3}\right)^{-\frac{1}{2}}$$

$$= \left(\frac{1}{81} + \frac{1}{27}\right)^{-\frac{1}{2}} = \left(\frac{1}{81} + \frac{3}{81}\right)^{-\frac{1}{2}}$$

$$= \left(\frac{4}{81}\right)^{-\frac{1}{2}} = \frac{1}{\left(\frac{4}{81}\right)^{\frac{1}{2}}} = \frac{1}{\sqrt{\frac{4}{81}}} = \frac{1}{\frac{2}{9}} = \frac{9}{2}$$

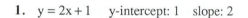

Problem set 7.3.3

1. $y = 2x + 1$ y-intercept: 1   slope: 2

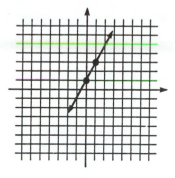

17. $x + y = 3$

$y = -x + 3$ y-intercept: 3   slope: -1

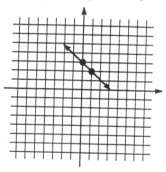

19. $2x + y = -1$

$y = -2x - 1$ y-intercept: -1 slope: -2

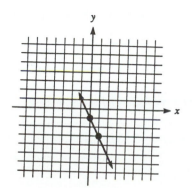

21. $2x + 3y = 6$

$3y = -2x + 6$

$y = -\frac{2}{3}x + \frac{6}{3}$

$y = -\frac{2}{3}x + 2$ y-intercept: 2   slope: $-\frac{2}{3}$

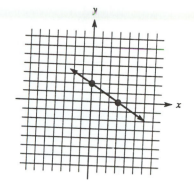

29. $2x + 5y > 10$

$5y > -2x + 10$

$y > -\frac{2}{5}x + 2$

The graph consists of the half-plane above the line whose equation is $y = -\frac{2}{5}x + 2$. Although the line is not included, its y-intercept is 2 and its slope is $-\frac{2}{5}$.

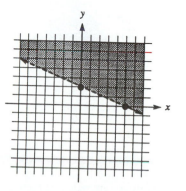

31. $x - 2y > 0$

$\quad -2y > -x$

$\quad y < \frac{1}{2}x$

$\quad y = \frac{1}{2}x \quad$ y-intercept: 0   slope: $\frac{1}{2}$

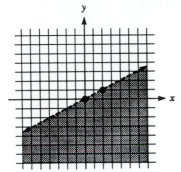

33. $4x - 3y \geq 12$

$\quad -3y \geq -4x + 12$

$\quad y \leq \frac{4}{3}x - 4$

$\quad y = \frac{4}{3}x - 4 \quad$ y-intercept: -4 slope: $\frac{4}{3}$

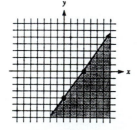

37. $y = 2x + 3$ slope: 2 $y = 2x + 17$ slope: 2

With equal slopes, the lines are parallel.

39. $y = 2x + 3$ slope: 2

$\quad y = -\frac{1}{2}x + 5$ slope: $-\frac{1}{2}$

Since slopes are negative reciprocals, the lines are perpendicular.

41. $2x + 3y = 5 \qquad\qquad 2x - 3y = 11$

$\quad 3y = -2x + 5 \qquad\quad -3y = -2x + 11$

$\quad y = -\frac{2}{3}x + \frac{5}{3} \qquad\quad y = \frac{2}{3}x - \frac{11}{3}$

$\quad$ slope: $-\frac{2}{3} \qquad\qquad\quad$ slope: $\frac{2}{3}$

Since the slopes are neither equal nor negative reciprocals, the lines intersect and are not perpendicular.

45. $y = 2 \; (y = 0x + 2) \;$ slope: 0

$\quad y = 4 \; (y = 0x + 4) \;$ slope: 0

With equal slopes, the lines are parallel.

47. $x = 0$ is the equation of the y-axis.

$\quad y = 0$ is the equation of the x-axis.

The axes are perpendicular.

49. $y = 2x - 1 \;$ slope: 2

slope of line whose equation we must write:

2

$\quad y - 3 = 2(x - 1)$

$\quad y = 2x + 1 \qquad \left[\text{or} \; 2x - y = -1 \right]$

$\quad -2x + y = 1$

51. $y = 2x - 1 \;$ slope: 2

slope of line whose equation we must write:

$-\frac{1}{2}$

$\quad y + 3 = -\frac{1}{2}(x - 1) \;$ Point-slope form

$\quad y + 3 = -\frac{1}{2}x + \frac{1}{2}$

$\quad y = -\frac{1}{2}x - \frac{5}{2} \;$ Slope-intercept form

$\quad 2y = 2\left(-\frac{1}{2}x - \frac{5}{2}\right)$

$\quad 2y = -x - 5$

$\quad x + 2y = -5 \;$ Standard form

55. $-\frac{1}{3}x + y = 4$

$y = \frac{1}{3}x + 4 \quad \left[\text{slope: } \frac{1}{3}\right]$

slope of line whose equation we must write:
-3

$y + 3 = -3(x - 2)$ Point-slope form

$y = -3x + 3$ Slope-intercept form

$3x + y = 3$ Standard form

59. $2x - 4y = 12$

$-4y = -2x + 12$

$y = \frac{1}{2}x - 3 \quad \left[\text{slope: } \frac{1}{2}\right]$

slope of line whose equation we must write:
-2

$y + 3 = -2(x - 1)$ Point-slope form

$y = -2x - 1$ Slope-intercept form

$2x + y = -1$ Standard form

63. Line passing through $(1, -5)$ and $(0, -6)$:

$m = \frac{-6+5}{0-1} = \frac{-1}{-1} = 1$

slope of line whose equation we must write:
1

$(x_1, y_1) = (2, 4)$

$y - 4 = 1(x - 2)$ Point-slope form

$y = x + 2$ Slope-intercept form

$-x + y = 2$ Standard form

65. Line passing through $\left(8, -\frac{3}{2}\right)$ and $\left(0, \frac{5}{2}\right)$:

$m = \dfrac{\frac{5}{2} + \frac{3}{2}}{0 - 8} = \frac{4}{-8} = -\frac{1}{2}$

slope of line whose equation we must write:
2

$(x_1, y_1) = (-3, -7)$

$y + 7 = 2(x + 3)$ Point-slope form

$y = 2x - 1$ Slope-intercept form

$-2x + y = -1$ Standard form

67. slope of x-axis (horizontal line): $m = 0$

$y - 4 = 0(x + 2)$ Point-slope form

$y = 4$ Slope-intercept and Standard form

69. $3x - 2y = 4$

slope: $-2y = -3x + 4$

$y = \frac{3}{2}x - 2 \qquad m = \frac{3}{2}$

slope of line whose equation we must write:
$-\frac{2}{3}$

x-intercept of $3x - 2y = 4$

$3x - 2(0) = 4$

$x = \frac{4}{3}$

$(x_1, y_1) = \left(\frac{4}{3}, 0\right)$

$y - 0 = -\frac{2}{3}\left(x - \frac{4}{3}\right)$ Point-slope form

$y = -\frac{2}{3}x + \frac{8}{9}$ Slope-intercept form

$9y = -6x + 8$

$6x + 9y = 8$ Standard form

71. $Ax + By = C$

$By = -Ax + C$

$y = -\frac{A}{B}x + \frac{C}{B}$

slope: $-\frac{A}{B}$

75. $by = 8x - 1$

$y = \frac{8}{b}x - \frac{1}{b}$

slope: $\frac{8}{b} = -2$

$-2b = 8$

$b = -4$

77. (t, u) satisfies the equation.

$$At + Bu = C$$

(v, w) satisfies the equation.

$$Av + Bw = C$$

Thus,

$$At + Bu = Av + Bw$$
$$At - Av = Bw - Bu$$
$$A(t - v) = B(w - u)$$
$$\frac{A}{B}(t - v) = (w - u)$$
$$\frac{A}{B} = \frac{w - u}{t - v}$$

Second method:

slope of $Ax + By = C$ is $-\frac{A}{B}$ (see ex. 71)

slope is also $\dfrac{w - u}{v - t}$

Thus, $-\dfrac{A}{B} = \dfrac{w - u}{v - t}$

$$\frac{A}{B} = (-1)\frac{w - u}{v - t}$$
$$\frac{A}{B} = \frac{w - u}{-v + t}$$
$$\frac{A}{B} = \frac{w - u}{t - v}$$

79. m: $\dfrac{6 - 4}{a - 2 - 0} = \dfrac{2}{a - 2}$

$$y - 4 = \frac{2}{a - 2}(x - 0)$$

x-intercept (set $y = 0$) :

$$-4 = \frac{2}{a - 2}x$$
$$-4(a - 2) = 2x$$
$$-2(a - 2) = x$$

x-intercept is $-2(a - 2)$.

Given: x-intercept must be a.

$$-2(a - 2) = a$$
$$-2a + 4 = a$$
$$4 = 3a; \qquad \frac{4}{3} = a$$

80. $\dfrac{-26}{3 + 2i} \cdot \dfrac{3 - 2i}{3 - 2i} = \dfrac{-26(3 - 2i)}{9 - 4i^2}$

$$= \frac{-26(3 - 2i)}{9 - 4(-1)} = \frac{-26(3 - 2i)}{13}$$
$$= -2(3 - 2i) = -6 + 4i$$

81. $\left(-\dfrac{x^{-3}y^{-2}}{x^3 y^{-4}}\right)^{-2} = \left(-x^{-6}y^2\right)^{-2}$

$$= (-1)^{-2}\left(x^{-6}\right)^{-2}\left(y^2\right)^{-2}$$
$$= \frac{1}{(-1)^2}x^{12}y^{-4} = \frac{x^{12}}{y^4}$$

82. x: number of gallons of 7% solution

$6 - x$: number of gallons of 12% solution

Amount of alcohol in the 7% solution
+ Amount of alcohol in the 12% solution
= Amount of alcohol in the 6 gallons of the 10% solution

$$0.07x + 0.12(6 - x) = .10(6)$$
$$0.07x + 0.72 - 0.12x = 0.6$$
$$-0.05x = -0.12$$
$$x = 2.4$$

number of gallons of 7% solution: 2.4
number of gallons of 12% solution:
$6 - 2.4 = 3.6$

Problem set 7.4.1

1. $y = x^2 + 6x + 5$

x-intercepts (set y = 0)

$x^2 + 6x + 5 = 0$

$(x+5)(x+1) = 0$

$x = -5 \qquad x = -1$

y-intercept (set x = 0)

$y = 5$

vertex: $x = -\dfrac{b}{2a} = -\dfrac{6}{2} = -3$

$y = (-3)^2 + 6(-3) + 5 = -4$

$(-3, -4)$

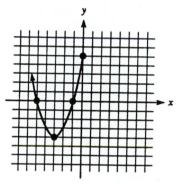

5. $y = -x^2 - 4x - 5$

x-intercepts: $-x^2 - 4x - 5 = 0$

$x^2 + 4x + 5 = 0$

$b^2 - 4ac = 16 - 4(1)(5) = -4 < 0$, so the equation has no real solutions. The graph has no x-intercepts.

y-intercept: $y = -5$

vertex: $x = -\dfrac{b}{2a} = -\dfrac{(-4)}{2(-1)} = \dfrac{4}{-2} = -2$

$y = -(-2)^2 - 4(-2) - 5 = -4 + 8 - 5 = -1$

vertex: $(-2, -1)$

Other values:

If $x = -3$, $y = -(-3)^2 - 4(-3) - 5 = -2$

If $x = -1$, $y = -(-1)^2 - 4(-1) - 5 = -2$

$(-3, -2)$ and $(-1, -2)$

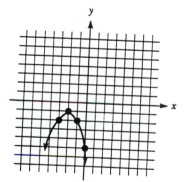

9. $y = 2x^2 - 3$

x-intercepts: $2x^2 - 3 = 0$

$$2x^2 = 3$$

$$x^2 = \frac{3}{2}$$

$$x = \pm\sqrt{\frac{3}{2}} \approx \pm 1.2$$

y-intercept: $y = -3$

vertex: $x = -\dfrac{b}{2a} = -\dfrac{0}{2(2)} = 0$ $(0, -3)$

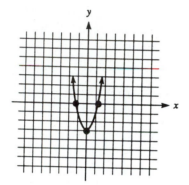

11. $y = x^2 - 4x + 4$

x-intercept: $x^2 - 4x + 4 = 0$

$$(x - 2)^2 = 0$$

$$x = 2$$

y-intercept: $y = 4$

vertex: $x = -\dfrac{b}{2a} = -\dfrac{(-4)}{2(1)} = 2$ $(2, 0)$

If $x = 4$, $y = 4^2 - 4 \cdot 4 + 4 = 4$ $(4, 4)$

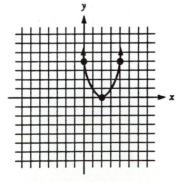

13. $y = -3x^2 + 6x - 1$

x-intercepts: $-3x^2 + 6x - 1 = 0$

$a = -3, b = 6, c = -1$

$$x = \frac{-b \pm \sqrt{b^2 - 4ac}}{2a}$$

$$= \frac{-6 \pm \sqrt{36 - 4(-3)(-1)}}{2(-3)}$$

$$= \frac{-6 \pm \sqrt{24}}{-6} = \frac{-6 \pm 2\sqrt{6}}{-6} = \frac{-3 \pm \sqrt{6}}{-3}$$

$x \approx 0.18$ $x \approx 1.82$

y-intercept: $y = -1$

vertex: $x = -\dfrac{b}{2a} = -\dfrac{6}{2(-3)} = 1$ $(1, 2)$

25. $v = -20,000x^2 + 1.28$

x	0	0.001	0.002	0.003	0.004
v	1.28	1.26	1.2	1.1	0.96

x	0.005	0.006	0.007	0.008
v	0.78	0.56	0.3	0

27. $y = -x^2 + 180x - 4500$

To maximum profit: $x = -\dfrac{b}{2a} = -\dfrac{180}{2(-1)} = 90$

90 units will maximize profit.

maximum profit :

$y = -90^2 + 180(90) - 4500$

$\quad = -8100 + 16200 - 4500$

$\quad = 3600$

maximum profit : $3600

x-intercepts : $-x^2 + 180x - 4500 = 0$

$\qquad\qquad\quad x^2 - 180x + 4500 = 0$

$\qquad\qquad\quad (x - 150)(x - 30) = 0$

$\qquad\qquad\quad x = 150 \qquad x = 30$

y-intercept : $y = -4500$

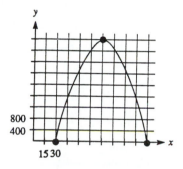

29. $P = -5I^2 + 80I$

$I = -\dfrac{b}{2a} = -\dfrac{80}{2(-5)} = -\dfrac{80}{-10} = 8$

$P = -5(8)^2 + 80(8) = 320$

A current of 8 amps produces a maximum power of 320 volts.

$-5I^2 + 80I = 0$

$-5I(I - 16) = 0$

$I = 0 \qquad I = 16$

31. $C = 20t^2 - 200t + 640$

$t = -\dfrac{b}{2a} = -\dfrac{(-200)}{2(20)} = 5$

Concentration is at a minimum after 5 days.

33. One number : x

Other number : $48 - x$

$y = x(48 - x) = -x^2 + 48x$

$x = -\dfrac{b}{2a} = -\dfrac{48}{2(-1)} = 24$

one number : 24

other number : $48 - 24 = 24$

35. $A = x(20 - x) = -x^2 + 20x$

$x = -\dfrac{b}{2a} = -\dfrac{20}{2(-1)} = 10$

maximum area : $10(20 - 10) = 100 \text{ ft}^2$

37.

LAKE

$240 - 2x$

x: measure of one side

$240 - 2x$: measure of other side

$A = x(240 - 2x) = -2x^2 + 240x$

$x = -\dfrac{b}{2a} = -\dfrac{240}{2(-2)} = 60$

Dimensions :   x: 60m

$\qquad\qquad\quad 240 - 2x:\ 240 - 2(60) = 120\text{m}$

maximum area : $(60)(120) = 7200 \text{ m}^2$

39. $y \geq x^2 + 6x + 5$

(see ex. 1)

Test $(0, 0)$:

$0^2 \geq 0^2 + 6 \cdot 0 + 5$

$0 > 5$, false

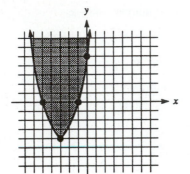

41. $y > x^2 + 4x + 3$

$y = x^2 + 4x + 3$

x-intercepts : $x = -3 \;\; x = -1$

y-intercept : $y = 3$

vertex : $(-2, \, -1)$

Test $(0, 0)$:

$0^2 > 0^2 + 4 \cdot 0 + 3$

$0 > 3$, false

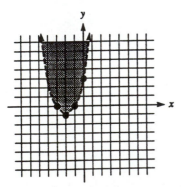

43. $x^{4a+1} - xy^{4a}$

$= x\left(x^{4a} - y^{4a}\right)$

$= x\left[\left(x^{2a}\right)^2 - \left(y^{2a}\right)^2\right]$

$= x\left(x^{2a} + y^{2a}\right)\left(x^{2a} - y^{2a}\right)$

$= x\left(x^{2a} + y^{2a}\right)\left[\left(x^a\right)^2 - \left(y^a\right)^2\right]$

$= x\left(x^{2a} + y^{2a}\right)\left(x^a + y^a\right)\left(x^a - y^a\right)$

44. $y(y+b)(y-b)\left[\dfrac{3}{y} + \dfrac{2}{y-b}\right]$

$\qquad = y(y+b)(y-b)\left[\dfrac{5}{y+b}\right]$

$3(y+b)(y-b) + 2y(y+b) = 5y(y-b)$

$3y^2 - 3b^2 + 2y^2 + 2by = 5y^2 - 5by$

$5y^2 - 3b^2 + 2by = 5y^2 - 5by$

$\qquad -3b^2 + 2by = -5by$

$\qquad\qquad -3b^2 = -7by$

$\qquad\qquad \dfrac{-3b^2}{-7b} = y$

$\qquad\qquad \dfrac{3b}{7} = y$

45. $4x - 5y = 7$

$\quad -5y = -4x + 7$

$\quad y = \dfrac{4}{5}x - \dfrac{7}{5} \qquad$ slope : $\dfrac{4}{5}$

slope of the line whose equation we must

write: $-\dfrac{5}{4}$

$y - 5 = -\dfrac{5}{4}(x + 4)$

$y - 5 = -\dfrac{5x}{4} - 5$

$\quad\quad y = -\dfrac{5x}{4}$

$\quad 4y = -5x$

$5x + 4y = 0$

Problem set 7.4.2

1. $(h, k) = (3, 2)$　　$r = 5$

$(x-h)^2 + (y-k)^2 = r^2$

$(x-3)^2 + (y-2)^2 = 5^2$

$(x-3)^2 + (y-2)^2 = 25$

5. $(h, k) = (-3, -1)$　　$r = \sqrt{3}$

$(x-h)^2 + (y-k)^2 = r^2$

$[x-(-3)]^2 + [y-(-1)]^2 = (\sqrt{3})^2$

$(x+3)^2 + (y+1)^2 = 3$

9. $(h, k) = (0, 0)$　　$r = 7$

$(x-h)^2 + (y-k)^2 = r^2$

$(x-0)^2 + (y-0)^2 = 7^2$

$x^2 + y^2 = 49$

11. $x^2 + y^2 = 16$

$(x-0)^2 + (y-0)^2 = 4^2$

$(x-h)^2 + (y-k)^2 = r^2$

center : $(h, k) = (0, 0)$

radius : $r = 4$

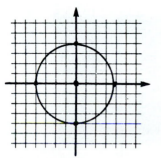

15. $(x+3)^2 + (y-2)^2 = 4$

$[x-(-3)]^2 + (y-2)^2 = 2^2$

$(x-h)^2 + (y-k)^2 = r^2$

center : $(h, k) = (-3, 2)$

radius : $r = 2$

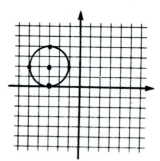

19. $x^2 + y^2 + 6x + 2y + 6 = 0$

$x^2 + 6x \underline{\quad} + y^2 + 2y \underline{\quad} = -6$

$\qquad\uparrow\qquad\qquad\qquad\uparrow$

$\left(\frac{1}{2}\right)(6) = 3 \qquad \left(\frac{1}{2}\right)(2) = 1$

$\qquad 3^2 = 9 \qquad\qquad 1^2 = 1$

$x^2 + 6x + 9 + y^2 + 2y + 1 = -6 + 9 + 1$

$(x+3)^2 + (y+1)^2 = 4$

center : $(-3, -1)$

radius : 2

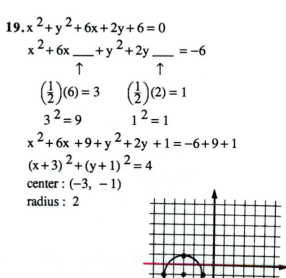

21. $x^2 + y^2 - 10x - 6y - 30 = 0$

$x^2 - 10x\_\_\_ + y^2 - 6y\_\_\_ = 30$

$\quad\uparrow\qquad\qquad\qquad\uparrow$

$\left(\frac{1}{2}\right)(-10) = -5 \qquad \left(\frac{1}{2}\right)(-6) = -3$

$(-5)^2 = 25 \qquad\qquad (-3)^2 = 9$

$x^2 - 10x + 25 + y^2 - 6y + 9 = 30 + 25 + 9$

$(x-5)^2 + (y-3)^2 = 64$

center : (5, 3)

radius : 8

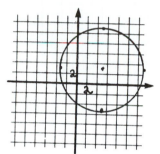

25. $x^2 - 2x + y^2 - 15 = 0$

$x^2 - 2x\_\_\_ + y^2 = 15$

$\qquad\uparrow$

$\left(\frac{1}{2}\right)(-2) = -1$

$(-1)^2 = 1$

$x^2 - 2x + 1 + y^2 = 15 + 1$

$(x-1)^2 + y^2 = 16$

center : (1, 0)

radius : 4

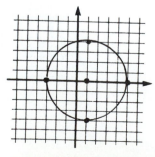

27. center : (0, 0)

radius : $r = \sqrt{(4-0)^2 + (3-0)^2} = \sqrt{25} = 5$

$(x-0)^2 + (y-0)^2 = 5^2$

$x^2 + y^2 = 25$

29. center : (0, 0)

radius : 4

$(x-0)^2 + (y-0)^2 = 4^2$

$x^2 + y^2 = 16$

31. center : (3, 7)

radius : $r = \sqrt{(6-3)^2 + (3-7)^2}$

$= \sqrt{9+16} = \sqrt{25} = 5$

$(x-3)^2 + (y-7)^2 = 5^2$

$(x-3)^2 + (y-7)^2 = 25$

35. $x^2 + y^2 \geq 9$

$x^2 + y^2 = 9 \quad$ center : (0, 0) $\quad$ radius : 3

Test point : (0, 0)

$0^2 + 0^2 \geq 9$

$0 \geq 9, \quad$ false

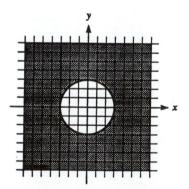

37. $x^2 + y^2 + 4y \geq 0$

$x^2 + y^2 + 4y = 0$

$x^2 + y^2 + 4y\underline{} = 0$

$\uparrow$

$\frac{1}{2}(4) = 2$

$2^2 = 4$

$x^2 + y^2 + 4y + 4 = 0 + 4$

$x^2 + (y + 2)^2 = 4$

center : $(0,\ -2)$　　radius : 2

Note: Don't use $(0,\ 0)$ as a test point since it lies on the circle.

Test : $(1,\ 1)$

$x^2 + y^2 + 4y \geq 0$

$1^2 + 1^2 + 4(1) \geq 0,\ \ $True

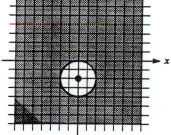

41.

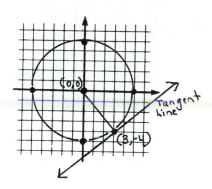

Tangent line : $(x_1,\ y_1) = (3,\ -4)$

slope of line through $(0,\ 0)$ and $(3,\ -4)$:

$m = \dfrac{-4-0}{3-0} = -\dfrac{4}{3}$

slope of tangent line : $\dfrac{3}{4}$

$\left(\text{negative reciprocal of} - \dfrac{4}{3}\right)$

Point-slope equation of tangent line:

$y - y_1 = m(x - x_1)$

$y + 4 = \dfrac{3}{4}(x - 3)$

42. $\dfrac{1-6x}{x} < 0$

$1 - 6x = 0 \qquad\qquad x = 0$

$\quad -6x = -1$

$\qquad x = \dfrac{1}{6}$

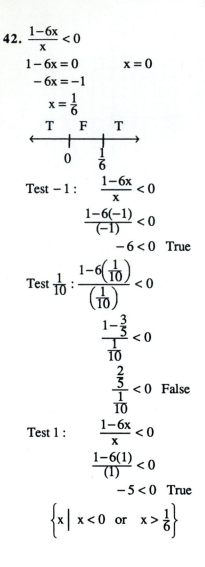

$\begin{array}{ccc} T & F & T \end{array}$

$\qquad\qquad 0 \qquad \dfrac{1}{6}$

Test -1 : $\quad \dfrac{1-6x}{x} < 0$

$\qquad\qquad \dfrac{1-6(-1)}{(-1)} < 0$

$\qquad\qquad\qquad -6 < 0 \quad \text{True}$

Test $\dfrac{1}{10}$: $\dfrac{1-6\left(\dfrac{1}{10}\right)}{\left(\dfrac{1}{10}\right)} < 0$

$\qquad\qquad \dfrac{1-\dfrac{3}{5}}{\dfrac{1}{10}} < 0$

$\qquad\qquad \dfrac{\dfrac{2}{5}}{\dfrac{1}{10}} < 0 \quad \text{False}$

Test 1 : $\quad \dfrac{1-6x}{x} < 0$

$\qquad\qquad \dfrac{1-6(1)}{(1)} < 0$

$\qquad\qquad\qquad -5 < 0 \quad \text{True}$

$\left\{ x \,\middle|\, x < 0 \ \text{ or } \ x > \dfrac{1}{6} \right\}$

43. $y = x^2 + 2x - 3$

x-intercepts: $x^2 + 2x - 3 = 0$

$\qquad\qquad (x+3)(x-1) = 0$

$\qquad\qquad\quad x = -3 \quad x = 1$

y-intercept : $y = 0^2 + 2\cdot\ 0 - 3$

$\qquad\qquad\qquad y = -3$

vertex : $x = -\dfrac{b}{2a} = -\dfrac{2}{2(1)} = -1$

$\qquad\qquad y = (-1)^2 + 2(-1) - 3 = -4$

$\qquad$ vertex : $(-1,\ -4)$

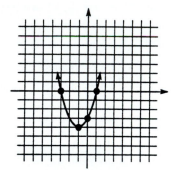

44. $x^6 - 5x^3 - 6$

$= \left(x^3 - 6\right)\left(x^3 + 1\right)$

$= \left(x^3 - 6\right)(x+1)\left(x^2 - x + 1\right)$

Problem set 7.4.3

1. $\dfrac{x^2}{9} + \dfrac{y^2}{4} = 1$

    y-intercepts (Let x = 0) : $\dfrac{y^2}{4} = 1$

                             $y^2 = 4$

                             $y = \pm 2$

    x-intercepts (Let y = 0) : $\dfrac{x^2}{9} = 1$

                             $x^2 = 9$

                             $x = \pm 3$

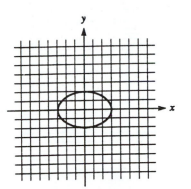

9. $25x^2 + 4y^2 = 100$

       $\dfrac{25x^2}{100} + \dfrac{4y^2}{100} = 1$

       $\dfrac{x^2}{4} + \dfrac{y^2}{25} = 1$

    y-intercepts : $\dfrac{y^2}{25} = 1$

                    $y^2 = 25$

                    $y = \pm 5$

    x-intercepts : $\dfrac{x^2}{4} = 1$

                    $x^2 = 4$

                    $x = \pm 2$

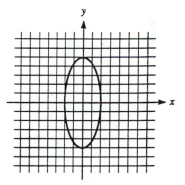

15. $x^2 + 2y^2 = 8$

$$\frac{x^2}{8} + \frac{2y^2}{8} = 1$$

$$\frac{x^2}{8} + \frac{y^2}{4} = 1$$

y-intercepts : $\frac{y^2}{4} = 1$

$$y^2 = 4$$

$$y = \pm 2$$

x-intercepts : $\frac{x^2}{8} = 1$

$$x^2 = 8$$

$$x = \pm\sqrt{8} = \pm 2\sqrt{2} \approx \pm 2.8$$

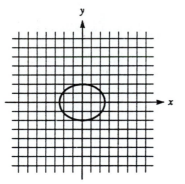

17. $\frac{x^2}{25} + \frac{y^2}{9} \leq 1$

$$\frac{x^2}{25} + \frac{y^2}{9} = 1 \qquad \text{y-intercepts :} \quad \frac{y^2}{9} = 1$$

$$y^2 = 9$$

$$y = \pm 3$$

$$\text{x-intercepts :} \quad \frac{x^2}{25} = 1$$

$$x^2 = 25$$

$$x = \pm 5$$

Test $(0, 0)$:

$$\frac{0^2}{25} + \frac{0^2}{9} \leq 1$$

$$0 \leq 1 \qquad \text{True}$$

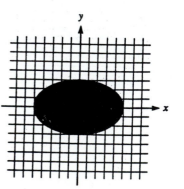

45. $4x + 6y = 2$

Equivalently : $\frac{1}{2}(4x + 6y) = \frac{1}{2}(2)$

$\qquad\qquad\quad 2x + 3y = 1$

The other equation must be $2x + 3y = 1$,
which is option (b).

47. $6y = 3\left[4 - 2y(y - 3)\right]$

$6y = 3\left[4 - 2y^2 + 6y\right]$

$6y = 12 - 6y^2 + 18y$

$6y^2 - 12y - 12 = 0$

$y^2 - 2y - 2 = 0$

$a = 1,\ b = -2,\ c = -2$

$y = \dfrac{-b \pm \sqrt{b^2 - 4ac}}{2a} = \dfrac{2 \pm \sqrt{4 - 4(1)(-2)}}{2(1)}$

$= \dfrac{2 \pm \sqrt{12}}{2} = \dfrac{2 \pm 2\sqrt{3}}{2} = 1 \pm \sqrt{3}$

$\{1 + \sqrt{3},\ 1 - \sqrt{3}\}$

48. $3(3y - 2)^{-\frac{2}{3}} + 5(3y - 2)^{-\frac{1}{3}} - 28 = 0$

$\qquad$ let $t = (3y - 2)^{-\frac{1}{3}}$

$3t^2 + 5t - 28 = 0$

$(3t - 7)(t + 4) = 0$

$3t - 7 = 0 \qquad$ or $\qquad t = -4$

If $\ t = \frac{7}{3}$:

$\qquad (3y - 2)^{-\frac{1}{3}} = \frac{7}{3}$

$\qquad \left[(3y - 2)^{-\frac{1}{3}}\right]^{-3} = \left(\frac{7}{3}\right)^{-3}$

$\qquad\qquad 3y - 2 = \frac{7^{-3}}{3^{-3}}$

$3y - 2 = \dfrac{3^3}{7^3}$

$3y - 2 = \dfrac{27}{343}$

$343(3y - 2) = 343\left(\dfrac{27}{343}\right)$

$1029y - 686 = 27$

$1029y = 713$

$y = \dfrac{713}{1029}$

If $t = -4$:

$(3y - 2)^{-\frac{1}{3}} = -4$

$\left[(3y - 2)^{-\frac{1}{3}}\right]^{-3} = (-4)^{-3}$

$3y - 2 = (-4)^{-3}$

$3y - 2 = \dfrac{1}{(-4)^3}$

$3y - 2 = \dfrac{1}{-64}$

$64(3y - 2) = 64\left(\dfrac{1}{-64}\right)$

$192y - 128 = -1$

$192y = 127$

$y = \dfrac{127}{192}$

$\left\{\dfrac{713}{1029},\ \dfrac{127}{192}\right\}$

49.

$$4y^2 - 9y + 9 - \dfrac{8}{2y + 3}$$

$$2y + 3\ \overline{\left|\ 8y^3 - 6y^2 - 9y + 19\right.}$$

$$\underline{8y^3 + 12y^2}$$

$$-18y^2 - 9y$$

$$\underline{-18y^2 - 27y}$$

$$18y + 19$$

$$\underline{18y + 27}$$

$$-8$$

Review Problems : Chapter 7

1.

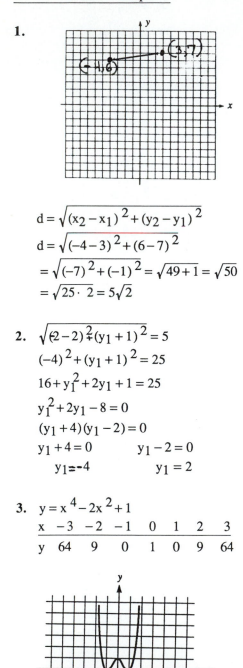

$$d = \sqrt{(x_2 - x_1)^2 + (y_2 - y_1)^2}$$

$$d = \sqrt{(-4-3)^2 + (6-7)^2}$$

$$= \sqrt{(-7)^2 + (-1)^2} = \sqrt{49+1} = \sqrt{50}$$

$$= \sqrt{25 \cdot 2} = 5\sqrt{2}$$

2. $\sqrt{(2-2)^2 + (y_1 + 1)^2} = 5$

$(-4)^2 + (y_1 + 1)^2 = 25$

$16 + y_1^2 + 2y_1 + 1 = 25$

$y_1^2 + 2y_1 - 8 = 0$

$(y_1 + 4)(y_1 - 2) = 0$

$y_1 + 4 = 0 \qquad\qquad y_1 - 2 = 0$

$\qquad y_1 = -4 \qquad\qquad\qquad y_1 = 2$

3. $y = x^4 - 2x^2 + 1$

x	−3	−2	−1	0	1	2	3
y	64	9	0	1	0	9	64

4. $y = \dfrac{5x}{110 - x}$

x	0	10	30	50	60	90	100
y	0	0.5	1.875	4.17	6	22.5	50

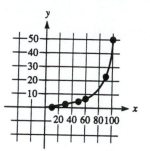

5. One 5-pound package : $6
Three 7-pound packages : 3($7) = $21
Two 19-pound packages : 2($12) = $24
Total cost: $51

6. a. 5 hours after noon : at 5 PM ; $-4°$
b. 8 hours after noon : at 8 PM ; $16°$
c. 4 and 6
    At 4 PM and 6 PM the temperature is 0°.
d. 12
    At noon the temperature is 12°.
e. At 7 PM : 4°   At 8 PM : 16°
    $\dfrac{\text{Amount of increase}}{\text{Original amount}} = \dfrac{12}{4} = 3$
      300% increase

7. $2x - 4y = -8$

    x-intercept : $2x = -8$

               $x = -4$

    y-intercept : $-4y = -8$

               $y = 2$

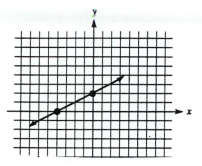

8. $x - 3y \leq 6$

    $x - 3y = 6$

    x-intercept : $x = 6$

    y-intercept : $-3y = 6$

               $y = -2$

    Test $(0,\ 0)$: $x - 3y \leq 6$

                    $0 - 3 \cdot 0 \leq 6$

                          $0 \leq 6$ True

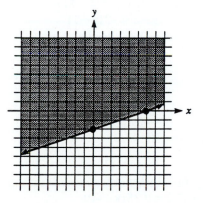

9. $x \geq 3$ and $y \leq 0$

    $x \geq 3$: Line parallel to the y-axis with
             x-intercept at 3 and portion of
             the plane to the right of this line.

    $y \leq 0$: x-axis and portion of plane below
             x-axis.

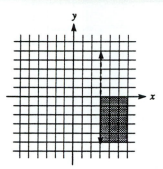

10. $2x - y > -4$

    $2x - y = -4$

    x-intercept : $2x = -4$

               $x = -2$

    y-intercept : $-y = -4$

               $y = 4$

    Test $(0,\ 0)$: $2(0) - 0 > -4$

                      $0 > -4$ True

    $x \geq 0$: y-axis and half-plane to the
             right of y-axis.

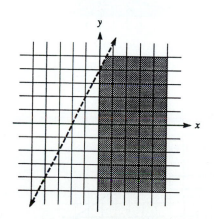

11. $|2x+1| \le 3$

$-3 \le 2x+1 \le 3$

$-4 \le 2x \le 2$

$-2 \le x \le 1$

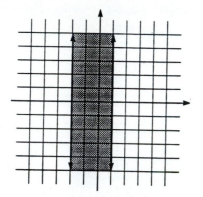

12. $|3y+2| > 8$

$3y+2 > 8$ or $3y+2 < -8$

$3y > 6$ $3y < -10$

$y > 2$ $y < -3\frac{1}{3}$

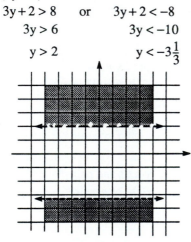

13. x-intercept : 2

slope : $\dfrac{-2}{3}$ $\dfrac{\text{horizontal change (rise)}}{\text{vertical change (run)}}$

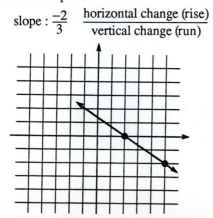

14. passing through $(1,-3)$

slope : $\dfrac{-2}{1}$

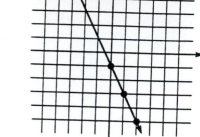

15. $(x_1, y_1) = (-2,\ 3)$

$m = -4$

Point-slope: $y - y_1 = m(x - x_1)$

$y - 3 = -4(x + 2)$

Slope-Intercept : $y = mx + b$

$y - 3 = -4x - 8$

$y = -4x - 5$

16. $(x_1, y_1) = (1,\ 3)$ or $\left[(x_1, y_1) = (-4,\ 18)\right]$

$m = \dfrac{18 - 3}{-4 - 1} = \dfrac{15}{-5} = -3$

Point-slope: $y - y_1 = m(x - x_1)$

$y - 3 = -3(x - 1)$

Slope-Intercept : $y = mx + b$

$y - 3 = -3x + 3$

$y = -3x + 6$

17. a. $m = \dfrac{54.8 - 47.2}{60 - 40} = \dfrac{7.6}{20} = 0.38$

$(x_1, y_1) = (40,\ 47.2)$

          or $\left[(x_1, y_1) = (60,\ 54.8)\right]$

$y - y_1 = m(x - x_1)$

$y - 47.2 = 0.38(x - 40)$

$y - 47.2 = 0.38x - 15.2$

$y = 0.38x + 32$

b. If $x = 80$

$y = 0.38(80) + 32$

$y = 62.4$

Crop yield : 62.4 bushels

18. $y = \frac{3}{4}x - 2$ $\left[\text{In the form } y = mx + b\right]$

y-intercept : $b = -2$

slope : $m = \frac{3}{4}$

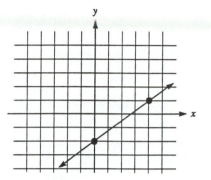

19. $3y - 2x \leq 6$

$3y \leq 2x + 6$

$y \leq \frac{2}{3}x + 2$

y-intercept : 2

slope : $\frac{2}{3}$

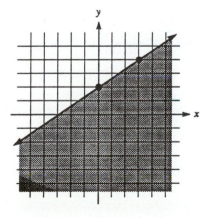

20. $(x_1, y_1) = (-1, -4)$

$2x - 3y = 6$

$-3y = -2x + 6$

$y = \frac{2}{3}x - 2$ slope : $\frac{2}{3}$

slope of line whose equation we must

write : $\frac{2}{3}$

$y - y_1 = m(x - x_1)$

$y + 4 = \frac{2}{3}(x + 1)$ Point-slope Form

$y + 4 = \frac{2}{3}x + \frac{2}{3}$

$y = \frac{2}{3}x - \frac{10}{3}$ Slope-Intercept Form

21. $(x_1, y_1) = (2, -3)$

$4y - 2x = -8$

$4y = 2x - 8$

$y = \frac{1}{2}x - 2$ slope : $\frac{1}{2}$

slope of line whose equation we must

write : -2 $\left(\text{negative reciprocal of } \frac{1}{2}\right)$

$y - y_1 = m(x - x_1)$

$y + 3 = -2(x - 2)$ Point-slope Form

$y + 3 = -2x + 4$

$y = -2x + 1$ Slope-Intercept Form

22. $(x_1, y_1) = (0, 0)$ (the origin)

slope of the line through (1, 4) and

(-2, 10) :

$m = \frac{10 - 4}{-2 - 1} = \frac{6}{-3} = -2$

slope of line whose equation we must

write : -2

$y - y_1 = m(x - x_1)$

$y - 0 = -2(x - 0)$ Point-Slope Form

$y = -2x$ Slope-Intercept Form

23. $(x_1, y_1) = (-3, 2)$

x-intercept 4 : (4, 0)

y-intercept -2 : (0, -2)

slope of line through (4, 0) and (0, -2) :

$m = \dfrac{-2-0}{0-4} = \dfrac{1}{2}$

slope of line whose equation we must

write : -2 $\left(\text{negative reciprocal of } \dfrac{1}{2}\right)$

$y - y_1 = m(x - x_1)$

$y - 2 = -2(x + 3)$ Point-Slope Form

$y - 2 = -2x - 6$

$y = -2x - 4$ Slope-Intercept Form

24. $y = x^2 + 5x + 4$

x-intercepts (set y = 0) : $x^2 + 5x + 4 = 0$

$(x + 4)(x + 1) = 0$

$x + 4 = 0$ $x + 1 = 0$

$x = -4$ $x = -1$

y-intercept (set x = 0) : $y = 0^2 + 5 \cdot 0 + 4$

$y = 4$

vertex : $x = -\dfrac{b}{2a} = -\dfrac{5}{2(1)} = -\dfrac{5}{2}$

$y = \left(-\dfrac{5}{2}\right)^2 + 5\left(-\dfrac{5}{2}\right) + 4$

$= \dfrac{25}{4} - \dfrac{25}{2} + 4 = \dfrac{25}{4} - \dfrac{50}{4} + \dfrac{16}{4} = -\dfrac{9}{4}$

vertex : $\left(-\dfrac{5}{2}, -\dfrac{9}{4}\right) = \left(-2\dfrac{1}{2}, -2\dfrac{1}{4}\right)$

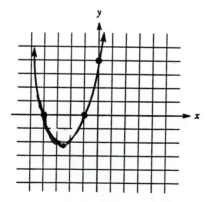

25. $y = -2x^2 + 12x - 10$

x-intercepts : $-2x^2 + 12x - 10 = 0$

$x^2 - 6x + 5 = 0$

$(x - 5)(x - 1) = 0$

$x = 5$ $x = 1$

y-intercept : $y = -10$

vertex : $x = -\dfrac{b}{2a} = -\dfrac{12}{2(-2)} = 3$

$y = -2 \cdot 3^2 + 12 \cdot 3 - 10$

$= -18 + 36 - 10 = -8$

vertex : (3, 8)

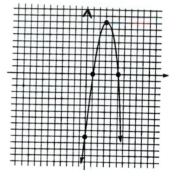

26. $C = 10t^2 - 100t + 320$

$t = -\dfrac{b}{2a} = -\dfrac{(-100)}{2(10)} = \dfrac{100}{20} = 5$

$C = 10(5)^2 - 100(5) + 320$

$= 250 - 500 + 320 = 70$

After 5 days the lowest concentration

of 70 bacteria per cm^3 will occur.

27. $h = -16t^2 + 256t$

$t = -\dfrac{b}{2a} = -\dfrac{256}{2(-16)} = \dfrac{-256}{-32} = 8$

Maximum height occurs after 8 seconds.

$h = -16(8)^2 + 256(8) = -1024 + 2048$

$ = 1024$

Maximum height is 1024 feet.

28. $x^2 + y^2 - 4x + 2y - 4 = 0$

$x^2 - 4x\underline{} + y^2 + 2y\underline{} = 4$

$\uparrow\uparrow$

$\left(\dfrac{1}{2}\right)(-4) = -2 \quad \left(\dfrac{1}{2}\right)(2) = 1$

$(-2)^2 = 4 \qquad\quad 1^2 = 1$

$x^2 - 4x + 4 + y^2 + 2y + 1 = 4 + 4 + 1$

$(x-2)^2 + (y+1)^2 = 9$

$(x-2)^2 + \left[y - (-1)\right]^2 = 3^2$

$(x-h)^2 + (y-k)^2 = r^2$

center : $(h, k) = (2, -1)$

radius : $r = 3$

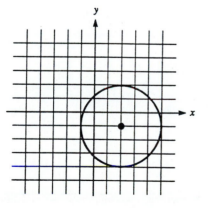

29. $\dfrac{x^2}{25} + \dfrac{y^2}{4} = 1$

x-intercepts : $\dfrac{x^2}{25} = 1$
(set y=0)

$x^2 = 25$

$x = \pm\sqrt{25} = \pm 5$

y-intercepts : $\dfrac{y^2}{4} = 1$
(set x=0)

$y^2 = 4$

$y = \pm\sqrt{4} = \pm 2$

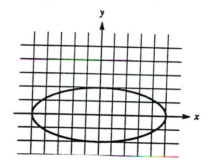

30. $9x^2 + 4y^2 = 36$

$\dfrac{9x^2}{36} + \dfrac{4y^2}{36} = 1$

$\dfrac{x^2}{4} + \dfrac{y^2}{9} = 1$

x-intercepts : $\dfrac{x^2}{4} = 1$
(set y=0)

$x^2 = 4$

$x = \pm 2$

y-intercepts : $\dfrac{y^2}{9} = 1$
(set x=0)

$y^2 = 9$

$y = \pm 3$

31. $\dfrac{x^2}{16} - \dfrac{y^2}{9} = 1$

$a^2 = 16 \quad b^2 = 9$

Asymptotes : use rectangle passing through
4 and -4 on x-axis and 3 and -3
on y-axis.

x-intercepts : $\dfrac{x^2}{16} = 1$

$x^2 = 16$

$x = \pm 4$

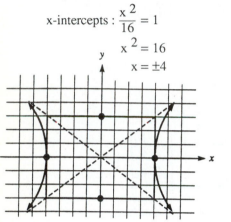

32. $x^2 - 4y^2 = -4$

$\dfrac{x^2}{-4} - \dfrac{4y^2}{-4} = 1$

$-\dfrac{x^2}{4} + \dfrac{4y^2}{4} = 1$

$\dfrac{y^2}{1} - \dfrac{x^2}{4} = 1$

Asymptotes : use rectangle passing through
1 and -1 on the y-axis and 2 and -2
on the x-axis.

y-intercepts : $\dfrac{y^2}{1} = 1$

$y^2 = 1; \quad y = \pm 1$

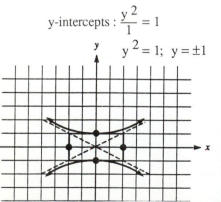

33. Intersection points :

$(1, 1), (0, 0), (-2, -8)$

check $(1, 1)$: $\quad y = x^3 \qquad y = 2x - x^2$

$\qquad\qquad\qquad 1 = 1^3 \qquad 1 = 2(1) - 1^2$

$\qquad\qquad\qquad 1 = 1 \qquad\quad 1 = 1$

check $(0, 0)$: $\quad y = x^3 \qquad y = 2x - x^2$

$\qquad\qquad\qquad 0 = 0^3 \qquad 0 = 2(0) - 0^2$

$\qquad\qquad\qquad 0 = 0 \qquad\quad 0 = 0$

check $(-2, -8)$:

$\qquad y = x^3 \qquad\qquad y = 2x - x^2$

$\qquad -8 = (-2)^3 \qquad -8 = 2(-2) - (-2)^2$

$\qquad -8 = -8 \qquad\quad -8 = -8$

$\qquad\qquad \{(1, 1), (0, 0), (-2, -8)\}$

34. $2x - y = 2$

$x + 2y = 11$

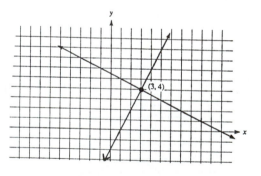

$\{(3, 4)\}$

35. $y = -x^2 + 2x$

x-intercepts : $-x^2 + 2x = 0$

$\quad\quad\quad\quad\quad -x(x-2) = 0$

$\quad\quad\quad\quad x = 0 \quad\quad x = 2$

y-intercept : $y = 0$

vertex : $x = -\dfrac{b}{2a} = -\dfrac{2}{2(-1)} = 1 \quad (1, 1)$

$x + y = 0$

Two points on line : $(0, 0), (1, -1)$

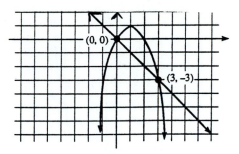

$\{(0, 0), (3, -3)\}$

36. $y = 2x + 3$

y-intercept : 3

slope : 2

$x^2 - 4x + y^2 = 0$

$x^2 - 4x \underline{\quad\quad} + y^2 = 0$

$\left(\dfrac{1}{2}\right)(-4) = -2$

$(-2)^2 = 4$

$x^2 - 4x + 4 + y^2 = 0 + 4$

$(x-2)^2 + y^2 = 4$

circle's center : $(2, 0)$; radius : 2

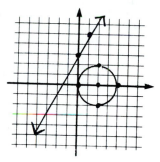

Graphs do not intersect.

$\varnothing$

37. $y = -x + 4$

y-intercept : 4

slope : -1

$3x + 3y = 12$

If $x = 0$: $3y = 12$

　　　　　$y = 4$　　(0, 4)

If $y = 0$: $3x = 12$

　　　　　$x = 4$　　(4, 0)

Graphs coincide. System is dependent. The same ordered pairs satisfy both equations.

38. $y = x^2 - x - 6$　　(parabola)

x-intercepts : $x^2 - x - 6 = 0$

　　　　　　　$(x - 3)(x + 2) = 0$

　　　　　　　$x = 3$　　$x = -2$

y-intercept : $y = -6$

vertex : $x = -\dfrac{b}{2a} = -\dfrac{(-1)}{2(1)} = \dfrac{1}{2}$

$y = \left(\dfrac{1}{2}\right)^2 - \left(\dfrac{1}{2}\right) - 6 = \dfrac{1}{4} - \dfrac{2}{4} - \dfrac{24}{4} = -\dfrac{25}{4}$

vertex: $\left(\dfrac{1}{2},\ -6\dfrac{1}{4}\right)$

$y = x - 3$;　　y-intercept : -3　　slope : 1

$\{(3, 0), (-1,\ -4)\}$

39. $4x^2 - y^2 = 36$

$$\frac{4x^2}{36} - \frac{y^2}{36} = 1$$

$$\frac{x^2}{9} - \frac{y^2}{36} = 1$$

Hyperbola : x-intercepts : ± 3

$$x^2 + 9y^2 = 9$$

$$\frac{x^2}{9} + \frac{9y^2}{9} = 1$$

$$\frac{x^2}{9} + \frac{y^2}{1} = 1$$

Ellipse : x-intercepts : ± 3

y-intercepts : ± 1

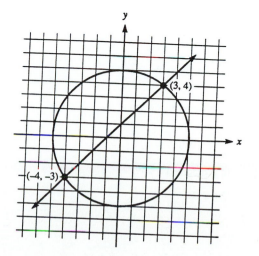

$\{(3,\ 0),\ (-3,\ 0)\}$

40. $y = x + 1$ y-intercept : 1   slope : 1

$x^2 + y^2 = 25$

circle's center : $(0,\ 0)$ radius : 5

$\{(3,\ 4),\ (-4,\ -3)\}$

Problem Set 8.1.1

1. $x = 2y - 5$

$x - 3y = 8$

$(2y - 5) - 3y = 8$

$-y - 5 = 8$

$-y = 13$

$y = -13$

$x = 2y - 5$

$x = 2(-13) - 5 = -31$

$\{(-31, -13)\}$

3. $4x + y = 5$

$2x - 3y = 13$

$y = 5 - 4x$

$2x - 3(5 - 4x) = 13$

$2x - 15 + 12x = 13$

$14x = 28$

$x = 2$

$y = 5 - 4x = 5 - 4(2) = -3$ $\{(2, -3)\}$

9. $3x - 2y = 14$

$2x + 3y = -8$

$3x = 2y + 14$

$x = \dfrac{2y + 14}{3}$

$2(\dfrac{2y + 14}{3}) + 3y = -8$

$\dfrac{4y + 28}{3} + 3y = -8$

$3(\dfrac{4y + 28}{3} + 3y) = 3(-8)$

$4y + 28 + 9y = -24$

$13y + 28 = -24$

$13y = -52$

$y = -4$

$x = \dfrac{2y + 14}{3} = \dfrac{2(-4) + 14}{3} = \dfrac{-8 + 14}{3} = \dfrac{6}{3} = 2$

$\{(2, -4)\}$

13. $y - 3x = 2$

$x = \dfrac{1}{4}y$

$y - 3(\dfrac{1}{4}y) = 2$

$y - \dfrac{3}{4}y = 2$

$4(y - \dfrac{3}{4}y) = 4(2)$

$4y - 3y = 8$

$y = 8$

$x = \dfrac{1}{4}y = \dfrac{1}{4}(8) = 2$

$\{(2, 8)\}$

17. $y - bx = 2$

$3bx + 2y = 1$

$y = 2 + bx$

$3bx + 2(2 + bx) = 1$

$3bx + 4 + 2bx = 1$

$5bx + 4 = 1$

$5bx = -3$

$x = -\dfrac{3}{5b}$

$y = 2 + bx = 2 + b(-\dfrac{3}{5b}) = 2 + (-\dfrac{3}{5}) = \dfrac{7}{5}$

$\{(-\dfrac{3}{5b}, \dfrac{7}{5})\}$

23. $x + 3y = 11$

$x - 5y = -13$

$x = 5y - 13$

$(5y - 13) + 3y = 11$

$8y = 24$

$y = 3$

$x = 5(3) - 13 = 2$

$y - y_1 = m(x - x_1)$ $(x_1, y_1) = (2, 3)$ $m = -2$

$y - 3 = -2(x - 2)$

$y - 3 = -2x + 4$

$y = -2x + 7$ Point - Slope Form

25. $x + 3y = 9$

$4x + 5y = 22$

$x = 9 - 3y$

$4(9 - 3y) + 5y = 22$

$36 - 12y + 5y = 22$

$-7y = -14$

$y = 2$

$x = 9 - 3(2) = 3$ $(x_1, y_1) = (3, 2)$

Slope of $4x - 2y = 7$:

$-2y = -4x + 7$

$y = 2x - \dfrac{7}{2}$ $m = 2$

Slope of line whose equation we must

　write : 2

$y - y_1 = m(x - x_1)$

$y - 2 = 2(x - 3)$

$y - 2 = 2x - 6$

$y = 2x - 4$ Point Slope Form

27. $4x = 12$ $3x = y + 3$ $x = 3$

$3(3) = y + 3$ $9 = y + 3$ $6 = y$

$(x_1, y_1) = (3, 6)$

Slope of $5x - 20y = 30$:

$-20y = -5x + 30$

$y = \dfrac{1}{4}x - \dfrac{3}{2}$ $m = \dfrac{1}{4}$

Slope of the line whose equation

we must write : -4(negative reciprocal of $\dfrac{1}{4}$)

$y - y_1 = m(x - x_1)$

$y - 6 = -4(x - 3)$

$y - 6 = -4x + 12$

$y = -4x + 18$ Point Slope Form

29. $\dfrac{3x + y}{5} = \dfrac{4}{1}$ $3x + y = 20$

$\dfrac{x - 3y}{4} = \dfrac{3}{1}$ $x - 3y = 12$

$x = 12 + 3y$

$3(12 + 3y) + y = 20$

$36 + 9y + y = 20$

$10y = -16$

$y = -\dfrac{16}{10} = -\dfrac{8}{5}$

$x = 12 + 3y = 12 + 3(-\dfrac{8}{5}) = 12 - \dfrac{24}{5}$

$= \dfrac{60}{5} - \dfrac{24}{5} = \dfrac{36}{5}$

$\{(\dfrac{36}{5}, -\dfrac{8}{5})\}$

33. x : width of border

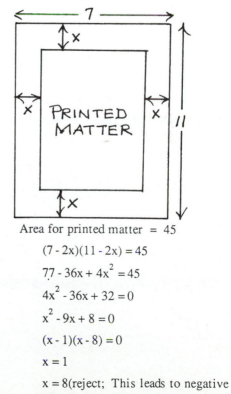

Area for printed matter = 45

$(7 - 2x)(11 - 2x) = 45$

$77 - 36x + 4x^2 = 45$

$4x^2 - 36x + 32 = 0$

$x^2 - 9x + 8 = 0$

$(x - 1)(x - 8) = 0$

$x = 1$

$x = 8$(reject; This leads to negative

dimensions for the printed page.)

Width of Border : 1 inch

34. $49y^2 + 70y + 9 < 0$

Consider $49y^2 + 70y + 9 = 0$

$(7y + 9)(7y + 1) = 0$

$7y + 9 = 0 \quad 7y + 1 = 0$

$y = -\dfrac{9}{7} \quad y = -\dfrac{1}{7}$

$\qquad\qquad$ F $\qquad\qquad$ T $\qquad\qquad$ F

$\qquad\qquad -\dfrac{9}{7} \qquad\qquad -\dfrac{1}{7}$

Test -2 : $\quad 49(-2)^2 + 70(-2) + 9 < 0$

$\qquad\qquad\qquad 65 < 0 \quad$ False

Test -1 : $\quad 49(-1)^2 + 70(-1) + 9 < 0$

$\qquad\qquad\qquad -12 < 0 \quad$ True

Test 0 : $\quad 49(0)^2 + 70(0) + 9 < 0$

$\qquad\qquad\qquad 9 < 0 \quad$ False

$\{y \,|\, -\dfrac{9}{7} < y < -\dfrac{1}{7}\}$

35. $\sqrt{2y + 3} - \sqrt{y + 1} = 1$

$\sqrt{2y + 3} = 1 + \sqrt{y + 1}$

$(\sqrt{2y + 3})^2 = (1 + \sqrt{y + 1})^2$

$2y + 3 = 1 + 2\sqrt{y + 1} + y + 1$

$2y + 3 = y + 2 + 2\sqrt{y + 1}$

$y + 1 = 2\sqrt{y + 1}$

$(y + 1)^2 = (2\sqrt{y + 1})^2$

$y^2 + 2y + 1 = 4(y + 1)$

$y^2 + 2y + 1 = 4y + 4$

$y^2 - 2y - 3 = 0$

$(y - 3)(y + 1) = 0$

$y - 3 = 0 \quad y + 1 = 0$

$y = 3 \quad y = -1$

Check 3 : $\qquad \sqrt{2(3) + 3} - \sqrt{3 + 1} = 1$

$3 - 2 = 1 \qquad 1 = 1$

Check -1 : $\qquad \sqrt{2(-1) + 3} - \sqrt{-1 + 1} = 1$

$1 - 0 = 1 \qquad 1 = 1 \qquad \{-1, 3\}$

Problem Set 8.1.2

1. $x^2 + 4y^2 = 29$

$x - 3y = 2$

$x = 2 + 3y$

$(2 + 3y)^2 + 4y^2 = 29$

$4 + 12y + 9y^2 + 4y^2 = 29$

$13y^2 + 12y - 25 = 0$

$(13y + 25)(y - 1) = 0$

$13y + 25 = 0$ or $y - 1 = 0$

$y = -\dfrac{25}{13}$ $y = 1$

If $y = -\dfrac{25}{13}$: $x = 2 + 3y = 2 + 3(-\dfrac{25}{13}) = \dfrac{26}{13} - \dfrac{75}{13}$

$= -\dfrac{49}{13}$

$(-\dfrac{49}{13}, -\dfrac{25}{13})$

If $y = 1$: $x = 2 + 3y = 2 + 3(1) = 5$

$(5, 1)$

$\{(-\dfrac{49}{13}, -\dfrac{25}{13}), (5, 1)\}$

3. $x^2 + y^2 = 40$

$3x - y + 20 = 0$

$y = 3x + 20$

$x^2 + (3x + 20)^2 = 40$

$x^2 + 9x^2 + 120x + 400 = 40$

$10x^2 + 120x + 360 = 0$

$x^2 + 12x + 36 = 0$

$(x + 6)^2 = 0$

$x = -6$

$y = 3(-6) + 20 = 2$

$\{(-6, 2)\}$

5. $2x + y = 12$

$x^2 + 6y = 39$

$y = 12 - 2x$

$x^2 + 6(12 - 2x) = 39$

$x^2 + 72 - 12x = 39$

$x^2 - 12x + 33 = 0$

$x = \dfrac{-b \pm \sqrt{b^2 - 4ac}}{2a} = \dfrac{12 \pm \sqrt{144 - 4(1)(33)}}{2}$

$= \dfrac{12 \pm \sqrt{12}}{2} = \dfrac{12 \pm 2\sqrt{3}}{2} 6 \pm \sqrt{3}$

If $x = 6 + \sqrt{3}$: $y = 12 - 2x = 12 - 2(6 + \sqrt{3})$

$= 12 - 12 - 2\sqrt{3} = -2\sqrt{3}$

$(6 + \sqrt{3}, -2\sqrt{3})$

If $x = 6 - \sqrt{3}$: $y = 12 - 2x = 12 - 2(6 - \sqrt{3})$

$= 2\sqrt{3}$

$(6 - \sqrt{3}, 2\sqrt{3})$

$\{(6 + \sqrt{3}, -2\sqrt{3}), (6 - \sqrt{3}, 2\sqrt{3})\}$

9. $x^2 + y^2 = 52$

$3x - 2y = 0$

$3x = 2y$

$x = \dfrac{2y}{3}$

$(\dfrac{2y}{3})^2 + y^2 = 52$

$\dfrac{4y^2}{9} + y^2 = 52$

$9(\dfrac{4y^2}{9} + y^2) = 9(52)$

$4y^2 + 9y^2 = 468$

$13y^2 = 468$ $y^2 = 36$ $y = \pm 6$

If $y = 6$: $x = \dfrac{2y}{3} = \dfrac{2(6)}{3} = 4$

If $y = -6$: $x = \dfrac{2y}{3} = \dfrac{2(-6)}{3} = -4$

$\{(4, 6), (-4, -6)\}$

15. $x - y = 2$

$x^2 - 3y^2 = 8$

$x = y + 2$

$(y + 2)^2 - 3y^2 = 8$

$y^2 + 4y + 4 - 3y^2 = 8$

$-2y^2 + 4y - 4 = 0$

$y^2 - 2y + 2 = 0$

$y = \dfrac{-b \pm \sqrt{b^2 - 4ac}}{2a} = \dfrac{2 \pm \sqrt{4 - 4(1)(2)}}{2(1)}$

$= \dfrac{2 \pm \sqrt{-4}}{2} = \dfrac{2 \pm 2i}{2} = 1 \pm i$

If $y = 1 + i$: $x = y + 2 = (1 + i) + 2 = 3 + i$

If $y = 1 - i$: $x = y + 2 = (1 - i) + 2 = 3 - i$

$\{(3 + i, 1 + i), (3 - i, 1 - i)\}$

17. $y - 4x = 0$

$y = x^2 + 5$

$(x^2 + 5) - 4x = 0$　　$x^2 - 4x + 5 = 0$

$x = \dfrac{4 \pm \sqrt{16 - 4(1)(5)}}{2} = \dfrac{4 \pm \sqrt{-4}}{2} = \dfrac{4 \pm 2i}{2} = 2 \pm i$

If $x = 2 + i$: $y = 4x = 4(2 + i) = 8 + 4i$

If $x = 2 - i$: $y = 4x = 4(2 - i) = 8 - 4i$

$\{(2 + i, 8 + 4i), (2 - i, 8 - 4i)\}$

23. $x^2 + y^2 = 100$　　$x + y = 10\sqrt{2}$　　$y = 10\sqrt{2} - x$

$x^2 + (10\sqrt{2} - x)^2 = 100$

$x^2 + 200 - 20\sqrt{2}x + x^2 = 100$

$2x^2 - 20\sqrt{2}x + 100 = 0$

$x^2 - 10\sqrt{2}x + 50 = 0$

$a = 1,\ b = -10\sqrt{2},\ c = 50$

$x = \dfrac{-b \pm \sqrt{b^2 - 4ac}}{2a} = \dfrac{10\sqrt{2} \pm \sqrt{200 - 4 \cdot 1 \cdot 50}}{2}$

$= \dfrac{10\sqrt{2} \pm \sqrt{0}}{2} = 5\sqrt{2}$

$y = 10\sqrt{2} - x = 10\sqrt{2} - 5\sqrt{2} = 5\sqrt{2}$

$\{(5\sqrt{2}), (5\sqrt{2})\}$

27. $2x^2 - y + 12 = 0$

$x^2 + 2y - 4 = 0$

$y = 2x^2 + 12$

$x^2 + 2(2x^2 + 12) - 4 = 0$

$5x^2 + 20 = 0$

$5x^2 = -20$

$x^2 = -4$

$x = \pm\sqrt{-4} = \pm 2i$

If $x = 2i$: $y = 2x^2 + 12 = 2(2i)^2 + 12$

$= 2(4i^2) + 12 = 8(-1) + 12 = 4$

If $x = -2i$: $y = 2x^2 + 12 = 2(-2i)^2 + 12$

$= 2(4i^2) + 12 = 4$

$\{(2i, 4), (-2i, 4)\}$

29. $x - 3y = -1$

$7 - 3x^2 + 2xy + y^2 = 0$

$x = 3y - 1$

$7 - 3(3y - 1)^2 + 2(3y - 1)y + y^2 = 0$

$7 - 27y^2 + 18y - 3 + 6y^2 - 2y + y^2 = 0$

$-20y^2 + 16y + 4 = 0$

$5y^2 - 4y - 1 = 0$

$(5y + 1)(y - 1) = 0$

$y = -\dfrac{1}{5}$　$y = 1$

If $y = -\dfrac{1}{5}$: $x = 3y - 1 = 3(-\dfrac{1}{5}) - 1 = -\dfrac{3}{5} - \dfrac{5}{5} = -\dfrac{8}{5}$

If $y = 1$: $x = 3y - 1 = 3(1) - 1 = 2$

$\{(-\dfrac{8}{5}, -\dfrac{1}{5}), (2, 1)\}$

33. $y^2 - 8x = 0$

$2x - y - 6 = 0$

$y = 2x - 6$

$(2x - 6)^2 - 8x = 0$

$4x^2 - 24x + 36 - 8x = 0$

$4x^2 - 32x + 36 = 0$

$x^2 - 8x + 9 = 0$

$x = \dfrac{-b \pm \sqrt{b^2 - 4ac}}{2a} = \dfrac{8 \pm \sqrt{64 - 4(1)(9)}}{2}$

$= \dfrac{8 \pm \sqrt{28}}{2}$

$= \dfrac{8 \pm 2\sqrt{7}}{2} = 4 \pm \sqrt{7}$

If $x = 4 + \sqrt{7}$: $y = 2x - 6 = 2(4 + \sqrt{7}) - 6$

$= 8 + 2\sqrt{7} - 6 = 2 + 2\sqrt{7}$

If $x = 4 - \sqrt{7}$: $y = 2x - 6 = 2(4 - \sqrt{7}) - 6$

$= 8 - 2\sqrt{7} - 6 = 2 - 2\sqrt{7}$

$\{(4 + \sqrt{7}, 2 + 2\sqrt{7}), (4 - \sqrt{7}, 2 - 2\sqrt{7})\}$

35. $2y = -1 - x$

$\sqrt{x^2 + y^2} + y = 1$

$x = -1 - 2y$

$\sqrt{(-1 - 2y)^2 + y^2} + y = 1$

$\sqrt{1 + 4y + 4y^2 + y^2} + y = 1$

$\sqrt{5y^2 + 4y + 1} = 1 - y$

$(\sqrt{5y^2 + 4y + 1})^2 = (1 - y)^2$

$5y^2 + 4y + 1 = 1 - 2y + y^2$

$4y^2 + 6y = 0$

$2y(2y + 3) = 0$

$2y = 0$ or $2y + 3 = 0$

$y = 0 \quad y = -\dfrac{3}{2}$

If $y = 0$: $x = -1 - 2y = -1 - 2(0) = -1$

If $y = -\dfrac{3}{2}$: $x = -1 - 2(-\dfrac{3}{2}) = 2$

We must check for extraneous solutions.

Check: $(-1, 0)$

$2y = -1 - x \qquad\qquad \sqrt{x^2 + y^2} + y = 1$

$2(0) = -1 - (-1) \qquad \sqrt{(-1)^2 + 0^2} + 0 = 1$

$0 = 0 \qquad\qquad\qquad \sqrt{1} = 1$

Check: $(2, -\dfrac{3}{2})$

$2y = -1 - x \qquad\qquad \sqrt{x^2 + y^2} + y = 1$

$2(-\dfrac{3}{2}) = -1 - 2 \qquad \sqrt{2^2 + (-\dfrac{3}{2})^2} + (-\dfrac{3}{2}) = 1$

$-3 = -3 \qquad\qquad\qquad \sqrt{\dfrac{25}{4}} - \dfrac{3}{2} = 1$

$\qquad\qquad\qquad\qquad\qquad \dfrac{5}{2} - \dfrac{3}{2} = 1$

$\{(-1, 0), (2, -\dfrac{3}{2})\}$

37. $xy = 4$

$x^2 + y^2 = 10$

$x = \dfrac{4}{y}$

$(\dfrac{4}{y})^2 + y^2 = 10$

$\dfrac{16}{y^2} + y^2 = 10$

$y^2(\dfrac{16}{y^2} + y^2) = y^2(10)$

$16 + y^4 = 10y^2$

$y^4 - 10y^2 + 16 = 0$

Let $t = y^2$

$t^2 - 10t + 16 = 0$

$(t - 8)(t - 2) = 0$

$t = 8$ or $t = 2$

$y^2 = 8$ or $y^2 = 2$

$y = \pm\sqrt{8}$ $y = \pm\sqrt{2}$

$y = \pm 2\sqrt{2}$

$x = \dfrac{4}{y}$

If $y = 2\sqrt{2}$: $x = \dfrac{4}{2\sqrt{2}} = \dfrac{2}{\sqrt{2}} = \dfrac{2}{\sqrt{2}} \cdot \dfrac{\sqrt{2}}{\sqrt{2}}$

$= \dfrac{2\sqrt{2}}{2} = \sqrt{2}$

If $y = -2\sqrt{2}$: $x = \dfrac{4}{-2\sqrt{2}} = -\sqrt{2}$

If $y = \sqrt{2}$: $x = \dfrac{4}{\sqrt{2}} = \dfrac{4}{\sqrt{2}} \cdot \dfrac{\sqrt{2}}{\sqrt{2}} = \dfrac{4\sqrt{2}}{2}$

$= 2\sqrt{2}$

If $y = -\sqrt{2}$: $x = \dfrac{4}{-\sqrt{2}} = -2\sqrt{2}$

$\{(\sqrt{2}, 2\sqrt{2}), (-\sqrt{2}, -2\sqrt{2}), (2\sqrt{2}, \sqrt{2}),$

$(-2\sqrt{2}, -\sqrt{2})\}$

39. $xy = ab$

$y + 2x = b + 2a$

$y = b + 2a - 2x$

$x(b + 2a - 2x) = ab$

$bx + 2ax - 2x^2 = ab$

$- 2x^2 + (2a + b)x - ab = 0$

$A = -2 \quad B = 2a + b \quad C = -ab$

$x = \dfrac{-B \pm \sqrt{B^2 - 4AC}}{2A}$

$= \dfrac{-(2a + b) \pm \sqrt{(2a + b)^2 - 4(-2)(-ab)}}{-4}$

$= \dfrac{-2a - b \pm \sqrt{4a^2 + 4ab + b^2 - 8ab}}{-4}$

$= \dfrac{-2a - b \pm \sqrt{4a^2 - 4ab + b^2}}{-4}$

$= \dfrac{-2a - b \pm \sqrt{(2a - b)^2}}{-4}$

$= \dfrac{-2a - b \pm (2a - b)}{-4}$

$x = \dfrac{-2a - b + 2a - b}{-4}$ or $x = \dfrac{-2a - b - 2a + b}{-4}$

$= \dfrac{-2b}{-4} = \dfrac{b}{2}$ $= \dfrac{-4a}{-4} = a$

If $x = \dfrac{b}{2}$: $y = b + 2a - 2x = b + 2a - 2(\dfrac{b}{2})$

$= b + 2a - b = 2a$

If $x = a$: $y = b + 2a - 2x = b + 2a - 2a = b$

$\{(\dfrac{b}{2}, 2a), (a, b)\}$

43. $x^3 + y^3 = -19$

$\qquad x + y = -1$

$\qquad \dfrac{x^3 + y^3}{x + y} = \dfrac{-19}{-1}$

$\qquad \dfrac{(x + y)(x^2 - xy + y^2)}{(x + y)} = 19$

$\qquad x^2 - xy + y^2 = 19$

$\qquad x = -y - 1$

$\qquad (-y - 1)^2 - (-y - 1)y + y^2 = 19$

$\qquad y^2 + 2y + 1 + y^2 + y + y^2 = 19$

$\qquad 3y^2 + 3y + 1 = 19$

$\qquad 3y^2 + 3y - 18 = 0$

$\qquad y^2 + y - 6 = 0$

$\qquad (y + 3)(y - 2) = 0$

$\qquad y = -3 \ \text{ or } \ y = 2$

$\qquad$ If $y = -3$: $x = -y - 1 = -(-3) - 1 = 2$

$\qquad$ If $y = 2$: $\ x = -y - 1 = -2 - 1 = -3$

$\qquad \{(2, -3), (-3, 2)\}$

45. $5x^2 + xy - 4 = 0$

$\qquad x^2 y^2 + 7xy + 6 = 0$

$\qquad (xy + 1)(xy + 6) = 0$

$\qquad xy + 1 = 0 \ \text{ or } \ xy + 6 = 0$

$\qquad xy = -1 \qquad\qquad xy = -6$

$\qquad x = -\dfrac{1}{y} \qquad\qquad x = -\dfrac{6}{y}$

Substitute $-\dfrac{1}{y}$ for x in the first equation.

$\qquad 5(-\dfrac{1}{y})^2 + (-\dfrac{1}{y})y - 4 = 0$

$\qquad \dfrac{5}{y^2} - 1 - 4 = 0$

$\qquad \dfrac{5}{y^2} - 5 = 0$

$\qquad y^2(\dfrac{5}{y^2} - 5) = y^2(0)$

$\qquad 5 - 5y^2 = 0$

$\qquad 5 = 5y^2$

$\qquad 1 = y^2$

$\qquad y = \pm 1$

If $y = 1$: $x = -\dfrac{1}{y} = -\dfrac{1}{1} = -1$ $\qquad$ $(-1, 1)$

If $y = -1$: $x = -\dfrac{1}{y} = \dfrac{-1}{(-1)} = 1$ $\qquad$ $(1, -1)$

Next, substitute $-\dfrac{6}{y}$ for x in the first equation.

$\qquad 5(-\dfrac{6}{y})^2 + (-\dfrac{6}{y})y - 4 = 0$

$\qquad \dfrac{180}{y^2} - 6 - 4 = 0$

$\qquad \dfrac{180}{y^2} - 10 = 0$

$\qquad y^2(\dfrac{180}{y^2} - 10) = y^2(0)$

$\qquad 180 - 10y^2 = 0$

$\qquad 180 = 10y^2$

$\qquad 18 = y^2$

$\qquad y = \pm\sqrt{18} = \pm\sqrt{9 \cdot 2} = \pm 3\sqrt{2}$

If $y = 3\sqrt{2}:$ $x = -\dfrac{6}{y} = \dfrac{-6}{3\sqrt{2}} = \dfrac{-2}{\sqrt{2}}$

$= \dfrac{-2}{\sqrt{2}} \cdot \dfrac{\sqrt{2}}{\sqrt{2}} = \dfrac{-2\sqrt{2}}{2} = -\sqrt{2}$

If $y = -3\sqrt{2}:$ $x = -\dfrac{6}{y} = \dfrac{-6}{-3\sqrt{2}} = \dfrac{2}{\sqrt{2}}$

$= \dfrac{2}{\sqrt{2}} \cdot \dfrac{\sqrt{2}}{\sqrt{2}} = \sqrt{2}$

$\{(-1,1), (1,-1), (-\sqrt{2}, 3\sqrt{2}), (\sqrt{2}, -3\sqrt{2})\}$

46. $\dfrac{x^{x+1} y^{2n-1}}{(x^n y^{1-n})^3} = \dfrac{x^{n+1} y^{2n-1}}{x^{3n} y^{3-3n}}$

$= x^{n+1-3n} y^{2n-1-(3-3n)} = x^{1-2n} y^{5n-4}$

$\left[\text{Equivalently}: \dfrac{y^{5n-4}}{x^{2n-1}} \right]$

47. $(x_1, y_1) = (1, 4)$

Slope of $2x - 5y + 7 = 0$:

$-5y = -2x - 7$

$y = \dfrac{2}{5}x + \dfrac{7}{5}$ $m = \dfrac{2}{5}$

Slope of the line whose equation

we must write : $\dfrac{2}{5}$

$y - y_1 = m(x - x_1)$

$y - 4 = \dfrac{2}{5}(x - 1)$ Point Slope Equation

48. $\sqrt{3 - 3y} = 3 + \sqrt{3y + 2}$

$(\sqrt{3 - 3y})^2 = (3 + \sqrt{3y + 2})^2$

$3 - 3y = 9 + 6\sqrt{3y + 2} + 3y + 2$

$3 - 3y = 3y + 11 + 6\sqrt{3y + 2}$

$-8 - 6y = 6\sqrt{3y + 2}$

$4 + 3y = -3\sqrt{3y + 2}$

$(4 + 3y)^2 = (-3\sqrt{3y + 2})^2$

$16 + 24y + 9y^2 = 9(3y + 2)$

$16 + 24y + 9y^2 = 27y + 18$

$9y^2 - 3y - 2 = 0$

$(3y - 2)(3y + 1) = 0$

$3y - 2 = 0$ $3y + 1 = 0$

$y = \dfrac{2}{3}$ or $y = -\dfrac{1}{3}$

Check $\dfrac{2}{3}$: $\sqrt{3 - 3(\dfrac{2}{3})} = 3 + \sqrt{3(-\dfrac{2}{3}) + 2}$

$\sqrt{3 - 2} = 3 + \sqrt{-2 + 2}$

1   3

$\dfrac{2}{3}$ is extraneous.

Check $-\dfrac{1}{3}$: $\sqrt{3 - 3(-\dfrac{1}{3})} = 3 + \sqrt{3(-\dfrac{1}{3}) + 2}$

$\sqrt{3 + 1} = 3 + \sqrt{-1 + 2}$

2   4

$-\dfrac{1}{3}$ is extraneous.

$\varnothing$

Problem Set 8.2.1

1.

$x + y = 7$

$\underline{x - y = 3}$

Add: $2x = 10$

$x = 5$

$x + y = 7$

$5 + y = 7$

$y = 2$

$\{(5, 2)\}$

5. $x - 2y = 5$

$5x - y = -2$ (multiply by -2)

$x - 2y = 5$

$\underline{-10x + 2y = 4}$

Add: $-9x = 9$

$x = -1$

$5x - y = -2$

$5(-1) - y = -2$

$-5 - y = -2$

$-3 = y$

$\{(-1, -3)\}$

9. $3x - 7y = 1$ (multiply by 2)

$2x - 3y = -1$ (multiply by -3)

$6x - 14y = 2$

$\underline{-6x + 9y = 3}$

Add: $-5y = 5$

$y = -1$

$3x - 7y = 1$

$3x - 7(-1) = 1$

$3x + 7 = 1$

$3x = -6$

$x = -2$

$\{(-2, -1)\}$

13. $2y = 5 - 5x$

$9x - 15 = -3y$

$5x + 2y = 5$ (multiply by 3)

$9x + 3y = 15$ (multiply by -2)

$15x + 6y = 15$

$\underline{-18x - 6y = -30}$

Add: $-3x = -15$

$x = 5$

$2y = 5 - 5(5)$

$2y = -20$

$y = -10$

$\{(5, -10)\}$

15. $9x + \dfrac{4}{3}y = 5$

$4x - \dfrac{1}{3}y = 5$ (multiply by 4)

$9x + \dfrac{4}{3}y = 5$

$16x - \dfrac{4}{3}y = 20$

Add: $25x = 25$

$x = 1$

$9x + \dfrac{4}{3}y = 5$

$9(1) + \dfrac{4}{3}y = 5$

$3(9 + \dfrac{4}{3}y) = 3(5)$

$27 + 4y = 15$

$4y = -12$

$y = -3$

$\{(1, -3)\}$

17. $\frac{1}{8}y + \frac{1}{5}x = 5$ (multiply by 40)

$\frac{1}{3}y + \frac{1}{2}x = 13$ (multiply by 6)

$5y + 8x = 200$ (multiply by 2)

$2y + 3x = 78$ (multiply by -5)

$10y + 16x = 400$

$\underline{-10y - 15x = -390}$

Add: $x = 10$

　　$\frac{1}{8}y + \frac{1}{5}x = 5$

$\frac{1}{8}y + \frac{1}{5}(10) = 5$

$\frac{1}{8}y + 2 = 5$

$8(\frac{1}{8}y + 2) = 8(5)$

$y + 16 = 40$

$y = 24$

$\{(10, 24)\}$

19. $2x = 6 - 3y$

$9x = 8y + 4$

$2x + 3y = 6$ (multiply by 9)

$9x - 8y = 4$ (multiply by -2)

$18x + 27y = 54$

$\underline{-18x + 16y = -8}$

Add: $43y = 46$

$y = \frac{46}{43}$

$2x + 3y = 6$ (multiply by 8)

$9x - 8y = 4$ (multiply by 3)

$16x + 24y = 48$

$\underline{27x - 24y = 12}$

Add: $43x = 60$

$x = \frac{60}{43}$

$\{(\frac{60}{43}, \frac{46}{43})\}$

21. $\frac{x}{10} + \frac{y}{5} = \frac{1}{2}$ (multiply by 10)

$\frac{x}{4} - \frac{y}{3} = -\frac{5}{12}$ (multiply by 12)

$x + 2y = 5$ (multiply by 2)

$3x - 4y = -5$

$2x + 4y = 10$

$\underline{3x - 4y = -5}$

Add: $5x = 5$

$x = 1$

$x + 2y = 5$

$1 + 2y = 5$

$2y = 4$

$y = 2$

$\{(1, 2)\}$

23. $2y + 2x = -2 - 4y$

$7y + 27 = 3x + y$

$2x + 6y = -2$

$-3x + 6y = -27$ (multiply by -1)

$2x + 6y = -2$

$\underline{3x - 6y = 27}$

Add: $5x = 25$

$x = 5$

$2y + 2x = -2 - 4y$

$2y + 10 = -2 - 4y$

$6y = -12$

$y = -2$

$\{(5, -2)\}$

29. $1/2(x-y) - 1/6(x-4y) = 4$ (multiply by 6)

$1/9(x+y) - 1/6(x-2y) = 11/9$ (multiply by 18)

$3(x-y) - (x-4y) = 24$

$2(x+y) - 3(x-2y) = 22$

$3x - 3y - x + 4y = 24$

$2x + 2y - 3x + 6y = 22$

$2x + y = 24$

$-x + 8y = 22$ (multiply by 2)

$2x + y = 24$

$\underline{-2x + 16y = 44}$

Add: $17y = 68$

$y = 68/17 = 4$

$2x + y = 24$

$2x + 4 = 24$

$2x = 20$

$x = 10$

$\{(10, 4)\}$

33. $3x + 3y = 17 - A$

$4x + 3y = 18 - A$ (multiply by -1)

$3x + 3y = 17 - A$

$\underline{-4x - 3y = -18 + A}$

Add: $-x = -1$

$x = 1$

$3x + 3y = 17 - A$

$3(1) + 3y = 17 - A$

$3y = 14 - A$

$y = \dfrac{14 - A}{3}$

$\left\{\left(1, \dfrac{14 - A}{3}\right)\right\}$

35. $Ax + By = C$

$\underline{-Ax + By = C}$

Add: $2By = 2C$

$y = \dfrac{2C}{2B} = \dfrac{C}{B}$

Since $A = B$ and $C = 3A$

$y = \dfrac{C}{B} = \dfrac{3A}{A} = 3$

$Ax + By = C$

$Ax + B(3) = C$

$Ax = C - 3B$

$x = \dfrac{C - 3B}{A}$

Since $A = B$ and $C = 3A$

$x = \dfrac{C - 3B}{A} = \dfrac{3A - 3A}{A} = \dfrac{0}{A} = 0$

$\{(0, 3)\}$

37. a. $2x - 3y = 1$ (multiply by -3)

$3x - 5y = -2$ (multiply by 2)

$-6x + 9y = -3$

$\underline{6x - 10y = -4}$

Add: $-y = -7$

$y = 7$

$2x - 3y = 1$

$2x - 3(7) = 1$

$2x = 22$

$x = 11$

$\{(11, 7)\}$

b. $A(2x - 3y - 1) + B(3x - 5y + 2) = 0$

$A(22 - 21 - 1) + B(33 - 35 + 2) = 0$

$A(0) + B(0) = 0$

$0 = 0$

39. $ax + by = c$

$\underline{ax - by = d}$

Add: $2ax = c + d$

$x = \dfrac{c + d}{2a}$　　$ax + by = c$

$ax - by = d$ (multiply by -1)

$ax + by = c$　　$-ax + by = -d$

Add: $2by = c - d$　　$y = \dfrac{c - d}{2b}$　　$\{(\dfrac{c+d}{2a}, \dfrac{c-d}{2b})\}$

41. Number of hours by carpenter

　　working alone: x

Number of hours by apprentice

　　working alone: $x + 5$

Portion of job done by the carpenter

　　in 6 hours + Portion of the job done

　　by apprentice in 6 hours $= 1$ whole job

$\dfrac{6}{x} + \dfrac{6}{x+5} = 1$

$x(x+5)\left[\dfrac{6}{x} + \dfrac{6}{x+5}\right] = x(x+5) \cdot 1$

$6x + 30 + 6x = x^2 + 5x$

$0 = x^2 - 7x - 30$

$0 = (x - 10)(x + 3)$

$x = 10$ or $x = -3$ (reject)

Carpenter: 10 hours

Apprentice: 15 hours

42. $(y-1)(y-2)\left[\dfrac{3}{y-1} - \dfrac{4}{y-2}\right]$

$= (y-1)(y-2)\left[3 - \dfrac{2y}{y-2}\right]$

$3(y-2) - 4(y-1) = 3(y-1)(y-2) - 2y(y-1)$

$3y - 6 - 4y + 4 = 3y^2 - 9y + 6 - 2y^2 + 2y$

$-y - 2 = y^2 - 7y + 6$

$0 = y^2 - 6y + 8$　　$0 = (y - 4)(y - 2)$

$y - 4 = 0$ or $y - 2 = 0$

$y = 4 \quad y = 2$　　$\{2, 4\}$

43. $9y^2 + 12y + 1 = 0$

$y^2 + \dfrac{12}{9}y + \dfrac{1}{9} = 0$

$y^2 + \dfrac{4}{3}y \underline{} = -\dfrac{1}{9}$

$(\dfrac{1}{2})(\dfrac{4}{3}) = (\dfrac{2}{3})$

$(\dfrac{2}{3})^2 = \dfrac{4}{9}$

$y^2 + \dfrac{4}{3}y + \dfrac{4}{9} = -\dfrac{1}{9} + \dfrac{4}{9}$

$(y + \dfrac{2}{3})^2 = \dfrac{1}{3}$

$y + \dfrac{2}{3} = \pm\sqrt{\dfrac{1}{3}}$

$y + \dfrac{2}{3} = \pm\dfrac{1}{\sqrt{3}} \cdot \dfrac{\sqrt{3}}{\sqrt{3}}$

$y + \dfrac{2}{3} = \pm\dfrac{\sqrt{3}}{3}$

$y = -\dfrac{2}{3} \pm \dfrac{\sqrt{3}}{\sqrt{3}}$

$y = \dfrac{-2 \pm \sqrt{3}}{3}$

$\{(\dfrac{-2+\sqrt{3}}{3}, \dfrac{-2-\sqrt{3}}{3})\}$

Problem Set 8.2.2

1. $x^2 + y^2 = 13$

$\underline{x^2 - y^2 = 5}$

Add: $2x^2 = 18$

$x^2 = 9$

$x = \pm 3$

If $x = 3$: $x^2 + y^2 = 13$

$9 + y^2 = 13$

$y^2 = 4$

$y = \pm 2$ (3, 2) (3, -2)

If $x = -3$: $(-3)^2 + y^2 = 13$

$y = \pm 2$ (-3, 2) (-3, -2)

$\{(3, 2)(3, -2)(-3, 2)(-3, -2)\}$

3. $x^2 - y^2 = 11$ (multiply by -2)

$2x^2 - 5y^2 = 7$

$-2x^2 + 2y^2 = -22$

$\underline{2x^2 - 5y^2 = 7}$

Add: $-3y^2 = -15$

$y^2 = 5$

$y = \pm \sqrt{5}$

If $y = \sqrt{5}$: $x^2 - 5 = 11$

$x^2 = 16$

$x = \pm 4$

Similarly. if $y = -\sqrt{5}$: $x = \pm 4$

$\{(4, \sqrt{5}), (-4, \sqrt{5}), (4, -\sqrt{5}), (-4, -\sqrt{5})\}$

7. $y = x^2 + 4$

$x^2 + y^2 = 16$

$x^2 - y = -4$ (multiply by -1)

$x^2 + y^2 = 16$

$-x^2 + y = 4$

$\underline{x^2 + y^2 = 16}$

Add: $y + y^2 = 20$

$y^2 + y - 20 = 0$

$(y + 5)(y - 4) = 0$

$y = -5$ or $y = 4$

If $y = -5$: $y = x^2 + 4$

$x^2 + 4 = -5$

$x^2 = -9$

$x = \pm \sqrt{-9}$

$x = \pm 3i$

If $y = 4$: $y = x^2 + 4$

$4 = x^2 + 4$

$x^2 = 0$

$x = 0$

$\{(3i, -5), (-3i, -5), (0, 4)\}$

11. $x^2 - 16y^2 = 8$ (multiply by 3)

$y^2 - 3x^2 = 23$

$3x^2 - 48y^2 = 24$

$\underline{-3x^2 + y^2 = 23}$

Add: $-47y^2 = 47$

$y^2 = -1 \qquad y = \pm\sqrt{-1} = \pm i$

If $y = i$: $x^2 - 16y^2 = 8$

$x^2 - 16i^2 = 8$

$x^2 - 16(-1) = 8$

$x^2 = -8 \qquad x = \pm\sqrt{-8}$

$x = \pm 2\sqrt{2i}$

If $y = -i$: $x^2 - 16(-i)^2 = 8$

$x^2 - 16i^2 = 8$

$x = \pm 2\sqrt{2i}$

$\{(2\sqrt{2i}, i), (-2\sqrt{2i}, i), (2\sqrt{2i}, -i), (-2\sqrt{2i}, -i)\}$

15. $4x^2 - y = 3$ (multiply by -2)

$8x^2 - y^2 = -9$

$-8x^2 + 2y = -6$

$\underline{8x^2 - y^2 = -9}$

Add: $2y - y^2 = -15$

$-y^2 + 2y + 15 = 0$

$y^2 - 2y - 15 = 0$

$(y - 5)(y + 3) = 0$

$y = 5$ or $y = -3$

If $y = 5$: $4x^2 - y = 3$

$4x^2 - 5 = 3$

$4x^2 = 8$

$x^2 = 2$

$x = \pm\sqrt{2}$

If $y = -3$: $4x^2 - y = 3$

$4x^2 - (-3) = 3$

$4x^2 = 0$

$x = 0$

$\{(\sqrt{2}, 5), (-\sqrt{2}, 5), (0, -3)\}$

17. $y^2 + 3xy = 1$ (multiply by 4)

$y^2 + 4xy = 2$ (multiply by -3)

$4y^2 + 12xy = 4$

$\underline{-3y^2 - 12xy = -6}$

Add: $y^2 = -2$

$y = \pm\sqrt{-2}$

$y = \pm\sqrt{2i}$

If $y = \sqrt{2i}$: $y^2 + 3xy = 1$

$(\sqrt{2i})^2 + 3x(\sqrt{2i}) = 1$

$2i^2 + 3\sqrt{2i}x = 1$

$-2 + 3\sqrt{2i}x = 1$

$3\sqrt{2i}x = 3$

$x = \dfrac{3}{3\sqrt{2i}} = \dfrac{1}{\sqrt{2i}} \cdot \dfrac{\sqrt{2i}}{\sqrt{2i}} = \dfrac{\sqrt{2i}}{2i^2} = \dfrac{\sqrt{2i}}{2(-1)}$

$= \dfrac{-\sqrt{2i}}{2}$

If $y = -\sqrt{2i}$: $y^2 + 3xy = 1$

$(-\sqrt{2i})^2 + 3x(-\sqrt{2i}) = 1$

$2i^2 - 3\sqrt{2i}x = 1$

$-2 - 3\sqrt{2i}x = 1$

$-3\sqrt{2i}x = 3$

$x = \dfrac{-1}{\sqrt{2i}} \cdot \dfrac{\sqrt{2i}}{\sqrt{2i}} = \dfrac{-\sqrt{2i}}{2i^2} = \dfrac{\sqrt{2i}}{2}$

$\left\{ \left(-\dfrac{\sqrt{2i}}{2}, \sqrt{2i}\right), \left(\dfrac{\sqrt{2i}}{2}, -\sqrt{2i}\right) \right\}$

21. $y = x^2 + a$

$x^2 + y^2 = a^2$

$-x^2 + y = a$

$\underline{x^2 + y^2 = a^2}$

Add: $y + y^2 = a + a^2$

$y^2 + y + (-a^2 - a) = 0$

$y = \dfrac{-1 \pm \sqrt{1 - 4(1)(-a^2 - a)}}{2}$

$= \dfrac{-1 \pm \sqrt{1 + 4a^2 + 4a}}{2}$

$= \dfrac{-1 \pm \sqrt{4a^2 + 4a + 1}}{2}$

$= \dfrac{-1 \pm \sqrt{(2a + 1)^2}}{2}$

$= \dfrac{-1 \pm (2a + 1)}{2}$

$y = \dfrac{-1 + 2a + 1}{2}$ or $y = \dfrac{-1 - (2a + 1)}{2}$

$y = \dfrac{2a}{2}$ $y = \dfrac{-2a - 2}{2}$

$y = a$ $y = -a - 1$

If $y = a$: $x^2 + y^2 = a^2$

$x^2 + a^2 = a^2$

$x^2 = 0$

$x = 0$

If $y = -a - 1$: $x^2 + y^2 = a^2$

$x^2 + (-a - 1)^2 = a^2$

$x^2 + a^2 + 2a + 1 = a^2$

$x^2 = -2a - 1$

$x = \pm \sqrt{-2a - 1}$

$x = \pm \sqrt{(-1)(2a + 1)}$

$x = \pm \sqrt{2a + 1}\, i$

$\{(0, a), (\sqrt{2a + 1}\, i, -a - 1), (-\sqrt{2a + 1}\, i, -a - 1)\}$

25. $x^2 + xy + y^2 = 37$

$\underline{x^2 \quad + y^2 = 25}$

Subtract : $xy = 12$

$x = \dfrac{12}{y}$ $x^2 + y^2 = 25$ $(\dfrac{12}{y})^2 + y^2 = 25$

$\dfrac{144}{y^2} + y^2 = 25$

$y^2(\dfrac{144}{y^2} + y^2) = y^2(25)$

$144 + y^4 = 25y^2$

$y^4 - 25y^2 + 144 = 0$

Let $t = y^2$

$t^2 - 25t + 144 = 0$

$(t - 16)(t - 9) = 0$

$t = 16$ or $t = 9$

$y^2 = 16$ $y^2 = 9$

$y = \pm 4$ $y = \pm 3$

If $y = \pm 4$: $x^2 + y^2 = 25$

$x^2 + 16 = 25$

$x^2 = 9$

$x = \pm 3$

If $y = \pm 3$: $x^2 + y^2 = 25$

$x^2 + 9 = 25$

$x^2 = 16$

$x = \pm 4$

Check $(3, 4)$ in $x^2 + xy + y^2 = 37$

$9 + 12 + 16 = 37$

$37 = 37$

Check $(-3, 4)$: $9 - 12 + 16 \quad 37$

Check $(3, -4)$: $9 - 12 + 16 \quad 37$

Check $(-3, -4)$: $9 + 12 + 16 = 37$

Check $(4, 3)$: $16 + 12 + 9 = 37$

Check $(4, -3)$: $16 - 12 + 9 \quad 37$

Check $(-4, 3)$: $16 - 12 + 9 \quad 37$

Check $(-4, -3)$: $16 + 12 + 9 = 37$

$\{(3, 4), (-3, -4), (4, 3), (-4, -3)\}$

26. $\dfrac{2y}{y-3} + \dfrac{3y+9}{(2y+3)(y-3)} + \dfrac{1}{2y+3} = 0$

$(2y+3)(y-3)\left[\dfrac{2y}{y-3} + \dfrac{3y+9}{(2y+3)(y-3)} + \dfrac{1}{2y+3}\right]$

$\qquad = (2y+3)(y-3)\cdot 0$

$2y(2y+3) + 3y + 9 + y - 3 = 0$

$4y^2 + 10y + 6 = 0$

$2y^2 + 5y + 3 = 0$

$(2y+3)(y+1) = 0$

$2y + 3 = 0 \quad y + 1 = 0$

$y = -\dfrac{3}{2}$(reject; causes division by zero

$\qquad$ in original equation) $y = -1$

$\{-1\}$

27. x : number of trees above the basic 20

$\qquad$ Yield = (number of trees)(number

$\qquad\qquad\qquad$ of grapefruits for each tree)

$\qquad$ Y = (20 + x)(300 - 10x)

$\qquad$ Y = 6000 - 200x + 300x - 10x^2$

$\qquad$ Y = -10x^2 + 100x + 6000$

$\qquad$ Maximum Y occurs when $x = -\dfrac{b}{2a}$.

$\qquad x = \dfrac{-100}{2(-10)} = \dfrac{-100}{-20} = 5$

$\qquad$ Plant $20 + 5 = 25$ trees

28. $d = \sqrt{(x_2 - x_1)^2 + (y_2 - y_1)^2}$

$\qquad = \sqrt{(2+6)^2 + (1+3)^2}$

$\qquad = \sqrt{64 + 16} = \sqrt{80} = \sqrt{16\cdot 5} = 4\sqrt{5}$

Problem Set 8.2.3

1. $x + 2y - 3 = 0$

$12 = 8y + 4x$

$x + 2y = 3$

$4x + 8y = 12$

$\dfrac{1}{4} = \dfrac{2}{8} = \dfrac{3}{12}$

Dependent

$\{(x, y) \mid x + 2y - 3 = 0\}$

or $\{(x, y) \mid 12 = 8y + 4x\}$

3. $0.2x - 0.5y = 0.1$ (multiply by 10)

$0.4x - y = -0.2$ (multiply by 10)

$2x - 5y = 1$

$4x - 10y = -2$

$\dfrac{2}{4} = \dfrac{-5}{-10} \quad \dfrac{1}{-2}$

Inconsistent

$\varnothing$

9. $2x^2 - xy = y^2$

$x = y$

$2x^2 - x(x) = x^2$

$2x^2 - x^2 = x^2$

$x^2 = x^2$

$0 = 0$

Dependent

$\{(x, y) \mid x = y\}$

17. $3(x - 3) - 2y = 0$

$2(x - y) = -x - 3$

$3x - 9 - 2y = 0$

$2x - 2y = -x - 3$

$3x - 2y = 9$

$3x - 2y = -3$

$\dfrac{3}{3} = \dfrac{-2}{-2} \quad \dfrac{9}{-3}$

Inconsistent

$\varnothing$

19. $\dfrac{x - 1}{2} = \dfrac{y - 2}{8}$

$12x - 3y = 6$

$8(x - 1) = 2(y - 2)$

$12x - 3y = 6$

$8x - 2y = 4$

$12x - 3y = 6$

$\dfrac{8}{12} = \dfrac{-2}{-3} = \dfrac{4}{6}$

Dependent $\{(x, y) \mid 12x - 3y = 6\}$

21. $6x - 9y = 3$

$4x - 6y = a$

$\dfrac{6}{4} = \dfrac{-9}{-6} = \dfrac{3}{a} \qquad \dfrac{3}{2} = \dfrac{3}{2} = \dfrac{3}{a}$

$a = 2$

23. $ax + by = c$

$bx + ay = c$

Inconsistent : $\dfrac{a}{b} = \dfrac{b}{a} \quad 1$

$a^2 = b^2$ and $\dfrac{b}{a} \quad 1$

$\qquad\qquad b \quad a$

If $a^2 = b^2$ and $b \quad a$, then :

$a^2 - b^2 = 0$

$(a + b)(a - b) = 0$

$a + b = 0 \quad a - b = 0$

$a = -b \quad a = b$ (reject since $b \quad a$)

$a = -b$

24. $\dfrac{2 - y^2}{y^2 + 3y - 4} - \dfrac{-y - 1}{y + 4}$

$= \dfrac{2 - y^2}{(y + 4)(y - 1)} - \dfrac{-y - 1}{y + 4} \cdot \dfrac{(y - 1)}{(y - 1)}$

$= \dfrac{2 - y^2 - (-y - 1)(y - 1)}{(y + 4)(y - 1)} = \dfrac{2 - y^2 + y^2 - 1}{(y + 4)(y - 1)}$

$= \dfrac{1}{(y + 4)(y - 1)}$

25. $(x+4)^3 + (x-1)^3$

$\quad = [(x+4)+(x-1)][(x+4)^2 - (x+4)(x-1)$

$\qquad + (x-1)^2]$

$\quad = (2x+3)(x^2 + 8x + 16 - x^2 - 3x + 4 + x^2$

$\qquad - 2x + 1)$

$\quad = (2x+3)(x^2 + 3x + 21)$

26. x : number of days working together

Portion of the job done in x days by

$\qquad$ Archie + Portion of the job done in

$\qquad$ x days by Edith = 1 whole job

$\dfrac{x}{15} + \dfrac{x}{10} = 1$

$30(\dfrac{x}{15} + \dfrac{x}{10}) = 30(1)$

$2x + 3x = 30$

$5x = 30$

$x = 6$

Working together : 6 days

Problem Set 8.3.1

5. x :   First number

     y :   second number

     $x + y = -1$

     $y = x^2 - 1$

     $x + (x^2 - 1) = -1$

     $x^2 + x = 0$

     $x(x + 1) = 0$

     $x = 0$ or $x = -1$

     If $x = 0$: $y = x^2 - 1 = 0^2 - 1 = -1$

     If $x = -1$: $y = (-1)^2 - 1 = 0$

     First Number :   0 and second number :   -1

     or First Number :   -1 and

                         second number :   0

7.   Numbers : x, y

     $x + y = 6$

     $2x = 10 - 2y$

     $x + y = 6$

     $2x + 2y = 10$

     $\dfrac{1}{2} = \dfrac{1}{2} \quad \dfrac{6}{10}$ System is inconsistent.

     No numbers satisfy the given conditions.

13. Numbers : x, y

     $xy = 4$

     $x^2 + y^2 = 8$

     $x = \dfrac{4}{y}$

     $(\dfrac{4}{y})^2 + y^2 = 8 \qquad \dfrac{16}{y^2} + y^2 = 8$

     $y^2(\dfrac{16}{y^2} + y^2) = y^2 \bullet 8 \qquad 16 + y^4 = 8y^2$

     $y^4 - 8y^2 + 16 = 0$

     Let $t = y^2$

     $t^2 - 8t + 16 = 0 \qquad (t - 4)^2 = 0 \qquad t = 4$

     $y^2 = 4$

     $y = \pm 2$

     If $y = 2$: $x = \dfrac{4}{y} = \dfrac{4}{2} = 2$

     If $y = -2$: $x = \dfrac{4}{y} = \dfrac{4}{-2} = -2$

     Numbers :   2 and 2 or - 2 and - 2

17. t :   tens' digit

     u :   units' digit

     Number : $10t + u$

     Number with digits reversed : $10u + t$

     $t + u = 14$

     $10u + t = 2(10t + u) - 23$

     $t + u = 14$

     $19t - 8u = 23$

     $t = 14 - u$

     $19(14 - u) - 8u = 23$

     $266 - 19u - 8u = 23$

     $-27u = -243$

     $u = 9$

     $t = 14 - u = 14 - 9 = 5$

     Number : $10t + u = 10(5) + 9 = 59$

23. t : tens' digit

u : units' digit

$(10t + u)(t + u) = 205$

$u + 3t = 13$

$10t^2 + 11tu + u^2 = 205$

$u = 13 - 3t$

$10t^2 + 11t(13 - 3t) + (13 - 3t)^2 = 205$

$10t^2 + 143t - 33t^2 + 169 - 78t + 9t^2 = 205$

$-14t^2 + 65t - 36 = 0$

$14t^2 - 65t + 36 = 0$

$(14t - 9)(t - 4) = 0$

$t = \dfrac{9}{14}$(reject; t must be a natural

number) $t = 4$

$u = 13 - 3t$

$u = 13 - 3(4) = 1$

Number : $10t + u = 10(4) + 1 = 41$

25. Numbers : x, y

$x + y = 7$

$\dfrac{1}{x} + \dfrac{1}{y} = \dfrac{7}{12}$

$x = 7 - y$

$\dfrac{1}{7 - y} + \dfrac{1}{y} = \dfrac{7}{12}$

$12y(7 - y)[\dfrac{1}{7 - y} + \dfrac{1}{y}] = 12y(7 - y)(\dfrac{7}{12})$

$12y + 84 - 12y = 49y - 7y^2$

$7y^2 - 49y + 84 = 0$

$y^2 - 7y + 12 = 0$

$(y - 4)(y - 3) = 0$

$y = 4$ or $y = 3$

If $y = 4$: $x = 7 - y = 7 - 4 = 3$

If $y = 3$: $x = 7 - y = 7 - 3 = 4$

Numbers : 3 and 4

27. Number : $\dfrac{x}{y}$

$\dfrac{x - 1}{y + 2} = \dfrac{1}{3}$

$\dfrac{x}{y + 1} = \dfrac{1}{2}$

$3(x - 1) = y + 2 \qquad 2x = y + 1$

$3x - y = 5$

$\underline{-2x + y = -1}$

Add : $x = 4$

$2x = y + 1 \qquad\qquad 8 = y + 1 \qquad\qquad 7 = y$

Number : $\dfrac{x}{y} = \dfrac{4}{7}$

28. $\dfrac{(-3)(-2)(-1)}{-1 - (-2)} \div \dfrac{6[(-5)(-1) - (-7)]}{[5 + (-2)][(-2) + (-1)]}$

$= \dfrac{-6}{1} \div \dfrac{6[5 + 7]}{(3)(-3)} \qquad = -6 \div \dfrac{72}{-9}$

$= -6 \div -8 \qquad = \dfrac{-6}{1} \cdot \dfrac{1}{-8} = \dfrac{3}{4}$

29. $y^2 + 1 \overline{\smash{\big)}\ y^7 + 2y^6 + y^5 + 2y^4 + 0y^3 + y^2 + 0y + 1}$

$\qquad\qquad\qquad\dfrac{y^5 + 2y^4 + 1}{}$

$\underline{y^7 \qquad\quad + y^5}$

$\quad 2y^6 \qquad\quad + 2y^4$

$\underline{\quad 2y^6 \qquad\quad + 2y^4}$

$\qquad\qquad\qquad\quad y^2 \qquad + 1$

$\qquad\qquad\qquad\underline{\quad y^2 \qquad + 1}$

$\qquad\qquad\qquad\qquad\qquad 0$

30. $\dfrac{-\frac{a+b}{a-b} - \frac{a-b}{a+b}}{1 - \frac{a^2+b^2}{(a+b)^2}} \cdot \dfrac{(a+b)^2(a-b)}{(a+b)^2(a-b)}$

$= \dfrac{(a+b)(a+b)^2 - (a+b)(a-b)^2}{(a+b)^2(a-b) - (a^2 + b^2)(a-b)}$

$= \dfrac{(a+b)[(a+b)^2 - (a-b)^2]}{(a-b)[(a+b)^2 - (a^2 + b^2)]}$

$= \dfrac{(a+b)[(a+b) + (a-b)][(a+b) - (a-b)]}{(a-b)[a^2 + 2ab + b^2 - a^2 - b^2]}$

$= \dfrac{(a+b)(2a)(2b)}{(a-b)(2ab)} = \dfrac{2(a+b)}{a-b}$

Problem Set 8.3.2

1.   Length : x   Width : y

   Perimeter is 20 ft. $2x + 2y = 20$

   Area is 21 ft.^{2} $xy = 21$

   $x + y = 10$ $(y = 10 - x)$

   $xy = 21$

   $x(10 - x) = 21$

   $10x - x^2 = 21$

   $0 = x^2 - 10x + 21$

   $0 = (x - 7)(x - 3)$

   $x = 7$ or $x = 3$

   If $x = 7$: $y = 10 - x = 10 - 7 = 3$

   If $x = 3$: $y = 10 - x = 10 - 3 = 7$

   Dimensions : 7ft by 3ft

5.   Legs : x, y

   Perimeter is 56 : $x + y + 25 = 56$

   By Pythagorean Theorem : $x^2 + y^2 = (25)^2$

   $x + y = 31$ $(x = 31 - y)$

   $x^2 + y^2 = 625$

   $(31 - y)^2 + y^2 = 625$

   $961 - 62y + y^2 + y^2 = 625$

   $2y^2 - 62y + 336 = 0$

   $y^2 - 31y + 168 = 0$

   $(y - 24)(y - 7) = 0$

   $y = 24$ or $y = 7$

   If $y = 24$: $x = 31 - y = 31 - 24 = 7$

   If $y = 7$: $x = 31 - 7 = 24$

   Legs :   24 inches and 7 inches

15. Base : x

   Altitude : y

   $\dfrac{1}{2}xy = 16$

   $\dfrac{1}{2}(x + 2)(y - 3) = 5$

   $xy = 32$

   $(x + 2)(y - 3) = 10$

   $xy = 32$ $(y = \dfrac{32}{x})$

   $xy - 3x + 2y = 16$

   $x(\dfrac{32}{x}) - 3x + 2(\dfrac{32}{x}) = 16$

   $32 - 3x + \dfrac{64}{x} = 16$

   $x(32 - 3x + \dfrac{64}{x}) = x(16)$

   $32x - 3x^2 - 64 = 16x$

   $0 = 3x^2 - 16x - 64$

   $0 = (3x + 8)(x - 8)$

   $x = -\dfrac{8}{3}$(Reject)

   $x = 8$

   If $x = 8$: $y = \dfrac{32}{x} = \dfrac{32}{8} = 4$

   Base : 8 feet   Altitude : 4 feet

17. Dimensions : x, y

   Perimeter is 46. $2x + 2y = 46$

   $x + y = 23$ $(x = 23 - y)$ Diagonal is 17.

   Pythagorean Theorem : $x^2 + y^2 = 289$

   $(23 - y)^2 + y^2 = 289$

   $529 - 46y + y^2 + y^2 = 289$

   $2y^2 - 46y + 240 = 0$

   $y^2 - 23y + 120 = 0$

   $(y - 8)(y - 15) = 0$ $y = 8$ or $y = 15$

   If $y = 8$: $x = 23 - y = 23 - 8 = 15$

   If $y = 15$: $x = 23 - 15 = 8$

   Dimensions : 15 feet by 8 feet

19. Length : x   Width : y

$x = 2y - 1$

$y + 3 = x - 1$ (Length = Width)

$y + 3 = (2y - 1) - 1$ $y + 3 = 2y - 2$ $5 = y$

If $y = 5$: $x = 2y - 1 = 2(5) - 1 = 9$

Length : 9 meters

Width : 5 meters

21. EB : x   AE : y

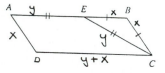

Note : $DC = x + y$ because opposite sides of

    a parallelogram have equal measures.

Perimeter of ABCD is 50.

$2(x + y) + 2x = 50$

Perimeter of AECD is 39.

$y + y + (y + x) + x = 39$

$4x + 2y = 50$

$2x + 3y = 39$ (multiply by - 2)

$4x + 2y = 50$

$\underline{-4x - 6y = -78}$

$-4y = -28$

$y = 7$

$2x + 3y = 39$

$2x + 21 = 39$

$x = 9$

AE : $y = 7m$

EB : $x = 9m$

DC : $x + y = 7 + 9 = 16m$

23.

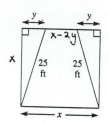

Perimeter of trapezoid is 84.

$x + 50 + (x - 2y) = 84$

Since the larger quadrilateral is a square,

    the side on the left is also x.

By the Pythagorean Theorem :

$x^2 + y^2 = (25)^2$

$2x - 2y + 50 = 84$

$x^2 + y^2 = 625$

$x - y = 17$ $(x = y + 17)$

$x^2 + y^2 = 625$

$(y + 17)^2 + y^2 = 625$

$2y^2 + 34y - 336 = 0$

$y^2 + 17y - 168 = 0$

$(y + 24)(y - 7) = 0$

$y = -24$ (reject) $y = 7$

If $y = 7$: $x = y + 17 = 7 + 17 = 24$

$x = 24, y = 7$

26. $(y - 1)(y - 2)\left[\dfrac{3}{y - 1} - \dfrac{4}{y - 2}\right]$

$= (y - 1)(y - 2)\left[3 - \dfrac{2y}{y - 2}\right]$

$3(y - 2) - 4(y - 1) = 3(y - 1)(y - 2) - 2y(y - 1)$

$3y - 6 - 4y + 4 = 3y^2 - 9y + 6 - 2y^2 + 2y$

$-y - 2 = y^2 - 7y + 6$

$0 = y^2 - 6y + 8$

$0 = (y - 4)(y - 2)$

$y - 4 = 0$ or $y - 2 = 0$

$y = 4$ $y = 2$ (reject; causes division

    by zero in original equation)

{4}

27. Algebraic Methods :

$3x + y = 3$

$y = x^2 - 5x + 4$

$3x + (x^2 - 5x + 4) = 3$

$x^2 - 2x + 1 = 0$

$(x - 1)^2 = 0$

$x = 1$

$y = x^2 - 5x + 4 = 1^2 - 5 \cdot 1 + 4 = 0$

$\{(1, 0)\}$

Graphic Methods :

$3x + y = 3$

Line : $y = -3x + 3$ y-intercept : 3

slope : -3

$y = x^2 - 5x + 4$

Parabola : x-intercepts : $x^2 - 5x + 4 = 0$

$(x - 4)(x - 1) = 0$

$x = 4$ $x = 1$

y-intercept : $y = 0^2 - 5 \cdot 0 + 4 = 4$

vertex : $x = -\dfrac{b}{2a} = -\dfrac{(-5)}{2(1)} = \dfrac{5}{2}$

If $x = \dfrac{5}{2}$, $y = (\dfrac{5}{2})^2 - 5(\dfrac{5}{2}) + 4$

$= \dfrac{25}{4} - \dfrac{50}{4} + \dfrac{16}{4} = -\dfrac{9}{4}$

$(2.5, -2.25)$

Intersection of line and parabola occurs at

$(1, 0)$.

$\{(1, 0)\}$

28. $3x - 2y \le 6$

$3x - 2y = 6$

x-intercept : $3x = 6$

$x = 2$

y-intercept : $-2y = 6$

$y = -3$

Test $(0, 0)$: $3(0) - 2(0) \le 6$

$0 \le 6$ True

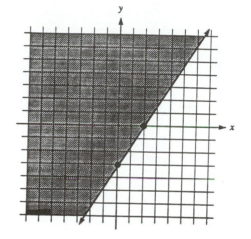

Problem Set 8.3.3

1. x : Tutor's wage
y : Grader's wage
$70x + 50y = 622.50$
$90x + 40y = 719$
$7x + 5y = 62.25$ (multiply by -4)
$9x + 4y = 71.9$ (multiply by 5)
$-28x - 20y = -249$
$\underline{45x + 20y = 359.5}$
Add: $17x = 110.5$
$x = 6.5$
$7(6.5) + 5y = 62.25$
$y = 3.35$
Tutor's wage : \$6.50
Grader's wage : \$3.35

3. Daily pay for Louis : x
Daily pay for Lestat : y
$8x + 10y = 640$
$12x = 9y$
$4x + 5y = 320$
$\underline{4x - 3y = 0}$
$8y = 320$
$y = 40$
$8x + 10(40) = 640$
$x = 30$
Louis : \$30/daily   Lestat : \$40/daily

7. x : speed of plane in still air
y : speed of wind
$RT = D$
$(x + y)5 = 3000$
$(x - y)6 = 3000$
$x + y = 600$
$\underline{x - y = 500}$
$2x = 1100$
$x = 550$
$550 + y = 600$
$y = 50$
Plane's speed : 550 km/h
Wind speed : 50 km/h

9. x : Steve's speed in still water
y : speed of current
$RT = D$
$2.5(x + y) = D$
$5(x - y) = D$
$2.5(x + y) = 5(x - y)$
$2.5x + 2.5y = 5x - 5y$
$7.5y = 2.5x$
$3y = x$
Since $x = 3y$, Steve's rate in still water (x)
        is three times as fast as the current (y).

15. x : Tail length
y : Body length
$x = 6 + \frac{1}{4}y$
$y = (6 + x) + 3$
$4x = 24 + y$
$y = x + 9$
$4x - y = 24$
$\underline{-x + y = 9}$
Add: $3x = 33$
$x = 11$
$y = x + 9 = 11 + 9 = 20$
Lobster's length : head + body + tail
$= 6 + 20 + 11 = 37$ inches

19. $y = ax^2 + bx - 1$
(1,4) satisfies the equation. $4 = a(1)^2 + b(1) - 1$
(-2,1) satisfies the equation. $1 = a(-2)^2 + b(-2) - 1$
$5 = a + b$ (multiply by 2)
$2 = 4a - 2b$
$2a + 2b = 10$
$\underline{4a - 2b = 2}$
Add: $6a = 12$
$a = 2$
$a + b = 5$
$2 + b = 5$
$b = 3$
$y = 2x^2 + 3x - 1$

21. $S = S_0 + V_0 t - 16t^2$

When $t = 5$, $S = 10,000$:

$10,000 = S_0 + 5V_0 - 16(5)^2$

$10,400 = S_0 + 5V_0$

When $t = 10$, $S = 8,550$:

$8,550 = S_0 + 10V_0 - 16(10)^2$

$10,150 = S_0 + 10V_0$

$S_0 + 5V_0 = 10,400$

$\underline{S_0 + 10V_0 = 10,150}$

$-5V_0 = 250$

$V_0 = -50$

$10,150 = S_0 + 10V_0$

$10,150 = S_0 + 10(-50)$

$S_0 = 10,650$

Initial Height : $10,650$ feet (S_0)

Initial Velocity : -50 feet / second (V_0)

23. $y = ax^2 + bx + 32$

After one - half a second, height is 36.

When $x = \dfrac{1}{2}$, $y = 36$

After two seconds, ball hits the ground, so height is 0.

When $x = 2$, $y = 0$.

$(\dfrac{1}{2}, 36) : \ 36 = a(\dfrac{1}{2})^2 + b(\dfrac{1}{2}) + 32$

$36 = \dfrac{1}{4}a + \dfrac{1}{2}b + 32$

$4 = \dfrac{1}{4}a + \dfrac{1}{2}b$

$16 = a + 2b$

$(2, 0) : \ 0 = a(2)^2 + b(2) + 32$

$0 = 4a + 2b + 32$

$-32 = 4a + 2b$

$a + 2b = 16$

$\underline{4a + 2b = -32}$

$-3a = 48$

$a = -16$

$a + 2b = 16$

$-16 + 2b = 16$

$2b = 32$

$b = 16$

$a = -16, \ b = 16$

25. c : Ricky's distance

d : Fred's distance

$c + d = 9 \ \ (d = 9 - c)$

$80c = 100d$

$80c = 100(9 - c)$

$80c = 900 - 100c$

$180c = 900$

$c = 5$

$5 + d = 9$

$d = 4$

Ricky's distance : 5 feet

Fred's distance : 4 feet

27. x : number of blenders purchased by the merchant

y : merchant's cost of each blender

Total cost to merchant = $\$720 \ \ \ xy = 720$

Merchant's profit = cost of 8 blenders.

$\quad P = 8y$

Profit : Money brought in from x blenders

$\qquad\qquad\qquad\qquad$ - cost of x blenders

Profit : $40x - xy$

Thus, $40x - xy = 8y$

System is :

$xy = 720$

$40x - xy = 8y$

$x = \dfrac{720}{y}$

$40(\dfrac{720}{y}) - (\dfrac{720}{y})y = 8y$

$\dfrac{28,800}{y} - 720 = 8y$

$28,800 - 720y = 8y^2$

$8y^2 - 720y + 28,800 = 0$

$y^2 - 90y + 3600 = 0$

$(y - 30)(y + 120) = 0$

$y = 30 \quad y = -120$ (reject)

$x = \dfrac{720}{y} = \dfrac{720}{30} = 24$

Merchant purchased and sold 24 blenders.

29. $I = \dfrac{E}{R + \frac{r}{n}} \cdot \dfrac{n}{n}$

$I = \dfrac{En}{Rn + r}$

$(Rn + r)I = (Rn + r)(\dfrac{En}{Rn + r})$

$IRn + Ir = En$

$Ir = En - IRn$

$Ir = (E - IR)n$

$\dfrac{Ir}{E - IR} = n$

30. $(5^a x^{2a} y^{-3a})^{\frac{2}{a}} (x^{3a} y^{-4a})^{-\frac{3}{a}}$

$= 5^2 x^4 y^{-6} x^{-9} y^{12}$

$= 25x^{-5} y^6 = \dfrac{25y^6}{x^5}$

31. $\dfrac{2}{3} \sqrt[3]{\dfrac{27}{4}} = \dfrac{2}{3} \cdot \dfrac{3}{\sqrt[3]{4}} = \dfrac{2}{\sqrt[3]{4}} \cdot \dfrac{\sqrt[3]{4^2}}{\sqrt[3]{4^2}}$

$= \dfrac{2\sqrt[3]{4^2}}{\sqrt[3]{4^3}} = \dfrac{2\sqrt[3]{16}}{4} = \dfrac{\sqrt[3]{16}}{2} = \dfrac{\sqrt[3]{8.2}}{2}$

$= \dfrac{2\sqrt[3]{2}}{2} = \sqrt[3]{2}$

Problem Set 8.4.1

1. (1) $x + y + 2z = 11$

  (2) $x + y + 3z = 14$

  (3) $x + 2y - z = 5$

  $(1) + -1(2)$:

  $x + y + 2z = 11$

  $\underline{-x - y - 3z = -14}$

  $-z = -3$

  $z = 3$

  $(1) + -1(3)$:

  $x + y + 2z = 11$

  $\underline{-x - 2y + z = -5}$

  $-y + 3z = 6$

  $-y + 3(3) = 6$

  $-y = -3$

  $y = 3$

   $x + y + 2z = 11$

  $x + 3 + 6 = 11$

  $x = 2$

  $\{(2, 3, 3)\}$

3. (1) $4x - y + 2z = 11$

  (2) $x + 2y - z = -1$

  (3) $2x + 2y - 3z = -1$

  Eliminate y.

  $2(1) + (2)$: $8x - 2y + 4z = 22$

  $\underline{\qquad x + 2y - z = -1}$

  $9x \qquad + 3z = 21$

  Equivalently : $3x + z = 7$

  $-1(2) + (3)$: $-x - 2y + z = 1$

  $\underline{\qquad 2x + 2y - 3z = -1}$

  $x \qquad - 2z = 0$

  $3x + z = 7$ (multiply by 2)

   $x - 2z = 0$

   $6x + 2z = 14$

   $\underline{x - 2z = 0}$

   Add : $7x = 14$

  $x = 2$

  $3x + z = 7$

  $6 + z = 7$

  $z = 1$

  $4x - y + 2z = 11$

  $4(2) - y + 2(1) = 11$

  $-y + 10 = 11$

  $y = -1$

  $\{(2, -1, 1)\}$

5. (1) $3x + 5y + 2z = 0$

(2) $12x - 15y + 4z = 12$

(3) $6x - 25y - 8z = 8$

Eliminate z.

$-2(1) + (2): \quad -6x - 10y - 4z = 0$

$\underline{\qquad 12x - 15y + 4z = 12 \qquad}$

$\qquad 6x - 25y \qquad = 12$

$2(2) + (3): \quad 24x - 30y + 8z = 24$

$\underline{\qquad 6x - 25y - 8z = 8 \qquad}$

$\qquad 30x - 55y = 32$

$6x - 25y = 12$ (multiply by -5)

$-30x + 125y = -60$

$\underline{30x - 55y = 32}$

$70y = -28$

$y = \dfrac{-28}{70} = \dfrac{-2}{5}$

$6x - 25y = 12$

$6x - 25(-\dfrac{2}{5}) = 12$

$6x + 10 = 12$

$6x = 2$

$x = \dfrac{1}{3}$

$3x + 5y + 2z = 0$

$3(\dfrac{1}{3}) + 5(-\dfrac{2}{5}) + 2z = 0$

$1 - 2 + 2z = 0$

$2z = 1$

$z = \dfrac{1}{2}$

$\{(\dfrac{1}{3}, -\dfrac{2}{5}, \dfrac{1}{2})\}$

9. (1) $2x + y = 2$

(2) $x + y - z = 4$

(3) $3x + 2y + z = 0$

Eliminate z.

(1): $2x + y = 2$ (multiply by -3)

(2) + (3): $4x + 3y = 4$

$-6x - 3y = -6$

$\underline{4x + 3y = 4}$

Add: $-2x = -2$

$x = 1$

$2x + y = 2$

$2 + y = 2$

$y = 0$

$3x + 2y + z = 0$

$3 + 0 + z = 0$

$z = -3$

$\{(1, 0, -3)\}$

11. (1) $x + y = -4$

(2) $y - z = 1$

(3) $2x + y + 3z = -21$

Eliminate z.

(1): $x + y = -4$

$3(2) + (3): 2x + 4y = -18$

Equivalently: $x + 2y = -9$

$x + y = -4$

$\underline{x + 2y = -9}$

$-y = 5$

$y = -5$

$x + y = -4$

$x - 5 = -4$

$x = 1$

$y - z = 1$

$-5 - z = 1$

$-6 = z$

$\{(1, -5, -6)\}$

13. $6x - y + 3z = 9$

$\dfrac{1}{4}x - \dfrac{1}{2}y - \dfrac{1}{3}z = -1$ (multiply by 12)

$-x + \dfrac{1}{6}y - \dfrac{2}{3}z = 0$ (multiply by 6)

(1) $6x - y + 3z = 9$

(2) $3x - 6y - 4z = -12$

(3) $-6x + y - 4z = 0$

Eliminate x.

(1) + (2) : $-z = 9$

$z = -9$

2(2) + (3) : $-11y - 12z = -24$

$-11y - 12(-9) = -24$

$-11y = -132$

$y = 12$

$6x - y + 3z = 9$

$6x - 12 + 3(-9) = 9$

$6x - 39 = 9$

$6x = 48$

$x = 8$

$\{(8, 12, -9)\}$

15. (1) $2x + y + 4z = 4$

(2) $x - y + z = 6$

(3) $x + 2y + 3z = 5$

Eliminate x.

(1) + -2(2) : $3y + 2z = -8$

$-1(2) + (3) :$ $3y + 2z = -1$

$\dfrac{3}{3} = \dfrac{2}{2}$ $\dfrac{-8}{-1}$

System is inconsistent.

$\varnothing$

17. Eliminate x.

$-3(1) + (2) :$ $0 = 0$

$-2(1) + (3) :$ $0 = 0$

System is dependent

$\{(x, y, z) \mid x - 4y + z = -5\}$

19. Eliminate z.

(1): $x + y = 4$

(2) + -1(3): $x - y = 0$

$x + y = 4$

$\underline{x - y = 0}$

Add: $2x = 4$

$x = 2$

$x + y = 4 \quad 2 + y = 4 \quad y = 2$

$x + z = 4 \quad 2 + z = 4 \quad z = 2$

$\{(2, 2, 2)\}$

25. (1) $x + y = a$

(2) $y + z = b$

(3) $x + z = c$

Eliminate z.

(1): $x + y = a$

(2) + -1(3): $y - x = b - c$

$x + y = a$

$\underline{-x + y = b - c}$

Add: $2y = a + b - c$

$y = \dfrac{a + b - c}{2}$

$x + y = a$

$x + \dfrac{a + b - c}{2} = a$

$2x + a + b - c = 2a$

$2x = a - b + c$

$x = \dfrac{a - b + c}{2}$

$y + z = b$

$\dfrac{a + b - c}{2} + z = b$

$a + b - c + 2z = 2b$

$2z = -a + b + c$

$z = \dfrac{-a + b + c}{2}$

$\{(\dfrac{a - b + c}{2}, \dfrac{a + b - c}{2}, \dfrac{-a + b + c}{2})\}$

27. (1) $x - y + 2z - 2w = -1$ (2) $x - y - z + w = -4$

(3) $-x + 2y - 2z - w = -7$

(4) $2x + y + 3z - w = 6$

Eliminate x.

(1) + (3): $y - 3w = -8$ (1)′

(2) + (3): $y - 3z = -11$ (2)′

2(3) + (4): $5y - z - 3w = -8$ (3)′

Now eliminate w.

(2)′: $y - 3z = -11$

$-1(1)′ + (3)′$: $4y - z = 0$ (multiply by -3)

$y - 3z = -11 \qquad -12y + 3z = 0$

Add: $-11y = -11$

$y = 1$

$4y - z = 0$

$4(1) - z = 0$

$4 = z$

$y - 3w = -8$

$1 - 3w = -8$

$-3w = -9$

$w = 3$

$x - y + 2z - 2w = -1$

$x - 1 + 2(4) - 2(3) = -1$

$x + 1 = -1$

$x = -2$

$\{(-2, 1, 4, 3)\}$

29. $0x^2 + 1x + 0 = Ax^2 + (B - 4A)x + (4A - 2B + C)$

Equating coefficients:

$A = 0 \qquad B - 4A = 1$

$4A - 2B + C = 0$

$B - 4A = 1 \qquad B - 4(0) = 1 \qquad B = 1$

$4A - 2B + C = 0$

$4(0) - 2(1) + C = 0 \qquad C = 2$

$A = 0, B = 1, C = 2$

31. $-12x^5 + 9x^3 + 10 = (6B + 4C)x^5$
$+ (3A + 3B)x^3 + (2A - 3C)$

Equating coefficients :

$6B + 4C = -12$ (multiply by $\frac{1}{2}$)

$3A + 3B = 9$ (multiply by $\frac{1}{3}$)

$2A - 3C = 10$

(1) $3B + 2C = -6$

(2) $A + B = 3$

(3) $2A - 3C = 10$

Eliminate A.

(1) : $3B + 2C = -6$ (multiply by 3)

$-2(2) + (3)$: $-2B - 3C = 4$ (multiply by 2)

             $9B + 6C = -18$

             $\underline{-4B - 6C = 8}$

Add : $5B = -10$

          $B = -2$

$A + B = 3$

$A + (-2) = 3$

$A = 5$

$3B + 2C = -6$

$3(-2) + 2C = -6$

$-6 + 2C = -6$

$2C = 0$

$C = 0$

$A = 5, B = -2, C = 0$

32. $10\sqrt{\dfrac{3}{5}} + 8\sqrt{\dfrac{1}{12}} - \dfrac{5}{\sqrt{75}}$

$= \dfrac{10\sqrt{3}}{\sqrt{5}} + \dfrac{8}{2\sqrt{3}} - \dfrac{5}{5\sqrt{3}}$

$= \dfrac{10\sqrt{3}}{\sqrt{5}} \cdot \dfrac{\sqrt{5}}{\sqrt{5}} + \dfrac{4}{\sqrt{3}} \cdot \dfrac{\sqrt{3}}{\sqrt{3}} - \dfrac{1}{\sqrt{3}} \cdot \dfrac{\sqrt{3}}{\sqrt{3}}$

$= \dfrac{10\sqrt{15}}{5} + \dfrac{4\sqrt{3}}{3} - \dfrac{\sqrt{3}}{3}$

$= 2\sqrt{15} + \dfrac{3\sqrt{3}}{3}$

$= 2\sqrt{15} + \sqrt{3}$

33. $\sqrt{\dfrac{x}{y} + \dfrac{y}{x} + 2} - \sqrt{\dfrac{y}{x}}$

$= \sqrt{\dfrac{x^2 + y^2 + 2xy}{xy}} - \sqrt{\dfrac{y}{x}}$

$= \sqrt{\dfrac{(x + y)^2}{xy}} - \sqrt{\dfrac{y}{x}}$

$= \dfrac{(x + y)}{\sqrt{xy}} - \dfrac{\sqrt{y}}{\sqrt{x}} \cdot \dfrac{\sqrt{y}}{\sqrt{y}}$

$= \dfrac{x + y - y}{\sqrt{xy}}$

$= \dfrac{x}{\sqrt{xy}} \cdot \dfrac{\sqrt{xy}}{\sqrt{xy}}$

$= \dfrac{x\sqrt{xy}}{xy}$

$= \dfrac{\sqrt{xy}}{y}$

34. $y < x + 1$

For $y = x + 1$: y-intercept $= 1$ slope $= 1$

Test $(0, 0)$

$y < x + 1$

$0 < 0 + 1$

$0 < 1$ True

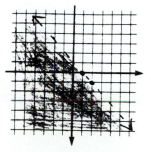

Problem Set 8.4.2

3. x : number of nickels

   y : number of dimes

   z : number of quarters

   $x + y + z = 17$

   $15x + 10y + 25z = 200$ (multiply by $\frac{1}{5}$)

   $x + y = z + 9$

   (1) $x + y + z = 17$

   (2) $x + 2y + 5z = 40$

   (3) $x + y - z = 9$

   Eliminate x.

   (1) - (2) : $-y - 4z = -23$

   (1) - (3) : $2z = 8$

   $z = 4$

   $-y - 4z = -23$

   $-y - 4(4) = -23$

   $7 = y$

   $x + y + z = 17$

   $x + 7 + 4 = 17$

   $x = 6$

   Number of nickels : $x = 6$ dimes : $y = 7$

   quarters : $z = 4$

5. $y = ax^2 + bx + c$

   $(-1, 1) : 1 = a(-1)^2 + b(-1) + c$

   $(3, 1) : 1 = a(3)^2 + b(3) + c$

   $(4, -4) : -4 = a(4)^2 + b(4) + c$

   (1) $a - b + c = 1$

   (2) $9a + 3b + c = 1$

   (3) $16a + 4b + c = -4$

   Eliminate c.

   (2) - (1) : $8a + 4b = 0$ (multiply by $\frac{1}{4}$)

   (3) - (1) : $15a + 5b = -5$ (multiply by $\frac{1}{5}$)

   $2a + b = 0$

   $\underline{3a + b = -1}$

   $-a = 1$

   $a = -1$

   $8a + 4b = 0$

   $8(-1) + 4b = 0$

   $4b = 8$

   $b = 2$

   $a - b + c = 1$

   $-1 - 2 + c = 1$

   $-3 + c = 1$

   $c = 4$

   $y = ax^2 + bx + c$

   $y = -x^2 + 2x + 4$

13. x : first angle

y : second angle

z : third angle

$x + y + z = 180$

$x + y = z$

$x - y = \dfrac{2}{3}z$ (multiply by 3)

(1) $x + y + z = 180$

(2) $x + y - z = 0$

(3) $3x - 3y - 2z = 0$

Eliminate z.

(1) + (2) : $2x + 2y = 180$ (multiply by $\dfrac{1}{2}$)

2(1) + (3) : $5x - y = 360$

$x + y = 90$

$\underline{5x - y = 360}$

Add : $6x = 450$

$x = 75$

$x + y = 90$

$75 + y = 90$

$y = 15$

$x + y + z = 180$

$75 + 15 + z = 180$

$90 + z = 180$

$z = 90$

Angles measure $75°$, $15°$, and $90°$.

15. x : rare birds (three legs each)

y : canaries

z : puppies

Pets have 20 heads : $x + y + z = 20$

Pets have 56 legs : $3x + 2y + 4z = 56$

$z = x + y - 10$

(1) $x + y + z = 20$

(2) $3x + 2y + 4z = 56$

(3) $x + y - z = 10$

Eliminate z.

(1) + (3) : $2x + 2y = 30$ (multiply by -3)

4(3) + (2) : $7x + 6y = 96$

$-6x - 6y = -90$

$\underline{7x + 6y = 96}$

Add : $x = 6$

$2x + 2y = 30$

$2(6) + 2y = 30$

$2y = 18$

$y = 9$

$x + y + z = 20$

$6 + 9 + z = 20$

$z = 5$

6 rare birds, 9 canaries, 5 puppies

19. x : amount invested at 10%

    y : amount invested at 12%

    z : amount invested at 15%

    $x + y + z = 17,000$

    $.10x + .12y + .15z = 2110$ (multiply by 10)

    $y = x + z - 1,000$

    (1) $x + y + z = 17,000$

    (2) $10x + 12y + 15z = 211,000$

    (3) $x - y + z = 1,000$

    Eliminate y.

    $(1) + (3)$: $2x + 2z = 18,000$

    $\quad -12(1) + (2)$: $-2x + 3z = 7,000$

    ──────────────────────

    Add : $5z = 25,000$

    $z = 5,000$

    $2x + 2z = 18,000$

    $2x + 2(5,000) = 18,000$

    $2x + 10,000 = 18,000$

    $2x = 8,000$

    $x = 4,000$

    $x + y + z = 17,000$

    $4,000 + y + 5,000 = 17,000$

    $y = 8,000$

    $4,000$ at 10%, $8,000$ at 12%,

        $5,000$ at 15%

23. Truck I : x cubic yards

    Truck II : y cubic yards

    Truck III : z cubic yards

    Day 1 : $4x + 3y + 5z = 78$ (1)

    Day 2 : $5x + 4y + 4z = 81$ (2)

    Day 3 : $3x + 5y + 3z = 69$ (3)

    Eliminate z.

    $3(2) + -4(3)$: $3x - 8y = -33$

    $4(1) + -5(2)$: $-9x - 8y = -93$

    ──────────────────────

                 $12x = 60$

    $x = 5$

    $3x - 8y = -33$

    $3(5) - 8y = -33$

    $-8y = -48$

    $y = 6$

    $4x + 3y + 5z = 78$

    $4(5) + 3(6) + 5z = 78$

    $5z = 40$ $z = 8$

    Truck I : $5yd^3$ Truck II : $6yd^3$

        Truck III : $8yd^3$

25. x : number of adult tickets

    y : number of tickets for adult with 1 child

    z : number of tickets for adult with 2

        children

    $415 was collected : $20x + 30y + 35z$

            $= 415$ (multiply by $\frac{1}{5}$)

    28 passenger total : $x + 2y + 3z = 28$

    15 adults : $x + y + z = 15$

    (1) $4x + 6y + 7z = 83$

    (2) $x + 2y + 3z = 28$

    (3) $x + y + z = 15$

    Eliminate x.

    $(2) - (3)$: $y + 2z = 13$ (multiply by -2)

    $-4(3) + (1)$: $2y + 3z = 23$

    $-2y - 4z = -26$

    $2y + 3z = 23$

    ──────────────────────

    Add : $-z = -3$

    $z = 3$

    $y + 2z = 13$

    $y + 2(3) = 13$

    $y = 7$

    $x + y + z = 15$

    $x + 7 + 3 = 15$

    $x = 5$

    5 adult tickets, 7 adult with one child,

    3 adult with 2 children

27. Slope of line through $(4, 2)$ and $(-2, 3)$:

    $m = \dfrac{3 - 2}{-2 - 4} = \dfrac{1}{-6}$

    slope of desired line $= 6$ (negative

    reciprocal of $-\dfrac{1}{6}$)

    $(x_1, y_1) = (1, 5)$

    $m = 6$

    $y - y_1 = m(x - x_1)$

    $y - 5 = 6(x - 1)$

    $y - 5 = 6x - 6$

    $y = 6x - 1$ slope - intercept form

28. $\dfrac{7\sqrt{2}}{2\sqrt{2}-1} \cdot \dfrac{2\sqrt{2}+1}{2\sqrt{2}+1} + \dfrac{\sqrt{2}}{\sqrt{2}-1} \cdot \dfrac{\sqrt{2}+1}{\sqrt{2}+1}$

$= \dfrac{28+7\sqrt{2}}{8-1} + \dfrac{2+\sqrt{2}}{2-1}$

$= \dfrac{28+7\sqrt{2}}{7} + 2 + \sqrt{2}$

$= 4 + \sqrt{2} + 2 + \sqrt{2}$

$= 6 + 2\sqrt{2}$

29. $\dfrac{3}{y+1} - \dfrac{5}{y} = \dfrac{19}{y^2+y}$

$y(y+1)\left[\dfrac{3}{y+1} - \dfrac{5}{y}\right] = y(y+1)\left[\dfrac{19}{y(y+1)}\right]$

$3y - 5(y+1) = 19$

$3y - 5y - 5 = 19$

$-2y = 24$

$y = -12$

$\{-12\}$

Problem Set 8.5.1

1. $\begin{vmatrix} 5 & 7 \\ 2 & 3 \end{vmatrix} = (5)(3) - (2)(7) = 15 - 14 = 1$

7. $\begin{vmatrix} -5 & -1 \\ -2 & -7 \end{vmatrix} = (-5)(-7) - (-2)(-1) = 35 - 2 = 33$

9. $\begin{vmatrix} \frac{1}{2} & \frac{1}{2} \\ \frac{1}{8} & -\frac{3}{4} \end{vmatrix} = (\frac{1}{2})(-\frac{3}{4}) - (\frac{1}{8})(\frac{1}{2}) = -\frac{3}{8} - \frac{1}{16} = -\frac{6}{16} - \frac{1}{16}$

$= \dfrac{-7}{16}$

11. $x + y = 7$

$x - y = 3$

$D = \begin{vmatrix} 1 & 1 \\ 1 & -1 \end{vmatrix} = -1 - 1 = -2$

$D_x = \begin{vmatrix} 7 & 1 \\ 3 & -1 \end{vmatrix} = -7 - 3 = -10$

$D_y = \begin{vmatrix} 1 & 7 \\ 1 & 3 \end{vmatrix} = 3 - 7 = -4$

$x = \dfrac{D_x}{D} = \dfrac{-10}{-2} = 5$

$y = \dfrac{D_y}{D} = \dfrac{-4}{-2} = 2$

$\{(5, 2)\}$

17. $x + 2y = 3$

$5x + 10y = 15$

$D = \begin{vmatrix} 1 & 2 \\ 5 & 10 \end{vmatrix} = 10 - 10 = 0$

$D_x = \begin{vmatrix} 3 & 2 \\ 15 & 10 \end{vmatrix} = 30 - 30 = 0$

$D_y = \begin{vmatrix} 1 & 3 \\ 5 & 15 \end{vmatrix} = 15 - 15 = 0$

Since $D_x = D_y = D = 0$,

the system is dependent.

$\{(x, y) \mid x + 2y = 3\}$

21. $2x = 3y + 2$

$5x = 51 - 4y$

$2x - 3y = 2$

$5x + 4y = 51$

$D = \begin{vmatrix} 2 & -3 \\ 5 & 4 \end{vmatrix} = 8 + 15 = 23$

$D_x = \begin{vmatrix} 2 & -3 \\ 51 & 4 \end{vmatrix} = 8 + 153 = 161$

$D_y = \begin{vmatrix} 2 & 2 \\ 5 & 51 \end{vmatrix} = 102 - 10 = 92$

$x = \dfrac{161}{23} = 7 \quad y = \dfrac{92}{23} = 4$

$\{(7, 4)\}$

23. $3x = 2 - 3y$

$2y = 3 - 2x$

$3x + 3y = 2$

$2x + 2y = 3$

$D = \begin{vmatrix} 3 & 3 \\ 3 & 2 \end{vmatrix} = 6 - 6 = 0$

$D_x = \begin{vmatrix} 2 & 3 \\ 3 & 2 \end{vmatrix} = 4 - 9 = -5$

Since $D = 0$ and $D_x \ne 0$, the system

is inconsistent.

$\varnothing$

27. $3x + 2y = 4$

$x = 5$

$3x + 2y = 4$

$1x - 0y = 5$

$D = \begin{vmatrix} 3 & 2 \\ 1 & 0 \end{vmatrix} = 0 - 2 = -2$

$D_x = \begin{vmatrix} 4 & 2 \\ 5 & 0 \end{vmatrix} = 0 - 10 = -10$

$D_y = \begin{vmatrix} 3 & 4 \\ 1 & 5 \end{vmatrix} = 15 - 4 = 11$

$x = \dfrac{-10}{-2} = 5 \quad y = \dfrac{11}{-2} = -\dfrac{11}{2}$

$\{(5, -\dfrac{11}{2})\}$

29. $\begin{vmatrix} 3 & 2 \\ y & 4 \end{vmatrix} = 0$

$(3)(4) - (y)(2) = 0$

$12 - 2y = 0$

$12 = 2y$

$6 = y$

$\{6\}$

35. $\begin{vmatrix} a_1 & Kb_1 \\ a_2 & Kb_2 \end{vmatrix} = Ka_1\,b_2 - Ka_2\,b_1$

$K\begin{vmatrix} a_1 & b_1 \\ a_2 & b_2 \end{vmatrix} = K(a_1\,b_2 - a_2\,b_1) = Ka_1\,b_2 - Ka_2\,b_1$

41. $a_1\,x + b_1\,y = c_1$

$a_2\,x + b_2\,y = c_2$

$x = \dfrac{\begin{vmatrix} c_1 & b_1 \\ c_2 & b_2 \end{vmatrix}}{\begin{vmatrix} a_1 & b_1 \\ a_2 & b_2 \end{vmatrix}}$

$y = \dfrac{\begin{vmatrix} a_1 & c_1 \\ a_2 & c_2 \end{vmatrix}}{\begin{vmatrix} a_1 & b_1 \\ a_2 & b_2 \end{vmatrix}}$

$Ka_1\,x + Kb_1\,y = Kc_1$

$a_2\,x + b_2\,y = c_2$

$x = \dfrac{\begin{vmatrix} Kc_1 & Kb_1 \\ c_2 & b_2 \end{vmatrix}}{\begin{vmatrix} Ka_1 & Kb_1 \\ a_2 & b_2 \end{vmatrix}} = \dfrac{K\begin{vmatrix} c_1 & b_1 \\ c_2 & b_2 \end{vmatrix}}{K\begin{vmatrix} a_1 & b_1 \\ a_2 & b_2 \end{vmatrix}} = \dfrac{\begin{vmatrix} c_1 & b_1 \\ c_2 & b_2 \end{vmatrix}}{\begin{vmatrix} a_1 & b_1 \\ a_2 & b_2 \end{vmatrix}}$

$y = \dfrac{\begin{vmatrix} Ka_1 & Kc_1 \\ a_2 & c_2 \end{vmatrix}}{\begin{vmatrix} Ka_1 & Kb_1 \\ a_2 & b_2 \end{vmatrix}} = \dfrac{K\begin{vmatrix} a_1 & c_1 \\ a_2 & b_2 \end{vmatrix}}{K\begin{vmatrix} a_1 & b_1 \\ a_2 & b_2 \end{vmatrix}} = \dfrac{\begin{vmatrix} a_1 & c_1 \\ a_2 & c_2 \end{vmatrix}}{\begin{vmatrix} a_1 & b_1 \\ a_2 & b_2 \end{vmatrix}}$

Both systems have identical values for x and y.

43. $y = m_1\,x + d_1$

$y = m_1\,x + d_2$

a, Lines are parallel because they have equal slopes $(m_1 = m_2)$ and different y-intercepts $(d_1 \neq d_2)$.

b. $-m_1\,x + y = d_1$

$-m_1\,x + y = d_2$

$\begin{vmatrix} a_1 & b_1 \\ a_2 & b_2 \end{vmatrix} = \begin{vmatrix} -m_1 & 1 \\ -m_1 & 1 \end{vmatrix} = -m_1 + m_1 = 0$

45. $6\sqrt{\dfrac{1}{3}} + \dfrac{5}{2}\sqrt{108} + 2\sqrt[4]{9}$

$= \dfrac{6}{\sqrt{3}} \cdot \dfrac{\sqrt{3}}{\sqrt{3}} + \dfrac{5}{2}\sqrt{36 \cdot 3} + 2(3^2)^{\frac{1}{4}}$

$= \dfrac{6\sqrt{3}}{3} + \dfrac{5}{2}(6\sqrt{3}) + 2(3)^{\frac{1}{2}}$

$= 2\sqrt{3} + 15\sqrt{3} + 2\sqrt{3}$

$= 19\sqrt{3}$

46. $x^2 + y^2 - 4x + 6y + 9 = 0$

$x^2 - 4x \,\underline{\quad}\, + y^2 + 6y \,\underline{\quad}\, = -9$

$\dfrac{1}{2}(-4) = -2 \qquad\qquad \dfrac{1}{2}(6) = 3$

$(-2)^2 = 4 \qquad\qquad 3^2 = 9$

$x^2 - 4x + 4 + y^2 + 6y + 9 = -9 + 4 + 9$

$(x - 2)^2 + (y + 3)^2 = 4$

$(x - h)^2 + (y - k)^2 = r^2$

Center : $(h, k) = (2, -3)$

radius : 2

47. Denominator : x

Numerator : x + 1

Fraction : $\dfrac{x + 1}{x}$

$\dfrac{x + 1}{x} + \dfrac{x}{x + 1} = \dfrac{61}{30}$

$30x(x + 1)\left[\dfrac{x + 1}{x} + \dfrac{x}{x + 1}\right] = 30x(x + 1)\left[\dfrac{61}{30}\right]$

$30x^2 + 60x + 30 + 30x^2 = 61x^2 + 61x$

$60x^2 + 60x + 30 = 61x^2 + 61x$

$0 = x^2 + x - 30$

$0 = (x + 6)(x - 5)$

x = -6 (reject; we are told fraction
         is positive)

x = 5

Fraction : $\dfrac{x + 1}{x} = \dfrac{6}{5}$

Problem Set 8.5.2

1.
$$\begin{vmatrix} 1 & -1 & 2 \\ 2 & 1 & 3 \\ 0 & -2 & 1 \end{vmatrix} = 1\begin{vmatrix} 1 & 3 \\ -2 & 1 \end{vmatrix} - 2\begin{vmatrix} -1 & 2 \\ -2 & 1 \end{vmatrix} + 0\begin{vmatrix} -1 & 2 \\ 1 & 3 \end{vmatrix}$$
$$= 1(1+6) - 2(-1+4)$$
$$= 7 - 6 = 1$$

5. Expand about the elements in the last column.
$$\begin{vmatrix} 4 & 1 & 0 \\ 1 & -1 & -1 \\ -2 & -1 & 0 \end{vmatrix} = 1\begin{vmatrix} 4 & 1 \\ -2 & -1 \end{vmatrix} = -4 + 2 = -2$$

9. Expand about the elements in the middle column.
$$\begin{vmatrix} -4 & 0 & 3 \\ 6 & -2 & -1 \\ 8 & 0 & 4 \end{vmatrix} = -2\begin{vmatrix} -4 & 3 \\ 8 & 4 \end{vmatrix} = -2(-16-24)$$
$$= -2(-40) = 80$$

11. Expand about elements in the first row.
$$\begin{vmatrix} 2 & 0 & 0 \\ 3 & -1 & -1 \\ 2 & -2 & 4 \end{vmatrix} = 2\begin{vmatrix} -1 & -1 \\ -2 & 4 \end{vmatrix} = 2(-4-2) = -12$$

13.
$$\begin{vmatrix} a_1 & a_1 & c_1 \\ a_2 & a_2 & c_2 \\ a_3 & a_3 & c_3 \end{vmatrix} = a_1\begin{vmatrix} a_2 & c_2 \\ a_3 & c_3 \end{vmatrix} - a_1\begin{vmatrix} a_2 & c_2 \\ a_3 & c_3 \end{vmatrix} + c_1\begin{vmatrix} a_2 & a_2 \\ a_3 & a_3 \end{vmatrix}$$
$$= a_1(a_2 c_3 - a_3 c_2) - a_1(a_2 c_3 - a_3 c_2)$$
$$\quad + c_1(a_2 a_3 - a_3 a_2)$$
$$= a_1 a_2 c_3 - a_1 a_3 c_2 - a_1 a_2 c_3 + a_1 a_3 c_2 + c_1(0)$$
$$= 0$$

15.
$$\begin{vmatrix} Ka_1 & b_1 & c_1 \\ Ka_2 & b_2 & c_2 \\ Ka_3 & b_3 & c_3 \end{vmatrix} = Ka_1\begin{vmatrix} b_2 & c_2 \\ b_3 & c_2 \end{vmatrix} - Ka_2\begin{vmatrix} b_1 & c_1 \\ b_3 & c_3 \end{vmatrix}$$
$$\quad + Ka_3\begin{vmatrix} b_1 & c_1 \\ b_2 & c_2 \end{vmatrix}$$
$$= K\left[a_1\begin{vmatrix} b_2 & c_2 \\ b_3 & c_2 \end{vmatrix} - a_2\begin{vmatrix} b_1 & c_1 \\ b_3 & c_3 \end{vmatrix} + a_3\begin{vmatrix} b_1 & c_1 \\ b_2 & c_2 \end{vmatrix} \right]$$
$$= K\begin{vmatrix} a_1 & b_1 & c_1 \\ a_2 & b_2 & c_2 \\ a_3 & b_3 & c_3 \end{vmatrix}$$

17.
$$\begin{vmatrix} a_1 & b_1 + Ka_1 & c_1 \\ a_2 & b_2 + Ka_2 & c_2 \\ a_3 & b_3 + Ka_3 & c_3 \end{vmatrix}$$
$$= a_1\begin{vmatrix} b_2 + Ka_2 & c_2 \\ b_3 + Ka_3 & c_3 \end{vmatrix} - a_2\begin{vmatrix} b_1 + Ka_1 & c_1 \\ b_3 + Ka_3 & c_3 \end{vmatrix}$$
$$\quad + a_3\begin{vmatrix} b_1 + Ka_1 & c_1 \\ b_2 + Ka_2 & c_2 \end{vmatrix}$$
$$= a_1(b_2 c_3 + Ka_2 c_3 - b_3 c_2 - Ka_3 c_2)$$
$$- a_2(b_1 c_3 + Ka_1 c_3 - b_3 c_1 - Ka_3 c_1)$$
$$+ a_3(b_1 c_2 + Ka_1 c_2 - b_2 c_1 - Ka_2 c_1)$$
$$= a_1 b_2 c_3 + Ka_1 a_2 c_3 - a_1 b_3 c_2 - Ka_1 a_3 c_2$$
$$- a_2 b_1 c_3 - Ka_1 a_2 c_3 + a_2 b_3 c_1 - Ka_2 a_3 c_1$$
$$+ a_3 b_1 c_2 + Ka_1 a_3 c_2 - a_3 b_2 c_1 - Ka_2 a_3 c_1$$
$$= a_1 b_2 c_3 - a_1 b_3 c_2 - a_2 b_1 c_3 + a_2 b_3 c_1$$
$$\quad + a_3 b_1 c_2 - a_3 b_2 c_1$$
$$= \begin{vmatrix} a_1 & b_1 & c_1 \\ a_2 & b_2 & c_2 \\ a_3 & b_3 & c_3 \end{vmatrix}$$

19.
$$\begin{vmatrix} 3 & 5 & -2 \\ 1 & 2 & -1 \\ x & -3 & 4x \end{vmatrix} = -12$$
$$3\begin{vmatrix} 2 & -1 \\ -3 & 4x \end{vmatrix} - 5\begin{vmatrix} 1 & -1 \\ x & 4x \end{vmatrix} - 2\begin{vmatrix} 1 & 2 \\ x & -3 \end{vmatrix} = -12$$
$$3(8x-3) - 5(4x+x) - 2(-3-2x) = -12$$
$$24x - 9 - 25x + 6 + 4x = -12$$
$$3x - 3 = -12$$
$$3x = -9$$
$$x = -3$$

21.
$$\begin{vmatrix} x-1 & x & -1 \\ x & x+1 & 1 \\ x+2 & x-1 & 2 \end{vmatrix} = 15$$
$$(x-1)\begin{vmatrix} x+1 & 1 \\ x-1 & 2 \end{vmatrix} - x\begin{vmatrix} x & -1 \\ x-1 & 2 \end{vmatrix} + (x+2)\begin{vmatrix} x & -1 \\ x+1 & 1 \end{vmatrix}$$
$$= 15$$
$$(x-1)[2(x+1) - (x-1)] - x[2x + (x-1)]$$
$$\quad + (x+2)[x + (x+1)] = 15$$
$$(x-1)(x+3) - x(3x-1) + (x+2)(2x+1) = 15$$
$$x^2 + 2x - 3 - 3x^2 + x + 2x^2 + 5x + 2 = 15$$
$$8x - 1 = 15 \quad\quad 8x = 16 \quad\quad x = 2$$

23 a. The equation of the line is

$$y - y_1 = \frac{y_2 - y_1}{x_2 - x_1}(x - x_1).$$

Multiply both sides by $(x_2 - x_1)$.

$$(x_2 - x_1)(y - y_1) = (y_2 - y_1)(x - x_1)$$

$$x_2 y - x_2 y_1 - x_1 y + x_1 y_1 = y_2 x - y_2 x_1 - xy_1 + x_1 y_1$$

$$x_2 y - x_2 y_1 - x_1 y = y_2 x - y_2 x_1 - xy_1 \quad (i)$$

We must now show that the equation involving the determinant is equivalent to the above equation, namely equation (i).

$$\begin{vmatrix} x & y & 1 \\ x_1 & y_1 & 1 \\ x_2 & y_2 & 1 \end{vmatrix} = 0$$

Expanding about the third column :

$$1\begin{vmatrix} x_1 & y_1 \\ x_2 & y_2 \end{vmatrix} - 1\begin{vmatrix} x & y \\ x_2 & y_2 \end{vmatrix} + 1\begin{vmatrix} x & y \\ x_1 & y_1 \end{vmatrix} = 0$$

$$x_1 y_2 - x_2 y_1 - xy_2 + yx_2 + xy_1 - yx_1 = 0$$

We can rearrange the terms to show that this equation is identical to equation (i).

$$x_2 y - x_2 y_1 - x_1 y = y_2 x - y_2 x_1 - xy_1$$

b.

$$\begin{vmatrix} x & y & 1 \\ 3 & 2 & 1 \\ -1 & 0 & 1 \end{vmatrix} = 0$$

$$x\begin{vmatrix} 2 & 1 \\ 0 & 1 \end{vmatrix} - 3\begin{vmatrix} y & 1 \\ 0 & 1 \end{vmatrix} - 1\begin{vmatrix} y & 1 \\ 2 & 1 \end{vmatrix} = 0$$

$$x(2 - 0) - 3(y - 0) - 1(y - 2) = 0$$

$$2x - 3y - y + 2 = 0$$

$$2x - 4y + 2 = 0$$

$$2x + 2 = 4y$$

$$y = \frac{2x + 2}{4}$$

$$y = \frac{1}{2}x + \frac{1}{2}$$

24. $\frac{1}{2}n(n + 1) = 78$

$$n(n + 1) = 156$$

$$n^2 + n - 156 = 0$$

$$(n - 12)(n + 13) = 0$$

$$n = 12 \quad n = -13 \text{ (reject)}$$

12 consecutive integers must be added.

25. $\dfrac{x + 2}{x - 2} \geq 6$

$$\frac{x + 2}{x - 2} - 6 \geq 0$$

$$\frac{x + 2}{x - 2} - \frac{6(x - 2)}{(x - 2)} \geq 0$$

$$\frac{x + 2 - 6x + 12}{x - 2} \geq 0$$

$$\frac{-5x + 14}{x - 2} \geq 0$$

$$-5x + 14 = 0 \quad x - 2 = 0$$

$$x = \frac{14}{5} \quad x = 2$$

Test 0 : $\dfrac{x + 2}{x - 2} \geq 6$

$$\frac{0 + 2}{0 - 2} \geq 6$$

$$-1 \geq 6 \text{ False}$$

Test $2\frac{1}{2}$: $\dfrac{2\frac{1}{2} + 2}{2\frac{1}{2} - 2} \geq 6$

$$\frac{\frac{9}{2}}{\frac{1}{2}} \geq 6$$

$$9 \geq 6 \text{ True}$$

Test 3 : $\dfrac{3 + 2}{3 - 1} \geq 6$

$$\frac{5}{2} \geq 6 \text{ False}$$

$$\{x \mid 2 < x \leq \frac{14}{5}\}$$

26. $y^2 - 7y + 12 + ay - 4a$

$$= (y - 4)(y - 3) + a(y - 4)$$

$$= (y - 4)(y - 3 + a)$$

Problem Set 8.5.3

1. $x + y + 2z = 11$

$x + y + 3z = 14$

$x + 2y - z = 5$

$$D = \begin{vmatrix} 1 & 1 & 2 \\ 1 & 1 & 3 \\ 1 & 2 & -1 \end{vmatrix} = -1$$

$$Dx = \begin{vmatrix} 11 & 1 & 2 \\ 14 & 1 & 3 \\ 5 & 2 & -1 \end{vmatrix} = -2$$

$$Dy = \begin{vmatrix} 1 & 11 & 2 \\ 1 & 14 & 3 \\ 1 & 5 & -1 \end{vmatrix} = -3$$

$$Dz = \begin{vmatrix} 1 & 1 & 11 \\ 1 & 1 & 14 \\ 1 & 2 & 5 \end{vmatrix} = -3$$

$x = \dfrac{Dx}{D} = \dfrac{-2}{-1} = 2 \quad y = \dfrac{Dy}{D} = \dfrac{-3}{-1} = 3$

$z = \dfrac{Dz}{D} = \dfrac{-3}{-1} = 3$

$\{(2, 3, 3)\}$

5. $3x + 5y + 2z = 0$

$12x - 15y + 4y = 12$

$6x - 25y - 8z = 8$

$$D = \begin{vmatrix} 3 & 5 & 2 \\ 12 & -15 & 4 \\ 6 & -25 & -8 \end{vmatrix} = 840$$

$$Dx = \begin{vmatrix} 0 & 5 & 2 \\ 12 & -15 & 4 \\ 8 & -25 & -8 \end{vmatrix} = 280$$

$$Dy = \begin{vmatrix} 3 & 0 & 2 \\ 12 & 12 & 4 \\ 6 & 8 & -8 \end{vmatrix} = -336$$

$$Dz = \begin{vmatrix} 3 & 5 & 0 \\ 12 & -15 & 12 \\ 6 & -25 & 8 \end{vmatrix} = 420$$

$x = \dfrac{Dx}{D} = \dfrac{280}{840} = \dfrac{1}{3}$

$y = \dfrac{Dy}{D} = \dfrac{-336}{840} = \dfrac{-2}{5}$

$z = \dfrac{Dz}{D} = \dfrac{420}{840} = \dfrac{1}{2}$

$\{(\dfrac{1}{3}, -\dfrac{2}{5}, \dfrac{1}{2})\}$

11. $x + y + 0z = -4$

$0x + y - z = 1$

$2x + y + 3z = -21$

$$D = \begin{vmatrix} 1 & 1 & 0 \\ 0 & 1 & -1 \\ 2 & 1 & 3 \end{vmatrix} = 2$$

$$Dx = \begin{vmatrix} -4 & 1 & 0 \\ 1 & 1 & -1 \\ -21 & 1 & 3 \end{vmatrix} = 2$$

$$Dy = \begin{vmatrix} 1 & -4 & 0 \\ 0 & 1 & -1 \\ 2 & -21 & 3 \end{vmatrix} = -10$$

$$Dz = \begin{vmatrix} 1 & 1 & -4 \\ 0 & 1 & 1 \\ 2 & 1 & -21 \end{vmatrix} = -12$$

$x = \dfrac{Dx}{D} = \dfrac{2}{2} = 1$

$y = \dfrac{Dy}{D} = \dfrac{-10}{2} = -5$

$z = \dfrac{Dz}{D} = \dfrac{-12}{2} = -6$

$\{(1, -5, -6)\}$

13. Multiply the second equation by 12 and the third by 6

$6x - y + 3z = 9$

$3x - 6y - 4z = -12$

$-6x + y - 4z = 0$

$$D = \begin{vmatrix} 6 & -1 & 3 \\ 3 & -6 & -4 \\ -6 & 1 & -4 \end{vmatrix} = 33$$

$$Dx = \begin{vmatrix} 9 & -1 & 3 \\ -12 & -6 & -4 \\ 0 & 1 & -4 \end{vmatrix} = 264$$

$$Dy = \begin{vmatrix} 6 & 9 & 3 \\ 3 & -12 & -4 \\ -6 & 0 & -4 \end{vmatrix} = 396$$

$$Dz = \begin{vmatrix} 6 & -1 & 9 \\ 3 & -6 & -12 \\ -6 & 1 & 0 \end{vmatrix} = -297$$

$x = \dfrac{Dx}{D} = \dfrac{264}{33} = 8 \quad y = \dfrac{Dy}{D} = \dfrac{396}{33} = 12$

$z = \dfrac{Dz}{D} = \dfrac{-297}{33} = -9$

$\{(8, 12, -9)\}$

15. $D = \begin{vmatrix} 2 & 1 & 4 \\ 1 & -1 & 1 \\ 1 & 2 & 3 \end{vmatrix} = 0$

$Dx = \begin{vmatrix} 4 & 1 & 4 \\ 6 & -1 & 1 \\ 5 & 2 & 3 \end{vmatrix} = 35$

Since $D = 0$ and $Dx \ne 0$, the system is inconsistent.

$\varnothing$

17. $D = \begin{vmatrix} 1 & -4 & 1 \\ 3 & -12 & 3 \\ 2 & -8 & 2 \end{vmatrix} = 0$

$Dx = \begin{vmatrix} -5 & -4 & 1 \\ -15 & -12 & 3 \\ -10 & -8 & 2 \end{vmatrix} = 0$

$Dy = \begin{vmatrix} 1 & -5 & 1 \\ 3 & -15 & 3 \\ 2 & -10 & 2 \end{vmatrix} = 0$

$Dz = \begin{vmatrix} 1 & -4 & -5 \\ 3 & -12 & -15 \\ 2 & -8 & -10 \end{vmatrix} = 0$

Since $D = Dx = Dy = Dz = 0$, the system is dependent. $\{(x, y, z) \mid x - 4y + z = -5\}$

19. $x + y + 0z = 4$
$x + 0y + z = 4$
$0x + y + z = 4$

$D = \begin{vmatrix} 1 & 1 & 0 \\ 1 & 0 & 1 \\ 0 & 1 & 1 \end{vmatrix} = -2$

$Dx = \begin{vmatrix} 4 & 1 & 0 \\ 4 & 0 & 1 \\ 4 & 1 & 0 \end{vmatrix} = -4$

$Dy = \begin{vmatrix} 1 & 4 & 1 \\ 1 & 4 & 1 \\ 0 & 4 & 1 \end{vmatrix} = -4$

$Dz = \begin{vmatrix} 1 & 1 & 4 \\ 1 & 0 & 4 \\ 0 & 1 & 4 \end{vmatrix} = -4$

$x = \dfrac{-4}{-2} = 2 \quad y = \dfrac{-4}{-2} = 2 \quad z = \dfrac{-4}{-2} = 2$

$\{(2, 2, 2)\}$

25. $x + y + 0z = a$
$0x + y + z = b$
$x + 0y + z = c$

$D = \begin{vmatrix} 1 & 1 & 0 \\ 0 & 1 & 1 \\ 1 & 0 & 1 \end{vmatrix} = 2 \quad Dx = \begin{vmatrix} a & 1 & 0 \\ b & 1 & 1 \\ c & 0 & 1 \end{vmatrix} = a - b + c$

$Dy = \begin{vmatrix} 1 & a & 0 \\ 0 & b & 1 \\ 1 & c & 1 \end{vmatrix} = a + b - c$

$Dz = \begin{vmatrix} 1 & 1 & a \\ 0 & 1 & b \\ 1 & 0 & c \end{vmatrix} = -a + b + c$

$\left\{ \left(\dfrac{a-b+c}{2}, \dfrac{a+b-c}{2}, \dfrac{-a+b+c}{2} \right) \right\}$

26. x: number of years after 1990
$4000 + 600x = 8000 - 400x$
$1000x = 4000$
$x = 4$
Same population in $1990 + 4 = 1994$

27.

$$y^2 + 5 \overline{\smash{\big)}\, 3y^3 - 2y^2 + y + 1} \qquad 3y - 2 + \dfrac{11 - 14y}{y^2 + 5}$$

$\underline{3y^3 \qquad + 15y}$

$-2y^2 - 14y + 1$

$\underline{-2y^2 \qquad -10}$

$-14y + 11$

28. $\left(x - 2 - \dfrac{4}{x+1} \right) \div \left(x - 1 - \dfrac{3}{x+1} \right)$

$= \left[\dfrac{x(x+1)}{(x+1)} - \dfrac{2(x+1)}{(x+1)} - \dfrac{4}{x+1} \right]$

$\div \left[\dfrac{x(x+1)}{(x+1)} - \dfrac{1(x+1)}{(x+1)} - \dfrac{3}{x+1} \right]$

$= \left(\dfrac{x^2 + x - 2x - 2 - 4}{x+1} \right) \div \left(\dfrac{x^2 + x - x - 1 - 3}{x+1} \right)$

$= \dfrac{x^2 - x - 6}{x+1} \div \dfrac{x^2 - 4}{x+1}$

$= \dfrac{(x-3)(x+2)}{x+1} \div \dfrac{(x+2)(x-2)}{x+1}$

$= \dfrac{(x-3)(x+2)}{x+1} \cdot \dfrac{x+1}{(x+2)(x-2)}$

$= \dfrac{x-3}{x-2}$

Review Problems : Chapter Eight

1. $2x - y = 2$ (multiply by 2)

   $x + 2y = 11$

   $4x - 2y = 4$

   $\underline{x + 2y = 11}$

   Add : $5x = 15$ $x = 3$

   $x + 2y = 11$

   $3 + 2y = 11$

   $2y = 8$ $y = 4$

   $\{(3, 4)\}$

2. $y = 4 - x$

   $3x + 3y = 12$

   Method 1. Substitution

   $3x + 3(4 - x) = 12$

   $3x + 12 - 3x = 12$

   $12 = 12$

   $0 = 0$ System is dependent.

   $\{(x.y) \mid y = 4 - x\}$

   Method 2. $x + y = 4$

   $3x + 3y = 12$

   $\dfrac{1}{3} = \dfrac{1}{3} = \dfrac{4}{12}$

   System is dependent.

3. $5x + 3y = -3$ (multiply by 2)

   $2x + 7y = -7$ (multiply by - 5)

   $10x + 6y = -6$

   $\underline{-10x - 35y = 35}$

   Add : $-29y = 29$

   $y = -1$

   $5x + 3y = -3$

   $5x + 3(-1) = -3$

   $5x - 3 = -3$

   $5x = 0$

   $x = 0$

   $\{(0, -1)\}$

4. $5y = x^2 - 1$

   $x - y = 1$

   Substitution :

   $x = y + 1$

   $5y = (y + 1)^2 - 1$

   $5y = y^2 + 2y + 1 - 1$

   $0 = y^2 - 3y$

   $0 = y(y - 3)$

   $y = 0$ $y - 3 = 0$

   $y = 0$ $y = 3$

   If $y = 0$: $x = y + 1 = 0 + 1 = 1$

   If $y = 3$: $x = y + 1 = 3 + 1 = 4$

   $\{(1, 0), (4, 3)\}$

5. $y = x^2 - 2x - 1$

   $y - x = 3$

   Substitution : $y = x + 3$

   $x + 3 = x^2 - 2x - 1$

   $0 = x^2 - 3x - 4$

   $0 = (x - 4)(x + 1)$

   $x - 4 = 0$ or $x + 1 = 0$

   $x = 4$ $x = -1$

   If $x = 4$: $y = x + 3 = 4 + 3 = 7$

   If $x = -1$: $y = x + 3 = -1 + 3 = 2$

   $\{(4.7), (-1, 2)\}$

6. $x^2 + y^2 = 2$

   $x + y = 0$

   Substitution : $x = -y$

   $(-y)^2 + y^2 = 2$

   $2y^2 = 2$

   $y^2 = 1$

   $y = \pm 1$

   If $y = 1$: $x = -y = -1$

   If $y = -1$: $x = -y = -(-1) = 1$

   $\{(-1, 1), (1, -1)\}$

7. $x - 2y + 3 = 0$

$2x - 4y + 7 = 0$

$x - 2y = -3$

$2x - 4y = -7$

$\dfrac{1}{2} = \dfrac{-2}{-4} \ne \dfrac{-3}{-7}$

The system is inconsistent.

$\varnothing$

8. $x^2 + y^2 = 4$

$2x - y = 0$

Substitution : $y = 2x$

$x^2 + (2x)^2 = 4$

$x^2 + 4x^2 = 4$

$5x^2 = 4$

$x^2 = 4/5$

$x = \pm \sqrt{\dfrac{4}{5}}$

$x = \pm \dfrac{2}{\sqrt{5}} \cdot \dfrac{\sqrt{5}}{\sqrt{5}}$

$x = \pm \dfrac{2\sqrt{5}}{5}$

If $x = \dfrac{2\sqrt{5}}{5}$: $y = 2x = 2(\dfrac{2\sqrt{5}}{5}) = \dfrac{4\sqrt{5}}{5}$

If $x = -\dfrac{2\sqrt{5}}{5}$: $y = 2x = 2\,(-\dfrac{2\sqrt{5}}{5}) = -\dfrac{4\sqrt{5}}{5}$

$\{(\dfrac{2\sqrt{5}}{5}, \dfrac{4\sqrt{5}}{5}),\ (-\dfrac{2\sqrt{5}}{5},\ -\dfrac{4\sqrt{5}}{5})\}$

9. $x^2 + 2y^2 = 4$

$x - y - 1 = 0$

Substitution : $x = y + 1$

$(y + 1)^2 + 2y^2 = 4$

$y^2 + 2y + 1 + 2y^2 = 4$

$3y^2 + 2y - 3 = 0$

$a = 3, b = 2, c = -3$

$y = \dfrac{-b \pm \sqrt{b^2 - 4ac}}{2a} = \dfrac{-2 \pm \sqrt{4 - 4(3)(-3)}}{6}$

$= \dfrac{-2 \pm \sqrt{40}}{6} = \dfrac{-2 \pm 2\sqrt{10}}{6}$

$= \dfrac{-1 \pm \sqrt{10}}{3}$

If $y = \dfrac{-1 + \sqrt{10}}{3}$: $x = y + 1 = \dfrac{-1 + \sqrt{10}}{3} + \dfrac{3}{3}$

$= \dfrac{2 + \sqrt{10}}{3}$

If $y = \dfrac{-1 - \sqrt{10}}{3}$: $x = y + 1 = \dfrac{-1 - \sqrt{10}}{3} + \dfrac{3}{3}$

$= \dfrac{2 - \sqrt{10}}{3}$

$\{(\dfrac{2 + \sqrt{10}}{3}, \dfrac{-1 + \sqrt{10}}{3}), (\dfrac{2 - \sqrt{10}}{3}, \dfrac{-1 - \sqrt{10}}{3})\}$

10. $2x^2 + y^2 = 24$

$x^2 + y^2 = 15$ (multiply by -1)

$2x^2 + y^2 = 24$

$\underline{-x^2 - y^2 = 15}$

Add : $x^2 = 9$

$x = \pm 3$

If $x = 3$: $x^2 + y^2 = 15$ $9 + y^2 = 15$ $y^2 = 6$

$y = \pm\sqrt{6}$

If $x = -3$: $x^2 + y^2 = 15$ $9 + y^2 = 15$ $y^2 = 6$

$y = \pm\sqrt{6}$

$\{(3, \sqrt{6}), (3, -\sqrt{6}), (-3, \sqrt{6}), (-3, -\sqrt{6})\}$

11. $xy - 4 = 0$

$y - x = 0$

Substitution : $y = x$

$x(x) - 4 = 0$

$x^2 = 4$

$x = \pm 2$

If $x = 2$: $y = x = 2$

If $x = -2$: $y = x = -2$

$\{(2, 2), (-2, -2)\}$

12. $x^2 + y^2 = 2$

$y = x^2$

Substitution : $x^2 + (x^2)^2 = 2$

$x^4 + x^2 - 2 = 0$

Let $t = x^2$

$t^2 + t - 2 = 0$

$(t + 2)(t - 1) = 0$

$t + 2 = 0$ or $t - 1 = 0$

$t = -2$ $t = 1$

$x^2 = -2$ $x^2 = 1$

$x = \pm\sqrt{-2}$ $x = \pm\sqrt{1}$

$x = \pm\sqrt{2}i$ $x = \pm 1$

If $x = \sqrt{2}i$: $y = x^2 = (\sqrt{2}i)^2 = 2i^2 = 2(-1)$

$= -2$

If $x = -\sqrt{2}i$: $y = x^2 = (-\sqrt{2}i)^2 = 2i^2 = 2(-1)$

$= -2$

If $x = 1$: $y = x^2 = 1^2 = 1$

If $x = -1$: $y = x^2 = (-1)^2 = 1$

$\{(\sqrt{2}i, -2), (-\sqrt{2}i, -2), (1, 1), (-1, 1)\}$

13. $y^2 = 4x$

$x - 2y + 3 = 0$

Substitution : $x = 2y - 3$

$y^2 = 4(2y - 3)$

$y^2 = 8y - 12$

$y^2 - 8y + 12 = 0$

$(y - 6)(y - 2) = 0$

$y = 6$ or $y = 2$

If $y = 6$: $x = 2y - 3 = 2(6) - 3 = 9$

If $y = 2$: $x = 2y - 3 = 2(2) - 3 = 1$

$\{(9, 6), (1, 2)\}$

14. $\dfrac{2x + y}{3} - \dfrac{x + 2y}{2} = \dfrac{23}{6}$ (multiply by 6)

$x = 2 + \dfrac{3x - 4y}{5}$ (multiply by 5)

$4x + 2y - 3x - 6y = 23$

$5x = 10 + 3x - 4y$

$x - 4y = 23$

$\underline{2x + 4y = 10}$

Add : $3x = 33$

$x = 11$

$x - 4y = 23$

$11 - 4y = 23$

$-12 = 4y$

$-3 = y$

$\{(11, -3)\}$

15. (1) $3x - y + 4z = 4$
(2) $4x + 4y - 3z = 3$
(3) $2x + 3y + 2z = -4$

Eliminate y.
$4(1) + (2):$ $12x - 4y + 16z = 16$
                 $4x + 4y - 3z = 3$
        Add: $16x + 13z = 19$
$3(1) + (3):$ $9x - 3y + 12z = 12$
                 $2x + 3y + 2z = -4$
        Add: $11x + 14z = 8$
$16x + 13z = 19$ (multipy by 11)
$11x + 14z = 8$ (multiply by -16)
$176x + 143z = 209$
$-176x - 224z = -128$
Add: $-81z = 81$
$z = -1$
$11x + 14z = 8$
$11x + 14(-1) = 8$
$11x = 22$ $x = 2$
$3x - y + 4z = 4$
$3(2) - y + 4(-1) = 4$
$-y + 2 = 4$
$-y = 2$
$y = -2$
$\{(2, -2, -1)\}$

16. (1) $x - z + 2 = 0$
(2) $y + 3z = 11$
(3) $x + y + z = 6$
Eliminate y. (1): $x - z = -2$
$-1(2) + (3):$ $-y - 3z = -11$
                 $x + y + z = 6$
        Add: $x \quad - 2z = -5$
$x - z = -2$
$x - 2z = -5$ (multiply by -1)
$x - z = -2$
$-x + 2z = 5$
Add: $z = 3$
$x - z = -2$ $x - 3 = -2$ $x = 1$
$y + 3z = 11$
$y + 3(3) = 11$
$y = 2$
$\{(1, 2, 3)\}$

17. $y = x^2 + 2$
$x^2 + y^2 = 4$
$y - x^2 = 2$
$y^2 + x^2 = 4$
Add: $y + y^2 = 6$
$y^2 + y - 6 = 0$
$(y + 3)(y - 2) = 0$
$y + 3 = 0$ or $y - 2 = 0$
$y = -3$ $y = 2$
If $y = -3$: $y = x^2 + 2$ $-3 = x^2 + 2$ $-5 = x^2$
$x = \pm\sqrt{-5} = \pm\sqrt{5}i$
If $y = 2$: $y = x^2 + 2$
$2 = x^2 + 2$
$0 = x^2$
$0 = x$

$\{(\sqrt{5}i, -3), (-\sqrt{5}i, -3), (0, 2)\}$

18. Numbers: x, y
$x^2 + y^2 = 25$
$x - 3y = -5$
Substitution: $x = 3y - 5$
$(3y - 5)^2 + y^2 = 25$
$9y^2 - 30y + 25 + y^2 = 25$
$10y^2 - 30y = 0$
$10y(y - 3) = 0$
$10y = 0$ or $y - 3 = 0$
$y = 0$ or $y = 3$
If $y = 0$: $x = 3y - 5 = 3(0) - 5 = -5$
If $y = 3$: $x = 3y - 5 = 3(3) - 5 = 4$
Numbers are -5 and 0 or 4 and 3.

19. t : tens' digit

u : units' digit

$10t + u$: the number

$10u + t$: the number with the digits

reversed

$t + u = 7$ $10u + t = 10t + u - 9$

$t + u = 7$

$9t - 9u = 9$ (multiply by $\frac{1}{9}$)

$t + u = 7$

$\underline{t - u = 1}$

Add : $2t = 8$

$t = 4$

$t + u = 7$

$4 + u = 7$

$u = 3$

Number : $10t + u = 10(4) + 3 = 43$

20. Dimensions : x, y

Perimeter is 38 : $2x + 2y = 38$ (multiply

by $\frac{1}{2}$)

$x + y = 19$

Area is 84 : $xy = 84$

Substitution : $y = 19 - x$

$x(19 - x) = 84$

$19x - x^2 = 84$

$0 = x^2 - 19x + 84$

$0 = (x - 12)(x - 7)$

$x = 12$ or $x = 7$

If $x = 12$: $y = 19 - x = 19 - 12 = 7$

If $x = 7$: $y = 19 - x = 19 - 7 = 12$

Dimensions : 12 yards, 7 yards

21. Legs : x, y

Perimeter is 36 : $x + y + 15 = 36$

$x + y = 21$

Pythagorean Theorem : $x^2 + y^2 = 15^2$

Substitution : $x = 21 - y$

$(21 - y)^2 + y^2 = 15^2$

$441 - 42y + y^2 + y^2 = 225$

$2y^2 - 42y - 216 = 0$

$y^2 - 21y - 108 = 0$

$(y - 9)(y - 12) = 0$

$y - 9 = 0$ or $y - 12 = 0$

$y = 9$ $y = 12$

If $y = 9$: $x = 21 - y = 21 - 9 = 12$

If $y = 12$: $x = 21 - y = 21 - 12 = 9$

Legs : 9 meters, 12 meters

22. Measure of two angles : x, y

$x = 3y - 80$

Angle sum is $180°$: $x + y + 80 = 180$

$x - 3y = -80$

$x + y = 100$ (multiply by -1)

$x - 3y = -80$

$\underline{-x - y = -100}$

Add : $-4y = -180$

$y = 45$

$x + y = 100$

$x + 45 = 100$

$x = 55$

Measure of two angles : $45°$ and $55°$

23. $x:$ cost of one pen

$y:$ cost of one pad

$8x + 6y = 16.10$

$3x + 2y = 5.85$ (multiply by -3)

$8x + 6y = 16.10$

$\underline{-9x - 6y = -17.55}$

Add: $-x = -1.45$

$x = 1.45$

$3x + 2y = 5.85$

$3(1.45) + 2y = 5.85$

$4.35 + 2y = 5.85$

$2y = 1.5$

$y = 0.75$

Pens: \$1.45 Pads: 75c

24. $x:$ speed of plane in still air

$y:$ speed of wind

$RT = D$

$(2h\ 15\ min = 2\frac{1}{4}h = \frac{9}{4}h)$

$(1h\ 20\ min = 1\frac{1}{3}h = \frac{4}{3}h)$

$\frac{9}{4}(x - y) = 360$ (multiply by 4)

$\frac{4}{3}(x + y) = 360$ (multiply by 3)

$9x - 9y = 1440$ (multiply by $\frac{1}{9}$)

$4x + 4y = 1080$ (multiply by $\frac{1}{4}$)

$x - y = 160$

$\underline{x + y = 270}$

Add: $2x = 430$

$x = 215$

$x + y = 270$

$215 + y = 270$

$y = 55$

Plane's speed in still air: 215m/h

Wind speed: 55m/h

25. $y = mx + b$

$(1, 6): \ 6 = m(1) + b$

$(-1, 12): \ -12 = m(-1) + b$

$m + b = 6$

$\underline{-m + b = -12}$

Add: $2b = -6$

$b = -3$

$m + b = 6$

$m - 3 = 6$

$m = 9$

$m = 9,\ b = -3: \ y = 9x - 3$

26. $x:$ age of older home

$y:$ age of newer home

$x - 20 = 10(y - 20)$

$x + 60 = 2(y + 60)$

$x - 20 = 10y - 200$

$x + 60 = 2y + 120$

$x - 10y = -180$

$x - 2y = 60$ (multiply by -1)

$x - 10y = -180$

$\underline{-x + 2y = -60}$

Add: $-8y = -240$

$y = 30$

$x - 2y = 60$

$x - 2(30) = 60$

$x = 120$

Older home: 120 years old

Newer home: 30 years old

27. $x:$ number of people upstairs

$y:$ number of people downstairs

$x + 1 = 10(y - 1)$

$y + 8 = x - 8$

$x - 10y = -11$

$\underline{-x + y = -16}$

Add: $-9y = -27$

$y = 3$

$x - 10y = -11$

$x - 10(3) = -11$

$x = 19$

Upstairs: 19 people

Downstairs: 3 people

28. h : hundreds' digit

t : tens' digit

u : units' digit

$h + t + u = 12$

$t = 3u + 1$

$2h + u = 8$

(1) $h + t + u = 12$

(2) $t - 3u = 1$

(3) $2h + u = 8$

Eliminate t.

(3) : $2h + u = 8$

$-1(1) + (2) : -h - t - u = -12$

$\underline{\hspace{3em} t - 3u = 1 \hspace{2em}}$

Add: $-h - 4u = -11$

$2h + u = 8$

$-h - 4u = -11$ (multiply by 2)

$2h + u = 8$

$\underline{-2h - 8u = -22}$

Add: $-7u = -14$

$u = 2$

$2h + u = 8$

$2h + 2 = 8$

$2h = 6$

$h = 3$

$t - 3u = 1$

$t - 3(2) = 1$

$t = 7$

Number : $100h + 10t + u = 100(3) + 10(7) + 2$

$= 372$

29. $y = ax^2 + bx + c$

$(1, 2) : 2 = a \cdot 1^2 + b \cdot 1 + c$

$a + b + c = 2$

$(-1, 6) : 6 = a(-1)^2 + b(-1) + c$

$a - b + c = 6$

$(2, 3) : 3 = a \cdot 2^2 + b \cdot 2 + c$

$4a + 2b + c = 3$

(1) $a + b + c = 2$

(2) $a - b + c = 6$

(3) $4a + 2b + c = 3$

Eliminate c.

$(1) + -1(2) : 2b = -4$

$b = -2$

$(1) + -1(3) : -3a - b = -1$

$-3a - (-2) = -1$

$-3a = -3$

$a = 1$

$a + b + c = 2$

$1 + (-2) + c = 2$

$-1 + c = 2$

$c = 3$

$a = 1, b = -2, c = 3 : y = x^2 - 2x + 3$

30. x : number of nickels

y : number of dimes

z : number of quarters

$x + y + z = 85$

$.05x + .10y + .25z = 6.25$ (multiply by 100)

$x = 3y$

$x + y + z = 85$

$5x + 10y + 25z = 625$ (multiply by $\frac{1}{5}$)

    $x = 3y$

(1) $x + y + z = 85$

(2) $x + 2y + 5z = 125$

(3) $x - 3y = 0$

Eliminate z.

(3) : $x - 3y = 0$

$-5(1) + (2)$: $-5x - 5y - 5z = -425$

                 $x + 2y + 5z = 125$

    Add : $-4x - 3y = -300$

$x - 3y = 0$ (multiply by -1)

$-4x - 3y = -300$

$-x + 3y = 0$

$-4x - 3y = -300$

Add : $-5x = -300$

$x = 60$

$x - 3y = 0$

$60 - 3y = 0$

$60 = 3y$

$20 = y$

$x + y + z = 85$

$60 + 20 + z = 85$

$z = 5$

60 nickels, 20 dimes, 5 quarters

31. CB : x

CD : y

CA : z

Note : Since diagonals of a parallelogram bisect, $BE = ED = \dfrac{x}{2}$ and $CE = EA = \dfrac{z}{2}$.

Perimeter of CDE = 12 : $y + \dfrac{x}{2} + \dfrac{z}{2} = 12$

     (multiply by 2)

Perimeter of CDA = 20 : $x + y + z = 20$

Perimeter of CDB = 18 : $x + y + x = 18$

(1) $x + 2y + z = 24$

(2) $x + y + z = 20$

(3) $2x + y = 18$

Eliminate z.

(3) : $2x + y = 18$

(1) + -1(2) : $y = 4$

$2x + 4 = 18$

$2x = 14$

$x = 7$

$x + y + z = 20$

$7 + 4 + z = 20$

$z = 9$

CB : 7ft   CD : 4ft   CA : 9ft

32. $\begin{vmatrix} 3 & 2 \\ -1 & 5 \end{vmatrix} = (3)(5) - (-1)(2) = 15 - (-2) = 17$

33. $\begin{vmatrix} -2 & -3 \\ -4 & -8 \end{vmatrix} = (-2)(-8) - (-4)(-3) = 16 - 12 = 4$

34. Expand about elements in the first column.

$\begin{vmatrix} 2 & 4 & -3 \\ -1 & 7 & -4 \\ 1 & -6 & -2 \end{vmatrix}$

$= 2 \begin{vmatrix} 7 & -4 \\ -6 & -2 \end{vmatrix} + 1 \begin{vmatrix} 4 & -3 \\ -6 & -2 \end{vmatrix} + 1 \begin{vmatrix} 4 & -3 \\ 7 & -4 \end{vmatrix} =$

$2(-14 - 24) + 1(-8 - 18) + 1(-16 + 21)$

$= 2(-38) + 1(-26) + 1(5)$

$= -76 - 26 + 5$

$= -97$

35. Expand about elements in the last column.

$\begin{vmatrix} 4 & 7 & 0 \\ -5 & 6 & 0 \\ 3 & 2 & -4 \end{vmatrix} = -4 \begin{vmatrix} 4 & 7 \\ -5 & 6 \end{vmatrix}$

$= -4(24 + 35)$

$= -4(59)$

$= -236$

36. $2x - y = 2$

$x + 2y = 11$

$D = \begin{vmatrix} 2 & -1 \\ 1 & 2 \end{vmatrix} = 4 - (-1) = 5$

$Dx = \begin{vmatrix} 2 & -1 \\ 11 & 2 \end{vmatrix} = 4 - (-11) = 15$

$Dy = \begin{vmatrix} 2 & 2 \\ 1 & 11 \end{vmatrix} = 22 - 2 = 20$

$x = \dfrac{Dx}{D} = \dfrac{15}{5} = 3$

$y = \dfrac{Dy}{D} = \dfrac{20}{5} = 4$

$\{(3, 4)\}$

37. $4x = 12 - 3y$

$2x - 6 = 5y$

$4x + 3y = 12$

$2x - 5y = 6$

$D = \begin{vmatrix} 4 & 3 \\ 2 & -5 \end{vmatrix} = -20 - 6 = -26$

$Dx = \begin{vmatrix} 12 & 3 \\ 6 & -5 \end{vmatrix} = -60 - 18 = -78$

$Dy = \begin{vmatrix} 4 & 12 \\ 2 & 6 \end{vmatrix} = 24 - 24 = 0$

$x = \dfrac{Dx}{D} = \dfrac{-78}{-26} = 3 \quad y = \dfrac{Dy}{D} = \dfrac{0}{-26} = 0$

$\{(3, 0)\}$

38. $4x + 2y + 3z = 9$

$3x + 5y + 4z = 19$

$9x + 3y + 2z = 3$

$$D = \begin{vmatrix} 4 & 2 & 3 \\ 3 & 5 & 4 \\ 9 & 3 & 2 \end{vmatrix}$$

$$= 4\begin{vmatrix} 5 & 4 \\ 3 & 2 \end{vmatrix} - 3\begin{vmatrix} 2 & 3 \\ 3 & 2 \end{vmatrix} + 9\begin{vmatrix} 2 & 3 \\ 5 & 4 \end{vmatrix}$$

$$= 4(10 - 12) - 3(4 - 9) + 9(8 - 15)$$

$$= -8 + 15 - 63 = -56$$

$$Dx = \begin{vmatrix} 9 & 2 & 3 \\ 19 & 5 & 4 \\ 3 & 3 & 2 \end{vmatrix}$$

$$= 9\begin{vmatrix} 5 & 4 \\ 3 & 2 \end{vmatrix} - 19\begin{vmatrix} 2 & 3 \\ 3 & 2 \end{vmatrix} + 3\begin{vmatrix} 2 & 3 \\ 5 & 4 \end{vmatrix}$$

$$= 9(10 - 12) - 19(4 - 9) + 3(8 - 15)$$

$$= -18 + 95 - 21 = 56$$

$$Dy = \begin{vmatrix} 4 & 9 & 3 \\ 3 & 19 & 4 \\ 9 & 3 & 2 \end{vmatrix}$$

$$= 4\begin{vmatrix} 19 & 4 \\ 3 & 2 \end{vmatrix} - 3\begin{vmatrix} 9 & 3 \\ 3 & 2 \end{vmatrix} + 9\begin{vmatrix} 9 & 3 \\ 19 & 4 \end{vmatrix}$$

$$= 4(38 - 12) - 3(18 - 9) + 9(36 - 57)$$

$$= 104 - 27 - 189 = -112$$

$$Dz = \begin{vmatrix} 4 & 2 & 9 \\ 3 & 5 & 19 \\ 9 & 3 & 3 \end{vmatrix}$$

$$= 4\begin{vmatrix} 5 & 19 \\ 3 & 3 \end{vmatrix} - 3\begin{vmatrix} 2 & 9 \\ 3 & 3 \end{vmatrix} + 9\begin{vmatrix} 2 & 9 \\ 5 & 19 \end{vmatrix}$$

$$= 4(15 - 57) - 3(6 - 27) + 9(38 - 45)$$

$$= -168 + 63 - 63 = -168$$

$$x = \frac{Dx}{D} = \frac{56}{-56} = -1$$

$$y = \frac{Dy}{D} = \frac{-112}{-56} = 2$$

$$z = \frac{Dz}{D} = \frac{-168}{-56} = 3$$

$$\{(-1, 2, 3)\}$$

39. $2x + 0y + z = -4$

$0x - 2y + z = -4$

$2x - 4y + z = -20$

$$D = \begin{vmatrix} 2 & 0 & 1 \\ 0 & -2 & 1 \\ 2 & -4 & 1 \end{vmatrix}$$

$$= 2\begin{vmatrix} -2 & 1 \\ -4 & 1 \end{vmatrix} + 2\begin{vmatrix} 0 & 1 \\ -2 & 1 \end{vmatrix}$$

$$= 2(-2 + 4) + 2(0 + 2) = 4 + 4 = 8$$

$$Dx = \begin{vmatrix} -4 & 0 & 1 \\ -4 & -2 & 1 \\ -20 & -4 & 1 \end{vmatrix}$$

$$= -2\begin{vmatrix} -4 & 1 \\ -20 & 1 \end{vmatrix} + 4\begin{vmatrix} -4 & 1 \\ -4 & 1 \end{vmatrix}$$

$$= -2(-4 + 20) + 4(-4 + 4)$$

$$= -32$$

$$Dy = \begin{vmatrix} 2 & -4 & 1 \\ 0 & -4 & 1 \\ 2 & -20 & 1 \end{vmatrix}$$

$$= -2\begin{vmatrix} -4 & 1 \\ -20 & 1 \end{vmatrix} + 2\begin{vmatrix} -4 & 1 \\ -4 & 1 \end{vmatrix}$$

$$= 2(-4 + 20) + 2(-4 + 4)$$

$$= 32$$

$$Dz = \begin{vmatrix} 2 & 0 & -4 \\ 0 & -2 & -4 \\ 2 & -4 & -20 \end{vmatrix}$$

$$= 2\begin{vmatrix} -2 & -4 \\ -4 & -20 \end{vmatrix} + 2\begin{vmatrix} 0 & -4 \\ -2 & -4 \end{vmatrix}$$

$$= 2(40 - 16) + 2(0 - 8)$$

$$= 48 - 16 = 32$$

$$x = \frac{Dx}{D} = \frac{-32}{8} = -4$$

$$y = \frac{Dy}{D} = \frac{32}{8} = 4$$

$$z = \frac{Dz}{D} = \frac{32}{8} = 4$$

$$\{(-4, 4, 4)\}$$

Problem Set 9.1.1

19. $f(70)=(70/12.3)^3 \approx 184.32$

The threshold weight for a man between
40 and 49 years old who is 70 inches tall
is approximately 184 pounds.

21. a. $g(x)=0.4x+0.88$

b. $g(2)=0.4(2)+0.88=1.68$

A subject kept isolated for 2 hours takes
1.68 minutes to get through a maze.

23. a. $f(x)=4x-148$

b. $f(60)=4(60)-148=92$

At 60° a cricket chirps 92 times a minute.

c. $4x-148=164$

$4x=312$

$x=78$

If the cricket chirps 164 times a
minute,the temperature is 78°.

d. $f(t+h)=4(t+h)-148=4t+4h-148$

e. $f(t+h)-f(t)/h=4t+4h-148-(4t-148)/h$

$=4t+4h-148-4t+148/h$

$=4h/h$

$=4$

25. a. $f(x)=40x-x^2$

b. $f(8)=40(8)-8^2$

$=320-64$

$=256$

After 8 days there are 256 flu cases.

c. $40x-x^2=175$

$0=x^2-40x+175$

$0=(x-35)(x-5)$

$x-35=0$ or $x-5=0$

$x=35$ $x=5$

If there are 175 flu cases,this
occurs 5 days after the outbreak and
again 35 days after the outbreak.

d. $f(x+h)=40(x+h)-(x+h)^2$

$=40x+40h-x^2-2xh-h^2$

e. $f(x+h)-f(x)/h$

$=40x+40h-x^2-2xh-h^2-(40x-x^2)/h$

$=40x+40h-x^2-2xh-h^2-40x+x^2/h$

$=40h-2xh-h^2/h$

$=40-2x-h$

27. a. $G(I)=-5I^2+80I$

b. $G(3)=-5(3)^2+80(3)$

$=-45+240=195$

A current of 3 amps produces a power of 195 volts.

c. $-5I^2+80I=75$

$0=5I^2-80I+75$

$0=I^2-16I+15$

$0=(I-15)(I-1)$

$I=15 \qquad I=1$

A power of 75 volts is produced when the current is 1 amp or 15 amps.

d. $G(x^2+h)-G(x^2)/h$

$=-5(x^2+h)^2+80(x^2+h)-[-5(x^2)^2+80x^2]/h$

$=-5x^4-10x^2h-5h^2+80x^2+80h+5x^4-80x^2/h$

$=-10x^2h-5h^2+80h/h$

$=-10x^2-5h+80$

29. a. $F(t)=t^3-3t^2+5t$

b. $F(3)=3^3-(3)(3^2)+(5)(3)$

$=27-27+15$

$=15$

15 people catch a cold 3 weeks after January 1.

c. $F(5)=5^3-(3)(5^2)+(5)(5)$

$=125-75+25$

$=75$

75 people catch a cold 5 weeks after January 1.

d. January 8: 1 week after January 1

January 22: 3 weeks after January 1

$F(3)-F(1)$

e. $F(u^3)-F(u^2)$

$=(u^3)^3-3(u^3)^2+5u^3-[(u^2)^3-3(u^2)^2+5u^2]$

$=u^9-3u^6+5u^3-u^6+3u^4-5u^2$

$=u^9-4u^6+3u^4+5u^3-5u^2$

$f(x)=x^2+2x-1$ and $g(x)=2x+3$

31. $g(1)=2(1)+3=5$

$f[g(1)]=f(5)$

$=5^2+(2)(5)-1$

$=25+10-1$

$=34$

35. g(a)=2a+3

f[g(a)]=f(2a+3)

=(2a+3)²+2(2a+3)-1

=4a²+12a+9+4a+6-1

=4a²+16a+14

37. g(b+1)=2(b+1)+3

=2b+2+3

=2b+5

f[g(b+1)]=f(2b+5)

=(2b+5)²+2(2b+5)-1

=4b²+20b+25+4b+10-1

=4b²+24b+34

f(x)=2x+3,g(x)=x-1,h(x)=2x²+x-3

39. (f+g)(x)=f(x)+g(x)

=(2x+3)+(x-1)

=3x+2

41. (g-f)(x)=g(x)-f(x)

=(x-1)-(2x+3)

=-x-4

43. (fg)(x)=f(x)g(x)

=(2x+3)(x-1)

=2x²+x-3

45. (h/f)(x)=h(x)/f(x)

=2x²+x-3/2x+3

=(2x+3)(x-1)/2x+3

=x-1(x≠-3/2)

47. (f-h)(-1)=f(-1)-h(-1)

=[2(-1)+3]-[2(-1)²+(-1)-3]

=1-(-2)=3

49. (f+g)(x+h)=f(x+h)+g(x+h)

=[2(x+h)+3]+[(x+h)-1]

=(2x+2h+3)+(x+h-1)

=3x+3h+2

53. (h+fg)(x)=h(x)+f(x)g(x)

=(2x²+x-3)+(2x+3)(x-1)

=(2x²+x-3)+(2x²+x-3)

=4x²+2x-6

55. f(x)=17

f(0)=17,f(2)=17,f(-3)=17

f(a+h)-f(a)/h=17-17/h

=0/h

=0

61. f(x)=3x+7

f(2x)-2f(x)+3=3(2x)+7-2(3x+7)+3

=6x+7-6x-14+3

=-4

63. f(x)=3x/2x-1

f(a-3/a+3)

=[3(a-3/a+3)/2(a-3/a+3)-1](a+3/a+3)

=3(a-3)/2(a-3)-(a+3)

=3a-9/a-9

67. $f(x)=ax^2+bx+c$

$g(x)=f(x+1)-f(x)$

$=a(x+1)^2+b(x+1)+c-(ax^2+bx+c)$

$=ax^2+2ax+a+bx+b+c-ax^2-bx-c$

$=2ax+a+b$

$g(x)$ is a linear function, in the form

$g(x)=mx+B$

with $m=2a$ and $B=a+b$.

(slope:$2a$,y-intercept:$a+b$)

69. Given:$f(0)=1$

$f(x+2)=f(x)+7$

$f(2)=f(0+2)$

$=f(0)+7$

$=1+7$

$=8$

$f(4)=f(2+2)$

$=f(2)+7$

$=8+7$

$=15$

70. $x+1=2y$

$x^2-xy+y^2-7=0$

Substitution:$x=2y-1$

$(2y-1)^2-(2y-1)y+y^2-7=0$

$4y^2-4y+1-2y^2+y+y^2-7=0$

$3y^2-3y-6=0$

$y^2-y-2=0$

$(y-2)(y+1)=0$

$y-2=0$ or $y+1=0$

$y=2$ $y=-1$

If $y=2$:$x=2y-1$

$=2(2)-1$

$=3$

If $y=-1$:$x=2y-1$

$=2(-1)-1$

$=-3$

$\{(3,2),(-3,-1)\}$

71. $|2x-3|<7$

$-7<2x-3<7$

$-4<2x<10$

$-2<x<5$

$\{x|-2<x<5\}$

72. $x+y+z=5$

$2x+7y+0z=6$

$3x+0y-5z=-1$

Expand about the third column.

$$D= \begin{vmatrix} 1 & 1 & 1 \\ 2 & 7 & 0 \\ 3 & 0 & -5 \end{vmatrix} = 1 \begin{vmatrix} 2 & 7 \\ 3 & 0 \end{vmatrix} - 5 \begin{vmatrix} 1 & 1 \\ 2 & 7 \end{vmatrix}$$

$=1(0-21)-5(7-2)$

$=-21-25$

$=-46$

$$Dx= \begin{vmatrix} 5 & 1 & 1 \\ 6 & 7 & 0 \\ -1 & 0 & -5 \end{vmatrix} = 1 \begin{vmatrix} 6 & 7 \\ -1 & 0 \end{vmatrix} - 5 \begin{vmatrix} 5 & 1 \\ 6 & 7 \end{vmatrix}$$

$=1(0+7)-5(35-6)$

$=7-145$

$=-138$

$$Dy= \begin{vmatrix} 1 & 5 & 1 \\ 2 & 6 & 0 \\ 3 & -1 & -5 \end{vmatrix} = 1 \begin{vmatrix} 2 & 6 \\ 3 & -1 \end{vmatrix} - 5 \begin{vmatrix} 1 & 5 \\ 2 & 6 \end{vmatrix}$$

$=1(-2-18)-5(6-10)$

$=-20+20$

$=0$

$$Dz= \begin{vmatrix} 1 & 1 & 5 \\ 2 & 7 & 6 \\ 3 & 0 & -1 \end{vmatrix} \text{ (Expand about third row.)}$$

$$=3 \begin{vmatrix} 1 & 5 \\ 7 & 6 \end{vmatrix} -1 \begin{vmatrix} 1 & 1 \\ 2 & 7 \end{vmatrix}$$

$=3(6-35)-1(7-2)$

$=-87-5$

$=-92$

$x=Dx/D=-138/-46=3$

$y=Dy/D=0/-46=0$

$z=Dz/D=-92/-46=2$

$\{(3,0,2)\}$

Problem Set 9.1.2

1. **a.** One number: x
   Other number: 100-x
   $P(x)=x(100-x)$
   $P(x)=-x^2+100x$

   b. $P(15)=-15^2+100(15)$
   $=-225+1500$
   $=1275$

   c. Maximum when $x=-b/2a$
   $a=-1 \quad b=100$
   $x=-100/2(-1)=50$
   Maximum $P(x)$ is $P(50)=-50^2+100(50)$
   $\qquad\qquad =-2500+5000=2500$

3. Numbers: x and y
   $x+2y=2$
   $2y=24-x$
   $y=24-x/2$
   Product=xy
   $P(x)=x(24-x/2)=24x-x^2/2$

9. Revenue=(number of units sold)(price
   per unit)
   $R(x)=x(200-0.04x)$
   $\qquad R(x)=200x-0.04x^2$
   $R(400)=(200)(400)-0.04(400)^2$
   $=80,000-6,400$
   $=73,600$

11. x:width(in feet)
    2x+3:length(in feet)
    $C(x)=92x+11[x+(2x+3)+(2x+3)]$
    $=92x+11(5x+6)$
    $=147x+66$

13. Time through cities: x
    Time on expressway: 6-x
    $d=rt$
    $d(x)=35x+60(6-x)$
    $=35x+360-60x$
    $=-25x+360$

15. height: x
    radius: 2x
    $V=\pi r^2h$
    $V(x)=\pi(2x)^2x$
    $V(x)=4\pi x^3$
    $V(4)=4\pi(4)^3=256\pi$

17. $V=10$
    $x^2y=10$
    $y=10/x^2$
    Cost=Cost of bottom+Cost of 4 sides
    $Cost=15x^2+(6)(4)(xy)$
    $=15x^2+(24)(x)(10/x^2)$
    $C(x)=15x^2+240/x$

19. Length of path A: A

$6^2+(8-x)^2=A^2$

$100-16x+x^2=A^2$

$A=\sqrt{x^2-16x+100}$

Cost=40(length of path A)+

    20(length of path B)

Cost=$40\sqrt{x^2-16x+100}+20x$

21. Feet of fencing=5x+2y

$xy=1000$

$y=1000/x$

$F(x)=5x+2(1000/x)$

$F(x)=5x+2000/x$

23. $(x-1/x-2)-[3x-2/(x+2)(x-2)]=3/x+2$

Multiply each side by $(x+2)(x-2)$.

$x^2+x-2-3x+2=3x-6$

$x^2-2x=3x-6$

$x^2-5x+6=0$

$(x-3)(x-2)=0$

x=3 or x=2(reject;causes division

by zero)

{3}

24. $54x^3+2y^3=2(27x^3+y^3)$

$=2[(3x)^3+y^3]$

$=2(3x+y)[(3x)^2-(3x)y+y^2]$

$=2(3x+y)(9x^2-3xy+y^2)$

25. Slope of line through (2,0) and (3,-2)

$=-2-0/3-2=-2$

Slope of desired line=1/2(negative

    reciprocal of-2)

$(x_1,y_1)=(-1,3)$

m=1/2

$y-y_1=m(x-x_1)$

$y-3=1/2(x+1)$

$y-3=1/2x+1/2$

$y=1/2x+7/2$ Slope-intercept form

Problem Set 9.1.3

9. $x^2/4 + y^2/9 = 1$

Ellipse

x-intercepts(y=0): $x^2/4 = 1$

$x^2 = 4$

$x = \pm 2$

y-intercepts(x=0): $y^2/9 = 1$

$y^2 = 9$

$y = \pm 3$

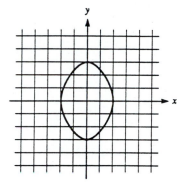

Domain = { x|-2≤x≤2 }

Range = { y|-3≤y≤3 }

11. $x^2/4 - y^2/9 = 1$

Hyperbola

x-intercepts: $x^2/4 = 1$

$x^2 = 4$

$x = \pm 2$

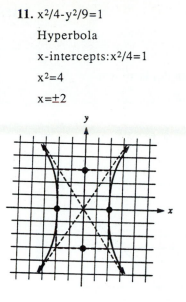

Domain = { x|x≤-2 or x≥2 }

Range = { y|y∈R }

13. $f(x)=x^2-25/x-5$

$=(x+5)(x-5)/x-5$

$=x+5(x\neq5)$

Since $x\neq5$, an open dot is shown

on the line above 5.

$f(x)=x+5$

slope=1   y-intercept=5

y is a function of x.

$$\text{Domain}=\{x|x<5\text{or } x>5\}$$

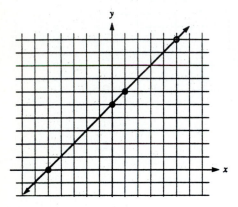

Since $x\neq5$,the range will

never have a value of

$f(5)=5+5=10.$

Range=$\{y|y<10$ or $y>10\}$

15. $y=x^3-8/x-2$

$=(x-2)(x^2+2x+4)/x-2$

$=x^2+2x+4$ $x\neq2$

$y=x^2+2x+4$

Parabola: x-intercepts(set y=0):

$x^2+2x+4=0$

$b^2-4ac=4-4(1)(4)<0$

No real roots; no x-intercepts

y-intercept(set x=0):   y=4

Vertex:

$x=-b/2a=-2/2(1)=-1$

$(-1,3)$

Other value:If x=-2,

$y=(-2)^2+2(-2)+4=4$

$(-2,4)$

Since $x\neq2,y\neq2^2+(2)(2)+4$

$y\neq12.$

An open dot is shown at$(2,12)$.

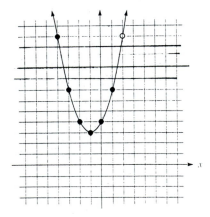

y is a function of x.

Domain=$\{x|x<2\text{or } x>2\}$

Range=$\{y|y\geq3\}$

17. $f(x)=\sqrt{x-3}$

x	3	4	7
f(x)	0	1	2

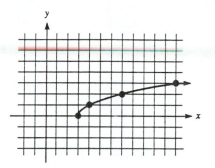

y is a function of x.

Domain=$\{x|x\geq3\}$

Range=$\{y|y\geq0\}$

19. $f(x)=\sqrt{x-3}$ if x≥0

         -x+3   if x<3

Add to the graph of ex.17 the portion of the line y=-x+3 (slope:-1,y-intercept:3) to the left of x=3.

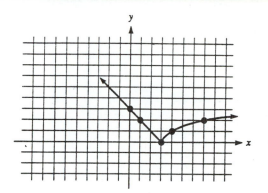

y is a function of x.

Domain=$\{x|x\in R\}$

Range=$\{y|y\geq0\}$

49. $f(x)=3x+2/5x^2-20$

        $=3x+2/5(x^2-4)$

        $=3x+2/5(x+2)(x-2)$

        Domain=$\{x|x\neq-2 \text{ and } x\neq2\}$

51. $f(x)=5x-2/6x^2+x-2$

        $=5x-2/(3x+2)(2x-1)$

        $3x+2\neq0 \text{ and } 2x-1\neq0$

        $x\neq-2/3$ $x\neq1/2$

        Domain=$\{x|x\neq-2/3 \text{ and } x\neq1/2\}$

55. $f(x)=\sqrt{2x+1}$

$2x+1\geq 0$

$2x\geq -1$

$x\geq -1/2$

Domain=$\{x|x\geq -1/2\}$

59. $f(x)=1/\sqrt{2x+1}$

$2x+1>0 \quad 2x+1\neq 0$ because

the expression is in the denominator.)

$x>-1/2$

Domain=$\{x|x>-1/2\}$

61. $f(x)=\sqrt{6x^2+13x-5}$

$6x^2+13x-5\geq 0$

$6x^2+13x-5=0$

$(3x-1)(2x+5)=0$

$x=1/3 \quad x=-5/2$

```
  T       F        T
←——+——————+———————+——→
  -5/2           1/3
```

Test -3: $6x^2+13x-5\geq 0$

$\quad 6(-3)^2+13(-3)-5\geq 0$

$\quad 10\geq 0$True

Test 0: $6(0)^2+13(0)-5\geq 0$

$\quad -5\geq 0$False

Test 1: $6(1)^2+13(1)-5\geq 0$

$\quad 14\geq 0$True

Domain=$\{x|x\leq -5/2 \text{ or } x\geq 1/3\}$

63. $f(x)=\dfrac{5}{\sqrt{6x^2-x-2}}$

$6x^2-x-2>0$

$6x^2-x-2=0$

$(3x-2)(2x+1)=0$

$3x-2=0 \quad 2x+1=0$

$x=2/3 \qquad x=-1/2$

```
    T        F         T
←———+————————+—————————+——→
   -1/2             2/3
```

Test -1: $6x^2-x-2>0$

$\qquad 6(-1)^2-(-1)-2>0$

$\qquad\qquad 5>0$ True

Test 0: $6(0)^2-0-2>0$

$\qquad\qquad -2>0$ False

Test 1: $6(1)^2-1-2>0$

$\qquad\qquad 3>0$ True

Domain=$\{x| \; x<-1/2 \; \text{ or } \; x>2/3\}$

69.

$$f(x)=\begin{cases} 2.4 & \text{if } 0<x\leq 1 \\ 2.74 & \text{if } 2<x\leq 3 \\ 3.18 & \text{if } 3<x\leq 4 \\ 3.61 & \text{if } 4<x\leq 5 \end{cases}$$

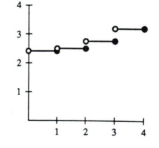

71. x²-2x-3/x²+x+1÷ x²-9/x³-1

=[(x-3)(x+1)/x²+x+1][(x-1)(x²+x

+1)/(x-3)(x+3)]

=(x+1)(x-1)/x+3

72. (1/y+2)+[y+3/(y+1)(y+2)]

-[3y-1/(y+1)(y-1)]

=[1/y+2][(y-1)(y+1)/(y-1)(y+1)]

+[y+3/(y+1)(y+2)][(y-1)/(y-1)]

-[3y-1/(y+1)(y-1)][(y+2)/(y+2]

=(y-1)(y+1)+(y+3)(y-1)-(3y-1)(y+2)/

(y+2)(y-1)(y+1)

=y²-1+y²+2y-3-3y²-5y+2/(y+2)(y-1)(y+1)

=-y²-3y-2/(y+2)(y-1)(y+1)

=-(y²+3y+2)/(y+2)(y-1)(y+1)

=-(y+2)(y+1)/(y+2)(y-1)(y+1)

=-1/y-1[Equivalently: 1/1-y]

73. [3√2-2√3/4√2-3√3][4√2+3√3/4√2+3√3]

=12(2)+9√6-8√6-6(3)/16(2)-9(3)

=6+√6/5

Problem Set 9.2

1. $f(x)=3x, g(x)=2x+5$

   $f[g(x)]=3[g(x)]=3(2x+5)=6x+15$

   $g[f(x)]=2[f(x)]+5=2(3x)+5=6x+5$

3. $f(x)=3x+4, g(x)=5x-2$

   $f[g(x)]=3[g(x)]+4=3(5x-2)+4=15x-2$

   $g[f(x)]=5[f(x)]-2=5(3x+4)-2=15x+18$

5. $f(x)=2x+5, g(x)=x^2+2$

   $f[g(x)]=2[g(x)]+5=2(x^2+2)+5=2x^2+9$

   $g[f(x)]=[f(x)]^2+2=(2x+5)^2+2=4x^2+20x+27$

7. $f(x)=5, g(x)=1-2x^3$

   $f[g(x)]=5$

   $g[f(x)]=1-2[f(x)]^3=1-2(5)^3=1-250=-249$

13. $f(x)=x+5/3, g(x)=3x-5$

   $f[g(x)]=g(x)+5/3=(3x-5)+5/3=3x/3=x$

   $g[f(x)]=3[f(x)]-5=3(x+5/3)-5=x+5-5=x$

15. $P=f(t)=10,000+100t^2$

   $C=g(p)=0.5p+2$

   a. Air pollution is a function
      of time.

   b. $g[f(6)]$:

   $f(6)=10,000+100(6)^2=13,600$

   $g[f(6)]=g(13,600)=0.5(13,600)+2=6,802$

   6,802 parts per million

   c. $f(10)=10,000+100(10)^2=20,000$

   $g[f(10)]=g(20,000)=0.5(20,000)+2$

   $\qquad =10,002$

   In 10 years, the carbon monoxide
   level will be 10,002 parts per million.

   d. $g[f(t)]=0.5[f(t)]+2$

   $=0.5(10,000+100t^2)+2$

   $=5,000+50t^2+2=50t^2+5,002$

19. $f(x)=3x+5$

   $y=3x+5$

   Inverse: $x=3y+5$

   $x-5=3y$

   $x-5/3=y$

   $f^{-1}(x)=x-5/3$

   $f[f^{-1}(x)]=3[f^{-1}(x)]+5=3(x-5/3)+5=x-5+5=x$

   $f^{-1}[f(x)]=f(x)-5/3=(3x+5)-5/3=3x/3=x$

25. $f(x)=1/2x+3$

   $y=1/2x+3$

   Inverse: $x=1/2y+3$

   $2x=y+6$

   $2x-6=y$

   $f^{-1}(x)=2x-6$

   $f[f^{-1}(x)]=1/2[f^{-1}(x)]+3=1/2(2x-6)+3$

   $\quad=x-3+3=x$

   $f^{-1}[f(x)]=2[f(x)]-6=2(1/2x+3)-6=x+6-6=x$

27. $f(x)=3/4x-2/3$

   $y=3/4x-2/3$

   Inverse: $x=3/4y-2/3$

   $12x=12(3/4y-2/3)$

   $12x=9y-8$

   $12x+8=9y$

   $12x+8/9=y$

   $f^{-1}(x)=12x+8/9$

   $f^{-1}(x)=12x/9+8/9$

   $f^{-1}(x)=4/3x+8/9$

   $f[f^{-1}(x)]=3/4[f^{-1}(x)]-2/3$

   $\quad=3/4[4/3x+8/9]-2/3$

   $\quad=x+2/3-2/3=x$

   $f^{-1}[f(x)]=4/3[f(x)]+8/9=4/3[3/4x-2/3]+8/9$

   $\quad=x-8/9+8/9=x$

29. $f(x)=\sqrt{2x+3}$

   $y=\sqrt{2x+3}$

   Inverse: $x=\sqrt{2y+3}$

   $x^2=2y+3$

   $x^2-3=2y$

   $x^2-3/2=y$

   $f^{-1}(x)=x^2-3/2$

   $f[f^{-1}(x)]=\sqrt{2[f^{-1}(x)]+3}=\sqrt{2(x^2-3/2)+3}$

   $\quad=\sqrt{x^2-3+3}=\sqrt{x^2}=x$

   $f^{-1}[f(x)]=[f(x)]^2-3/2=(\sqrt{2x+3})^2-3/2$

   $\quad\quad=2x+3-3/2=2x/2=x$

35. $f(x)=x^5-1$

   $y=x^5-1$

   Inverse: $x=y^5-1$

   $x+1=y^5$

   $\sqrt[5]{(x+1)}=y$

   $f^{-1}(x)=\sqrt[5]{(x+1)}$

   $f[f^{-1}(x)]=[f^{-1}(x)]^5-1=[\sqrt[5]{(x+1)}]^5-1$

   $\quad=x+1-1=x$

   $f^{-1}[f(x)]=\sqrt[5]{f(x)+1}$

   $\quad\quad=\sqrt[5]{(x^5-1)+1}=\sqrt[5]{(x^5)}=x$

37. $f(x)=mx+b$

$y=mx+b$

Inverse:$x=my+b$

$x-b=my$

$x-b/m=y$

$f^{-1}(x)=x-b/m$

$f[f^{-1}(x)]=m[f^{-1}(x)]+b=m(x-b/m)+b$

$\qquad =x-b+b=x$

$f^{-1}[f(x)]=f(x)-b/m=(mx+b)-b/m=mx/m=x$

39. $f(x)=4x+5$

$g(x)=1/4x+5$

$f[g(x)]=4[g(x)]+5=4(1/4x+5)+5$

$=4/4x+5+5(4x+5)/(4x+5)$

$\qquad =4+20x+25/4x+5$

$=20x+29/4x+5\neq x$

Since $f[g(x)]\neq x$, g is not the inverse of f.

51. Function; All vertical lines intersect the
graph no more than once.

Inverse is a function(one-to-one);All
horizontal lines intersect the graph no
more than once.

53. $y=x^2-1$ Inverse:$x=y^2-1$

$y^2=x+1$

$y=\pm\sqrt{(x+1)}$ (not a function)

x	0	3
y	±1	±2

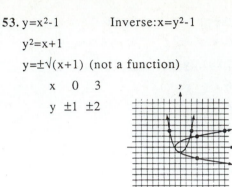

57. $y=x^3$

x	-2	-1	0	1	2
y	-8	-1	0	1	8

inverse:$x=y^3$

$y=\sqrt[3]{x}$ (function)

x	-8	-1	0	1	8
y	-2	-1	0	1	2

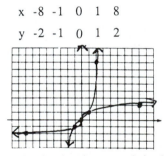

61. Ordered pairs on original function

$(-3,1),(-1,0),(1,1)$

Ordered pairs on inverse:

$(1,-3),(0,-1),(1,1)$

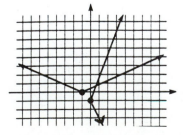

63.   Ordered pairs on original function:

(0,1),(1,2),(2,4)

Ordered pairs on inverse:

(1,0),(2,1),(4,2)

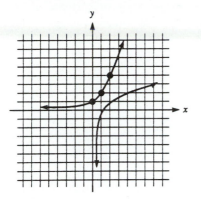

67. $g(x)=5x-2$

$[g(x)]^2=(5x-2)^2=25x^2-20x+4$

$g(x^2)=5x^2-2$

$g[g(x)]=5[g(x)]-2=5(5x-2)-2=25x-12$

69. $f(x)=x-1$

$f[g(x)]=g(x)-1=x^2$

$g(x)=x^2+1$

71. $f(x)=3x-2, g(x)=x^5$

$f[g(x)-f(x)]=3[g(x)-f(x)]-2$

$=3[x^5-(3x-2)]-2=3(x^5-3x+2)-2$

$=3x^5-9x+4$

$f[g(x)]-f[f(x)]=3[g(x)]-2-[3[f(x)]-2]$

$=3x^5-2-[3(3x-2)-2]$

$=3x^5-2-(9x-8)=3x^5-9x+6$

73. $f(x)=2x-5, g(x)=3x+b$

$f[g(x)]=2[g(x)]-5=2(3x+b)-5=6x+2b-5$

$g[f(x)]=3[f(x)]+b=3(2x-5)+b=6x-15+b$

Given: $f[g(x)]=g[f(x)]$

$6x+2b-5=6x-15+b$

$2b-5=-15+b$

$b=-10$

75. $y^{2/3}+7y^{1/3}+12=0$

Let $t=y^{1/3}$

$t^2+7t+12=0$

$(t+4)(t+3)=0$

$t+4=0$ or $t+3=0$

$t=-4$ $t=-3$

$y^{1/3}=-4$ $y^{1/3}=-3$

$(y^{1/3})^3=(-4)^3$ $(y^{1/3})^3=(-3)^3$

$y=-64$ $y=-27$

$\{-64,-27\}$

76. $y=x^2-4x+3$

x-intercepts(set y=0):$0=x^2-4x+3$

$0=(x-3)(x-1)$

$x=3 \qquad x=1$

y-intercept(set x=0):$y=3$

vertex:$x=-b/2a=-(-4)/2(1)=4/2=2$

$y=2^2-(4)(2)+3=-1$

Vertex:$(2,-1)$

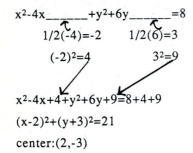

77. $x^2+y^2-4x+6y-8=0$

$x^2-4x\underline{\hspace{1cm}}+y^2+6y\underline{\hspace{1cm}}=8$

$1/2(-4)=-2 \qquad 1/2(6)=3$

$(-2)^2=4 \qquad\qquad 3^2=9$

$x^2-4x+4+y^2+6y+9=8+4+9$

$(x-2)^2+(y+3)^2=21$

center:$(2,-3)$

Problem Set 9.3.1

1. $f(x)=3^x$

x	f(x)
-2	1/9
-1	1/3
0	1
1	3
2	9

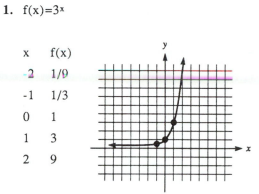

3. $f(x)=(1/3)^x$

x	f(x)
-2	9
-1	3
0	1
1	1/3
2	1/9

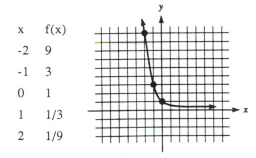

9. $f(x)=2^{-x}$

x	f(x)
-3	8
-2	4
-1	2
0	1
1	1/2
2	1/4

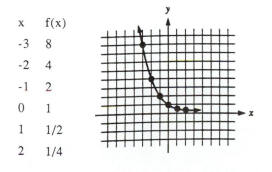

11. $f(x)=2^{x+1}$

x	f(x)
-3	1/4
-2	1/2
-1	1
0	2
1	4
2	8

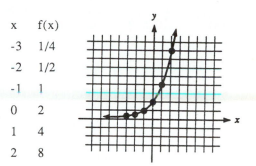

15. $f(x)=2^{2x}$

x	f(x)
-2	1/16
-1	1/4
0	1
1	4
2	16

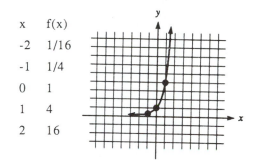

19. $f(x)=(3/4)^x$

x	f(x)
-3	$1/(3/4)^3=64/27\cong2.4$
-2	16/9=1 7/9
-1	4/3=1 1/3
0	1
1	3/4
2	9/16

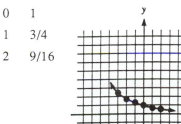

23. $f(x)=(1/2)^{-x}$

x	f(x)
-3	$(1/2)^3=1/8$
-2	1/4
-1	1/2
0	1
1	$(1/2)^{-1}$
	$=2$
2	4
3	8

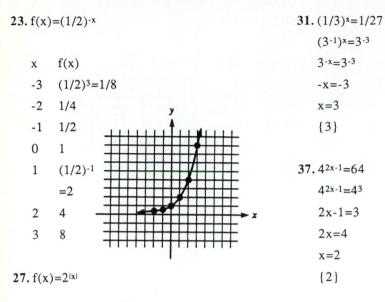

27. $f(x)=2^{|x|}$

x	f(x)
-3	8
-2	4
-1	2
0	1
1	2
2	4
3	8

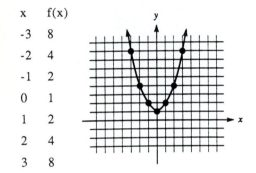

29. $2^x=32$

$2^x=2^5$

$x=5$

$\{5\}$

31. $(1/3)^x=1/27$

$(3^{-1})^x=3^{-3}$

$3^{-x}=3^{-3}$

$-x=-3$

$x=3$

$\{3\}$

37. $4^{2x-1}=64$

$4^{2x-1}=4^3$

$2x-1=3$

$2x=4$

$x=2$

$\{2\}$

39. $16^x=32$

$(2^4)^x=2^5$

$2^{4x}=2^5$

$4x=5$

$x=5/4$

$\{5/4\}$

43. $2^x+6=38$

$2^x=32$

$2^x=2^5$

$x=5$

$\{5\}$

45. $32^x=1/8$

$(2^5)^x=2^{-3}$

$2^{5x}=2^{-3}$

$5x=-3$

$x=-3/5$

$\{-3/5\}$

51. $(5^{2x})(5^{4x})=125$

$5^{6x}=5^3$

$6x=3$

$x=3/6=1/2$

$\{1/2\}$

55. $6^3=(2x-1)^3$

$6=2x-1$

$7=2x$

$7/2=x$

$\{7/2\}$

57. $5^{x^2-12}=25^{2x}$

$5^{x^2-12}=(5^2)^{2x}$

$5^{x^2-12}=5^{4x}$

$x^2-12=4x$

$x^2-4x-12=0$

$(x-6)(x+2)=0$

$x-6=0$ or $x+2=0$

$x=6$ $x=-2$

$\{-2,6\}$

61. $f(x)=(10^6)(2^x)$

a.

x	f(x)
-3	$(1/8)(10^6)$
-2	$(1/4)(10^6)$

Note:Since x represents time, the graph only shows the portion of the function for $x\geq0$.

x	f(x)
-1	$(1/2)(10^6)$
0	$(1)(10^6)$
1	$(2)(10^6)$
2	$(4)(10^6)$
3	$(8)(10^6)$

b. $f(5)/f(2)=(10^6)(2^5)/(10^6)(2^2)=2^3=8$

The count is 8 times as great.

65.   Graph   Calculator

a. $2^{1.15}$ 2 4/20=2.2 $2y^x1.15=2.21914$

b. $2^{2.65}$ 6 4/20 =6.2   6.27667

c. $2^{-1.5}$ 7/20=0.35   0.353553

d. $2^{-2.25}$ 4/20=0.2   0.210224

67. $f(x)=b^x$

$f(2)=b^2=1/4$

$b^2=(1/2)^2$

$b=1/2$

$f(x)=(1/2)^x$

$f(5)=(1/2)^5=1/32$

69. $f(x)=2^x+2^{-x}$

$g(x)=2^x-2^{-x}$

a. $f(x)+g(x)=(2^x+2^{-x})+(2^x-2^{-x})$

   $=(2)(2^x)=2^{x+1}$

b. $[f(x)][g(x)]=[(2^x+2^{-x})][(2^x-2^{-x})]$

   $=(2^x)(2^x)-(2^{-x})(2^{-x})$(Outside and

inside terms cancel.)

   $=2^{2x}-2^{-2x}$

c. $[f(x)]^2=(2^x+2^{-x})(2^x+2^{-x})$

$=2^{2x}+2^0+2^0+2^{-2x}=2^{2x}+2^{-2x}+2$

$[g(x)]^2=(2^x-2^{-x})(2^x-2^{-x})$

$=2^{2x}-2^0-2^0+2^{-2x}=2^{2x}+2^{-2x}-2$

$[f(x)]^2-[g(x)]^2=(2^{2x}+2^{-2x}+2)-(2^{2x}+2^{-2x}-2)$

   $=2-(-2)=4$

71. $3-4[2+(5-1)^2(6)]$

$=3-4[2+(16)(6)]$

$=3-4(2+96)$

$=3-4(98)$

$=3-392$

$=-389$

72. x:amount invested at 7%

12,000-x:amount invested at 5%

Interest earned at 7% + Interest earned

    at 5%=660

$.07x+.05(12,000-x)=660$

$.07x+600-.05x=660$

$.02x=60$

$x=3000$

Invested at7%: \$3,000

Invested at5%: 12,000-x

    $=12,000-3,000=\$9,000$

73.

$$x+7\,\overline{\smash{\big)}\,2x^3+11x^2-25x-28} \quad \frac{2x^2-3x-4}{}$$

$-7\rfloor$ 2   11   -25   -28

         -14   21   28

    2   -3   -4   0

$2x^3+11x^2-25x-28=(x+7)(2x^2-3x-4)$

Problem Set 9.3.2

1. $f(x)=10,000e^{x/4.6}$

   (a)$f(0)=10,000e^0=10,000$

      10,000 cells were present initially.

   b. $f(11.5)=10,000e^{11.5/4.6} \approx 121,825$

   [Calculator:$11.5 \div 4.6 = e^x \times 10,000 =$]

   Approximately 121,825 cells

3. $f(x)=6,164e^{0.00667x}$

   a. For 1650:

   $f(1650)=6,164e^{0.00667(1650)} \approx 371,000,000$

   Calculator:$0.00667 \times 1650 = e^x \times 6,164 =$

   b. The year 2000 is 20 years after 1980.

   $f(20)=4.2e^{0.02(20)} \approx 6.3$ billion

7. $A=Pe^{rx}$ $A=P(1+r/N)^{Nx}$

   $P=1$ and $x=2000-1700=300$

   a. $A=P(1+r/N)^{Nx}=1(1+0.03/1)^{(1)(300)}$

   $=(1.03)^{300} \approx 7098$

   [Calculator:$1.03 y^x 300 =$]

   Approximately \$7,098

   b. $A=P(1+r/N)^{Nx}=1(1+0.03/4)^{(4)(300)}$

   $=(1.0075)^{1200} \approx 7835$

   Approximately \$7,835

   c. $A=P(1+r/N)^{Nx}=1(1+0.03/52)^{(52)(300)}$

   $=(1.0005769)^{15,600}$

   ≈ 8079

   Approximately \$8,079

   d. $A=Pe^{rx}=1e^{(0.03)(300)} \approx 8103$

   Approximately \$8,103

9. $f(x)=0.8/1+e^{-0.2x}$

a. $f(0)=0.8/1+e^{-0.2(0)}=0.8/1+1=0.8/2=0.4$

0.4(or 40%)of the responses are correct prior to learning.

b. $f(10)=0.8/1+e^{-0.2(10)}$

$=0.8/1+e^{-2}=0.8/1+(0.1353)$

≈ 0.7

0.7 (or approximately 70%) of the responses are correct after 10 learning trials.

c. As x gets larger and larger,

$e^{-0.2x}=1/e^{0.2x}$ gets very close to zero.Thus, $f(x)=0.8/1+e^{-0.2x}$ gets close to $0.8/1+0=0.8$.

As continued learning takes place, 0.8(or 80%) of the responses will be correct.

11. $f(x)=1/\sqrt{2\pi}\ e^{-x^2/2}=0.4e^{-x^2/2}$

a. $f(0)\approx(0.4)e^0\approx0.4$

$f(1)\approx(0.4)e^{-1/2}\approx0.24$

$f(2)\approx(0.4)e^{-4/2}\approx0.05$

$f(-1)\approx(0.4)e^{-1/2}\approx0.24$

$f(-2)\approx(0.4)e^{-4/2}\approx0.05$

b. $\overset{\downarrow}{\sqrt{2\pi}\ e^{x^2/2}}$ has a denominator which gets extremely large.Since the numerator stays the same size (1), the expression approaches 0. This is shown by the graph getting closer and closer to the x-axis as x increases in size.

c. Again,the denominator gets large and the expression approaches 0. The graph gets closer to the x-axis as x decreases in size.

14. $f(x)=x^2+4x-3$

$f(b+h)-f(b)/h$

$=(b+h)^2+4(b+h)-3-(b^2+4b-3)/h$

$=b^2+2bh+h^2+4b+4h-3-b^2-4b+3/h$

$=2bh+h^2+4h/h=2b+h+4$

15. $y=10x-x^2/2$

$50=10x-x^2/2$

$100=20x-x^2$

$x^2-20x+100=0$

$(x-10)^2=0$

$x=10$

Density:10

16. x: Rider's speed

Time against the wind+Time with the

wind=3

$\qquad$ [RT=D;T=D/R]

40/x-10+40/x+10=3

(x-10)(x+10)[40/x-10+40/x+10]

$\qquad$ =(x-10)(x+10)(3)

40(x+10)+40(x-10)=3x²-300

$\qquad$ 40x+400+40x-400=3x²-300

$\qquad$ 0=3x²-80x-300

$\qquad$ 0=(x-30)(3x+10)

$\qquad$ x-30=0 or 3x+10=0

$\qquad$ x=30   x=-10/3(reject)

Rider's speed:30m/h

Problem Set 9. 4. 1

13. $\sqrt{100}=10$

     $100^{1/2}=10$

     $\log_{100}10=1/2$

15. $\sqrt[3]{64}=4$

     $(64)^{1/3}=4$

     $\log_{64}4=1/3$

17. $(\sqrt{81})^3=729$

     $[(81^{1/2})]^3=729$

     $81^{3/2}=729$

     $\log_{81}729=3/2$

23. $\sqrt{\sqrt{16}}=2$

     $(16^{1/2})^{1/2}=2$

     $16^{1/4}=2$

     $\log_{16}2=1/4$

39. $\log_3 9=x$

     $3^x=9$

     $x=2$

43. $\log_7\sqrt{7}=x$

     $7^x=\sqrt{7}$

     $7^x=(7)^{1/2}$

     $x=1/2$

47. $\log_2(1/32)=x$

     $2^x=1/32$

     $2^x=2^{-5}$

     $x=-5$

53. $\log_{10}1=x$

     $10^x=1$

     $x=0$

55. $\log_{10}(0.0001)=x$

     $10^x=0.0001$

     $10^x=10^{-4}$

     $x=-4$

57. $\log_{0.01}0.001=x$

     $0.01^x=0.001$

     $(10^{-2})^x=10^{-3}$

     $10^{-2x}=10^{-3}$

     $-2x=-3$

     $x=3/2$

59. $\log_{0.5}16=x$

     $(0.5)^x=16$

     $(1/2)^x=16$

     $(2^{-1})^x=2^4$

     $2^{-x}=2^4$

     $-x=4$

     $x=-4$

63. $10^{\log_{10}8}=x$

     In logarithmic form:

     $\log_{10}x=\log_{10}8$

     $x=8$

65. $\log_3(\log_7 7)$

Consider: $\log_7 7 = x$

$\qquad 7^x = 7$

$\qquad x = 1$

$\log_3(\log_7 7) = \log_3 1 = y$

$\qquad 3^y = 1$

$\qquad y = 0$

69. $\log_5 x = 3$

$5^3 = x$

$125 = x$

$\{125\}$

75. $\log_8 x = 2/3$

$8^{2/3} = x$

$(\sqrt[3]{8})^2 = x$

$2^2 = x$

$4 = x$

$\{4\}$

77. $\log_{125} x = -2/3$

$125^{-2/3} = x$

$1/125^{2/3} = x$

$1/(\sqrt[3]{125})^2 = x$

$1/5^2 = x$

$1/25 = x$

$\{1/25\}$

81. $\log_b 36 = 1/2$

$b^{1/2} = 36$

$\sqrt{b} = 36$

$(\sqrt{b})^2 = (36)^2$

$b = 1296$

$\{1296\}$

87. $\log_3(x-1) = 2$

$3^2 = x-1$

$9 = x-1$

$10 = x$

$\{10\}$

89. $\log_{10}(x^2 + 9x) = 1$

$10^1 = x^2 + 9x$

$0 = x^2 + 9x - 10$

$0 = (x+10)(x-1)$

$x+10 = 0 \quad \text{or} \quad x-1 = 0$

$x = -10 \qquad x = 1$

$\{-10, 1\}$

91. $\log_4(1/64) = -x^2 + x$

$4^{-x^2+x} = 1/64$

$4^{-x^2+x} = 4^{-3}$

$-x^2 + x = -3$

$0 = x^2 - x - 3$

$x = -b \pm \sqrt{b^2 - 4ac}/2a = 1 \pm \sqrt{(-1)^2 - 4(1)(-3)}/2(1)$

$= 1 \pm \sqrt{1 - (-12)}/2 = 1 \pm \sqrt{13}/2$

$\{1 + \sqrt{13}/2, 1 - \sqrt{13}/2\}$

95. $f(x)=4^x$

$y=4^x$

Inverse: $x=4^y$

$y=\log_4 x$

$f^{-1}(x)=\log_4 x$

$f(x)=4^x$		$f^{-1}(x)=\log_4 x$	
x	f(x)	x	$f^{-1}(x)$
-2	1/16	1/16	-2
-1	1/4	1/4	-1
0	1	1	0
1	4	4	1
2	16	16	2

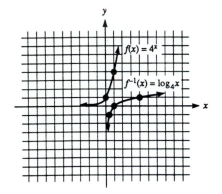

97. $f(x)=(1/3)^x$

$y=(1/3)^x$

Inverse: $x=(1/3)^y$

$y=\log_{1/3} x$

$f^{-1}(x)=\log_{1/3} x$

$f(x)=(1/3)^x$		$f^{-1}(x)=\log_{1/3} x$	
x	f(x)	x	$f^{-1}(x)$
-2	9	9	-2
-1	3	3	-1
0	1	1	0
1	1/3	1/3	1
2	1/9	1/9	2

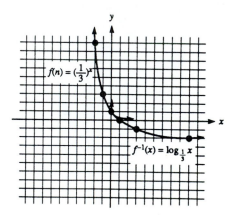

99. $f(x)=12\log_5(2x-5)$

　a. $f(15)=12\log_5[(2)(15)-5]$

　$=12\log_5 25$　$[\log_5 25=y; 5^y=25; y=2]$

　$=12(2)$

　$=24$

　b. $f(65)=12\log_5[(2)(65)-5]$

　$=12\log_5 125$　$[\log_5 125=y; 5^y=125; y=3]$

　$=12(3)$

　$=36$

　c. $f(315)=12\log_5[(2)(315)-5]$

　$=12\log_5 625$　$[\log_5 625=y; 5^y=625; y=4]$

　$=12(4)$

　$=48$

101. $D=10\log_{10}(I/10^{-12})$

　a. $D=10\log_{10}(10^{-8}/10^{-12})$

　　$=10\log_{10}[10^{-8-(-12)}]$

　$=10\log_{10}10^4=10(4)[\log_{10}10^4=y; 10^y$

　　　　　　　$=10^4; y=4]$

　$=40$

　b. $D=10\log_{10}(10^{-6}/10^{-12})$

　　$=10\log_{10}10^6=10(6)=60$

　$D=10\log_{10}(10^0/10^{-12})$

　　$=10\log_{10}(10^{12})=10(12)=120$

103. $\log_{25}5=x$

$25^x=5$

$(5^2)^x=5$

$5^{2x}=5^1$

$2x=1$

$x=1/2$

$\log_{1/16}8=y$

$(1/16)^y=8$

$(2^{-4})^y=2^3$

$2^{-4y}=2^3$

$-4y=3$

$y=-3/4$

$\log_9 1/27=z$

$9^z=1/27$

$(3^2)^z=3^{-3}$

$3^{2z}=3^{-3}$

$2z=-3$

$z=-3/2$

$\log_4 1=w$

$4^w=1$

$w=0$

Thus, $\log_{25}5-\log_{1/16}8/\log_9 1/27+\log_4 1$

$=1/2-(-3/4)/-3/2+0$

$=5/4/-3/2=-5/6$

105. $\log_5 1 = x$

$5^x = 1$

$5^x = 5^0$

$x = 0$

$\log_8[(4)(\sqrt[5]{16})] = y$

$8^y = (4)(\sqrt[5]{16})$

$(2^3)^y = (2^2)\sqrt[5]{2^4}$

$(2^3)^y = (2^2)(\sqrt[5]{2^4})$

$2^{3y} = 2^{2+4/5}$

$2^{3y} = 2^{14/5}$

$3y = 14/5$

$y = 14/15$

Thus, $\log_5 1 + \log_8(4)(\sqrt[5]{16})$

$\quad = 0 + 14/15 = 14/15$

107. $d = \sqrt{(x_2-x_1)^2+(y_2-y_1)^2} = \sqrt{(1+2)^2+(-3+9)^2}$

$\quad = \sqrt{3^2+6^2} = \sqrt{9+36} = \sqrt{45} = \sqrt{(9)(5)} = 3\sqrt{5}$

108. a. $m = y_2 - y_1/x_2 - x_1$

$\quad\quad = 104-98/323-129 = 6/194 = 3/97$

$(x_1, y_1) = (129, 98)$

$y - y_1 = m(x - x_1)$

$y - 98 = 3/97(x - 129)$

$y - 98 = 3/97x - 387/97$

$y = 3/97x - 387/97 + 98 \quad [98 = 9506/97]$

$y = 3/97x + 9119/97$

b. $y = 3/97(800) + 9119/97$

$y = 11519/97 \approx 119$

Approximately 119 people will die.

109. $2x+5y=11$(multiply by 2)

$3x-2y=-12$(multiply by 5)

$4x+10y=22$

$\underline{15x-10y=-60}$

Add: $19x=-38$

$x=-2$

$2x+5y=11$

$2(-2)+5y=11$

$-4+5y=11$

$5y=15$

$y=3$

$\{(-2,3)\}$

Problem Set 9. 4. 2

3. $\log_{10}(10)(100)=\log_{10}1000=x$

$\qquad 10^x=1000$

$\qquad x=3$

$\log_{10}10=y$

$10^y=10$

$y=1$

$\log_{10}100=z$

$10^z=100$

$z=2$

$\log_{10}(10)(100)=\log_{10}10+\log_{10}100$

$3=1+2$

$\qquad 3=3$

5. $\log_{3}(9)(1/3)=\log_{3}3=x$

$\qquad 3^x=3$

$\qquad x=1$

$\log_{3}9=y$

$3^y=9$

$y=2$

$\log_{3}1/3=z$

$3^z=1/3$

$z=-1$

$\log_{3}(9)(1/3)=\log_{3}9+\log_{3}1/3$

$1=2+(-1)$

$\qquad 1=1$

9. $\log_{e}(e^{17}/e^{4})=\log_{e}e^{13}=x$

$\qquad e^x=e^{13}$

$\qquad x=13$

$\log_{e}e^{17}=y$

$e^y=e^{17}$

$y=17$

$\log_{e}e^{4}=z$

$e^z=e^{4}$

$z=4$

$\log_{e}(e^{17}/e^{4})=\log_{e}e^{17}-\log_{e}e^{4}$

$13=17-4$

$13=13$

11. $\log_{5}5^3=\log_{5}125=x$

$\qquad 5^x=125$

$\qquad x=3$

$\log_{5}5=y$

$5^y=5$

$y=1$

$\log_{5}5^3=3\log_{5}5$

$3=3(1)$

$3=3$

15. $\log_{3}3x=\log_{3}3+\log_{3}x=1+\log_{3}x$

17. $\log_{b}x^2y=\log_{b}x^2+\log_{b}y=2\log_{b}x+\log_{b}y$

19. $\log_{10}x/100=\log_{10}x-\log_{10}100$

$\qquad\qquad =\log_{10}x-\log_{10}10^2$

$\qquad\qquad =\log_{10}x-2$

23. $\log_4 \sqrt{x}/16 = \log_4 \sqrt{x} - \log_4 16$

$\quad = \log_4 x^{1/2} - \log_4 4^2$

$\quad = 1/2\log_4 x - 2$

27. $\log_b x\sqrt{y}(^3\sqrt{z}) = \log_b x + \log_b \sqrt{y} + \log_b {}^3\sqrt{z}$

$\quad = \log_b x + \log_b y^{1/2} + \log_b z^{1/3}$

$\quad = \log_b x + 1/2\log_b y + 1/3\log_b z$

31. $\log_5 \sqrt{x/y} = \log_b(x/y)^{1/2} = 1/2[\log_b(x/y)]$

$\quad = 1/2(\log_b x - \log_b y)$

33. $\log_5 {}^3\sqrt{x^2 y/25} = \log_5(x^2 y/25)^{1/3}$

$\quad = 1/3\log_5(x^2 y/25)$

$\quad = 1/3[\log_5(x^2 y) - \log_5 25]$

$\quad = 1/3(\log_5 x^2 + \log_5 y - \log_5 5^2)$

$\quad = 1/3(2\log_5 x + \log_5 y - 2)$

$\quad [\text{Equivalently:} 2/3\log_5 x + 1/3\log_5 y - 2/3]$

35. $\log_b \sqrt{\sqrt{x}\, y^3} = \log_b(x^{1/2})^{1/2} y^3$

$\quad = \log_b x^{1/4} + \log_b y^3$

$\quad = 1/4\log_b x + 3\log_b y$

39. $3\log_2 x + 1/2\log_2(x+3) = \log_2 x^3 + \log_2(x+3)^{1/2}$

$\quad = \log_2[x^3(x+3)^{1/2}] = \log_2 x^3\sqrt{x+3}$

41. $\log_3(x^2-9) - \log_3(x-3) = \log_3 x^2 - 9/x - 3$

$\quad = \log_3(x-3)(x+3)/x - 3 = \log_3(x+3)$

43. $1/2\log_b x + 1/2\log_b y = \log_b x^{1/2} + \log_b y^{1/2}$

$\quad = \log_b x^{1/2} y^{1/2} = \log_b \sqrt{xy}$

49. $1/3\log_4 x + 2\log_4(3x+2)$

$\quad = \log_4 x^{1/3} + \log_4(3x+2)^2$

$\quad = \log_4 {}^3\sqrt{x}(3x+2)^2$

51. $1/2(\log_{10}x + \log_{10}y) = 1/2(\log_{10}xy)$

$\quad = \log_{10}(xy)^{1/2}$

$\quad = \log_{10}\sqrt{xy}$

55. $\log_b M = R$ and $\log_b N = S$

$\quad b^r = M$ and $b^s = N$

$\quad M/N = b^R/b^S = b^{R-S}$

Rewrite $M/N = b^{R-S}$ in logarithmic form.

$\quad \log_b(M/N) = R - S$

Substitute the original expressions

$\quad\quad$ for R and S.

$\quad \log_b(M/N) = \log_b M - \log_b N$

57. a. $\log_b(M)(1/M) = \log_b 1 = x$

$\quad\quad b^x = 1$

$\quad\quad x = 0$

$\quad$ **b.** $\log_b(M)(1/M) = \log_b M + \log_b 1/M$

$\quad\quad = 0$

$\quad \log_b 1/M = -\log_b M$

59. False

$\quad$ True: $\log_b \sqrt{x} = 1/2\log_b x$

63. False

$\quad \log_b \sqrt{xy/z} = \log_b(xy/z)^{1/2} = 1/2\log_b(xy/z)$

$\quad = 1/2[\log_b(xy) - \log_b z]$

$\quad\quad = 1/2(\log_b x + \log_b y - \log_b z)$

$\quad$ True: $\log_b \sqrt{xy/z} = 1/2(\log_b x + \log_b y - \log_b z)$

65. $\log_{10}100 = \log_{10}10^2 = 2$

$\quad \log_2 8 = \log_2 2^3 = 3$

$\quad (\log_{10}100)(\log_2 8) = (2)(3) = 6$

$\quad$ The statement is true.

71. R:influenza rate

D:number of people with the disease

N:number of people not ill with

    the disease

R=KDN

$99=K(1000)(99,000)$

$99=K(99,000,000)$

$0.000001=K$

$R=0.000001DN$

$R=(0.000001)(2000)(98,000)$

$R=196$

The rate is 196 cases per week.

72. x:width of walkway

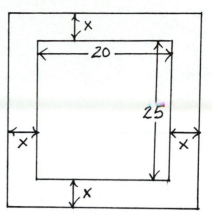

Area of walkway=250.

Area of large rectangle-area of inner

    rectangle=250

$(20+2x)(25+2x)-(20)(25)=250$

$500+90x+4x^2-500=250$

$4x^2+90x-250=0$

$2x^2+45x-125=0$

$(2x-5)(x+25)=0$

$2x-5=0$ or $x+25=0$

$x=5/2=2.5$ $x=-25$(reject)

Width of walkway:2.5yards

73. $3x^2+10x\geq8$

$3x^2+10x-8\geq0$

$(3x-2)(x+4)\geq0$

$(3x-2)(x+4)=0$

$x=2/3 \quad x=-4$

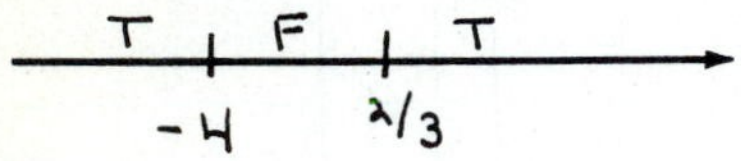

Test-5:$3(-5)^2+10(-5)-8\geq0$

$\quad\quad 17\geq0$True

Test0:$3(0)^2+10(0)-8\geq0$

$\quad\quad\quad -8\geq0$False

Test1:$3(1)^2+10(1)-8\geq0$

$\quad\quad\quad 5\geq0$True

$\quad\quad \{x\vert x\leq-4 \text{or } x\geq2/3\}$

Problem Set 9. 4. 3

3. **a.** We found $\log 0.2 = -0.69897$.

Now find x such that

$\log x = -0.69897$.

Possible key sequences:

0.69897 [+/-] [10^x]

0.69897 [+/-] [INV] [log]

$x = 0.2$

b. We found $\log 0.8 = -0.09691$.

Now find x such that

$\log x = -0.09691$.

Using a sequence like one of those

above:

$x = 0.8$

7. $R = \log I/I_o$

$\log I/I_o = 8$

[calculator: 8 [10^x]

or 8 [INV] [log]]

$I/I_o = 1(10^8)$

$I = 10^8 I_o$

10^8 times more intense

9. $\log(1-r) = 1/T \log W/P$

$\log(1-r) = 1/6 \log 3{,}000/12{,}000$

$\log(1-r) = 1/6 \log(0.25)$

$\log(1-r) = 1/6(-0.60206)$

$\log(1-r) = -0.10034$

$1 - r = 0.7937$

$1 - 0.7937 = r$

$0.206 = r$

Rate of depreciation: 20.6%

13. $\log C10^{8t} = \log C + \log 10^{8t}$

$= \log C + 8t$

17. $\log[A^3 10^{x^4}]/\ln[A^3 e^{x^3}]$

$= \log A^3 + \log 10^{x^4}/\ln A^3 + \ln e^{x^3}$

$= 3\log A + x^4/3\ln A + x^3$

19. $e^{\ln 7x^2 + \ln 2x} = e^{\ln(7x^2)(2x)} = e^{\ln 14x^3} = 14x^3$

21. $e^{\ln 14x^5 y^3 - \ln 2x^2 y} = e^{\ln 14x^5 y^3 / 2x^2 y} = e^{\ln 7x^3 y^2} = 7x^3 y^2$

23. $\log_b x = \ln x / \ln b$

   a. $\log_3 7 = \ln 7 / \ln 3 \approx 1.771$

   sequence: $7\,\boxed{\ln}\,+3\,\boxed{\ln}\,\boxed{=}$

   b. $\log_5 84 = \ln 84 / \ln 5 \approx 2.753$

   c. $\log_7 2.3 = \ln 2.3 / \ln 7 \approx 0.428$

   d. $\log_6 0.34 = \ln 0.34 / \ln 6 \approx -0.602$

   e. $\log_9 1,400 = \ln 1,400 / \ln 9 \approx 3.297$

   f. $\log_4 0.002 = \ln 0.002 / \ln 4 \approx -4.483$

25. a. $\log_b a = \log_a a / \log_a b = 1 / \log_a b$

   b. $\log_2 8 = x$

   $2^x = 8$

   $x = 3$

   $\log_8 2 = y$

   $8^y = 2$

   $(2^3)^y = 2^1$

   $2^{3y} = 2^1$

   $3y = 1$

   $y = 1/3$

   Thus, $\log_2 8 = 3$ and $\log_8 2 = 1/3$.

   $\log_2 8 = 1 / \log_8 2$

   $3 = 1/1/3$

       $3 = 3$

   c. $\log_3 81 = x$

   $3^x = 81$

   $x = 4$

   $\log_{81} 3 = y$

   $81^y = 3$

   $(3^4)^y = 3$

   $3^{4y} = 3^1$

   $4y = 1$

   $y = 1/4$

   Thus, $\log_3 81 = 4$ and $\log_{81} 3 = 1/4$.

   Substituting these values,

   $\log_3 81 = 1 / \log_{81} 3$.

27. $\log_{b^n}x=\log_b x/\log_b b^n=\log_b x/n$

Since $\log_{b^n}x=\log_b x/n$, then

$1/n\log_b x=\log_{b^n}x$.

Multiplying the

range values of $\log_b x$

by $1/n$ yields the range values

of $\log_{b^n}x$.

28. $f(x)=3x+17$

$y=3x+17$

Inverse: $x=3y+17$

$x-17=3y$

$x-17/3=y$

$f^{-1}(x)=x-17/3$

$f[f^{-1}(x)]=3[f^{-1}(x)]+17=3(x-17/3)+17$

$=x-17+17=x$

$f^{-1}[f(x)]=[f(x)]-17/3$

$\quad\quad =(3x+17)-17/3=3x/3=x$

29. (1) $x-5y-2z=6$

(2) $2x-3y+z=13$

(3) $3x-2y+4z=22$

Eliminate x.

$-2(1)+(2):7y+5z=1$ (multiply by-2)

$-3(1)+(3):13y+10z=4$

$\quad\quad -14y-10z=-2$

$\quad\quad \underline{13y+10z=4}$

$\quad\quad$ Add: $-y=2$

$\quad\quad$ $y=-2$

$\quad\quad 7y+5z=1$

$\quad\quad 7(-2)+5z=1$

$\quad\quad 5z=15$

$\quad\quad\quad z=3$

$\quad\quad x-5y-2z=6$

$\quad\quad x-5(-2)-2(3)=6$

$\quad\quad x+10-6=6$

$\quad\quad x+4=6$

$\quad\quad\quad x=2$

$\quad\quad\quad \{(2,-2,3)\}$

30. $x^2/25 - y^2/4 = 1$ $[x^2/a^2 - y^2/b^2 = 1]$

x-intercepts: $x^2/25 = 1$

$x^2 = 25$

$x = \pm 5$

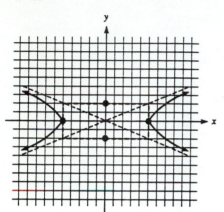

Problem Set 9. 5. 1

1. $\log(x+4)=\log x+\log 4$

   $\log(x+4)=\log(x)(4)$

   $x+4=4x$

   $4=3x$

   $4/3=x$

   $\{4/3\}$

7. $\log(x+4)-\log 2=\log(5x+1)$

   $\log x+4/2=\log(5x+1)$

   $x+4/2=5x+1$

   $\qquad x+4=10x+2$

   $\qquad 2=9x$

   $\qquad 2/9=x$

   $\qquad \{2/9\}$

9. $2\log x-\log 7=\log 112$

   $\log x^2-\log 7=\log 112$

   $\log x^2/7=\log 112$

   $x^2/7=112$

   $x^2=784$

   $\qquad x=\pm\sqrt{784}$

   $x=28$ or $x=-28$ (reject;produces the

        log of a negative number)

   $\qquad \{28\}$

11. $\log x+\log(x+3)=\log 10$

    $\log x(x+3)=\log 10$

    $x(x+3)=10$

    $x^2+3x-10=0$

    $(x+5)(x-2)=0$

    $x+5=0$ or $x-2=0$

    $x=-5$(reject; causes   or $x=2$

    log of a negative number)

    $\qquad \{2\}$

15. $\log_3(x+4)+\log_3(x+1)=2\log_3(x+3)$

    $\log_3(x+4)(x+1)=\log_3(x+3)^2$

    $(x+4)(x+1)=(x+3)^2$

    $\qquad\qquad x^2+5x+4=x^2+6x+9$

    $5x+4=6x+9$

    $\qquad\qquad -5=x$

    Reject-5

    (produces the log of a negative number)

    $\qquad \emptyset$

17. $\ln(1-x)-\ln(1+x)=\ln e$

    $\ln 1-x/1+x=\ln e$

    $1-x/1+x=e$

    $\qquad 1-x=e(1+x)$

    $\qquad 1-x=e+ex$

    $\qquad 1-e=x+ex$

    $\qquad 1-e=x(1+e)$

    $\qquad 1-e/1+e=x$

    Since $e\approx 2.72$, $1-e/1+e\approx-0.46$ and this value

    does not cause the natural log of a

    negative number in the equation.

    $\qquad \{1-e/1+e\}$

19. $\log(2x+1)+\log(x-2)=2\log x$

$\log(2x+1)(x-2)=\log x^2$

$(2x+1)(x-2)=x^2$

$2x^2-3x-2=x^2$

$x^2-3x-2=0$

$x=-b\pm\sqrt{b^2-4ac}/2a=3\pm\sqrt{9-4(1)(-2)}/2$

$=3\pm\sqrt{17}/2 \qquad 3-\sqrt{17}/2\approx-.56$ (reject)

$\{3+\sqrt{17}/2\}$

23. $\log x+\log 50=2$

$\log(x)(50)=2$

Exponential form: $10^2=50x$

$100=50x$

$2=x$

$\{2\}$

27. $\log_2(x+2)-\log_2(x-5)=3$

$\log_2 x+2/x-5=3$

Exponential form: $2^3=x+2/x-5$

$8=x+2/x-5$

$8(x-5)=x+2$

$8x-40=x+2$

$7x=42$

$x=6$

$\{6\}$

29. $\log_3(x-5)+\log_3(x+3)=2$

$\log_3(x-5)(x+3)=2$

Exponential form: $3^2=(x-5)(x+3)$

$9=x^2-2x-15$

$0=x^2-2x-24$

$0=(x-6)(x+4)$

$x-6=0$ or $x+4=0$

$x=6 \qquad x=-4$(reject)

$\{6\}$

31. $\log_6 x+\log_6(x-12)=2$

$\log_6 x(x-12)=2$

Exponential form: $6^2=x(x-12)$

$36=x^2-12x$

$0=x^2-12x-36$

$x=-b\pm\sqrt{b^2-4ac}/2a=12\pm\sqrt{144-4(1)(-36)}/2(1)$

$=12\pm\sqrt{288}/2=12\pm\sqrt{(144)(2)}/2=12\pm12\sqrt{2}/2$

$=6\pm6\sqrt{2}$

Reject $6-6\sqrt{2}\approx-2.5$, since this produces
the log of a negative number.

$\{6+6\sqrt{2}\}$

33. $\log_2(x-6)-\log_2 x+5=7-\log_2(x-4)$

$\log_2(x-6)-\log_2 x+\log_2(x-4)=2$

$\log_2(x-6)(x-4)/x=2$

Exponential form: $2^2=(x-6)(x-4)/x$

$4x=x^2-10x+24$

$0=x^2-14x+24$

$0=(x-12)(x-2)$

$x-12=0$ or $x-2=0$

$x=12$　$x=2$(reject;produces

log of a negative number in the

equation)

$\{12\}$

37. $x^2+y^2=34$

$\underline{x^2-y^2=16}$

Add: $2x^2=50$

$x^2=25$

$x=\pm5$

If $x=5$ or $x=-5$: $x^2+y^2=34$

$25+y^2=34$

$y^2=9$

$y=\pm3$

$\{(5,3),(5,-3),(-5,3),(-5,-3)\}$

$x^2+y^2=34$ circle:center(0,0)

radius $\sqrt{34}\approx5.8$

$x^2-y^2=16$

$x^2/16-y^2/16=1$ Hyperbola:x-intercepts

$x^2/16=1$

$x^2=16$

$x=\pm4$

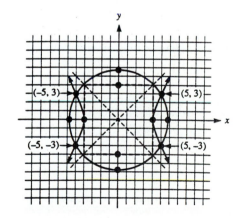

38. Slope of 4y-2x=8:

$\qquad$ 4y=2x+8

$\qquad\qquad$ y=2x+8/4

$\qquad\qquad$ y=1/2x+4

$\qquad\qquad\quad\uparrow$

$\qquad\qquad$ slope=1/2

Slope of the line whose equation we
must write=-2(negative reciprocal of1/2)

m=-2

(x_1,y_1)=(1,-3)

$y-y_1=m(x-x_1)$

y+3=-2(x-1) Point-Slope Form

y+3=-2x+2

y=-2x-1 Slope-Intercept Form

39. $p=-125t^2+1000t+10,000$

Maximum p occurs when t=-b/2a.

t=-b/2a=-1000/2(-125)=-1000/-250=4

After 4 hours the population will be
at a maximum.

$p=-125(4)^2+1000(4)+10,000$

=-2000+4000+10,000

=12,000

Maximum population: 12,000 bacteria

Problem Set 9.5.2

1. $10^x=2.91$
    $\log10^x=\log2.91$
    $x=\log2.91\approx0.46$

5. $e^x=2.7$
    $\ln e^x=\ln2.7$
    $x=\ln2.7\approx0.99$

9. $2e^{1.4x}=26$
    $e^{1.4x}=13$
    $\ln e^{1.4x}=\ln13$
    $1.4x=\ln13$
    $x=\ln13/1.4\approx1.83$
    Calculator sequence: $13\boxed{\ln}\boxed{\div}1.4\boxed{=}$

11. $3.4(10^{1.8x})=68$
    $10^{1.8x}=68/3.4$
    $10^{1.8x}=20$
    $\log10^{1.8x}=\log20$
    $1.8x=\log20$
    $x=\log20/1.8\approx0.72$

15. $0.7(10^{-1.3x})-21.7=16.24$
    $0.7(10^{-1.3x})=16.24+21.7$
    $0.7(10^{-1.3x})=37.94$
    $10^{-1.3x}=37.94/0.7$
    $10^{-1.3x}=54.2$
    $\log10^{-1.3x}=\log54.2$
    $-1.3x=\log54.2$
    $x=\log54.2/-1.3\approx1.33$
    calculator sequence:
    $54.2\boxed{\log}\boxed{\div}1.3\boxed{+/-}\boxed{=}$

17. $800-500e^{-0.5x}=733$
    $-500e^{-0.5x}=-67$
    $e^{-0.5x}=-67/-500$
    $e^{-0.5x}=0.134$
    $\ln e^{-0.5x}=\ln0.134$
    $-0.5x=\ln0.134$
    $x=\ln0.134/-0.5\approx4.02$

19. $A(1-e^{-x})=B$
    $A-Ae^{-x}=B$
    $A-B=Ae^{-x}$
    $A-B/A=e^{-x}$
    $\ln(A-B/A)=\ln e^{-x}$
    $\ln(A-B/A)=-x$
    $-\ln(A-B/A)=x$
    [Equivalently: $x=-\ln(1-B/A)$]

21. $(A+5)e^{-x/3}=A^2-25$

   $e^{-x/3}=A^2-25/A+5=(A+5)(A-5)/A+5$

   $e^{-x/3}=A-5$

   $\ln e^{-x/3}=\ln(A-5)$

   $-x/3=\ln(A-5)$

      $x=-3\ln(A-5)$

23. $400-Ae^{-x/k}=200$

   $-Ae^{-x/k}=-200$

   $Ae^{-x/k}=200$

   $e^{-x/k}=200/A$

   $\ln e^{-x/k}=\ln 200/A$

   $-x/k=\ln 200/A$

      $x=-k\ln 200/A$

25. $5^x=9$

   $\log 5^x=\log 9$

   $x\log 5=\log 9$

   $x=\log 9/\log 5\approx 1.37$

27. $7^{x-1}=13$

   $\log 7^{x-1}=\log 13$

   $(x-1)\log 7=\log 13$

   $x\log 7-\log 7=\log 13$

   $x\log 7=\log 13+\log 7$

   $x=\log 13+\log 7/\log 7\approx 2.32$

   calculator sequence:

      $13\boxed{\log}+7\boxed{\log}\boxed{=}\div 7\boxed{\log}\boxed{=}$

   Equivalently:$x=\log 13+\log 7/\log 7$

      $=\log(13)(7)/\log 7=\log 91/\log 7$

29. $3^{x^2}=11$

   $\log 3^{x^2}=\log 11$

   $x^2\log 3=\log 11$

   $x^2=\log 11/\log 3$

   $x=\pm\sqrt{\log 11/\log 3}\approx\pm 1.48$

   calculator sequence:$11\boxed{\log}\div 3\boxed{\log}\boxed{=}\boxed{\sqrt{x}}$

33. a. $3+9+27+81=120$

   Using the formula:

   $S=a-ar^n/1-r=3-3(3)^4/1-3=3-243/-2$

        $=-240/-2=120$

b. $S=a-ar^n/1-r$

   $S(1-r)=a-ar^n$

   $ar^n=a-S(1-r)$

   $r^n=a-S(1-r)/a$

   $r^n=1-S/a(1-r)$

   $r^n=1+S/a(r-1)$

   $\log r^n=\log[1+S/a(r-1)]$

   $n\log r=\log[1+S/a(r-1)]$

   $n=\log[1+S/a(r-1)]/\log r$

35. $(x^{-1}y^2)^3/(xy)^{1/2}=x^{-3}y^6/x^{1/2}y^{1/2}=x^{-3-1/2}y^{6-1/2}$

$=x^{-7/2}y^{11/2}=y^{11/2}/x^{7/2}$

37. $4x^2+25y^2=100$

$4x^2/100+25y^2/100=1$

$x^2/25+y^2/4=1$

36. $x^2+y^2=17$

$x-y=5$

Substitution Method:

$x=y+5$

$(y+5)^2+y^2=17$

$y^2+10y+25+y^2=17$

$2y^2+10y+8=0$

$y^2+5y+4=0$

$(y+4)(y+1)=0$

$y+4=0 \quad y+1=0$

$y=-4 \qquad y=-1$

If $y=-4$:$x=y+5=-4+5=1$

If $y=-1$:$x=y+5=-1+5=4$

$\{(1,-4),(4,-1)\}$

Graphic verification:

$x^2+y^2=17$Circle:center$(0,0)$

radius $\sqrt{17}\approx 4.1$

$x-y=5$ Line: x-intercept=5

y-intercept=-5

x-intercepts:$x^2/25=1$

$x^2=25$

$x=\pm5$

y-intercepts:$y^2/4=1$

$y^2=4$

$y=\pm2$

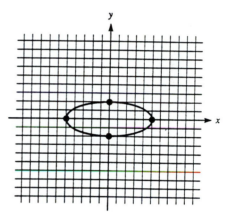

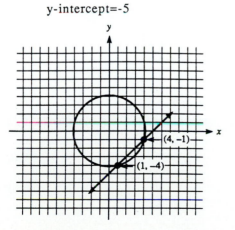

Problem Set 9. 5. 3

1. $A = Pe^{rx}$

$16{,}000 = 8{,}000e^{0.08x}$

$e^{0.08x} = 16{,}000/8{,}000$

$e^{0.08x} = 2$

$\ln e^{0.08x} = \ln 2$

$0.08x = \ln 2$

$x = \ln 2/0.08 \approx 8.7$

It will take 8.7 years.

3. $f(x) = 800 - 500e^{-0.5x}$

$800 - 500e^{-0.5x} = 733$

$-500e^{-0.5x} = -67$

$e^{-0.5x} = 0.134$

$\ln e^{-0.5x} = \ln 0.134$

$-0.5x = \ln 0.134$

$x = \ln 0.134/-0.5 \approx 4$

Approximately 4 months of training

7. a. $f(x) = Ae^{kx}$

$Ae^{kx} = 2A$

$e^{kx} = 2$

$\ln e^{kx} = \ln 2$

$kx = \ln 2$

$x = \ln 2/k$

b. $x = \ln 2/0.02 \approx 35$

It will take 35 years for the world's population to double.

9. $A = P(1 + r/N)^{Nx}$

$2600 = 1000(1 + 0.06/4)^{4x}$

$(1 + 0.015)^{4x} = 2.6$

$(1.015)^{4x} = 2.6$

$\log(1.015)^{4x} = \log 2.6$

$4x\log(1.015) = \log 2.6$

$x = \log 2.6/4\log(1.015) \approx 16$

16 years

11. $A = P(1 + r/N)^{Nx}$

$350 = 200(1 + r/1)^{1(5)}$

$(1 + r)^5 = 1.75$

$\log(1 + r)^5 = \log 1.75$

$5\log(1 + r) = \log 1.75$

$\log(1 + r) = \log 1.75/5$

$\log(1 + r) \approx 0.0486$

$1 + r \approx 1.118$

　　$[0.0486\boxed{10^x}$　or　$0.0486\boxed{INV}\boxed{log}]$

$r \approx 0.118$

Interest rate: 11.8%

15. $f(x)=20e^{kx}$

   a. $f(35)=60$

   $20e^{k(35)}=60$

   $e^{35k}=3$

   $\ln e^{35k}=\ln 3$

   $35k=\ln 3$

   $k=\ln 3/35 \approx 0.031$

   b. $f(x)=20e^{0.031x}$

   $100=20e^{0.031x}$

   $e^{0.031x}=5$

   $\ln e^{0.031x}=\ln 5$

   $0.031x=\ln 5$

   $x=\ln 5/0.031 \approx 51.9$

   Approximately 52 years

17. a. $f(x)=Ae^{kx}$

   Given: $f(0)=100$

   $f(10)=220$

   Use $f(0)=100$ to determine A.

   $f(0)=Ae^{k(0)}=100$

   $Ae^0=100$

   $A(1)=100$

   $A=100$

   $f(x)=100e^{kx}$

   Use $f(10)=220$ to determine k.

   $f(10)=100e^{10k}=220$

   $e^{10k}=2.2$

   $\ln e^{10k}=\ln 2.2$

   $10k=\ln 2.2$

   $k=\ln 2.2/10 \approx 0.079$

   The model is $f(x)=100e^{0.079x}$.

   The year 2000 is 30 years after 1970.

   Find $f(30)$.

   $f(30)=100e^{0.079(30)}$

   $=100e^{2.37}$ [2.37 $\boxed{e^x}$ $\boxed{x}$ 100 $\boxed{=}$]

   ≈ 1070

   For the year 2000, the GNP will

      be \$1070 billion.

   b. $m=y_2-y_1/x_2-x_1=220-100/10-0=12$

   $(x_1,y_1)=(0,100)$

   $y-y_1=m(x-x_1)$

   $y-100=12(x-0)$

   $y=12x+100$

   Year 2000: $x=30$

   $y=12(30)+100$

   $y=460$

   GNP: \$460 billion(with linear growth)

19. Distributive$[(2)(3+7)=(2)(3)+(2)(7)]$

Commutative of multiplication

$[(2)(3)+(2)(7)=(3)(2) +(7)(2)]$

20. x:pounds of 80% nut mixture to be used

Number of pounds　　Number of pounds

of pecans in 80%　+　pecans in 20%

pecan mixture　　　　pecan mixture

=Number of pounds of pecans

in final mixture.

$.8x+.2(50-x)=(.38)(50)$

$.8x+10-.2x=19$

$.6x=9$

$x=9/.6=15$

15pounds

21. $Pr/1-1/(1+r)^{36} \cdot (1+r)^{36}/(1+r)^{36}$

$=Pr(1+r)^{36}/(1+r)^{36}-1$

Review Problems: Chapter 9

1. $f(t)=t^3+2t^2-5t+3$

   a. $f(3)=3^3+2\cdot3^2-5\cdot3+3=27+18-15+3=33$

   $f(1)=1^3+2\cdot1^2-5\cdot1+3=1+2-5+3=1$

   $f(3)-f(1)=33-1=32$

   b. The particle moves 32 inches
   (from 1 inch to 33 inches) on the
   number line between 1 and 3 seconds.

2. $f(x)=2x^2-3x+1$

   $f(a+h)-f(a)/h$

   $=2(a+h)^2-3(a+h)+1-(2a^2-3a+1)/h$

   $=2a^2+4ah+2h^2-3a-3h+1-2a^2+3a-1/h$

   $=4ah+2h^2-3h/h$

   $=4a+2h-3$

3. $g(x)=7$

   $g(a+h)-g(a)/h=7-7/h=0/h=0$

4. d because for one of the values in
   the domain (12) there are two values
   in the range (13 and 19).

5. $f(x)=3x+2, g(x)=2x-1, h(x)=6x^2+x-2$

   a. $(f+g)(x)=f(x)+g(x)$

   $=(3x+2)+(2x-1)=5x+1$

   b. $(h-g)(-1)=h(-1)-g(-1)$

   $=6(-1)^2+(-1)-2-[2(-1)-1]$

   $=6-1-2-(-2-1)$

   $=3-(-3)=6$

   c. $(fh)(x)=f(x)h(x)$

   $=(3x+2)(6x^2+x-2)$

   $=18x^3+3x^2-6x+12x^2+2x-4$

   $=18x^3+15x^2-4x-4$

   d. $(h/g)(x)=h(x)/g(x)=6x^2+x-2/2x-1$

   $=(2x-1)(3x+2)/2x-1$

   $=3x+2 \quad (x\neq1/2)$

6. x: one side   y: adjacent side

   Perimeter=200

   $2x+2y=200$

   $2y=200-2x$

   $y=100-x$

   Area$=xy=x(100-x)$

   $A(x)=x(100-x)$

   [Equivalently:$A(x)=100x-x^2$]

7.　x:number of dimes

x/2:number of nickels

6(x/2)-4=3x-4:number of quarters

V(x)=0.10x+0.05(x/2)+0.25(3x-4)

V(x)=0.1x+0.025x+0.75x-1

V(x)=0.875x-1

8.　$f(x)=1/x^2-5x-24$

$x^2-5x-24\neq 0$

$(x-8)(x+3)\neq 0$

x-8≠0and x+3≠0

x≠8and x≠-3

Domain={xlx≠-3and x≠8}

9.　$g(x)=\sqrt{x^2-5x-24}$

$x^2-5x-24\geq 0$

$(x-8)(x+3)\geq 0$

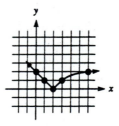

Test-4:$x^2-5x-24\geq 0$

$(-4)^2-5(-4)-24\geq 0$

12≥0True

Test0:$0^2-5\cdot 0-24\geq 0$

-24≥0False

Test9:$9^2-5\cdot 9-24\geq 0$

12≥0True

Domain={xlx≤-3and x≥8}

10.　$h(x)=1/\sqrt{x^2-5x-24}$

$x^2-5x-24>0$

Domain={xlx<-3and x>8}

11.　$f(x)=\begin{cases}\sqrt{x-2} & \text{if } x\geq 2\\ 2-x & \text{if } x<2\end{cases}$

$f(x)=\sqrt{x-2}$

x	2	3	6
f(x)	0	1	2

f(x)=-x+2 graphs as a line with

y-intercept = 2 and slope = -1.

However, only the portion of

the line to the left of x=2 is represented.

Domain={xlx∈R}

Range={yly≥0}

12. b because all vertical lines intersect the graph only once.

Domain=$\{x|-4<x<0$ or $x\geq3\}$

Range=$\{y|y\leq-1$ or $y=2\}$

13. Consider each option.

a. $4x+y^2=20$

$y^2=20-4x$

$y=\pm\sqrt{20-4x}$

Corresponding to certain values of x(such as x=0) there are two values of y($y=\pm\sqrt{20}$). y is not a function of x.

b. x=2

When x=2,y can take on any value. Corresponding to this one domain value (2) there are infinitely many range values.

y is not a function of x.

c. $xy-y=6$

$y(x-1)=6$

$y=6/x-1$

Corresponding to every value of x(except1), there is exactly one value for y.y is a function of x.

d. $x^2+y^2=4$

$y^2=4-x^2$

$y=\pm\sqrt{4-x^2}$

As with option a,y is not a function of x.

14. $P=f(t)=6,000+200t^2$

$C=g(P)=0.5P+1$

a. $f(5)=6,000+200(5)^2=6,000+5,000$

$=11,000$

$g[f(5)]=g(11,000)=0.5(11,000)+1=5,501$

In 5 years from now, the carbon monoxide level will be 5,501 parts per million.

b. $g[f(t)]=0.5[f(t)]+1$

$=0.5(6,000+200t^2)+1$

$=3,000+100t^2+1$

$=100t^2+3,001$

15. $f(x)=3x+2,g(x)=x^2+4x+1$

$f[g(x)]=3[g(x)]+2=3(x^2+4x+1)+2$

$=3x^2+12x+5$

$g[f(x)]=[f(x)]^2+4[f(x)]+1$

$=(3x+2)^2+4(3x+2)+1$

$=9x^2+12x+4+12x+8+1$

$=9x^2+24x+13$

16. $f(x)=4x-3$

$y=4x-3$

inverse:$x=4y-3$

$x+3=4y$

$x+3/4=y$

$f^{-1}(x)=x+3/4$

$f[f^{-1}(x)]=4[f^{-1}(x)]-3=4(x+3/4)-3=x+3-3=x$

$f^{-1}[f(x)]=f(x)+3/4=(4x-3)+3/4=4x/4=x$

17. $f(x)=\sqrt[3]{2x+1}$

$y=\sqrt[3]{2x+1}$

inverse: $x=\sqrt[3]{2y+1}$

$x^3=(\sqrt[3]{2y+1})^3$

$x^3=2y+1$

$x^3-1=2y$

$x^3-1/2=y$

$f^{-1}(x)=x^3-1/2$

$f[f^{-1}(x)]=\sqrt[3]{2[f^{-1}(x)]+1}=\sqrt[3]{2(x^3-1/2)+1}$

$=\sqrt[3]{x^3-1+1}=\sqrt[3]{x^3}=x$

$f^{-1}[f(x)]=[f(x)]^3-1/2=(\sqrt[3]{2x+1})^3-1/2$

$=2x+1-1/2=2x/2=x$

18. c because all horizontal lines intersect the graph only once. c has an inverse that is a function and c is one-to-one.

19. $2^{4x-2}=64$

$2^{4x-2}=2^6$

$4x-2=6$

$4x=8$

$x=2$

$\{2\}$

20. $3^{x^2+4x}=1/27$

$3^{x^2+4x}=3^{-3}$

$x^2+4x=-3$

$x^2+4x+3=0$

$(x+3)(x+1)=0$

$x+3=0$ or $x+1=0$

$x=-3$ $x=-1$

$\{-3,-1\}$

21. $f(x)=16 \cdot 2^{-x}$

a.

x	0	1	2	3
f(x)	16	8	4	2

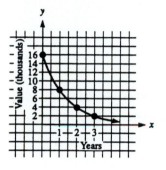

b. $f(3)/f(0)=16\cdot2^{-3}/16\cdot2^0=2^{-3}=1/8$

$f(3)/f(0)=1/8$

$f(3)=1/8f(0)$

At the end of 3 years the automobile is worth only 1/8 of what it was worth when it was new.

22. $f(x)=5000/1+9e^{-0.2x}$

a. $f(0)=5000/1+9e^{-0.2(0)}=5000/1+9=500$

500 fish were stocked in the lake.

b. $f(10)=5000/1+9e^{-0.2(10)}=5000/1+9e^{-2}$

$=5000/1+9(0.13533)$ $[2\boxed{+/-}\boxed{e^x}]$

≈ 2254

Approximately 2,254 fish

c. $f(20)=5000/1+9e^{-0.2(20)}=5000/1+9e^{-4}$

$=5000/1+9(0.018315)$

≈ 4292

Approximately 4,292 fish

d. Consider $9e^{-0.2x}=9/e^{0.2x}$. As x gets larger, the denominator gets larger, but the numerator (9) stays the same size. Thus, $9/e^{0.2x}$ gets closer and closer to 0. Consequently, the fish population approaches $5,000/1+0=5,000$ fish. 5,000 fish will eventually inhabit the lake.

23. $\log_3 1/27=x$

$3^x=1/27$

$3^x=3^{-3}$

$x=-3$

$\{-3\}$

24. $\log_6(3x+4)=2$

$6^2=3x+4$

$36=3x+4$

$32=3x$

$32/3=x$

$\{32/3\}$

25. $\log_{25}x=-3/2$

$(25)^{-3/2}=x$

$1/(25)^{3/2}=x$

$1/(\sqrt{25})^3=x$

$1/5^3=x$

$1/125=x$

$\{1/125\}$

e.

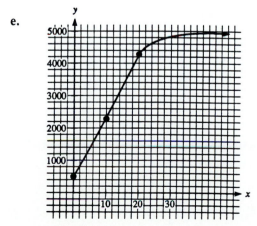

26. $f(x)=5^x$

$y=5^x$

inverse: $x=5^y$

$y=\log_5 x$

$f^{-1}(x)=\log_5 x$

$f(x)=5^x$		$f^{-1}(x)=\log_5 x$	
x	f(x)	x	$f^{-1}(x)$
-2	1/25	1/25	-2
-1	1/5	1/5	-1
0	1	1	0
1	5	5	1
2	25	25	2

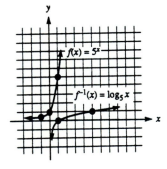

27. $\log_6 x^3 y^2/36=\log_6(x^3 y^2)-\log_6 6^2$

$\qquad =\log_6 x^3+\log_6 y^2-\log_6 6^2$

$\qquad =3\log_6 x+2\log_6 y-2\log_6 6$

$\qquad =3\log_6 x+2\log_6 y-2$

28. $\log_2\sqrt[5]{64/x^3}=\log_2(64/x^3)^{1/5}=1/5\log_2(64/x^3)$

$\qquad =1/5[\log_2 2^6-\log_2 x^3]$

$\qquad =1/5(6-3\log_2 x)=6/5-3/5\log_2 x$

29. $\log B/10^{7a}=\log B-\log 10^{7a}$

$\qquad =\log B-7a\ \ [\log 10^x=x]$

30. $\ln Ae^{9y^2}=\ln A+\ln e^{9y^2}$

$\qquad =\ln A+9y^2\ [\ln e^x=x]$

31. $1/2\ \log_6 x+5\log_6(3x+2)$

$\qquad =\log_6 x^{1/2}+\log_6(3x+2)^5$

$\qquad =\log_6 x^{1/2}(3x+2)^5$

$\qquad =\log_6\sqrt{x}(3x+2)^5$

32. $1/4(\log_b x+\log_b y)-2\log_b(2x+1)$

$\qquad =1/4\log_b x+1/4\log_b y-2\log_b(2x+1)$

$\qquad =\log_b x^{1/4}+\log_b y^{1/4}-\log_b(2x+1)^2$

$\qquad =\log_b x^{1/4}y^{1/4}-\log_b(2x+1)^2$

$\qquad =\log_b\ x^{1/4}y^{1/4}\ /\ (2x+1)^2$

$\qquad =\log_b\ \sqrt[4]{xy}\ /\ (2x+1)^2$

33. $\log(1-r)=1/T\log W/P$

$\log(1-r)=1/4\log\ 10{,}000/20{,}000$

$\log(1-r)=1/4\log 0.5$

$\log(1-r)=-0.0752575\ [0.5\ \log\ \div\ 4\ =]$

$1-r=0.84089$

$r=1-0.84089$

$r=0.15911\approx 15.9\%$

Approximately 15.9% (or 16%)

34. $10^{\log 7x+\log 4x}=10^{\log(7x)(4x)}$

$\qquad =10^{\log 28x^2}$

$\qquad =28x^2\ [10^{\log y}=y]$

35. $e^{\ln 8x^3-\ln 2x}=e^{\ln\ 8x^3/2x}=\ln 4x^2$

$\qquad =4x^2\ [e^{\ln y}=y]$

36. $\log_b x = \log_a x / \log_a b$

    a. $\log_3 7 = \log_{10} 7 / \log_{10} 3 = \log 7 / \log 3$

    b. $\log_3 7 = \log_e 7 / \log_e 3 = \ln 7 / \ln 3$

37. $\log_4(2x+1) = \log_4(x-3) + \log_4(x+5)$

    $\log_4(2x+1) = \log_4(x-3)(x+5)$

    $2x+1 = (x-3)(x+5)$

    $2x+1 = x^2 + 2x - 15$

    $16 = x^2$

    $x = \pm\sqrt{16}$

    $x = \pm 4$

    Reject -4 (produces log of a negative

        number, which is undefined).

    $\{4\}$

38. $\ln(x+1) - \ln 3 = \ln(1-2x)$

    $\ln x+1/3 = \ln(1-2x)$

    $x+1/3 = 1-2x$

    $x+1 = 3(1-2x)$

    $x+1 = 3-6x$

    $7x = 2$

    $x = 2/7$

    $\{2/7\}$

39. $\log_2(x-5) - 5 = \log_2(3x+2) - 7$

    $\log_2(x-5) - \log_2(3x+2) = -2$

    $\log_2 x-5 / 3x+2 = -2$

    Exponential form: $2^{-2} = x-5 / 3x+2$

    $1/4 = x-5 / 3x+2$

    $3x+2 = 4(x-5)$

    $3x+2 = 4x-20$

    $22 = x$

    $\{22\}$

40. $10^x = 72.3$

    $\log 10^x = \log 72.3$

    $x = \log 72.3 \approx 1.86$

41. $0.5 \cdot 10^{-0.2x} + 4.6 = 11.1$

    $0.5 \cdot 10^{-0.2x} = 6.5$

    $10^{-0.2x} = 13$

    $\log 10^{-0.2x} = \log 13$

    $-0.2x = \log 13$

    $x = \log 13 / -0.2 \approx -5.57$

    Calculator sequence:

        13   log $\div$ 0.2 +/- =

42. $5^{x-1} = 12$

    $\log 5^{x-1} = \log 12$

    $(x-1)\log 5 = \log 12$

    $x\log 5 - \log 5 = \log 12$

    $x\log 5 = \log 12 + \log 5$

    $x = \log 12 + \log 5 / \log 5 \approx 2.54$

43. $e^x = 47$

    $\ln e^x = \ln 47$

    $x = \ln 47 \approx 3.85$

44. $1.2e^{2.5x}-8.1=21.3$

$1.2e^{2.5x}=29.4$

$e^{2.5x}=24.5$

$\ln e^{2.5x}=\ln 24.5$

$2.5x=\ln 24.5$

$x=\ln 24.5 / 2.5 \approx 1.28$

45. $2^{x+1}=9$

$\ln 2^{x+1}=\ln 9$

$(x+1)\ln 2=\ln 9$

$x\ln 2+\ln 2=\ln 9$

$x\ln 2=\ln 9-\ln 2$

$x=\ln 9-\ln 2 / \ln 2 \approx 2.17$

46. $f(x)=Ae^{kx}$

$3A=Ae^{kx}$

$e^{kx}=3$

$\ln e^{kx}=\ln 3$

$kx=\ln 3$

$x=\ln 3 / k$

47. $f(x)=140e^{-0.1x}+70$

$140e^{-0.1x}+70=100$

$140e^{-0.1x}=30$

$e^{-0.1x}=3/14$

$e^{-0.1x}=0.21428$

$\ln e^{-0.1x}=\ln 0.21428$

$x=\ln 0.21428 / -0.1 \approx 15.4$

After 15.4 minutes

48. $f(x)=12 \cdot 2^{1.5x}$

a. $f(0)=12 \cdot 2^{1.5(0)}=12 \cdot 2^{0}=12(1)=12$

12 crocodiles

b. The population would reach

12(2)=24 crocodiles.

$12\cdot 2^{1.5x}=24$

$2^{1.5x}=2'$

$1.5x=1$

$x=1/1.5=2/3$

2/3 of a year.

Problem Set 10.1

1. $a_n = 2n - 1$ $a_1 = 2(1) - 1 = 1$

     $a_2 = 2(2) - 1 = 3$ $a_3 = 2(3) - 1 = 5$

     $a_4 = 2(4) - 1 = 7$ $1, 3, 5, 7, \ldots$

11. $a_n = 3^{-n}$ $a_1 = 3^{-1} = \dfrac{1}{3}$

     $a_2 = 3^{-2} = \dfrac{1}{3^2} = \dfrac{1}{9}$ $a_3 = 3^{-3} = \dfrac{1}{3^3} = \dfrac{1}{27}$

     $a_4 = 3^{-4} = \dfrac{1}{3^4} = \dfrac{1}{81}$ $\dfrac{1}{3}, \dfrac{1}{9}, \dfrac{1}{27}, \dfrac{1}{81}, \ldots$

15. $a_n = 1 + \dfrac{1}{n}$ $a_1 = 1 + \dfrac{1}{1} = 2$

     $a_2 = 1 + \dfrac{1}{2} = \dfrac{3}{2}$ $a_3 = 1 + \dfrac{1}{3} = \dfrac{4}{3}$

     $a_4 = 1 + \dfrac{1}{4} = \dfrac{5}{4}$ $2, \dfrac{3}{2}, \dfrac{4}{3}, \dfrac{5}{4}, \ldots$

17. $a_n = \dfrac{n(n+1)}{2}$ $a_1 = \dfrac{1(2)}{2} = 1$

     $a_2 = \dfrac{2(3)}{2} = 3$ $a_3 = \dfrac{3(4)}{2} = 6$

     $a_4 = \dfrac{4(5)}{2} = 10$ $1, 3, 6, 10, \ldots$

19. $a_n = (-1)^n n$ $a_1 = (-1)^1 \cdot 1 = -1$

     $a_2 = (-1)^2 \cdot 2 = 2$ $a_3 = (-1)^3 \cdot 3 = -3$

     $a_4 = (-1)^4 \cdot 4 = 4$ $-1, 2, -3, 4, \ldots$

23. $a_n = 3n - 4$ $a_{12} = 3(12) - 4 = 32$

27. $a_n = 3(2)^{2-n}$ $a_6 = 3(2)^{2-6}$

     $= 3(2)^{-4} = \dfrac{3}{2^4} = \dfrac{3}{16}$

29. $2, 1, \dfrac{2}{3}, \dfrac{1}{2}, \ldots$ $a_1 = 2 = \dfrac{2}{1}$

     $a_2 = 1 = \dfrac{2}{2}$ $a_3 = \dfrac{2}{3}$

     $a_4 = \dfrac{1}{2} = \dfrac{2}{4}$ Thus, $a_n = \dfrac{2}{n}$

33. $1, \dfrac{1}{4}, \dfrac{1}{9}, \dfrac{1}{16}, \ldots$ $a_1 = 1 = \dfrac{1}{1^2}$

     $a_2 = \dfrac{1}{4} = \dfrac{1}{2^2}$ $a_3 = \dfrac{1}{9} = \dfrac{1}{3^2}$

     $a_4 = \dfrac{1}{16} = \dfrac{1}{4^2}$ $a_n = \dfrac{1}{n^2}$

35. $\underset{a_1}{100,} \quad \underset{a_2}{200,} \quad \underset{a_3}{400,}$

    end of   end of

    first hour   second hour

    $\underset{a_4}{800,} \quad \underset{a_5}{1600,} \quad \underset{a_6}{3200}$

After 6 hours :   3200 bacteria

After n hours : $a_1 = 100 = 100 \cdot 2^{1-1}$

                $a_2 = 200 = 100 \cdot 2^{2-1}$

                $a_3 = 400 = 100 \cdot 2^{3-1}$

                $a_4 = 800 = 100 \cdot 2^{4-1}$

                $a_n = 100 \cdot 2^{n-1}$

After n hours : $100 \cdot 2^{n-1}$ bacteria

37. End of first year :   100,000

     $+ 0.05(100,000) = 105,000$

   End of 2 years :   105,000

     $+ 0.05(105,000) = 110,250$

   End of 3 years :   110,250

     $+ 0.05(110,250) = 115,762.50$

   End of 4 years :   115,762.50

     $+ 0.05(115,762.50)$

     $= 121,550.63$

   $105000, 110250, 11576.5, 121550.63, \ldots$

41. $a_n = (1 + \frac{1}{n})^n$

$a_{10} = \left(1 + \frac{1}{10}\right)^{10} = \left(\frac{11}{10}\right)^{10}$

≈ 2.59374

Calculator: $11 \div 10 = y^x \ 10 =$

$a_{100} = \left(1 + \frac{1}{100}\right)^{100} = \left(\frac{101}{100}\right)^{100}$

≈ 2.70481

$a_{1,000} = \left(1 + \frac{1}{1,000}\right)^{1,000}$

$= \left(\frac{1,001}{1,000}\right)^{1,000}$

≈ 2.71692

$a_{10,000} = \left(1 + \frac{1}{10,000}\right)^{10,000}$

$= \left(\frac{10,001}{10,000}\right)^{10,000}$

≈ 2.71815

$a_{100,000} = \left(1 + \frac{1}{100,000}\right)^{100,000}$

$= \left(\frac{100,001}{100,000}\right)^{100,000}$

≈ 2.71827

The terms approach the decimal approximation for e.

43. $a_1 = 1 = 2^1 - 1$

$a_2 = 3 = 2^2 - 1$

$a_3 = 7 = 2^3 - 1$

$a_4 = 15 = 2^4 - 1$

Thus $a_n = 2^n - 1$

$a_5 = 2^5 - 1 = 32 - 1 = 31$

$a_6 = 2^6 - 1 = 64 - 1 = 63$

$a_7 = 2^7 - 1 = 128 - 1 = 127$

$a_8 = 2^8 - 1 = 256 - 1 = 255$

$31, 63, 127, 255$

44. $2y^6 + 16 = 2(y^6 + 8)$

$= 2\left[(y^2)^3 + 2^3\right]$

$2(y^2 + 2)\left[(y^2)^2 - y^2 \cdot 2 + 2^2\right]$

$= 2(y^2 + 2)(y^4 - 2y^2 + 4)$

45. $\dfrac{x^2}{16} + \dfrac{y^2}{9} = 1$

Ellipse

x - intercepts: $\dfrac{x^2}{16} = 1$

$x^2 = 16$

$x = \pm 4$

y - intercepts: $\dfrac{y^2}{9} = 1$

$y^2 = 9$

$y = \pm 3$

46. $\log_4 x + \log_4 (x - 6) = 2$

$\log_4 x(x - 6) = 2$

Exponential form : $4^2 = x(x - 6)$

$16 = x^2 - 6x$

$0 = x^2 - 6x - 16$

$0 = (x - 8)(x + 2)$

$x - 8 = 0$ or $x + 2 = 0$

$x = 8$ $x = -2$

(reject; causes the log of a

negative number)

$\{8\}$

Problem Set 10.2

1. $\displaystyle\sum_{i=1}^{4} 3i = 3(1) + 3(2) + 3(3) + 3(4)$

$= 3 + 6 + 9 + 12 = 30$

3. $\displaystyle\sum_{i=2}^{6} (i^2 + 3) = (2^2 + 3) + (3^2 + 3)$

$+ (4^2 + 3) + (5^2 + 3) + (6^2 + 3)$

$= 7 + 12 + 19 + 28 + 39 = 105$

9. $\displaystyle\sum_{i=1}^{4}\left(-\frac{1}{2}\right)^i = \left(-\frac{1}{2}\right)^1 + \left(-\frac{1}{2}\right)^2$

$+ \left(-\frac{1}{2}\right)^3 + \left(-\frac{1}{2}\right)^4$

$= -\frac{1}{2} + \frac{1}{4} - \frac{1}{8} + \frac{1}{16}$

$= -\frac{8}{16} + \frac{4}{16} - \frac{2}{16} + \frac{1}{16} = \frac{-5}{16}$

13. $\displaystyle\sum_{i=3}^{5} \frac{2i-1}{i-1} = \frac{2(3)-1}{3-1}$

$+ \frac{2(4)-1}{4-1} + \frac{2(5)-1}{5-1}$

$= \frac{5}{2} + \frac{7}{3} + \frac{9}{4} = \frac{30+28+27}{12}$

$= \frac{85}{12}$

15. $\displaystyle\sum_{i=1}^{4} x^i = x + x^2 + x^3 + x^4$

21. $\displaystyle\sum_{i=1}^{4} ix^{i-1} = 1x^{1-1} + 2x^{2-1}$

$+ 3x^{3-1} + 4x^{4-1}$

$= 1 + 2x + 3x^2 + 4x^3$

27. $\displaystyle\bar{x} = \frac{\sum_{i=1}^{4} x_i}{4} = \frac{x_1 + x_2 + x_3 + x_4}{4}$

$= \frac{7.2 + 2.3 + 4.9 + 1.1}{4} = \frac{15.5}{4}$

$= 3.875$

33. $2 + 4 + 6 + 8 + 10$

$= 2(1) + 2(2) + 2(3) + 2(4) + 2(5)$

$= \displaystyle\sum_{i=1}^{5} 2i$

35. $5 + 10 + 17 + 26$

$= (2^2 + 1) + (3^2 + 1) + (4^2 + 1) + (5^2 + 1)$

$= \displaystyle\sum_{i=2}^{5} (i^2 + 1)$

37. $3 + 5 + 7 + 9 = [2(1) + 1]$

$+ [2(2) + 1] + [2(3) + 1] + [2(4) + 1]$

$= \displaystyle\sum_{i=1}^{4} (2i + 1)$

41. $1 + x + x^2 + x^3 + x^4 + x^5$

$= x^0 + x^1 + x^2 + x^3 + x^4 + x^5$

$= \displaystyle\sum_{i=0}^{5} x^i$

$\left[\text{Equivalently}: \displaystyle\sum_{i=1}^{6} x^{i-1}\right]$

49. $x - \dfrac{x^2}{2} + \dfrac{x^3}{3} - \dfrac{x^4}{4} + \dfrac{x^5}{5}$

Consider: $x + \dfrac{x^2}{2} + \dfrac{x^3}{3} + \dfrac{x^4}{4} + \dfrac{x^5}{5}$

$$= \sum_{i=1}^{5} \frac{x^i}{i}$$

Now we must introduce the alternating sign for the terms.

Since $\displaystyle\sum_{i=1}^{5} (-1)^i = -1 + 1 - 1 + 1 - 1,$

we see the signs are not alternating correctly. However,

$$\sum_{i=1}^{5} (-1)^{i+1} = (-1)^2 + (-1)^3 + (-1)^4$$
$$+ (-1)^5 + (-1)^6$$
$$= 1 - 1 + 1 - 1 + 1,$$

and this follows the $+ - + - +$ pattern of the original series. Thus,

$$x - \frac{x^2}{2} + \frac{x^3}{3} - \frac{x^4}{4} + \frac{x^5}{5}$$
$$= \sum_{i=1}^{5} (-1)^{i+1} \frac{x^i}{i}$$

53. $\displaystyle\sum_{i=2}^{5} \log i = \log 2 + \log 3$
$$+ \log 4 + \log 5$$
$$= \log 2 \cdot 3 \cdot 4 \cdot 5$$
$$= \log 120$$

55. $\displaystyle\sum_{i=1}^{4} i \log x = \log x + 2 \log x + 3 \log x + 4 \log x$
$$= \log x + \log x^2 + \log x^3 + \log x^4$$
$$= \log x \cdot x^2 \cdot x^3 \cdot x^4$$
$$= \log x^{10}$$

57. $\displaystyle\sum_{i=0}^{4} (ai + b) = 10$

$a(0) + b + a(1) + b + a(2) + b + a(3)$
$$+ b + a(4) + b = 10a + 5b = 10$$

$\displaystyle\sum_{i=1}^{4} (ai + b) = 14$

$a(1) + b + a(2) + b + a(3) + b$
$$+ a(4) + b = 14$$

$10a + 5b = 10$

$\underline{10a + 4b = 14}$

$b = -4$

$10a + 5b = 10$

$10a + 5(-4) = 10$

$10a - 20 = 10$

$10a = 30$

$a = 3$

$a = 3, b = -4$

59. $\displaystyle\sum_{i=1}^{n} ka_i = ka_1 + ka_2 + ka_3 + ka_4 + \ldots + ka_n$
$$= k(a_1 + a_2 + a_3 + a_4 + \ldots + a_n)$$
$$= k \sum_{i=1}^{n} a_i$$

61. $\displaystyle\sum_{i=1}^{n} a_i b_i = a_1 b_1 + a_2 b_2$
$$+ a_3 b_3 + a_4 b_4 + \ldots + a_n b_n$$

$\displaystyle\sum_{i=1}^{n} a_i \sum_{i=1}^{n} b_i = (a_1 + a_2 + a_3$
$$+ a_4 \ldots + a_n)(b_1 + b_2 + b_3$$
$$+ b_4 + \ldots b_n)$$

By the distributive property (distributing a_1 to each term in the second factor, a_2 to each term in the second factor, etc.), this is not equal to $a_1 b_1 + a_2 b_2 + a_3 b_3$
$+ a_4 b_4 + \ldots + a_n b_n$.

The given statement is not true.

63. $\log_3 xy\sqrt[3]{z} = \log_3 x +$

$\log_3 y + \log_3 z^{\frac{1}{3}}$

$= \log_3 x + \log_3 y + \dfrac{1}{3}\log_3 z$

64. $2x - 4y < 8$

consider : $\quad 2x - 4y = 8$

x - intercept : $\quad 2x = 8$

$x = 4$

y - intercept : $\quad -4y = 8$

$y = -2$

Test $(0,0)$: $\quad 2(0) - 4(0) < 8$

$0 < 8$ True

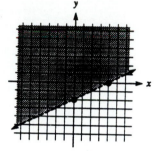

65. $5 - \sqrt{y+5} = \sqrt{y}$

$(5 - \sqrt{y+5})^2 = (\sqrt{y})^2$

$25 - 10\sqrt{y+5} + y + 5 = y$

$30 + y - 10\sqrt{y+5} = y$

$-10\sqrt{y+5} = -30$

$\sqrt{y+5} = 3$

$(\sqrt{y+5})^3 = 3^2$

$y + 5 = 9$

$y = 4$

check 4 : $\quad 5 - \sqrt{y+5} = \sqrt{y}$

$5 - \sqrt{9} = \sqrt{4}$

$5 - 3 = 2$

$2 = 2$

$\{4\}$

Problem Set 10.3.1

1. $2, 6, 10, 14, \ldots$

    $d = 6 - 2 = 4$

5. $a, a + 5b, a + 10d, \ldots$

    Common difference

    $= a + 5d - a = 5d$

9. $4, 7, 10, \ldots$ $d = 7 - 4 = 3$

    $a_n = a_1 + (n - 1)d$

    $a_{26} = 4 + (26 - 1) \cdot 3$

    $a_{26} = 4 + 25 \cdot 3$

    $a_{26} = 4 + 75$

    $a_{26} = 79$

13. $a_1 = 9, d = 2$

    $a_n = a_1 + (n - 1)d$

    $a_{16} = 9 + (16 - 1)(2)$

    $a_{16} = 9 + (15)(2)$

    $a_{16} = 9 + 30 = 39$

17. $7, 10, 13, 16, \ldots$ $d = 10 - 7 = 3$

    $a_n = a_1 + (n - 1)d$

    $a_n = 7 + (n - 1) \cdot 3$

    $a_n = 7 + 3n - 3$

    $a_n = 3_n + 4$

23. $a_1 = 2, d = 5$

    $a_n = a_1 + (n - 1)d$

    $a_n = 2 + (n - 1)5$

    $a_n = 2 + 5n - 5$

    $a_n = 5n - 3$

27. a_1

    $= 60,000$ (value during the

    first year)

    $d = -4,500$

    $a_n = a_1 + (n - 1)d$

    $a_n = 60,000 + (n - 1)(-4,500)$

    $a_n = 60,000 - 4,500n + 4,500$

    $a_n = -4,500n + 64,500$

29. $a_1 = 26,000, \ d = 3,500$

    $a_n = a_1 + (n - 1)d$

    Problem :   What value of n will

    result in $a_n = 99,500$?

    $99,500 = 26,000 + (n - 1)(3,500)$

    $99,500 = 26,000 + 3,500n - 3,500$

    $99,500 = 22,500 + 3,500n$

    $77,000 = 3,500n$

    $n = \dfrac{77,000}{3,500} = 22$

    It will take 22 years to reach

    the maximum salary.

33. Company A : $a_1 = 12,000, d = 800$

    $a_n = a_1 + (n - 1)d$

    Year ten : $a_{10} = 12,000 + (10 - 1)(800)$

    $= 19,200$

    Company B : $a_1 = 14,000, d = 500$

    Year ten : $a_{10} = 14,000 + (10 - 1)(500)$

    $= 18,500$

    Company A pays $19,200 - 18,500 = \$700$

           more.

35. $a_n = a_1 + (n-1)d$

$a_3 = 7:$ $a_3 = a_1 + (3-1)d = 7$

$a_1 + 2d = 7$

$a_8 = 17:$ $a_8 = a_1 + (8-1)d = 17$

$a_1 + 7d = 17$

$a_1 + 2d = 7$

$\underline{a_1 + 7d = 17}$

$-5d = -10$

$d = 2$

$a_1 + 2d = 7$

$a_1 + 2(2) = 7$

$a_1 = 3$

with $a_1 = 3$ and $d = 2:$

$a_n = a_1 + (n-1)d$

$a_n = 3 + (n-1)(2)$

$a_n = 2n + 1$

39. $3-x, \ x, \ \sqrt{9-2x}$

The common difference is obtained
by subtracting consecutive terms.

Common difference : $x - (3-x) = 2x - 3$

Common difference : $\sqrt{9-2x} - x$

Thus :

$2x - 3 = \sqrt{9-2x} - x$

$3x - 3 = \sqrt{9-2x}$

$(3x-3)^2 = (\sqrt{9-2x})^2$

$9x^2 - 18x + 9 = 9 - 2x$

$9x^2 - 16x = 0$

$x(9x - 16) = 0$

$x = 0$ $9x - 16 = 0$

$\qquad\qquad x = \dfrac{16}{9}$

0 is extraneous. Thus, $x = \dfrac{16}{9}$.

41. $x^2 + 4y^2 = 13$

$x^2 - y^2 = 8$ (multiply by -1)

$x^2 + 4y^2 = 13$

$\underline{-x^2 + y^2 = -8}$

Add: $5y^2 = 5$

$y^2 = 1$

$y = \pm 1$

If $y = \pm 1 :$ $x^2 - y^2 = 8$

$x^2 - 1 = 8$

$x^2 = 9$

$x = \pm 3$

$\{(3,1), (3,-1), (-3,1), (-3,-1)\}$

42. Resistance :   R

Current :   I

$R = \dfrac{K}{I^2}$

$50 = \dfrac{K}{(0.8)^2}$

$50 = \dfrac{K}{0.64}$

$(50)(0.64) = K$

$K = 32$

$R = \dfrac{32}{I^2}$

$R = \dfrac{32}{(0.5)^2}$

$R = \dfrac{32}{0.25}$

$R = 128$

128 ohms

43. $\dfrac{4x^2}{(x+y)(x-y)} + \dfrac{x+y}{x-y} \cdot \dfrac{(x+y)}{(x+y)}$

$\qquad\qquad - \dfrac{x-y}{x+y} \cdot \dfrac{(x-y)}{(x-y)}$

$= \dfrac{4x^2 + (x+y)(x+y) - (x-y)(x-y)}{(x+y)(x-y)}$

$= \dfrac{4x^2 + x^2 + 2xy + y^2 - x^2 + 2xy - y^2}{(x+y)(x-y)}$

$= \dfrac{4x^2 + 4xy}{(x+y)(x-y)} = \dfrac{4x(x+y)}{(x+y)(x-y)}$

$= \dfrac{4x}{x-y}$

Problem Set 10.3.2

1. $4, 10, 16, 22, \ldots \quad d = 6$

$$a_n = a_1 + (n-1)d$$

$$a_{20} = 4 + (20-1)6$$

$$a_{20} = 118$$

$$S_n = \frac{n}{2}(a_1 + a_n)$$

$$S_{20} = \frac{n}{2}(a_1 + a_{20})$$

$$= \frac{20}{2}(4 + 118)$$

$$= 1,220$$

5. $100 + 95 + 90 + \ldots + 10$

$$S_n = \frac{n}{2}(a_1 + a_n) \quad a_1 = 100 \text{ and } a_n = 10$$

We must find n (the number of terms

we are adding).

$$a_n = a_1 + (n-1)d \quad 100, 95, 90, \ldots, 10(d = -5)$$

$$10 = 100 + (n-1)(-5)$$

$$10 = 100 - 5n + 5$$

$$10 = 105 - 5n$$

$$5n = 95$$

$$n = 19$$

$$S_n = \frac{19}{2}(100 + 10)$$

$$S_n = 9.5(110)$$

$$S_n = 1045$$

7. $S_n = \frac{n}{2}(a_1 + a_n)$

$$S_{25} = \frac{25}{2}(a_1 + a_{25})$$

$$a_1 = -9, \ d = 5$$

We must find a_{25}.

$$a_n = a_1 + (n-1)d$$

$$a_{25} = -9 + (25-1)(5)$$

$$a_{25} = 111$$

$$S_{25} = \frac{25}{2}(-9 + 111)$$

$$S_{25} = 1275$$

11. $S_n = \frac{n}{2}(a_1 + a_n)$

$$S_{12} = \frac{12}{2}(\frac{1}{2} + a_{12})$$

Find a_{12}.

$$a_n = a_1 + (n-1)d$$

$$a_{12} = \frac{1}{2} + (12-1)(-\frac{1}{2})$$

$$a_{12} = \frac{1}{2} - \frac{11}{2} = -5$$

$$S_{12} = \frac{12}{2}(\frac{1}{2} - 5)$$

$$S_{12} = 6(-4.5)$$

$$S_{12} = -27$$

13. Company A : $a_1 = 15,000$ $d = 500$

$$S_{10} = \frac{n}{2}(a_1 + a_n) = \frac{10}{2}(15,000 + a_{10})$$

$$a_{10} = a_1 + (10 - 1)d$$

$$a_{10} = 15,000 + 9(500) = 19,500$$

$$S_{10} = \frac{10}{2}(15,000 + 19,500) = 172,500$$

Company B : $a_1 = 16,000$ $d = 400$

$$S_{10} = \frac{10}{2}(16,000 + a_{10})$$

$$a_{10} = 16,000 + 9(400) = 19,600$$

$$S_{10} = \frac{10}{2}(16,000 + 19,600) = 178,000$$

Company B will pay the greater amount ($178,000, as opposed to $172,500).

19. $\displaystyle\sum_{i=1}^{17} (5i + 3) = [5(1) + 3] + [5(2) + 3]$

$$+ [5(3) + 3] + \ldots + [5(17) + 3]$$

$$= 8 + 13 + 18 + \ldots + 88$$

$$S_n = \frac{n}{2}(a_1 + a_n)$$

$$S_{17} = \frac{17}{2}(a_1 + a_{17})$$

$$S_{17} = \frac{17}{2}(8 + 88) = 816$$

25. Find S_{12} given that $a_1 = 3$ and $a_{10} = 30$.

$$S_n = \frac{n}{2}(a_1 + a_n)$$

$$S_{12} = \frac{12}{2}(3 + a_{12}) \quad \text{Find } a_{12}.$$

Given : $a_{10} = 30$

$$a_n = a_1 + (n - 1)d$$

$$a_{10} = 3 + (10 - 1)d = 30$$

$$3 + 9d = 30$$

$$9d = 27$$

$$d = 3$$

$$a_{12} = a_1 + (12 - 1)d$$

$$a_{12} = 3 + 11(3) = 36$$

$$S_{12} = \frac{12}{2}(3 + 36) = 6(39) = 234$$

29. $2 + 4 + 6 + \ldots$

$$S_n = \frac{n}{2}(a_1 + a_n)$$

$$S_{30} = \frac{30}{2}(a_1 + a_{30}) \quad a_1 = 2, \ d = 2$$

Find a_{30}.

$$a_n = a_1 + (n - 1)d$$

$$a_{30} = 2 + (30 - 1)(2)$$

$$a_{30} = 60$$

$$S_{30} = \frac{30}{2}(2 + 60)$$

$$= 930$$

33. $22 + 24 + 26 + \ldots + 44 \quad d = 2$

$S_n = \dfrac{n}{2}(a_1 + a_n) \quad a_1 = 22, \ a_n = 44$

Find n.

$a_n = a_1 + (n-1)d$

$44 = 22 + (n-1)2$

$44 = 22 + 2n - 2$

$44 = 20 + 2n$

$24 = 2n$

$12 = n$

$S_n = \dfrac{12}{2}(22 + 44)$

$S_n = 6(66) = 396$

35. $73 + 79 + 85 + \ldots \quad a_1 = 73, \ d = 6$

$S_n = \dfrac{n}{2}(a_1 + a_n)$

$4,077 = \dfrac{n}{2}(73 + a_n) \quad$ Find n.

$a_n = a_1 + (n-1)d$

$a_n = 73 + (n-1)6$

$a_n = 73 + 6n - 6$

$a_n = 6n + 67$

$4,077 = \dfrac{n}{2}(73 + 6n + 67)$

$8154 = n(6n + 140)$

$8154 = 6n^2 + 140n$

$0 = 6n^2 + 140n - 8154$

$0 = 3n^2 + 70n - 4077$

$n = \dfrac{-b \pm \sqrt{b^2 - 4ac}}{2a}$

$= \dfrac{-70 \pm \sqrt{(70)^2 - 4(3)(-4077)}}{2(3)}$

$= \dfrac{-70 \pm \sqrt{53,824}}{6} = \dfrac{-70 \pm 232}{6}$

$n = \dfrac{-70 + 232}{6} = 27$

$n = \dfrac{-70 - 232}{6} = \dfrac{-302}{6} \quad$ (reject; the

number of rows must be a natural number)

27 rows

37. d: the fixed sum

$700,\ 700 + d,\ 700 + 2d, \ldots$

$S_n = \dfrac{n}{2}(a_1 + a_n)$

$S_8 = \dfrac{8}{2}(a_1 + a_8)$

$S_8 = 4(700 + a_8) = 6580$

$2800 + 4a_8 = 6580$

$4a_8 = 3780$

$a_8 = 945$

$a_n = a_1 + (n-1)d$

$a_8 = a_1 + (8-1)d$

$a_8 + 700 + 7d = 945$

$7d = 245$

$d = 35$

Fixed sum : $\$35$

41. $\displaystyle\sum_{i=1}^{n}(ai + b) = (a + b) + (2a + b)$

$+ (3a + b) + \ldots + (na + b)$

$S_n = \dfrac{n}{2}(a_1 + a_n)$

$a_1 = a + b$

$a_n = na + b$

$n = n$

$S_n = \dfrac{n}{2}(a + b + na + b)$

$S_n = \dfrac{n}{2}(a + na + 2b)$

$S_n = \dfrac{na}{2} + \dfrac{n^2 a}{2} + nb$

[Equivalently : $S_n = \dfrac{n(n+1)a}{2} + nb$]

43. $\displaystyle\sum_{i=1}^{2n+1} i = 1 + 2 + 3 + \ldots + (2n+1)$

$S_n = \dfrac{n}{2}(a_1 + a_n)$ $a_1 = 1,\ a_n = 2n + 1,\ n = 2n + 1$

$= \dfrac{2n+1}{2}(1 + 2n + 1)$

$= \dfrac{2n+1}{2} \cdot (2n + 2)$

$= \dfrac{2n+1}{2} \cdot \dfrac{2(n+1)}{1}$

$= (2n+1)(n+1)$

$\dfrac{1}{2n+1}\displaystyle\sum_{i=1}^{2n+1} i = \dfrac{1}{2n+1}(2n+1)(n+1)$

$= n + 1$

45. $(y + 1)(2y + 3) - 3(y + 2)(y + 1) = -3(y + 5)$

$2y^2 + 5y + 3 - 3y^2 - 9y - 6 = -3y - 15$

$-y^2 - 4y - 3 = -3y - 15$

$0 = y^2 + y - 12$

$0 = (y + 4)(y - 3)$

$y + 4 = 0$ or $y - 3 = 0$

$y = -4$ $y = 3$

$\{-4, 3\}$

46. $\begin{vmatrix} 5 & 2 & 34 \\ -1 & 3 & 22 \\ 0 & 0 & 4 \end{vmatrix} = 5\begin{vmatrix} 3 & 22 \\ 0 & 4 \end{vmatrix} + 1\begin{vmatrix} 2 & 34 \\ 0 & 4 \end{vmatrix}$

$= 5(12 - 0) + 1(8 - 0)$

$= 60 + 8 = 68$

47. (1) $4x + y - 2z = 8$

(2) $3x + 2y - z = 5$

(3) $-3y + z = 9$

Eliminate x.

(3) : $-3y + z = 9$

$-3(1) + 4(2)$: $-12x - 3y + 6z = -24$

$$\underline{12x + 8y - 4z = 20}$$

$$5y + 2z = -4$$

$-3y + z = 9$ (multiply by -2)

$5y + 2z = -4$

$6y - 2z = -18$

$\underline{5y + 2z = -4}$

Add: $11y = -22$

$y = -2$

$5y + 2z = -4$

$5(-2) + 2z = -4$

$-10 + 2z = -4$

$2z = 6$

$z = 3$

$3x + 2y - z = 5$

$3x + 2(-2) - 3 = 5$

$3x - 7 = 5$

$3x = 12$

$x = 4$

$\{(4, -2, 3)\}$

Problem Set 10.4.1

1. $81, 54, 36, \ldots$

$$\frac{54}{81} = 0.\bar{6}$$

$$\frac{36}{54} = 0.\bar{6}$$

There is a common ratio. The

sequence is geometric.

$$r = \frac{2}{3}$$

3. $1, 4, 9, 16, \ldots$

$$\frac{4}{1} = 4$$

$$\frac{9}{4} = 2.25$$

No common ratio; not geometric

$$4 - 1 = 3$$

$$9 - 4 = 5$$

No common difference; not arithmetic

Neither

5. $1, -3, 9, -27, \ldots$

$$-\frac{3}{1} = -3; \quad \frac{9}{-3} = -3; \quad \frac{-27}{9} = -3$$

Geometric with $r = -3$

15. $\sqrt{3}, 3, 3\sqrt{3}, \ldots$

$$\frac{3}{\sqrt{3}}$$

$$\frac{3\sqrt{3}}{3} = \sqrt{3}$$

Since $\dfrac{3}{\sqrt{3}} = \sqrt{3} [\dfrac{3}{\sqrt{3}} \cdot \dfrac{\sqrt{3}}{\sqrt{3}} = \dfrac{3\sqrt{3}}{3} = \sqrt{3}]$,

the sequence is geometric with $r = \sqrt{3}$.

19. $a_1 = 10, \; r = \dfrac{1}{2}$

$$a_1 = 10$$

$$a_2 = 10(\frac{1}{2}) = 5$$

$$a_3 = 5(\frac{1}{2}) = \frac{5}{2}$$

$$a_4 = \frac{5}{2}(\frac{1}{2}) = \frac{5}{4}$$

$$a_5 = \frac{5}{4}(\frac{1}{2}) = \frac{5}{8}$$

$$10, 5, \frac{5}{2}, \frac{5}{4}, \frac{5}{8}, \ldots$$

25. $a_1 = \dfrac{a^2}{b}, \; r = \dfrac{2b}{a}$

$$a_1 = \frac{a^2}{b}$$

$$a_2 = \frac{a^2}{b} \cdot \frac{2b}{a} = 2a$$

$$a_3 = 2a \cdot \frac{2b}{a} = 4b$$

$$a_4 = 4b \cdot \frac{2b}{a} = \frac{8b^2}{a}$$

$$a_5 = \frac{8b^2}{a} \cdot \frac{2b}{a} = \frac{16b^3}{a^2}$$

$$\frac{a^2}{b}, 2a, 4b, \frac{8b^2}{a}, \frac{16b^3}{a^2}, \ldots$$

27. $-3, -15, -75, \ldots$

$$r = \frac{-15}{-3} = 5$$

$$a_n = a_1 r^{n-1}$$

$$a_6 = (-3)(5)^{6-1}$$

$$a_6 = (-3)(5)^5 = -9,375$$

31. $250, 50, 10, \ldots$

$$r = \frac{50}{250} = \frac{1}{5}$$

$$a_n = a_1 r^{n-1}$$

$$a_6 = (250)\left(\frac{1}{5}\right)^{6-1}$$

$$a_6 = (250)\left(\frac{1}{5}\right)^{5} = \frac{250}{3125} = \frac{2}{25}$$

33. $\sqrt{2}, 2, 2\sqrt{2}, \ldots$

$$r = \frac{2}{\sqrt{2}} \cdot \frac{\sqrt{2}}{\sqrt{2}} = \frac{2\sqrt{2}}{2} = \sqrt{2}$$

$$a_n = a_1 r^{n-1}$$

$$a_7 = \sqrt{2}(\sqrt{2})^{7-1}$$

$$a_7 = \sqrt{2}(2^{\frac{1}{2}})^{6}$$

$$a_7 = \sqrt{2} \cdot 2^3 = 8\sqrt{2}$$

37. $c^7 d^6, c^6 d^4, c^5 d^2, \ldots$

$$r = \frac{c^6 d^4}{c^7 d^6} = \frac{1}{cd^2}$$

$$a_n = a_1 r^{n-1}$$

$$a_7 = (c^7 d^6)(\frac{1}{cd^2})^{7-1}$$

$$a_7 = (c^7 d^6)(\frac{1}{cd^2})^{6}$$

$$a_7 = \frac{c^7 d^6}{c^6 d^{12}} = cd^{-6} = \frac{c}{d^6}$$

39. $48, 12, 3, \ldots$

$$r = \frac{12}{48} = \frac{1}{4}$$

$$a_n = a_1 r^{n-1}$$

$$a_n = (48)(\frac{1}{4})^{n-1}$$

43. $3, \frac{3}{2}, \frac{3}{4}, \ldots$

$$r = \frac{\frac{3}{2}}{3} = \frac{1}{2}$$

$$a_n = a_1 r^{n-1}$$

$$a_n = 3(\frac{1}{2})^{n-1}$$

45. a. $200, 198, 196.02, 194.0598, \ldots$

b. $a_n = a_1 r^{n-1}$

$$a_n = 200(0.99)^{n-1}$$

47. $a_n = a_1 r^{n-1}$

$a_1 = 12,000$

$r = 0.75$

$a_n = 12,000(0.75)^{n-1}$

$a_4 = (12,000)(0.75)^{4-1} = (12,000)(0.75)^3$

[calculator : $0.75\ y^x\ 3\ =\ \times\ 12,000\ =$]

$a_4 = 5062.5$

Value :　$5062.50

49. $100000,\quad 120000,\quad 144000\ldots$

$a_n = a_1 r^{n-1}$　　　　$r = \dfrac{120,000}{100,000} = 1.2$

$a_n = (100,000)(1.2)^{n-1}$

In twenty years, $n = 5$.

$a_5 = (100,000)(1.2)^{5-1}$

$a_5 = 207,360$

53. Given: $a_1 = 6, a_5 = 96$

Find r.

$a_n = a_1 r^{n-1}$

$a_5 = 6r^{5-1}$

$a_6 = 6r^4$

$16 = r^4$

$r = \pm 2$

Two possible sequences: $6, 12, 24, 48, 96, \ldots$

or $6, -12, 24, -48, 96, \ldots$

In both, $a_1 = 6$ and $a_5 = 96$.

57. Given: $a_3 = 28, \; a_5 = 112$

Find a_1.

$a_n = a_1 r^{n-1}$

$a_3 = a_1 r^{3-1} = 28$

$a_1 r^2 = 28$

$a_5 = a_1 r^{5-1} = 112$

$a_1 r^4 = 112$

$\dfrac{a_1 r^4}{a_1 r^2} = \dfrac{112}{28}$

$r^2 = 4$

$r = \pm 2$

$a_1 (\pm 2)^2 = 28$

$4a_1 = 28$

$a_1 = 7$

61. Given: $a_4 = 24$ and $r = 2$

Find a_8.

$a_n = a_1 r^{n-1}$

$a_4 = a_1 (2)^{4-1}$

$24 = a_1 8$

$3 = a_1$

$a_8 = a_1 r^{8-1}$

$a_8 = 3(2)^7$

$a_8 = 384$

63. Company A: $20000, \; 21000, \; 22000, \ldots$

                      year 1   year 2   year 3

Arithmetic sequence:

$a_n = a_1 + (n-1)d$

Year 6: $a_6 = 20,000 + (6-1)1000$

            $= 20,000 + 5,000$

            $= \$25,000$

Company B: $20000, \; 21000, \; 22050, \ldots$

                      year 1   year 2   year 3

$$\begin{bmatrix} (21,000)(0.05) \\ + 21,000 \end{bmatrix}$$

Geometric Sequence:

$a_n = a_1 r^{n-1} \qquad r = \dfrac{21000}{20000} = 1.05$

Year 6: $a_6 = 20,000(1.05)^{6-1}$

$a_6 = \$25,525.60$

Company B will pay more in the sixth year (approximately \$526 more).

65. Since $a_1, a_2, a_3, \ldots, a_n$

is geometric with

common ratio r, $\dfrac{a_2}{a_1} = r.$

Consider: $\dfrac{1}{a_1}, \dfrac{1}{a_2}, \dfrac{1}{a_3}, \ldots, \dfrac{1}{a_n}$

Common ratio $= \dfrac{\frac{1}{a_2}}{\frac{1}{a_1}} = \dfrac{a_1}{a_2} = \dfrac{1}{r}$

69. Given: $\dfrac{a_2}{a_1} = r$ and $\dfrac{a_n}{a_{n-1}} = r$

Consider: $a_n, a_{n-1}, a_{n-2}, \ldots, a_1$

Common ratio $= \dfrac{a_{n-1}}{a_n} = \dfrac{1}{r}$

71. $\log 5x + 2 \log x$

$= \log 5x + \log x^2$

$= \log (5x \cdot x^2)$

$= \log 5x^3$

72. $8^{y-1} = 4^{y+2}$

$(2^3)^{y-1} = (2^2)^{y+2}$

$2^{3y-3} = 2^{2y+4}$

$3y - 3 = 2y + 4$

$y = 7$

$\{7\}$

73. $\dfrac{\sqrt{5}+\sqrt{3}}{\sqrt{5}-\sqrt{3}} \cdot \dfrac{\sqrt{5}+\sqrt{3}}{\sqrt{5}+\sqrt{3}} = \dfrac{5 + 2\sqrt{15} + 3}{5 - 3}$

$= \dfrac{8 + 2\sqrt{15}}{2} = 4 + \sqrt{15}$

Problem Set 10.4.2

1. $2, 6, 18, \ldots$

$$S_n = \frac{a_1 - a_1 r^n}{1 - r} \qquad a_1 = 2, \; r = \frac{6}{2} = 3$$

$$S_6 = \frac{2 - 2(3)^6}{1 - 3}$$

$$S_6 = \frac{2 - 1458}{-2}$$

$$S_6 = \frac{-1456}{-2} = 728$$

5. $-\frac{3}{2}, 3, -6, \ldots \qquad a_1 = -\frac{3}{2}, \; r = \frac{3}{\left(-\frac{3}{2}\right)} = -2$

$$S_n = \frac{a_1 - a_1 r^n}{1 - r}$$

$$S_7 = \frac{-\frac{3}{2} - \left(-\frac{3}{2}\right)(-2)^7}{1 - (-2)}$$

$$S_7 = \frac{-\frac{3}{2} + \frac{3}{2}(-128)}{3}$$

$$S_7 = \frac{-\frac{3}{2} - 192}{3} = \frac{-3 - 384}{6}$$

$$S_7 = \frac{-387}{6} = -64.5$$

7. $\displaystyle\sum_{i=0}^{6} 3^i = 3^0 + 3^1 + 3^2 + \ldots + 3^6$

$$= 1 + 3 + 9 + \ldots + 729$$

$a_1 = 1, \; r = 3, \; n = 7$ (There are seven terms.)

$$S_n = \frac{a_1 - a_1 r^n}{1 - r}$$

$$S_7 = \frac{1 - 1(3)^7}{1 - 3} = \frac{-2186}{-2} = 1093$$

11. $\displaystyle\sum_{i=1}^{5} 2^{i-1} = 2^{1-1} + 2^{2-1} + 2^{3-1} + 2^{4-1} + 2^{5-1}$

$$= 1 + 2 + 4 + 8 + 16$$

$a_1 = 1, \; r = 2 \; n = 5$ (Since we begin at
$i = 1, \;$ not $i = 0, \;$ we have five terms.)

$$S_n = \frac{a_1 - a_1 r^n}{1 - r}$$

$$S_5 = \frac{1 - 1(2)^5}{1 - 2} = \frac{-31}{-1} = 31$$

13. $\displaystyle\sum_{i=1}^{4} \left(-\frac{2}{3}\right)^i = \left(-\frac{2}{3}\right)^1 + \left(-\frac{2}{3}\right)^2 + \left(-\frac{2}{3}\right)^3 + \left(-\frac{2}{3}\right)^4$

$$= -\frac{2}{3} + \frac{4}{9} + \left(-\frac{2}{3}\right)^3 + \left(-\frac{2}{3}\right)^4$$

$$a_1 = -\frac{2}{3}, \; r = \frac{\frac{4}{9}}{-\frac{2}{3}} = -\frac{2}{3}, \; n = 4$$

$$S_n = \frac{a_1 - a_1 r^n}{1 - r}$$

$$S_4 = \frac{\left(-\frac{2}{3}\right) - \left(-\frac{2}{3}\right)\left(-\frac{2}{3}\right)^4}{1 - \left(-\frac{2}{3}\right)}$$

$$S_4 = \frac{-\frac{2}{3} + \frac{32}{243}}{\frac{5}{3}} \cdot \frac{243}{243}$$

$$S_4 = \frac{-162 + 32}{405} = \frac{-130}{405} = \frac{-26}{81}$$

15. $6, 5.4, 4.86, \ldots$

$$\begin{array}{ccc} a_1 & a_2 & a_3 \end{array}$$

$$a_1 = 6, \; r = \frac{5.4}{6} = 0.9, \; n = 6$$

$$S_n = \frac{a_1 - a_1 r^n}{1 - r}$$

$$S_6 = \frac{6 - 6(0.9)^6}{1 - 0.9}$$

$$S_6 = \frac{2.811354}{0.1}$$

$$S_6 = 28.11354$$

The bob will travel approximately
28 inches.

21. Parents :   2

Grandparents :   4

Great grandparents :   8

$\underset{\underset{a_1}{\uparrow}}{2} + \underset{\underset{a_2}{\uparrow}}{4} + \underset{\underset{a_3}{\uparrow}}{8} + \ldots + a_6$

$a_1 = 2, \ r = 2, \ n = 6$

$S_n = \dfrac{a_1 - a_1 r^n}{1 - r}$

$S_6 = \dfrac{2 - 2(2)^6}{1 - 2} = \dfrac{-126}{-1} = 126$

126 ancestors

23. Given : $S_8 = -170, \ r = -2$

Find a_1.

$S_n = \dfrac{a_1 - a_1 r^n}{1 - r}$

$S_8 = \dfrac{a_1 - a_1 (-2)^8}{1 - (-2)} = -170$

$\dfrac{a_1 - 256 a_1}{3} = -170$

$-255 a_1 = -510$

$a_1 = \dfrac{-510}{-255} = 2$

29. Given : $a_1 = 10 \quad a_{10} = 30$

Find r and S_{10}.

$a_n = a_1 r^{n-1}$

$a_{10} = a_1 r^9$

$30 = 10 r^9$

$3 = r^9$

$r = \sqrt[9]{3}$

$(r \approx 1.1298)$[calculator : $3 \ y^x \ 9 \ \dfrac{1}{x} \ = \ ]$

$S_n = \dfrac{a_1 - a_1 r^n}{1 - r}$

$S_{10} = \dfrac{10 - 10(3^{\frac{1}{9}})^{10}}{1 - \sqrt[9]{3}}$

$S_{10} = \dfrac{10 - 10(3.38949)}{1 - 1.1298}$

$S_{10} = \dfrac{-23.8949}{-0.1298} \approx 184$

33. Given : $r = 2, \ S_6 = 20,000$

Find a_1.

$S_n = \dfrac{a_1 - a_1 r^n}{1 - r}$

$S_6 = \dfrac{a_1 - a_1 (2)^6}{1 - 2} = 20,000$

$\dfrac{a_1 - 64 a_1}{-1} = 20,000$

$-63 a_1 = -20,000$

$a_1 = \dfrac{-20,000}{-63} \approx 317.46$

$\$317.46$

35. Company A :

Arithmetic sequence : $a_1 = 20,000$

$d = 1,000$

$S_n = \dfrac{n}{2}(a_1 + a_n)$

$S_6 = \dfrac{6}{2}(a_1 + a_6)$

Find a_6.

$a_n = a_1 + (n-1)d$

$a_6 = a_1 + (6-1)d$

$a_6 = 20,000 + 5(1,000)$

$a_6 = 25,000$

$S_6 = 3(a_1 + a_6)$

$S_6 = 3(20,000 + 25,000) = 135,000$

Company B :

Geometric sequence : $a_1 = 20,000$

$a_2 = 20,000(0.05) + 20,000$

$\quad = 21,000$

$r = \dfrac{21,000}{20,000} = 1.05$

$S_n = \dfrac{a_1 - a_1 r^n}{1 - r}$

$S_6 = \dfrac{20,000 - 20,000(1.05)^6}{1 - 1.05}$

$S_6 = \dfrac{-6802}{-0.05} = 136,040$

[Key sequence for Numerator of S_6 :

$1.05 \; y^x \; 6 \; = \; x \; 20,000 \; = \; \pm \; + \; 20,000 \; = \;$]

Over Six-Years : Company A : $135,000$

Company B : $136,040$

Company B produces the better total income.

37. Given : $a_1 = 1280, \; r = 1.25, \; S_n = 7380$

Find n.

$S_n = \dfrac{a_1 - a_1 r^n}{1 - r}$

$7380 = \dfrac{1280 - 1280(1.25)^n}{1 - 1.25}$

$7380 = \dfrac{1280 - 1280(1.25)^n}{-0.25}$

$-1845 = 1280 - 1280(1.25)^n$

$-3125 = -1280(1.25)^n$

$2.44 = (1.25)^n$

$\log 2.44 = \log (1.25)^n$

$\log 2.44 = n \log 1.25$

$\dfrac{\log 2.44}{\log 1.25} = n$

$n \approx 4$

4 years

39. $\displaystyle\sum_{i=0}^{n} 2^i = \underbrace{2^0 + 2^1 + 2^2 + 2^3 + \ldots + 2^n}_{n+1 \text{ terms}} = 63$

$a_1 = 2^0 = 1, \; r = 2, \; S_{n+1} = 63$

Find n.

$S_n = \dfrac{a_1 - a_1 r^n}{1 - r}$

$S_{n+1} = \dfrac{a_1 - a_1 r^{n+1}}{1 - r}$

$63 = \dfrac{1 - 1 \cdot 2^{n+1}}{1 - 2}$

$-63 = 1 - 2^{n+1}$

$2^{n+1} = 64$

$2^{n+1} = 2^6$

$n + 1 = 6$

$n = 5$

43. $\log(x+2) + \log(x-1) = \log 4$

$\log(x+2)(x-1) = \log 4$

$(x+2)(x-1) = 4$

$x^2 + x - 6 = 0$

$(x+3)(x-2) = 0$

$x+3 = 0$ or $x-2 = 0$

$x = -3(\text{reject})$ $x = 2$

$\{2\}$

44. $t:$ tens' digit

$u:$ units' digit

$t = u + 4$

$\dfrac{tu}{t+u} = \dfrac{3}{2}$

$\qquad\qquad 2tu = 3t + 3u$

$2(u+4)u = 3(u+4) + 3u$

$2u^2 + 8u = 3u + 12 + 3u$

$2u^2 + 8u = 6u + 12$

$2u^2 + 2u - 12 = 0$

$u^2 + u - 6 = 0$

$(u+3)(u-2) = 0$

$u + 3 = 0$ $u - 2 = 0$

$u = -3$ (reject; Digits can only

$\qquad\qquad$ be $0, 1, 2, \ldots, 9$)

or $u = 2$

$t = u + 4$

$t = 2 + 4 = 6$

Number : $10t + u = 10(6) + 2 = 62$

45. $\left(\dfrac{y^{\frac{3}{2}} y^{-\frac{3}{4}}}{y^{-\frac{5}{2}}}\right)^{-8} = \left(\dfrac{y^{\frac{3}{2} - \frac{3}{4}}}{y^{-\frac{5}{2}}}\right)^{-8} = \left(\dfrac{y^{\frac{3}{4}}}{y^{-\frac{5}{2}}}\right)^{-8} = \left(y^{\frac{3}{4} - \left(-\frac{5}{2}\right)}\right)^{-8}$

$= \left(y^{\frac{3}{4} + \frac{10}{4}}\right)^{-8} = \left(y^{\frac{13}{4}}\right)^{-8} = y^{\left(\frac{13}{4}\right)(-8)}$

$= y^{-26} = \dfrac{1}{y^{26}}$

Problem Set 10.4.3

1. $1 + \dfrac{1}{4} + \dfrac{1}{16} + \ldots$

$a_1 = 1$

$r = \dfrac{1}{4}$

$S = \dfrac{a_1}{1-r}$

$S = \dfrac{1}{1-\frac{1}{4}} = \dfrac{1}{\frac{3}{4}} = \dfrac{4}{3}$

3. $12 + 6 + 3 + \ldots$

$a_1 = 12, \ r = \dfrac{6}{12} = \dfrac{1}{2}$

$S = \dfrac{a_1}{1-r}$

$S = \dfrac{12}{1-\frac{1}{2}} = \dfrac{12}{\frac{1}{2}} = 24$

7. $5 + 10 + 15 + \ldots$

$a_1 = 5, \ r = \dfrac{10}{5} = 2$

Since r does not lie between -1 and 1, the infinite series has no sum.

9. $\dfrac{4}{3} + \dfrac{2}{9} + \dfrac{1}{27} + \ldots$

$a_1 = \dfrac{4}{3} \qquad r = \dfrac{\frac{2}{9}}{\frac{4}{3}} = \dfrac{2}{9} \cdot \dfrac{3}{4} = \dfrac{1}{6}$

$S = \dfrac{a_1}{1-r}$

$S = \dfrac{\frac{4}{3}}{1-\frac{1}{6}} = \dfrac{\frac{4}{3}}{\frac{5}{6}} = \dfrac{4}{3} \cdot \dfrac{6}{5} = \dfrac{8}{5}$

11. $1 - \dfrac{1}{2} + \dfrac{1}{4} - \dfrac{1}{8} + \ldots$

$a_1 = 1 \qquad r = \dfrac{-\frac{1}{2}}{1} = -\dfrac{1}{2}$

$S = \dfrac{a_1}{1-r}$

$S = \dfrac{1}{1-(-\frac{1}{2})} = \dfrac{1}{\frac{3}{2}} = \dfrac{2}{3}$

13. $0.\bar{5} = 0.5555\ldots = 0.5 + 0.05 + 0.005 + \ldots$

$a_1 = 0.5, \ r = 0.1$

$S = \dfrac{a_1}{1-r} = \dfrac{0.5}{1-0.1} = \dfrac{0.5}{0.9} = \dfrac{5}{9}$

15. $0.\overline{49} = 0.494949\ldots$

$\qquad = 0.49 + 0.0049 + 0.000049 + \ldots$

$a_1 = 0.49, \ r = 0.01$

$S = \dfrac{a_1}{1-r} = \dfrac{0.49}{1-0.01} = \dfrac{0.49}{0.99} = \dfrac{49}{99}$

19. $5.\overline{47}$

$0.\overline{47} = 0.474747\ldots$

$\qquad = 0.47 + 0.0047 + 0.000047 + \ldots$

$a_1 = 0.47, \ r = 0.01$

$S = \dfrac{a_1}{1-r} = \dfrac{0.47}{1-0.01} = \dfrac{0.47}{0.99} = \dfrac{47}{99}$

$5.\overline{47} = 5\dfrac{47}{99} = \dfrac{542}{99}$

23. $0.1\bar{2} = 0.12222\ldots$

Consider $0.0\bar{2} = 0.02222\ldots =$

$\qquad 0.02 + 0.002 + 0.0002 + \ldots$

$a_1 = 0.02, \ r = 0.1$

$S = \dfrac{a_1}{1-r} = \dfrac{0.02}{1-0.1} = \dfrac{0.02}{0.9} = \dfrac{2}{90} = \dfrac{1}{45}$

$0.1\bar{2} = 0.1 + \dfrac{1}{45} = \dfrac{1}{10} + \dfrac{1}{45}$

$\qquad = \dfrac{9}{90} + \dfrac{2}{90} = \dfrac{11}{90}$

25. The amount spent is given by :

$0.9(40) + (0.9)^2(40) + (0.9)^3(40) + \dots$

$a_1 = 0.9(40) = 36$

$r = 0.9$

$S = \dfrac{a_1}{1-r} = \dfrac{36}{1-0.9} = \dfrac{36}{0.1} = 360$

Additional spending : $360 billion

29. Perimeter of original square $= 4(40) = 160$

Perimeter of next - smaller square

$= 4(20) = 80$

Perimeter of next - smaller square

$= 4(10) = 40$

Perimeter of next - smaller square

$= 4(5) = 20$

Series : $160 + 80 + 40 + 20 + \dots$

$a_1 = 160, \ r = \dfrac{1}{2}$

$S = \dfrac{a_1}{1-r} = \dfrac{160}{1-\frac{1}{2}} = \dfrac{160}{\frac{1}{2}} = 320$

320 inches

31. Given : $r = 0.9, \ S = 20,000$

a_1 :   Number of flies released each day

$S = \dfrac{a_1}{1-r}$

$20,000 = \dfrac{a_1}{1-0.9}$

$20,000 = \dfrac{a_1}{0.1}$

$2,000 = a_1$

Release 2,000 flies each day.

33. Employee is paid (after taxes) :

$\$10,000 - \$1,000 = \$9,000$

$+ \quad \$1,000 - \$100 = \$900$

$+ \qquad \$100 - \$10 \ = \$90$

$+ \qquad \$10 - \$1 \quad = \$9$

etc. ad. infinitum

Employee's salary :

$9,000 + 900 + 90 + 9 + \dots$

$a_1 = 9,000 \quad r = 0.1$

$S = \dfrac{a_1}{1-r} = \dfrac{9,000}{1-0.01} = \dfrac{9,000}{0.9} = \$10,000$

Yes, the employee will get $10,000

after taxes with this continued process.

37. $x + x^2 + x^3 + \dots = \dfrac{1+x}{x}$

$a_1 = x, \ r = x, \ S = \dfrac{1+x}{x}$

$S = \dfrac{a_1}{1-r}$

$\dfrac{1+x}{x} = \dfrac{x}{1-x}$

$x^2 = (1+x)(1-x)$

$x^2 = 1 - x^2$

$2x^2 = 1$

$x^2 = \dfrac{1}{2}$

$x = \pm\sqrt{\dfrac{1}{2}} = \pm\dfrac{1}{\sqrt{2}} \cdot \dfrac{\sqrt{2}}{\sqrt{2}} = \pm\dfrac{\sqrt{2}}{2}$

39. $\sqrt[3]{54x^6y^7} = \sqrt[3]{27 \cdot 2(x^2)^3(y^2)^3 y}$

$= 3x^2y^2\sqrt[3]{2y}$

40. Algebraically :

$4x^2 + y^2 = 16$

$2x + y = 4$

Substitution :

$y = 4 - 2x$

$4x^2 + (4 - 2x)^2 = 16$

$4x^2 + 16 - 16x + 4x^2 = 16$

$8x^2 - 16x = 0$

$8x(x - 2) = 0$

$8x = 0$ or $x - 2 = 0$

$x = 0$ or $x = 2$

If $x = 0$: $y = 4 - 2x = 4 - 2(0) = 4$

If $x = 2$: $y = 4 - 2(2) = 4 - 4 = 0$

$\{(0, 4), (2, 0)\}$

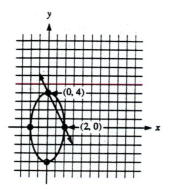

Graphically :

$4x^2 + y^2 = 16$

$\dfrac{4x^2}{16} + \dfrac{y^2}{16} = 1$

$\dfrac{x^2}{4} + \dfrac{y^2}{16} = 1$

Ellipse :   x - intercepts : $\dfrac{x^2}{4} = 1$

$x^2 = 4$

$x = \pm 2$

y - intercepts : $\dfrac{y^2}{16} = 1$

$y^2 = 16$

$y = \pm 4$

$2x + y = 4$

Line :   x - intercept : $2x = 4$

$x = 2$

y - intercept : $y = 4$

41. $f(x) = x^2 - 2x + 4$

$\dfrac{f(a + h) - f(a)}{h}$

$= \dfrac{(a + h)^2 - 2(a + h) + 4 - (a^2 - 2a + 4)}{h}$

$= \dfrac{a^2 + 2ah + h^2 - 2a - 2h + 4 - a^2 + 2a - 4}{h}$

$= \dfrac{2ah + h^2 - 2h}{h} = 2a + h - 2$

Problem Set 10.5

1. $3! = 3 \cdot 2 \cdot 1 = 6$

5. $\dfrac{10!}{8!2!} = \dfrac{10 \cdot 9}{2 \cdot 1} = 45$

9. $\dbinom{6}{3} = \dfrac{6!}{3!(6-3)!} = \dfrac{6!}{3!3!} = \dfrac{6 \cdot 5 \cdot 4}{3 \cdot 2 \cdot 1} = 20$

13. $\dbinom{6}{6} = \dfrac{6!}{6!(6-6)!} = \dfrac{6!}{6!0!} = \dfrac{1}{0!} = \dfrac{1}{1} = 1$

15. $(c+2)^5 = \dbinom{5}{0}c^5 + \dbinom{5}{1}c^4 \cdot 2 + \dbinom{5}{2}c^3 \cdot 2^2$

$+ \dbinom{5}{3}c^2 \cdot 2^3 + \dbinom{5}{4}c \cdot 2^4 + \dbinom{5}{5} \cdot 2^5$

$= c^5 + 5c^4(2) + 10c^3(4) + 10c^2(8) + 5c(16) + 32$

$= c^5 + 10c^4 + 40c^3 + 80c^2 + 80c + 32$

19. $\left(\dfrac{a}{2}+1\right)^4 = \dbinom{4}{0}\left(\dfrac{a}{2}\right)^4 + \dbinom{4}{1}\left(\dfrac{a}{2}\right)^3 \cdot 1$

$+ \dbinom{4}{2}\left(\dfrac{a}{2}\right)^2 \cdot 1^2 + \dbinom{4}{3}\left(\dfrac{a}{2}\right)^1 \cdot 1^3$

$+ \dbinom{4}{4} \cdot 1^4$

$= (1)\left(\dfrac{a^4}{16}\right) + 4\left(\dfrac{a^3}{8}\right) + 6\left(\dfrac{a^2}{4}\right) + 4\left(\dfrac{a}{2}\right) + 1$

$= \dfrac{a^4}{16} + \dfrac{a^3}{2} + \dfrac{3a^2}{2} + 2a + 1$

23. $(2x^2 - y^2)^3 = [2x^2 + (-y^2)]^3$

$= \dbinom{3}{0}(2x^2)^3 + \dbinom{3}{1}(2x^2)^2(-y^2) + \dbinom{3}{2}(2x^2)(-y^2)^2$

$+ \dbinom{3}{3}(-y^2)^3$

$= 1(8x^6) + 3(4x^4)(-y^2) + 3(2x^2)(y^4) + 1(-y^6)$

$= 8x^6 - 12x^4y^2 + 6x^2y^4 - y^6$

27. $(a^{\frac{1}{2}}+2)^4 = \dbinom{4}{0}(a^{\frac{1}{2}})^4 + \dbinom{4}{1}(a^{\frac{1}{2}})^3(2)$

$+ \dbinom{4}{2}(a^{\frac{1}{2}})^2(2)^2 + \dbinom{4}{3}(a^{\frac{1}{2}})(2)^3 + \dbinom{4}{4}(2)^4$

$= 1(a^2) + 4(a^{\frac{3}{2}})(2) + 6a(4) + 4a^{\frac{1}{2}}(8) + 1(16)$

$= a^2 + 8a^{\frac{3}{2}} + 24a + 32a^{\frac{1}{2}} + 16$

31. $(x^2 + x)^8 = \dbinom{8}{0}(x^2)^8 + \dbinom{8}{1}(x^2)^7 x$

$+ \dbinom{8}{2}(x^2)^6 x^2 + \dots$

$= 1x^{16} + 8(x^{14})x + 28(x^{12})x^2 + \dots$

$= x^{16} + 8x^{15} + 28x^{14} + \dots$

35. $(a - 2b)^8 = [a + (-2b)]^8$

$= \dbinom{8}{0}a^8 + \dbinom{8}{1}a^7(-2b) + \dbinom{8}{2}a^6(-2b)^2 + \dots$

$= 1a^8 + 8a^7(-2b) + 28a^6(4b^2) + \dots$

$= a^8 - 16a^7b + 112a^6b^2 - + \dots$

41. $\left(y + \dfrac{1}{y}\right)^7 = \dbinom{7}{0}y^7 + \dbinom{7}{1}y^6\left(\dfrac{1}{y}\right)$

$+ \dbinom{7}{2}y^5\left(\dfrac{1}{y}\right)^2 + \dots$

$= 1y^7 + 7y^6\left(\dfrac{1}{y}\right) + 21y^5\left(\dfrac{1}{y^2}\right) + \dots$

$= y^7 + 7y^5 + 21y^3 + \dots$

43. $(2a + b)^6$; 3rd term

$n = 6, \ r = 3$

$\dbinom{n}{r-1}a^{n-r+1}b^{r-1}$ becomes :

$\dbinom{6}{3-1}(2a)^{6-3+1}b^{3-1}$

$= \dbinom{6}{2}(2a)^4b^2 = 15(16a^4)b^2 = 240a^4b^2$

45. $(x+y)^{15}$; 7th term $n = 15$, $r = 7$

$$\binom{n}{r-1}a^{n-r+1}b^{r-1} = \binom{15}{7-1}x^{15-7+1}y^{7-1}$$

$$= \binom{15}{6}x^9 y^6 = 5{,}005x^9 y^6$$

47. $(c^5 + d^7)^9$; 3rd term $n = 9$, $r = 3$

$$\binom{n}{r-1}a^{n-r+1}b^{r-1} = \binom{9}{3-1}(c^5)^{9-3+1}(d^7)^{3-1}$$

$$= \binom{9}{2}(c^5)^7(d^7)^2 = 36c^{35}d^{14}$$

51. $\dfrac{(n+1)!}{n!} = \dfrac{(n+1)n!}{n!} = n+1$

53. $(1+i)^5 = \binom{5}{0}(1)^5 + \binom{5}{1}(1)^4(i) + \binom{5}{2}(1)^3(i)^2$

$$+ \binom{5}{3}(1)^2(i)^3 + \binom{5}{4}(1)(i)^4 + \binom{5}{5}(i)^5$$

$$= 1 + 5i + 10i^2 + 10i^3 + 5i^4 + i^5$$

$$i^2 = -1; \quad i^3 = i^2 \cdot i = (-1)i = -i;$$

$$i^4 = i^2 \cdot i^2 = (-1)(-1) = 1;$$

$$i^5 = i^4 \cdot i = 1i = i$$

$$(1+i)^5 = 1 + 5i + 10(-1) + 10(-i) + 5(1) + i$$

$$= 1 + 5i - 10 - 10i + 5 + i = -4 - 4i$$

55. $\left(\dfrac{3}{x} + \dfrac{x}{3}\right)^{10}$ There are 11 terms,

so the middle term is the 6th term.

$n = 10$, $r = 6$

$$\binom{n}{r-1}a^{n-r+1}b^{r-1} = \binom{10}{6-1}\left(\dfrac{3}{x}\right)^{10-6+1}\left(\dfrac{x}{3}\right)^{6-1}$$

$$= \binom{10}{5}\left(\dfrac{3}{x}\right)^5\left(\dfrac{x}{3}\right)^5 = \binom{10}{5} = 252$$

57. $(a+b)^{10}$ Since b is introduced in the second term, appears as b^2 in the third term, b^3 in the 4th term, etc., the term containing b^7 is the 8th term.

$$n = 10 \quad r = 8$$

$$\binom{n}{r-1}a^{n-r+1}b^{r-1} = \binom{10}{8-1}a^{10-8+1}b^{8-1}$$

$$= \binom{10}{7}a^3 b^7 = 120a^3 b^7$$

62. $f(x) = 2x - 1$, $g(x) = x^2 - 3x + 2$

$$f[g(x)] = 2[g(x)] - 1 = 2(x^2 - 3x + 2) - 1$$

$$= 2x^2 - 6x + 3$$

$$g[f(x)] = [f(x)]^2 - 3[f(x)] + 2$$

$$= (2x - 1)^2 - 3(2x - 1) + 2$$

$$= 4x^2 - 4x + 1 - 6x + 3 + 2$$

$$= 4x^2 - 10x + 6$$

63. $f(x) = 3x + 5$

$$y = 3x + 5$$

inverse : $x = 3y + 5$

$$x - 5 = 3y$$

$$\dfrac{x-5}{3} = y$$

$$f^{-1}(x) = \dfrac{x-5}{3}$$

$$f[f^{-1}(x)] = 3[f^{-1}(x)] + 5 = 3\left(\dfrac{x-5}{3}\right) + 5$$

$$= x - 5 + 5 = x$$

$$f^{-1}[f(x)] = \dfrac{f(x) - 5}{3} = \dfrac{3x + 5 - 5}{3} = \dfrac{3x}{3} = x$$

64. $f(x) = Ae^{0.04x}$

$A = 1,000 \quad f(x) = 4,000$

$1,000e^{0.04x} = 4,000$

$e^{0.04x} = 4$

$\ln e^{0.04x} = \ln 4$

$0.04x = \ln 4$

$x = \dfrac{\ln 4}{0.04} \approx 35$

Approximately 35 hours

Review Problems :   Chapter 10

1. $a_n = n^2 + 1$

$a_1 = 1^2 + 1 = 2$

$a_2 = 2^2 + 1 = 5$

$a_3 = 3^2 + 1 = 10$

$a_4 = 4^2 + 1 = 17$

$2, 5, 10, 17$

2. $a_1 = 1 = 1^2$

$a_2 = 4 = 2^2$

$a_3 = 9 = 3^2$

$a_4 = 16 = 4^2$

$a_n = n^2$

3. $\displaystyle\sum_{i=1}^{4} (2i^2 - 3) = (2 \cdot 1^2 - 3) + (2 \cdot 2^2 - 3)$

$+ (2 \cdot 3^2 - 3) + (2 \cdot 4^2 - 3)$

$= (-1) + (5) + (15) + (29)$

$= 48$

4. $\displaystyle\sum_{i=1}^{5} 6x^i = 6x + 6x^2 + 6x^3 + 6x^4 + 6x^5$

5. $\bar{x} = \dfrac{\sum_{i=1}^{5} x}{n}$

$= \dfrac{3.8 + 2.3 + 1.1 + 7.2 + 8.1}{5}$

$= \dfrac{22.5}{5} = 4.5$

6. $1 + 8 + 27 + 64 + 25 = 1^3 + 2^3 + 3^3 + 4^3 + 5^3$

$= \displaystyle\sum_{i=1}^{5} i^3$

7. $\dfrac{x+1}{x} + \dfrac{x+2}{x} + \dfrac{x+3}{x} + \dfrac{x+4}{x} =$

$\displaystyle\sum_{i=1}^{4} \dfrac{x+i}{x}$

8. $-7, -3, 1, 5, \ldots \qquad d = -3 - (-7) = 4$

$a_n = a_1 + (n - 1)d$

$a_{15} = -7 + (15 - 1)(4)$

$a_{15} = -7 + 14(4) = -7 + 56 = 49$

9. $5, 2, -1, -4, , \ldots$

$d = 2 - 5 = -3$

$a_n = a_1 + (n - 1)d$

$a_n = 5 + (n - 1)(-3)$

$a_n = 8 - 3n$

10. $9, 15, 21, \ldots$

$d = 15 - 9 = 6$

$a_n = a_1 + (n - 1)d$

$a_{15} = 9 + (15 - 1)(6)$

$a_{15} = 9 + 84 = 93$

93 oranges

11. $18000, 18850, 19700, \ldots$

$a_1 = 18,000 \qquad d = 850 \qquad a_n = 25,650$

Find n.

$a_n = a_1 + (n - 1)d$

$25,650 = 18,000 + (n - 1)(850)$

$25,650 = 18,000 + 850n - 850$

$25,650 = 17,150 + 850n$

$8,500 = 850n$

$10 = n$

10 years

12. $5, 12, 19, 26, \ldots$

$d = 12 - 5 = 7$

$S_n = \dfrac{n}{2}(a_1 + a_n)$

$S_{22} = \dfrac{22}{2}(a_1 + a_{22})$

Find a_{22}. $a_n = a_1 + (n - 1)d$

$a_{22} = 5 + (22 - 1)(7)$

$a_{22} = 5 + (21)(7)$

$a_{22} = 152$

$S_{22} = \dfrac{22}{2}(5 + 152)$

$S_{22} = 11(157) = 1,727$

13. Given : $n = 16$, $a_1 = 3$, $d = 5$

Find S_{16}.

$S_n = \dfrac{n}{2}(a_1 + a_n)$

$S_{16} = \dfrac{16}{2}(a_1 + a_{16})$

Find a_{16}. $a_n = a_1 + (n - 1)d$

$a_{16} = 3 + (16 - 1)(5)$

$a_{16} = 78$

$S_{16} = \dfrac{16}{2}(3 + 78)$

$S_{16} = 648$

14. $\displaystyle\sum_{i=1}^{16}(3i + 2) = (3 \cdot 1 + 2) + (3 \cdot 2 + 2)$

$+ (3 \cdot 3 + 2) + \ldots + (3 \cdot 16 + 2)$

$= 5 + 8 + 11 + \ldots + 50$

$a_1 = 5$, $n = 16$ (There are 16 terms),

$a_{16} = 50$

$S_n = \dfrac{n}{2}(a_1 + a_n)$

$S_{16} = \dfrac{16}{2}(a_1 + a_{16})$

$S_{16} = \dfrac{16}{2}(5 + 50)$

$S_{16} = 8(55) = 440$

15. Given : $a_1 = 68$, $a_5 = 59$

Find S_{17}.

$S_n = \dfrac{n}{2}(a_1 + a_n)$

$S_{17} = \dfrac{17}{2}(a_1 + a_{17})$

$a_n = a_1 + (n - 1)d$

$a_5 = 68 + (5 - 1)d = 59$

$68 + 4d = 59$

$4d = -9$

$d = -\dfrac{9}{4}$

Find a_{17}.

$a_{17} = a_1 + (17 - 1)d$

$a_{17} = 68 + 16(-\dfrac{9}{4})$

$a_{17} = 68 + (-36) = 32$

$S_{17} = \dfrac{17}{2}(68 + 32)$

$S_{17} = \dfrac{17}{2}(100) = 850$

16. d: number of additional bushels each day

$35, 35 + d, 35 + 2d, \ldots$

$a_1 = 35, \quad S_{14} = 854$

Find d.

$$S_n = \frac{n}{2}(a_1 + a_n)$$

$$S_{14} = \frac{14}{2}(a_1 + a_{14})$$

Find a_{14}. $a_n = a_1 + (n-1)d$

$$a_{14} = 35 + (14-1)d$$

$$a_{14} = 35 + 13d$$

$$S_{14} = \frac{14}{2}(35 + 35 + 13d) = 854$$

$$7(70 + 13d) = 854$$

$$490 + 91d = 854$$

$$91d = 364$$

$$d = 4$$

4 additional bushels of fruit each day

17. $a_1 = \dfrac{9}{25}$

$$a_2 = \frac{9}{25}\left(-\frac{5}{3}\right) = -\frac{3}{5}$$

$$a_3 = -\frac{3}{5}\left(-\frac{5}{3}\right) = 1$$

$$a_4 = 1\left(-\frac{5}{3}\right) = -\frac{5}{3}$$

$$\frac{9}{25}, -\frac{3}{5}, 1, -\frac{5}{3}$$

18. $\dfrac{1}{3}, \dfrac{1}{2}, \dfrac{3}{4}, \ldots$

$$a_1 = \frac{1}{3} \quad r = \frac{\frac{1}{2}}{\frac{1}{3}} = \frac{3}{2}$$

$$a_n = a_1 r^{n-1}$$

$$a_6 = \frac{1}{3}\left(\frac{3}{2}\right)^5 = \frac{3^4}{2^5} = \frac{81}{32}$$

19. $3, 3\sqrt{3}, 9, \ldots$

$$a_1 = 3 \quad r = \frac{3\sqrt{3}}{3} = \sqrt{3}$$

$$a_n = a_1 r^{n-1}$$

$$a_{10} = 3\left(\sqrt{3}\right)^{10-1} = 3\left(3^{\frac{1}{2}}\right)^9$$

$$= 3 \cdot 3^{\frac{9}{2}} = 3 \cdot 3^4 \cdot 3^{\frac{1}{2}}$$

$$a_{10} = 3^5 \cdot 3^{\frac{1}{2}} = 243\sqrt{3}$$

20. $5, -10, 20, \ldots$

$$a_1 = 5, \quad r = \frac{-10}{5} = -2$$

$$a_n = a_1 r^{n-1}$$

$$a_n = 5(-2)^{n-1}$$

21. $a_1 = 10,000 \quad r = 0.8$

$$a_n = a_1 r^{n-1}$$

$$a_5 = 10,000(0.8)^4$$

$$a_5 = 10,000(0.4096) = 4096$$

$4,096

22. Given : $a_1 = 20, \quad a_4 = \dfrac{5}{16}$

Find r.

$$a_n = a_1 r^{n-1}$$

$$a_4 = 20r^3 = \frac{5}{16}$$

$$r^3 = \frac{1}{64}$$

$$r = \sqrt[3]{\frac{1}{64}} = \frac{1}{4}$$

23. $7, -14, 28, \ldots$

$$a_1 = 7 \qquad r = \frac{-14}{7} = -2$$

$$S_n = \frac{a_1 - a_1 r^n}{1 - r}$$

$$S_6 = \frac{7 - 7(-2)^6}{1 - (-2)} = \frac{-441}{3} = -147$$

24. $\displaystyle\sum_{i=1}^{5} \frac{1}{2}(6)^{i-1} = \frac{1}{2}(6)^0 + \frac{1}{2}(6)^1 + \frac{1}{2}(6)^2$

$$+ \frac{1}{2}(6)^3 + \frac{1}{2}(6)^4$$

$a_1 = \frac{1}{2}, \; r = 6, \; n = 5 \text{(There are five terms.)}$

$$S_n = \frac{a_1 - a_1 r^n}{1 - r}$$

$$S_5 = \frac{\frac{1}{2} - \frac{1}{2}(6)^5}{1 - 6} = \frac{-3887.5}{-5} = 777.5$$

25. $4, 12, 36, \ldots \qquad r = \frac{12}{4} = 3$

$$S_n = \frac{a_1 - a_1 r^n}{1 - r}$$

$$S_6 = \frac{4 - 4(3)^6}{1 - 3} = \frac{-2912}{-2} = 1456$$

$\$1,456$

26. Given: $\; S_4 = -100, \; r = -3$

Find a_1.

$$S_n = \frac{a_1 - a_1 r^n}{1 - r}$$

$$S_4 = \frac{a_1 - a_1 (-3)^4}{1 - (-3)} = -100$$

$$\frac{a_1 - 81a_1}{4} = -100$$

$$-80a_1 = -400$$

$$a_1 = 5$$

27. $36 + 12 + 4 + \ldots$

$$r = \frac{12}{36} = \frac{1}{3}$$

$$S = \frac{a_1}{1 - r}$$

$$S = \frac{36}{1 - \frac{1}{3}} = \frac{36}{\frac{2}{3}} = 36 \cdot \frac{3}{2} = 54$$

28. $\dfrac{4}{3} - 1 + \dfrac{3}{4} - \ldots$

$$a_1 = \frac{4}{3} \qquad r = \frac{-1}{\frac{4}{3}} = -\frac{3}{4}$$

$$S = \frac{a_1}{1 - r}$$

$$S = \frac{\frac{4}{3}}{1 - (-\frac{3}{4})} = \frac{\frac{4}{3}}{\frac{7}{4}} = \frac{16}{21}$$

29. $0.\overline{36} = 0.363636\ldots$

$$= 0.36 + 0.0036 + 0.000036 + \ldots$$

$$a_1 = 0.36 \qquad r = \frac{0.0036}{0.36} = 0.01$$

$$S = \frac{a_1}{1 - r}$$

$$S = \frac{0.36}{1 - 0.01} = \frac{0.36}{0.99} = \frac{36}{99} = \frac{4}{11}$$

30. $a_1 = 3,000$

$$a_2 = 3,000 - (3,000)(0.25) = 2,250$$

$$r = \frac{2,250}{3,000} = 0.75$$

$$S = \frac{a_1}{1 - r}$$

$$S = \frac{3000}{1 - 0.75} = \frac{3000}{0.25} = 12,000$$

$\$12,000$

31. $\dbinom{9}{2} = \dfrac{9!}{2!(9-2)!} = \dfrac{9!}{2!7!} = \dfrac{9 \cdot 8}{2 \cdot 1} = 36$

32.
$$(x^2 + 3y)^4 = \binom{4}{0}(x^2)^4 + \binom{4}{1}(x^2)^3(3y)$$
$$+ \binom{4}{2}(x^2)^2(3y)^2 + \binom{4}{3}(x^2)^1(3y)^3$$
$$+ \binom{4}{4}(3y)^4$$
$$= 1x^8 + 4x^6(3y) + 6x^4(9y^2)$$
$$+ 4x^2(27y^3) + 1(81y^4)$$
$$= x^8 + 12x^6y + 54x^4y^2 + 108x^2y^3 + 81y^4$$

33.
$$(x^3 - 2)^5 = [x^3 + (-2)]^5$$
$$= \binom{5}{0}(x^3)^5 + \binom{5}{1}(x^3)^4(-2) + \binom{5}{2}(x^3)^3(-2)^2$$
$$+ \binom{5}{3}(x^3)^2(-2)^3 + \binom{5}{4}(x^3)^1(-2)^4$$
$$+ \binom{5}{5}(-2)^5$$
$$= 1x^{15} + 5x^{12}(-2) + 10x^9(4) + 10x^6(-8)$$
$$+ 5x^3(16) + 1(-32)$$
$$= x^{15} - 10x^2 + 40x^9 - 80x^6 + 80x^3 - 32$$

34.
$$(x + y)^{12} = \binom{12}{0}x^{12} + \binom{12}{1}x^{11}y$$
$$+ \binom{12}{2}x^{10}y^2 + \ldots$$
$$= x^{12} + 12x^{11}y + 66x^{10}y^2 + \ldots$$

35.
$$(2x - y)^6 = [2x + (-y)]^6 = \binom{6}{0}(2x)^6$$
$$+ \binom{6}{1}(2x)^5(-y) + \binom{6}{2}(2x)^4(-y)^2 + \ldots$$
$$= 1(64x^6) + 6(32x^5)(-y)$$
$$+ 15(16x^4)(y^2) + \ldots$$
$$= 64x^6 - 192x^5y + 240x^4y^2 - + \ldots$$

36. $(3c + d)^9$

7th term : $r = 7, \ n = 9$

$$r^{th} \text{ term } = \binom{n}{r-1}a^{n-r+1}b^{r-1}$$
$$7^{th} \text{ term } = \binom{9}{7-1}(3c)^{9-7+1}d^{7-1}$$
$$= \binom{9}{6}(3c)^3 d^6 = 84(27c^3)d^6$$
$$= 2268c^3d^6$$

37. $(a - 2b)^7$

4th term : $r = 4, \ n = 7$

$$r^{th} \text{ term } = \binom{n}{r-1}a^{n-r+1}b^{r-1}$$
$$4^{th} \text{ term } = \binom{7}{4-1}a^{7-4+1}(-2b)^{4-1}$$
$$= \binom{7}{3}a^4(-2b)^3$$
$$= 35a^4(-8b^3) = -280a^4b^3$$

1.
$$\frac{4y-2}{3} - \frac{y+2}{4} = \frac{7y-2}{12}$$

$$12\left[\frac{4y-2}{3} - \frac{y+2}{4}\right] = 12\left[\frac{7y-2}{12}\right]$$

$$4(4y-2) - 3(y+2) = 7y-2$$

$$16y - 8 - 3y - 6 = 7y - 2$$

$$13y - 14 = 7y - 2$$

$$6y = 12$$

$$y = 2$$

$$\{2\}$$

2. $6 - 4(2y-3) < 5(3-y)$

$$6 - 8y + 12 < 15 - 5y$$

$$18 - 8y < 15 - 5y$$

$$-3y < -3$$

$$y > 1$$

$$\{y \mid y > 1\}$$

3. $-7x > -14 \qquad 3x < 15$

$$x < 2 \qquad\qquad x < 5$$

$$\{x \mid x < 2\} \cap \{x \mid x < 5\} = \{x \mid x < 2\}$$

4. $\left|\dfrac{8-2y}{3}\right| = 4$

$$\frac{8-2y}{3} = 4 \text{ or } \frac{8-2y}{3} = -4$$

$$8 - 2y = 12 \qquad 8 - 2y = -12$$

$$-2y = 4 \qquad\quad -2y = -20$$

$$y = -2 \qquad\qquad y = 10$$

$$\{-2, 10\}$$

5. $|3y + 1| \geq 16$

$$3y + 1 \geq 16 \text{ or } 3y + 1 \leq -16$$

$$3y \geq 15 \qquad\qquad 3y \leq -17$$

$$y \geq 5 \qquad\qquad y \leq -\frac{17}{3}$$

$$\left\{y \mid y \geq 5 \text{ or } y \leq -\frac{17}{3}\right\}$$

6.
$$\frac{3}{9y+6} + \frac{2}{5y-1}$$

$$= \frac{4y}{15y^2 + 7y - 2}$$

$$\frac{3}{3(3y+2)} + \frac{2}{5y-1}$$

$$= \frac{4y}{(3y+2)(5y-1)}$$

$$3(3y+2)(5y-1)\left[\frac{3}{3(3y+2)} + \frac{2}{5y-1}\right]$$

$$= 3(3y+2)(5y-1)\left[\frac{4y}{(3y+2)(5y-1)}\right]$$

$$3(5y-1) + 2 \cdot 3(3y+2) = 3(4y)$$

$$15y - 3 + 18y + 12 = 12y$$

$$33y + 9 = 12y$$

$$9 = -21y$$

$$y = \frac{9}{-21} = -\frac{3}{7}$$

$$\left\{-\frac{3}{7}\right\}$$

7.
$$\frac{1}{y^2-4}+\frac{1}{5}=\frac{2}{y+2}$$

$$\frac{1}{(y+2)(y-2)}+\frac{1}{5}=\frac{2}{y+2}$$

$$5(y+2)(y-2)\left[\frac{1}{(y+2)(y-2)}+\frac{1}{5}\right]$$

$$=5(y+2)(y-2)\left[\frac{2}{y+2}\right]$$

$$5+(y+2)(y-2)=10(y-2)$$

$$5+y^2-4=10y-20$$

$$y^2+1=10y-20$$

$$y^2-10y+21=0$$

$$(y-7)(y-3)=0$$

$$y-7=0 \text{ or } y-3=0$$

$$y=7 \qquad y=3$$

$$\{3,7\}$$

8.
$$\sqrt{2y+5}-\sqrt{2y}=3$$

$$\sqrt{2y+5}=\sqrt{2y}+3$$

$$\left(\sqrt{2y+5}\right)^2=\left(\sqrt{2y}+3\right)^2$$

$$2y+5=2y+6\sqrt{2y}+9$$

$$-4=6\sqrt{2y}$$

$$-2=3\sqrt{2y}$$

$$(-2)^2=\left(3\sqrt{2y}\right)^2$$

$$4=9(2y)$$

$$4=18y$$

$$y=\frac{4}{18}=\frac{2}{9}$$

Check $\frac{2}{9}$:
$$\sqrt{2\left(\frac{2}{9}\right)+5}-\sqrt{2\left(\frac{2}{9}\right)}=3$$

$$\sqrt{\frac{4}{9}+\frac{45}{9}}-\sqrt{\frac{4}{9}}=3$$

$$\sqrt{\frac{49}{9}}-\sqrt{\frac{4}{9}}=3$$

$$\frac{7}{3}-\frac{2}{3}=3$$

$$\frac{5}{3}\neq 3$$

$\frac{2}{9}$ is extraneous.

$$\varnothing$$

9.
$$x^{\frac{2}{3}}-x^{\frac{1}{3}}-6=0$$

Let $t=x^{\frac{1}{3}}$

$$t^2-t-6=0$$

$$(t-3)(t+2)=0$$

$$t-3=0 \text{ or } t+2=0$$

$$t=3 \quad \text{ or } \quad t=-2$$

$$x^{\frac{1}{3}}=3 \qquad x^{\frac{1}{3}}=-2$$

$$\left(x^{\frac{1}{3}}\right)^3=3^3 \qquad \left(x^{\frac{1}{3}}\right)^3=(-2)^3$$

$$x=27 \qquad\qquad x=-8$$

$$\{-8,27\}$$

10. $\dfrac{1}{y+1} = 2 + \dfrac{2}{y-3}$

$(y+1)(y-3)[\dfrac{1}{y+1}]$

$\qquad = (y+1)(y-3)\left[2 + \dfrac{2}{y-3}\right]$

$y - 3 = 2(y+1)(y-3) + 2(y+1)$

$y - 3 = 2y^2 - 4y - 6 + 2y + 2$

$y - 3 = 2y^2 - 2y - 4$

$0 = 2y^2 - 3y - 1$

$a = 2, b = -3, c = -1$

$y = \dfrac{-b \pm \sqrt{b^2 - 4ac}}{2a}$

$\quad = \dfrac{-(-3) \pm \sqrt{(-3)^2 - 4(2)(-1)}}{2(2)}$

$\quad = \dfrac{3 \pm \sqrt{9 - (-8)}}{4} = \dfrac{3 \pm \sqrt{17}}{4}$

$\left\{ \dfrac{3 + \sqrt{17}}{4}, \dfrac{3 - \sqrt{17}}{4} \right\}$

11. $\dfrac{1}{y+2} - \dfrac{1}{3} = \dfrac{1}{y}$

$3y(y+2)\left[\dfrac{1}{y+2} - \dfrac{1}{3}\right] = 3y(y+2)\left(\dfrac{1}{y}\right)$

$3y - y(y+2) = 3(y+2)$

$3y - y^2 - 2y = 3y + 6$

$y - y^2 = 3y + 6$

$0 = y^2 + 2y + 6$

$a = 1, b = 2, c = 6$

$y = \dfrac{-b \pm \sqrt{b^2 - 4ac}}{2a}$

$\quad = \dfrac{-2 \pm \sqrt{2^2 - 4(1)(6)}}{2(1)}$

$\quad = \dfrac{-2 \pm \sqrt{-20}}{2}$

$\quad = \dfrac{-2 \pm \sqrt{4(5)(-1)}}{2}$

$\quad = \dfrac{-2 \pm 2\sqrt{5}\, i}{2}$

$\quad = -1 \pm \sqrt{5}\, i$

$\left\{ -1 + \sqrt{5}\, i, -1 - \sqrt{5}\, i \right\}$

12. $3x^2 + 8x + 5 < 0$

$(3x + 5)(x + 1) < 0$

Set each factor equal to zero.

$3x + 5 = 0 \quad x + 1 = 0$

$x = -\dfrac{5}{3} \qquad x = -1$

Test -2: $3x^2 + 8x + 5 < 0$

$3(-2)^2 + 8(-2) + 5 < 0$

$12 - 16 + 5 < 0$

$1 < 0$ False

Test $-1\dfrac{1}{2}$: $3x^2 + 8x + 5 < 0$

$3(-\dfrac{3}{2})^2 + 8(-\dfrac{3}{2}) + 5 < 0$

$\dfrac{27}{4} - 12 + 5 < 0$

$\dfrac{27 - 48 + 20}{4} < 0$

$-\dfrac{1}{4} < 0$ True

Test 0: $3x^2 + 8x + 5 < 0$

$3(0)^2 + 8(0) + 5 < 0$

$5 < 0$ False

$\{x \mid -\dfrac{5}{3} < x < -1\}$

13. $\dfrac{x-1}{x+3} \le 0$

$x - 1 = 0 \quad x + 3 = 0$

$x = 1 \qquad x = -3$

Test -4: $\dfrac{x-1}{x+3} \le 0$

$\dfrac{-4-1}{-4+3} \le 0$

$\dfrac{-5}{-1} \le 0$

$5 \le 0$ False

Test 0: $\dfrac{x-1}{x+3} \le 0$

$\dfrac{0-1}{0+3} \le 0$

$-\dfrac{1}{3} \le 0$ True

Test 2: $\dfrac{x-1}{x+3} \le 0$

$\dfrac{2-1}{2+3} \le 0$

$\dfrac{1}{5} \le 0$ False

$\{x \mid -3 < x \le 1\}$

14. $s = \dfrac{1}{2}gt^2$

$2s = gt^2$

$g = \dfrac{2s}{t^2}$

15. $5(y^2 - 1) = 4(y^2 + 2)$

$5y^2 - 5 = 4y^2 + 8$

$y^2 = 13$

$y = \pm\sqrt{13}$

$\left\{-\sqrt{13}, \sqrt{13}\right\}$

16. $6x + 3y = -1$ (multiply by 3)

        $9x + 5y = 1$ (multiply by -2)

                $18x + 9y = -3$

              $\underline{-18x - 10y = -2}$

Add : $-y = -5$

                  $y = 5$

         $6x + 3y = -1$

         $6x + 3(5) = -1$

         $6x + 15 = -1$

           $6x = -16$

            $x = \dfrac{-16}{6} = \dfrac{-8}{3}$

$$\left\{ \left(\dfrac{-8}{3}, 5 \right) \right\}$$

17. $2x^2 + y^2 = 7$ (multiply by 2)

        $x^2 - 2y^2 = -4$

          $4x^2 - 2y^2 = 14$

           $\underline{x^2 - 2y^2 = -4}$

Add : $5x^2 = 10$

             $x^2 = 2$

             $x = \pm\sqrt{2}$

If $x = \sqrt{2}$: $2x^2 + y^2 = 7$

            $2(\sqrt{2})^2 + y^2 = 7$

              $4 + y^2 = 7$

                $y^2 = 3$

                  $y = \pm\sqrt{3}$

$$\left(\sqrt{2}, \sqrt{3}\right) \quad \left(\sqrt{2}, -\sqrt{3}\right)$$

If $x = -\sqrt{2}$: $2x^2 + y^2 = 7$

            $2\left(-\sqrt{2}\right)^2 + y^2 = 7$

              $4 + y^2 = 7$

                $y^2 = 3$

                  $y = \pm\sqrt{3}$

$$\left(-\sqrt{2}, \sqrt{3}\right) \quad \left(-\sqrt{2}, -\sqrt{3}\right)$$

$$\left\{ \left(\sqrt{2}, \sqrt{3}\right), \left(\sqrt{2}, -\sqrt{3}\right), \right.$$
$$\left. \left(-\sqrt{2}, \sqrt{3}\right), \left(-\sqrt{2}, -\sqrt{3}\right) \right\}$$

18. $6x = 10 + y$

$3x^2 - xy = 3$

Substitution : $6x = 10 + y$

$6x - 10 = y$

$3x^2 - xy = 3$

$3x^2 - x(6x - 10) = 3$

$3x^2 - 6x^2 + 10x = 3$

$-3x^2 + 10x = 3$

$0 = 3x^2 - 10x + 3$

$0 = (3x - 1)(x - 3)$

$3x - 1 = 0$ or $x - 3 = 0$

$x = \dfrac{1}{3}$ $x = 3$

If $x = \dfrac{1}{3}$: $y = 6x - 10$

$y = 6(\dfrac{1}{3}) - 10 = 2 - 10 = -8$ $(\dfrac{1}{3}, -8)$

If $x = 3$: $y = 6x - 10$

$y = 6(3) - 10 = 18 - 10 = 8$ $(3, 8)$

$\{(\dfrac{1}{3}, -8), (3, 8)\}$

19.   (1) $x + y + z = 3$

(2) $x - y + z = 5$

(3) $2x - y + z = 6$

Eliminate z.

$-1(1) + (2) :$ $-2y = 2$

$y = -1$

$-1(1) + (3) :$ $x - 2y = 3$

$x - 2(-1) = 3$

$x + 2 = 3$

$x = 1$

$x + y + z = 3$

$1 + (-1) + z = 3$

$z = 3$ $\{(1, -1, 3)\}$

20. $5^{2x-1} = 125^{\frac{x}{2}}$

$5^{2x-1} = (5^3)^{\frac{x}{2}}$

$5^{2x-1} = 5^{\frac{3x}{2}}$

$2x - 1 = \dfrac{3x}{2}$

$4x - 2 = 3x$

$-2 = -x$

$2 = x$

$\{2\}$

21. $\log_2 x + \log_2 (2x - 3) = 1$

$\log_2 x(2x - 3) = 1$

$\log_2 (2x^2 - 3x) = 1$

Exponential form : $2^1 = 2x^2 - 3x$

$0 = 2x^2 - 3x - 2$

$0 = (2x + 1)(x - 2)$

$2x + 1 = 0$ or $x - 2 = 0$

$x = -\dfrac{1}{2}$ $x = 2$

Reject $-\dfrac{1}{2}$ since it causes the logarithm of a negative number in the original equation.

$\{2\}$

22. $4y^4 + 4y = 4y(y^3 + 1)$

$= 4y(y + 1)(y^2 - y \cdot 1 + 1^2)$

$= 4y(y + 1)(y^2 - y + 1)$

23. $x^3 - 2x^2 - 9x + 18$

$= x^2(x - 2) - 9(x - 2)$

$= (x - 2)(x^2 - 9)$

$= (x - 2)(x + 3)(x - 3)$

24. $(2x + y)^2 + 15(2x + y) + 36$

Let $z = 2x + y$.

$z^2 + 15z + 36$

$= (z + 12)(z + 3)$

$(2x + y)^2 + 15(2x + y) + 36$

$= (2x + y + 12)(2x + y + 3)$

25. $(-2x^{-2}y)^2(-4x^2y^{-2})^{-2}$

$= (-2)^2(x^{-2})^2y^2(-4)^{-2}(x^2)^{-2}(y^{-2})^{-2}$

$= 4x^{-4}y^2 \cdot \dfrac{1}{(-4)^2}x^{-4}y^4$

$= \dfrac{4}{16}x^{-8}y^6 = \dfrac{y^6}{4x^8}$

26. $\left(\dfrac{-30x^{4n-3}y^{1-2n}}{10x^{4n-2}y^{-3-2n}}\right)^{-2}$

$= (-3x^{4n-3-(4n-2)}y^{1-2n-(-3-2n)})$

$= (-3x^{-1}y^4)^{-2} = (-3)^{-2}(x^{-1})^{-2}(y^4)^{-2} = \dfrac{1}{(-3)^2}x^2y^{-8}$

$= \dfrac{x^2}{9y^8}$

27. $a^{\frac{5}{8}}b^{\frac{1}{2}}(a^{\frac{5}{2}}b^{-5})^{-\frac{3}{5}} = a^{\frac{5}{8}}b^{\frac{1}{2}}(a^{\frac{5}{2}})^{-\frac{3}{5}}(b^{-5})^{-\frac{3}{5}}$

$= a^{\frac{5}{8}}b^{\frac{1}{2}}a^{-\frac{3}{2}}b^3 = a^{\frac{5}{8}-\frac{3}{2}}b^{\frac{1}{2}+3} = a^{-\frac{7}{8}}b^{\frac{7}{2}} = \dfrac{b^{\frac{7}{2}}}{a^{\frac{7}{8}}}$

28. $(2x - y)(x + 3y)(x - 2y) = (2x - y)(x^2 + xy - 6y^2)$

$= 2x^3 + 2x^2y - 12xy^2$

$\underline{\qquad\quad - x^2y - xy^2 + 6y^3}$

$= 2x^3 + x^2y - 13xy^2 + 6y^3$

29.

$$3x - 4 \overline{)3x^3 - 19x^2 + 17x + 4}$$

quotient: $x^2 - 5x - 1$

$\underline{3x^3 - 4x^2}$

$-15x^2 + 17x$

$\underline{-15x^2 + 20x}$

$-3x + 4$

$\underline{-3x + 4}$

30. $\dfrac{3x^3 - 5x^2 + 2x - 1}{x - 2} = 3x^2 + x + 4 + \dfrac{7}{x - 2}$

2⌋	3	-5	2	-1
		6	2	8
	3	1	4	7

31. $\dfrac{y^3 - 7y^2 + 12y}{y^2 - y - 6} \div \dfrac{y^3 - 4y^2}{y^2 - 3y - 10}$

$= \dfrac{y(y^2 - 7y + 12)}{(y - 3)(y + 2)} \div \dfrac{y^2(y - 4)}{(y - 5)(y + 2)}$

$= \dfrac{y(y - 4)(y - 3)}{(y - 3)(y + 2)} \cdot \dfrac{(y - 5)(y + 2)}{y^2(y - 4)}$

$= \dfrac{y - 5}{y}$

32. $\dfrac{2y - 6}{3y^2 - 14y - 5} - \dfrac{y - 3}{y^2 - 5y}$

$= \dfrac{2y - 6}{(3y + 1)(y - 5)} - \dfrac{y - 3}{y(y - 5)}$

The LCD is $y(3y + 1)(y - 5)$.

$= \dfrac{2y - 6}{(3y + 1)(y - 5)} \cdot \dfrac{y}{y} - \dfrac{y - 3}{y(y - 5)} \cdot \dfrac{(3y + 1)}{(3y + 1)}$

$= \dfrac{y(2y - 6) - (y - 3)(3y + 1)}{y(3y + 1)(y - 5)}$

$= \dfrac{2y^2 - 6y - 3y^2 + 8y + 3}{y(3y + 1)(y - 5)}$

$= \dfrac{-y^2 + 2y + 3}{y(3y + 1)(y - 5)}$

33. $\dfrac{1 - \frac{14y - 45}{y^2}}{\frac{y}{9} - \frac{9}{y}} \cdot \dfrac{9y^2}{9y^2}$

$= \dfrac{9y^2 - 9(14y - 45)}{y^3 - 81y}$

$= \dfrac{9y^2 - 126y + 405}{y^3 - 81y}$

$= \dfrac{9(y^2 - 14y + 45)}{y(y^2 - 81)}$

$= \dfrac{9(y - 9)(y - 5)}{y(y + 9)(y - 9)}$

$= \dfrac{9(y - 5)}{y(y + 9)} \left[\text{Equivalently}: \quad \dfrac{9y - 45}{y^2 + 9y} \right]$

34. $\sqrt[3]{4x^2 y^5} \cdot \sqrt[3]{4xy^2 z^2}$

$= \sqrt[3]{16x^3 y^7 z^2}$

$= \sqrt[3]{8 \cdot 2x^3 (y^2)^3 yz^2}$

$= 2xy^2 \sqrt[3]{2yz^2}$

35. $7\sqrt{18x^5} - 3x\sqrt{2x^3}$

$= 7\sqrt{9 \cdot 2(x^2)^2 x} - 3x\sqrt{2x^2 x}$

$= 7 \cdot 3x^2 \sqrt{2x} - 3x(x)\sqrt{2x}$

$= 21x^2 \sqrt{2x} - 3x^2 \sqrt{2x}$

$= 18x^2 \sqrt{2x}$

36. $\dfrac{1 + \sqrt{3}}{3\sqrt{3} - 1} \cdot \dfrac{3\sqrt{3} + 1}{3\sqrt{3} + 1}$

$= \dfrac{3\sqrt{3} + 1 + 3(3) + \sqrt{3}}{9(3) - 1}$

$= \dfrac{4\sqrt{3} + 10}{26}$

$= \dfrac{2\sqrt{3} + 5}{13}$

37. $(2 + 3i)(4 - 5i)$

$= 8 - 10i + 12i - 15i^2$

$= 8 + 2i - 15(-1)$

$= 23 + 2i$

38. $\dfrac{6}{3 + 5i} \cdot \dfrac{3 - 5i}{3 - 5i} = \dfrac{6(3 - 5i)}{9 - 25i^2}$

$= \dfrac{6(3 - 5i)}{9 - 25(-1)} = \dfrac{6(3 - 5i)}{34}$

$= \dfrac{3(3 - 5i)}{17} = \dfrac{9 - 15i}{17}$

$= \dfrac{9}{17} - \dfrac{15}{17}i$

39. $(1, -4)$ and $(-5, 8)$

$m = \dfrac{y_2 - y_1}{x_2 - x_1} = \dfrac{8 - (-4)}{-5 - 1} = \dfrac{12}{-6} = -2$

$(x_1, y_1) = (1, -4)$

[Equivalently : $(x_1, y_1) = (-5, 8)$]

Point - Slope Equation : $y - y_1 = m(x - x_1)$

$y - (-4) = -2(x - 1)$

$y + 4 = -2(x - 1)$

Slope - Intercept Equation : $y = mx + b$

$y + 4 = -2(x - 1)$

$y + 4 = -2x + 2$

$y = -2x - 2$

40. $(3, -2) = (x_1, y_1)$

Slope of $-\frac{1}{4}x + y = 5$: $\quad y = \frac{1}{4}x + 5$

$\qquad\qquad\qquad\qquad$ Slope is $\frac{1}{4}$.

Slope of the line whose equation we must

write $= -4$ (negative reciprocal of $\frac{1}{4}$)

Point - Slope Equation : $\quad y - y_1 = m(x - x_1)$

$\quad x_1 = 3, y_1 = -2, m = -4$: $\quad y + 2 = -4(x - 3)$

Slope - Intercept Equation : $\quad y = mx + b$

$\qquad\qquad\qquad\qquad\qquad y + 2 = -4x + 12$

$\qquad\qquad\qquad\qquad\qquad\quad y = -4x + 10$

41. $f(x) = 2x - 1, g(x) = x^2 - 3x - 1$

$f[g(x)] = 2[g(x)] - 1 = 2(x^2 - 3x - 1) - 1$

$\qquad\quad = 2x^2 - 6x - 3$

$f(\frac{1}{2}) = 2(\frac{1}{2}) - 1 = 1 - 1 = 0$

$g[f(\frac{1}{2})] = g(0) = 0^2 - 3 \cdot 0 - 1 = -1$

42. $f(x) = 3x^2 - 5x - 2$

$\dfrac{f(a + h) - f(a)}{h}$

$= \dfrac{3(a + h)^2 - 5(a + h) - 2 - (3a^2 - 5a - 2)}{h}$

$= \dfrac{3a^2 + 6ah + 3h^2 - 5a - 5h - 2 - 3a^2 + 5a + 2}{h}$

$= \dfrac{6ah + 3h^2 - 5h}{h} = 6a + 3h - 5$

43. $f(x) = \frac{1}{2}x - 5$

$\qquad y = \frac{1}{2}x - 5$

inverse : $\quad x = \frac{1}{2}y - 5$

$\qquad\qquad 2x = y - 10$

$\qquad\quad 2x + 10 = y$

$\qquad\quad f^{-1}(x) = 2x + 10$

$f[f^{-1}(x)] = \frac{1}{2}[f^{-1}(x)] - 5$

$\qquad\quad = \frac{1}{2}(2x + 10) - 5 = x + 5 - 5 = x$

$f^{-1}[f(x)] = 2[f(x)] + 10$

$\qquad\quad = 2(\frac{1}{2}x - 5) + 10 = x - 10 + 10 = x$

44. $f(x) = \dfrac{2}{3x^2 - 5x}$

$\qquad 3x^2 - 5x \neq 0$

$\qquad x(3x - 5) \neq 0$

$\qquad x \neq 0$ and $3x - 5 \neq 0$

$\qquad\qquad\qquad x \neq \frac{5}{3}$

Domain $= \{x \mid x \neq 0$ and $x \neq \frac{5}{3}\}$

45. $f(x) = \sqrt{12 - 4x}$

$\qquad 12 - 4x \geq 0$

$\qquad -4x \geq -12$

$\qquad x \leq 3$

Domain $= \{x \mid x \leq 3\}$

46. $f(x) = \log(x - 5)$

$\qquad x - 5 > 0$

$\qquad x > 5$

Domain $= \{x \mid x > 5\}$

47. $\log_2 \dfrac{1}{64} = y$

$$2^y = \dfrac{1}{64}$$

$$2^y = 2^{-6}$$

$$y = -6$$

48. $\log_b 2 = \dfrac{1}{3}$

$$b^{\frac{1}{3}} = 2$$

$$(b^{\frac{1}{3}})^3 = 2^3$$

$$b = 8$$

49. $\begin{vmatrix} 2 & 4 & 0 \\ 5 & 0 & -1 \\ -2 & 1 & -1 \end{vmatrix}$ Expand about the middle column.

$$= -4 \begin{vmatrix} 5 & -1 \\ -2 & -1 \end{vmatrix} - 1 \begin{vmatrix} 2 & 0 \\ 5 & -1 \end{vmatrix}$$

$$= -4(-5 - 2) - 1(-2 - 0) = -4(-7) - 1(-2)$$

$$= 28 + 2 = 30$$

50. $\dfrac{1}{4}\log_b x + 2\log_b (x^2 - 3) - \dfrac{1}{3}\log_b z$

$$= \log_b x^{\frac{1}{4}} + \log_b (x^2 - 3)^2 - \log_b z^{\frac{1}{3}}$$

$$= \log_b x^{\frac{1}{4}}(x^2 - 3)^2 - \log_b z^{\frac{1}{3}}$$

$$= \log_b \dfrac{x^{\frac{1}{4}}(x^2 - 3)^2}{z^{\frac{1}{3}}}$$

$$= \log_b \dfrac{\sqrt[4]{x}(x^2 - 3)^2}{\sqrt[3]{z}}$$

51. $3(7 + 4) = 3 \cdot 7 + 3 \cdot 4$ Distributive property

52. Prime numbers between 3 and 17 :

$$5, 7, 11, 13$$

Both 7 and 11 yield remainders of 3 when divided by 4.

$$\{7, 11\}$$

53. $\dfrac{1}{B} = \dfrac{1}{A} - \dfrac{1}{C}$

$$ABC(\dfrac{1}{B}) = ABC(\dfrac{1}{A} - \dfrac{1}{C})$$

$$AC = BC - AB$$

$$AC + AB = BC$$

$$A(C + B) = BC$$

$$A = \dfrac{BC}{C + B}$$

54. $(4, 3)$ to $(2, -1)$

$$d = \sqrt{(x_2 - x_1)^2 + (y_2 - y_1)^2}$$

$$= \sqrt{(2 - 4)^2 + (-1 - 3)^2}$$

$$= \sqrt{(-2)^2 + (-4)^2} = \sqrt{4 + 16}$$

$$= \sqrt{20} = \sqrt{4 \cdot 5} = 2\sqrt{5}$$

55. $\dfrac{5}{x^{\frac{1}{2}}} - 3x^{\frac{1}{2}} = \dfrac{5}{x^{\frac{1}{2}}} - 3x^{\frac{1}{2}} \cdot \dfrac{x^{\frac{1}{2}}}{x^{\frac{1}{2}}} = \dfrac{5 - 3x}{x^{\frac{1}{2}}}$

$$= \dfrac{5 - 3x}{\sqrt{x}} \cdot \dfrac{\sqrt{x}}{\sqrt{x}} = \dfrac{5\sqrt{x} - 3x\sqrt{x}}{x}$$

56. $x^2 + y^2 - 4x + 6y - 12 = 0$

$x^2 - 4x + 4 + y^2 + 6y + 9 = 12 + 4 + 9$

$\frac{1}{2}(-4) = -2 \qquad \frac{1}{2}(6) = 3$

$(-2)^2 = 4 \qquad 3^2 = 9$

$(x-2)^2 + (y+3)^2 = 25$

$(x-2)^2 + [y-(-3)]^2 = 5^2$

$[(x-h)^2 + (y-k)^2 = r^2$

center: (h, k) radius: $r]$

center: $(2, -3)$

radius: 5

57.

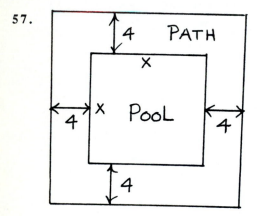

Area of path = Area of large square

- Area of inner square = $(x+8)(x+8) - x^2$

$A(x) = x^2 + 16x + 64 - x^2$

$A(x) = 16x + 64$

58. $(2x - y^3)^5 = [2x + (-y^3)]^5$

$= \binom{5}{0}(2x)^5 + \binom{5}{1}(2x)^4(-y^3) + \binom{5}{2}(2x)^3(-y^3)^2$

$+ \binom{5}{3}(2x)(-y^3)^3 + \binom{5}{4}(2x)^1(-y^3)^4$

$+ \binom{5}{5}(-y^3)^5$

$= (1)(32x^5) + (5)(16x^4)(-y^3)$

$+ (10)(8x^3)(y^6)$

$+ (10)(4x^2) \cdot (y^9) + (5)(2x)(y^{12})$

$+ (1) \cdot (y^{15})$

$= 32x^5 - 80x^4y^3 + 80x^3y^6 - 40x^2y^9$

$+ 10xy^{12} - y^{15}$

59. $2, 6, 10, \ldots$

$a_1 = 2, d = 6 - 2 = 4$

$S_n = \frac{n}{2}(a_1 + a_n)$

$S_{30} = \frac{30}{2}(a_1 + a_{30})$

Find a_{30}.

$a_n = a_1 + (n-1)d$

$a_{30} = 2 + (30-1)(4)$

$a_{30} = 2 + (29)(4) = 118$

$S_{30} = \frac{30}{2}(2 + 118)$

$S_{30} = (15)(120) = 1,800$

60. Given: $a_{16} = -30, d = -3$

Find a_1.

$a_n = a_1 + (n-1)d$

$a_{16} = a_1 + (16-1)d$

$-30 = a_1 + 15(-3)$

$-30 = a_1 - 45$

$15 = a_1$

61. $\frac{1}{2}, 2, 8, \ldots$

$$a_1 = \frac{1}{2} \qquad r = \frac{2}{\frac{1}{2}} = 4$$

$$S_n = \frac{a_1 - a_1 r^n}{1 - r}$$

$$S_8 = \frac{\frac{1}{2} - \frac{1}{2}(4)^8}{1 - 4} = \frac{\frac{1}{2} - \frac{1}{2}(65536)}{-3}$$

$$= \frac{-32767.5}{-3} = 10,922.5$$

62. $0.\overline{450} = 0.450450450\ldots$

$$= 0.450 + 0.000450$$

$$+ 0.000000450 + \ldots$$

$$a_1 = 0.450 \qquad r = \frac{0.000450}{0.450} = 0.001$$

$$S = \frac{a_1}{1 - r}$$

$$S = \frac{0.450}{1 - 0.001} = \frac{0.450}{0.999} = \frac{450}{999} = \frac{50}{111}$$

63. Given: $a_1 = 8, r = -\frac{1}{2}$

Find a_5.

$$a_n = a_1 r^{n-1}$$

$$a_5 = (8)(-\frac{1}{2})^{5-1}$$

$$a_5 = (8)(-\frac{1}{2})^4 = 8(\frac{1}{16}) = \frac{1}{2}$$

64. $y = -\frac{2}{3}x + 1$

Line: $[y = mx + b \quad y\text{-intercept} = b$

slope $= m]$

$$y\text{-intercept} = 1 \qquad slope = -\frac{2}{3}$$

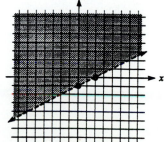

65. $x - 2y < 2$

$x - 2y = 2$: x-intercept: $x - 2(0) = 2$

$$x = 2$$

y-intercept: $0 - 2y = 2$

$$y = -1$$

Test point: $(0, 0)$

$$x - 2y < 2$$

$$0 - 2(0) < 2$$

$$0 < 2 \text{ True}$$

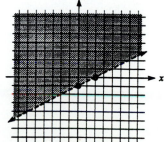

66. $y = x^2 - 4x - 5$

Parabola :

x - intercepts : $0 = x^2 - 4x - 5$

$0 = (x - 5)(x + 1)$

$x - 5 = 0 \quad\quad x + 1 = 0$

$x = 5 \quad\quad\quad x = -1$

y - intercept : $y = 0^2 - 4 \cdot 0 - 5$

$y = -5$

vertex : $x = -\dfrac{b}{2a} = \dfrac{-(-4)}{2(1)} = \dfrac{4}{2} = 2$

$y = 2^2 - 4 \cdot 2 - 5 = -9$

vertex : $(2, -9)$

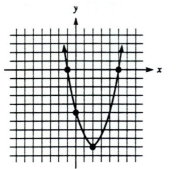

67. $\dfrac{x^2}{4} + \dfrac{y^2}{9} = 1$

Ellipse : $[\dfrac{x^2}{a^2} + \dfrac{y^2}{b^2} = 1]$

x - intercepts : $\dfrac{x^2}{4} = 1$

$x^2 = 4$

$x = \pm 2$

y - intercepts : $\dfrac{y^2}{9} = 1$

$y^2 = 9$

$y = \pm 3$

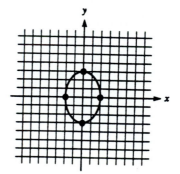

68. $\dfrac{x^2}{4} - \dfrac{y^2}{9} = 1$

Hyperbola : $\left[\dfrac{x^2}{a^2} - \dfrac{y^2}{b^2} = 1\right]$

x - intercepts : $\dfrac{x^2}{4} = 1$

$$x^2 = 4$$

$$x = \pm 2$$

There are no y - intercepts.

$$\dfrac{-y^2}{9} = 1$$

$$y^2 = -9$$

$$y = \pm\sqrt{-9}$$

No solution in the real number system. Since $a^2 = 4(a = \pm 2)$ and $b^2 = 9(b = \pm 3)$, the rectangle whose diagonals form the asymptotes passes through 2 and - 2 on the x - axis and 3 and - 3 on the y - axis.

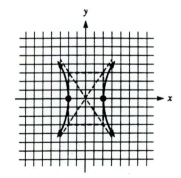

69. x : first integer

x + 2 : second consecutive even integer

x + 4 : third consecutive even integer

$$(x + 2) + (x + 4) = 3x + 2$$

70. x : Al's present age

$$x + 13 = 4(x - 17)$$

71. x : angle

supplement : 180 - x

complement : 90 - x

$$180 - x + 2(90 - x) = 207$$

72. x : greatest angle

x - 1 : next - smaller angle

x - 2 : smallest angle

$$x + (x - 1) + (x - 2) = 180$$

73. x : plane's rate of speed during the first three hours

$$\begin{array}{ccc} \text{Distance} & & \text{Distance} \\ \text{covered for the} & + & \text{covered for the} \\ \text{first 3 hours} & & \text{last 2 hours} \end{array} = 2540$$

$$[RT = D]$$

$$3x + 2(x - 30) = 2540$$

74. number of dimes : x

number of quarters : 2x - 1

number of nickels : 2(2x - 1) - 1 = 4x - 3

Value of dimes + Value of quarters + Value of nickels + Value of pennies = 145

$$10x + 25(2x - 1) + 5(4x - 3) = 145$$

75. $2Ax + By = -20$

(-2, 2) :   If $x = -2, y = 2$.

$2A(-2) + B(2) = -20$

$-4A + 2B = -20$

$Ax - 2By = 10$

(-2, 2) :   If $x = -2, \ y = 2$.

$A(-2) - 2B(2) = 10$

$-2A - 4B = 10$

System : $-4A + 2B = -20$

$-2A - 4B = 10$

76. x :   Boat's rate of speed in still water

y :   Rate of current

$RT = D$

With current :   45 minutes $= \dfrac{45}{60}$ of

an hour $= \dfrac{3}{4}$ h to cover 6 kilometers :

$(x + y) \cdot \dfrac{3}{4} = 6$

Against current : $1\dfrac{1}{2}$ hours to cover

6 kilometers : $(x - y) \cdot \dfrac{3}{2} = 6$

System : $\dfrac{3}{4}(x + y) = 6$

$\dfrac{3}{2}(x - y) = 6$

77. t :   tens' digit

u :   units' digit

$10t + u$:   the number

$10u + t$:   the number with its digits

reversed

$t + u = 8$

$10u + t = 10t + u + 18$

[Equivalently : $t + u = 8 \quad -9t + 9u = 18$]

78.

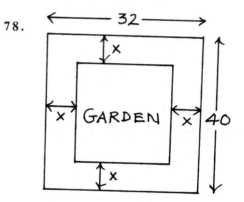

Area of garden $= 560$

$(32 - 2x)(40 - 2x) = 560$

79. x :   amount invested at 6%

$15,000 - x$:   amount invested at 8%

Income from the 6% investment +

Income from the 8% investment $= 1090$

[Income = Interest • Amount Invested]

$0.06x + 0.08(15,000 - x) = 1090$

80. x :   one number

x - 4 :   other number

$$\frac{1}{x} + \frac{1}{x-4} = \frac{10}{21}$$

Using two variables :

x :   one number

y :   other number

x - y = 4

$$\frac{1}{x} + \frac{1}{y} = \frac{10}{21}$$

Substitution :   x - 4 = y

$$\frac{1}{x} + \frac{1}{x-4} = \frac{10}{21}$$

81. x :   time working together

The fractional part of the job done by
Bill in x days + the fractional part of
the job done by Julie in x days =
one whole job

$$\frac{x}{4} + \frac{x}{6} = 1$$

[Equivalently :   3x + 2x = 12]

82. x :   walking speed

x + 4 :   running speed

Time to run 17 km = Time to walk
9 km

$$[RT = D; \ T = \frac{D}{R}]$$

$$\frac{17}{x+4} = \frac{9}{x}$$

83. $\dfrac{\text{Wages}}{\text{Hours worked}}$: $\dfrac{83.50}{7.5} = \dfrac{962}{x}$

$$\left[\text{Equivalently :} \quad \frac{7.5}{83.50} = \frac{x}{962}; \ \frac{7.5}{x} = \frac{83.50}{962} \right]$$

84. $F = \dfrac{KV^2}{L}$

$$5 = \frac{K(5)^2}{1}$$

$$5 = 25K$$

$$\frac{5}{25} = K$$

$$K = \frac{1}{5}$$

$$F = \frac{\frac{1}{5}V^2}{L} \left[\text{Equivalently :} \quad F = \frac{0.2V^2}{L} \right]$$

85.

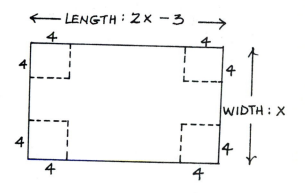

Volume of box = 532

(V = LWH)

(2x - 3 - 8)(x - 8)4 = 532

Equivalently :   4(2x - 11)(x - 8) = 532

86.

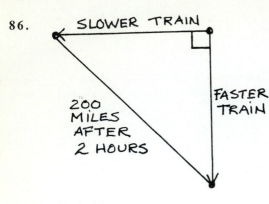

Slower train's speed :   x

Faster train's speed :   x + 20

[RT = D]

Distance of slower train in 2 hours :   2x

Distance of faster train in 2 hours :

$2(x + 20) = 2x + 40$

By the Pythagorean Theorem :

$(2x)^2 + (2x + 40)^2 = (200)^2$

87. x :   number of $8 tickets sold

250 - x :   number of $12 tickets sold

Income from $8 tickets + Income from

$12 tickets = $2200

$8x + 12(250 - x) = 2200$

88. x :   number of grams of the 80% solution

Amount of   Amount of   Amount of

antifreeze   antifreeze   antifreeze

in the 80% + in the 12% = in the 60%

solution   solution   mixture

$0.8x + (0.12)(175) = (0.6)(x + 175)$

89. x :   number of people in the smaller

group

x + 4 :   number of people in the larger

group

Amount paid by   Amount paid by

each person in the - each person in   = $20

smaller group   the larger group

$\left[\begin{array}{c} \text{Amount paid by each person} \\ = \dfrac{\text{Total cost}}{\text{Number of people}} \end{array} \right.$

$\dfrac{19,200}{x} - \dfrac{19,200}{x+4} = 20$

$\left[\text{Equivalently} : \quad \dfrac{19,200 - 20}{x} = \dfrac{19,200}{x+4} \right]$

90. x :   edge of open box

x + 1 :   edge of closed box

Surface area of open box(with five square

faces) $= 5x^2$

Surface area of closed box (with six square

faces) $= 6(x + 1)^2$

$5x^2 = 6(x + 1)^2 - 51$

91. Perimeter = 30

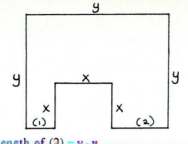

Length of (1) + Length of (2) = y - x

Perimeter = $y + y + y + x + x + x + (y - x) =$

$4y + 2x = 30$

    Area = 27.

$y^2 - x^2 = 27$

System : $4y + 2x = 30$
              $y^2 - x^2 = 27$

92. Area of triangle + Area of square = 37

$\frac{1}{2}(2y)(4) + x^2 = 37$

$4y + x^2 = 37$

By the Pythagorean Theorem :

$4^2 + y^2 = x^2$

$16 + y^2 = x^2$

System : $4y + x^2 = 37$
              $16 + y^2 = x^2$